Frontiers of Laser Spectroscopy of Gases

NATO ASI Series

Advanced Science Institutes Series

A Series presenting the results of activities sponsored by the NATO Science Committee, which aims at the dissemination of advanced scientific and technological knowledge, with a view to strengthening links between scientific communities.

The Series is published by an international board of publishers in conjunction with the NATO Scientific Affairs Division

A Life Sciences	Plenum Publishing Corporation
B Physics	London and New York
C Mathematical and Physical Sciences	Kluwer Academic Publishers Dordrecht, Boston and London
D Behavioural and Social Sciences	
E Applied Sciences	
F Computer and Systems Sciences	Springer-Verlag
G Ecological Sciences	Berlin, Heidelberg, New York, London,
H Cell Biology	Paris and Tokyo

Frontiers of Laser Spectroscopy of Gases

edited by

Antonio C. P. Alves
University of Coimbra, Portugal

John M. Brown
University of Oxford, U.K.

and

J. Michael Hollas
University of Reading, U.K.

Kluwer Academic Publishers

Dordrecht / Boston / London

Published in cooperation with NATO Scientific Affairs Division

Proceedings of the NATO Advanced Study Institute on
Frontiers of Laser Spectroscopy of Gases
Vimeiro, Portugal
30 March – 10 April 1987

Library of Congress Cataloging in Publication Data

```
Frontiers of laser spectroscopy of gases / edited by A.C.P. Alves,
  J.M. Brown, and J.M. Hollas.
      p.   cm. -- (NATO ASI series. Series C, Mathematical and
  physical sciences ; vol. 234)
      "Contains the lectures presented at the NATO Advanced Study
  Institute on 'Frontiers of Laser Spectroscopy of Gases' held in the
  Hotel Golf Mar, Vimeiro, near Torres Vedras, Portugal, from 30 March
  to 10 April 1987"--Pref.
      "Published in cooperation with NATO Scientific Affairs Division."
      Includes index.
      ISBN 978-94-010-7849-8          ISBN 978-94-009-3003-2 (eBook)
      DOI 10.1007/ 978-94-009-3003-2
      1. Laser spectroscopy--Congresses.   I. Alves, A. C. P.
  II. Brown, J. M.   III. Hollas, J. Michael (John Michael)   IV. NATO
  Advanced Study Institute on "Frontiers of Laser Spectroscopy of
  Gases" (1987 : Vimeiro, Lisbon, Portugal) V. North Atlantic Treaty
  Organization. Scientific Affairs Division.   VI. Series: NATO ASI
  series. Series C, Mathematical and physical sciences ; no. 234.
  QC454.L3F76 1988
  535.5'8--dc19                                          88-12042
                                                             CIP
```

ISBN 978-94-010-7849-8

Published by Kluwer Academic Publishers,
P.O. Box 17, 3300 AA Dordrecht, The Netherlands.

Kluwer Academic Publishers incorporates the publishing programmes of
D. Reidel, Martinus Nijhoff, Dr W. Junk, and MTP Press.

Sold and distributed in the U.S.A. and Canada
by Kluwer Academic Publishers,
101 Philip Drive, Norwell, MA 02061, U.S.A.

In all other countries, sold and distributed
by Kluwer Academic Publishers Group,
P.O. Box 322, 3300 AH Dordrecht, The Netherlands.

CONTENTS

MULTIPHOTON IONIZATION SPECTROSCOPY WITH PHOTOELECTRON
AND MASS SPECTRAL ANALYSIS, S.D. COLSON

MULTIPHOTON IONIZATION MASS SPECTROMETRY OF BIOMOLECULES,
J. GROTEMEYER AND E.W. SCHLAG

UNIMOLECULAR DECAY OF ENERGY-SELECTED POLYATOMIC MOLECULAR
IONS BY RESONANCE-ENHANCED MULTIPHOTON IONIZATION IN A
REFLECTRON MASS SPECTROMETER, A. KIERMEIER, H. KÜHLEWIND,
H.J. NEUSSER, AND E.W. SCHLAG

INFRARED SPECTROSCOPY OF MOLECULAR IONS, T. OKA

DYNAMICS OF THE PHOTODISSOCIATION OF SMALL MOLECULES,
P. ANDRESEN

PREFACE

This volume contains the lectures presented at the NATO Advanced Study
Institute on "Frontiers of Laser Spectroscopy of Gases" held in the
Hotel Golf Mar, Vimeiro, near Torres Vedras, Portugal from 30 March to
10 April 1987. The objective of the meeting was to take stock of the
recent technological developments involving lasers and to assess their
impact on spectroscopy. The whole range of wavelengths from the far
infrared through to the extreme ultraviolet was covered. In addition,
specific applications to both atoms and molecules were described.
Indeed, one of the most successful and pleasant aspects of the Institute
was the joint participation of atomic physicists and molecular spectroscopists, who meet all too rarely these days. The Institute also
succeeded in covering a wide time span from the very earliest days of
lasers to some of the very latest developments in both lasers and their
applications to spectroscopy.

There were 14 invited lecturers, giving a total of 40 lectures, and
89 other participants at the Institute. Each of the invited lecturers
has contributed a chapter to this volume. In addition, on Thursday 2nd
April a special one-day session was held in the Chemistry Department at
the University of Coimbra to mark the retirement of Professor Dr. F.
Pinto-Coelho. Three speakers, Professors Carrington, Mills and
Winnewisser, gave excellent talks on this occasion and these are also
reproduced as chapters in this volume. The Institute was fortunate in
the participation of Dr K. M. Evenson, one of the pioneers of far-
infrared spectroscopy. He was persuaded to make a presentation on
developments in this area and has also contributed a chapter.

All who were present at the A.S.I. can testify to its successful
outcome. This was fuelled by the enthusiasm for the subject matter of
all the participants, both young and old, and was helped in no small
way by a fine setting for the meeting on the cliffs overlooking the
Atlantic Ocean. The Portuguese were warm and generous hosts and the
one-day visit to the lovely mediaeval city of Coimbra was particularly
memorable.

It is a pleasure to express our thanks to the NATO Scientific
Affairs Division, whose financial support made this Institute possible.
In addition, we have received generous contributions from the following
companies and institutions:

Imperial Chemical Industries
Unilever
Spectra-Physics
Newport Corporation
Burleigh Instruments
U.S. Air Force
U.S. Army
U.S. Office of Naval Research

Junta Nacional de Investigação Científica e Tecnológica
Instituto Nacional de Investigação Científica
Fundação Calouste Gulbenkian
INVOTÁN-Portugal
Secretaria do Estado do Ensino Superior Reitoria da Universidade de
  Coimbra

and also support from:

Banco Pinto e Sotto Mayor
Banco Português do Atlântico
Região de Turismo do Oeste

We are very grateful to all of these for their help in making the A.S.I.
such a success both scientifically and otherwise.

                                          A.C.P. Alves
                                          J.M. Brown
                                          J.M. Hollas

EMERITUS PROFESSOR FERNANDO PINTO COELHO
A brief outline of an outstanding career

Born on 18 April 1912 in Funchal, região autónoma da Madeira, Portugal,
Fernando Pinto Coelho is the son of Augusto Pinto Coelho and Maria
Lidia Costa Pinto Coelho.

He graduated with distinction from both the Curso Complementar do
Liceu, Funchal, (1930), and with the degree of Licenciatura in Chemical
and Physical Sciences (Ciências Físico-Químicas) from the University of
Coimbra (1934).

He was awarded his doctorate in Ciências Físico-Químicas by the
University of Coimbra in 1944 for his thesis on structural studies of
triterpenic alcohols.

Dr. Pinto Coelho was first appointed Assistant Lecturer at the
University of Coimbra (Department of Chemistry) in 1934, and subse-
quently sat open examination (Concurso de Provas Públicas) to become
Professor Extraordinário (Associate Professor) of the 2° Grupo da 2ª
Secção (Química) of the Faculty of Sciences, now Faculty of Science and
Technology, in 1950, and Professor Catedrático (Full Professor) in 1956.
Both these Concursos and the award of his doctorate were approved
unanimously.

On reaching retirement age in 1984 he was appointed Professor
Catedrático Jubilado (Emeritus Professor) by the University of Coimbra.

Professor Pinto Coelho was elected to membership of the Academy of
Sciences of Lisbon in 1979.

In 1938 Dr. Pinto Coelho was invited to work in the laboratories
(Imperial College of Science and Technology) of Sir Ian Heilbron and
E.R.H.Jones on triterpenes. The advent of world war II did not allow the
full accomplishment of the planned research project. He subsequently
spent the period 1953-54 at the University of Cambridge, England, with
Dr. A.G. Maddock, working on radiochemical studies of organometallic
compounds (vitamin B12), the Radiochemical Laboratory of the Chemistry
Department of the University of Cambridge and the Atomic Energy Research
Establishment at Harwell (U.K.), and has since visited and collaborated
with universities and research laboratories in England, Angola,
Mocambique and Acores.

Professor Pinto Coelho has published more than 25 scientific papers,
three books of a pedagogical nature, and has presented many papers at
national and international scientific meetings and conferences.

He still is active as a teacher and researcher within the Academy
of Sciences of Lisbon and is in collaboration with the Escola Superior
de Tecnologia de Tomar.

We were honoured to present a special scientific session held in
the Auditorio da Universidade de Coimbra, on the 2 April 1987, on
"Lasers and Some Important Applications" dedicated to Professor Doctor
Fernando Pinto Coelho.

A.C.P. Alves, The Organizing Committee

DIRECTORS

Alves, Prof. A.C.P., Universidade de Coimbra, Portugal
Hollas, Dr. J.M., University of Reading, U.K.

UK ORGANIZING COMMITTEE

Brown, Dr. J.M., University of Oxford, U.K.
Davies, Dr. P.B., University of Cambridge, U.K.
Dixon, Prof. R.N., University of Bristol, U.K.
Hollas, Dr. J.M., University of Reading, U.K.
Sarre, Dr. P.J., University of Nottingham, U.K.

PORTUGUESE ORGANIZING COMMITTEE

Alves, Prof. A.C.P., Universidade de Coimbra, Portugal
Formosinho, Prof. S.J., Universidade de Coimbra, Portugal
Mariano, Dr. J.S., Universidade de Coimbra, Portugal

LECTURERS

Andresen, Prof. Dr. P., Max Planck Institut, Göttingen, F.R. Germany
Baird, Dr. P.E.G., University of Oxford, U.K.
Barron, Prof. L.D., University of Glasgow, U.K.
Cagnac, Prof. B., Université Pierre et Marie Curie, Paris, France
Colson, Prof. S.D., Yale University, New Haven, U.S.A.
Dixon, Prof. R.N., University of Bristol, U.K.
Leach, Dr. S.D., Université de Paris Sud, Orsay, France
Miller, Prof. T.A., Ohio State University, Columbus, U.S.A.
Oka, Prof. T., University of Chicago, U.S.A.
Ramsay, Dr. D.A., Herzberg Institute of Astrophysics, Ottawa, Canada
Schlag, Prof. Dr. E.W., Technischen Universität München, Garching,
  F.R. Germany
Stoicheff, Prof. B.P., University of Toronto, Canada
Urban, Prof. Dr. W., Universität Bonn, F.R. Germany
Wallenstein, Prof. Dr. R., Universität Bielefeld, F.R. Germany

COIMBRA SPECIAL SESSION LECTURERS

Carrington, Prof. A., University of Oxford, U.K.
Mills, Prof. I.M., University of Reading, U.K.
Winnewisser, Prof. Dr. G., Universität zu Köln, F.R. Germany

OTHER PARTICIPANTS

Alugbin, Dr. D., Exxon Res. and Eng. Co., Annandale, U.S.A.
Arqueros, Dr. F., Universidad Complutense, Madrid, Spain

Bartels, Dr. M., University of Edinburgh, U.K.
Bermejo Plaza, Dr. D., Inst. de Estructura de la Materia, Madrid, Spain
Bin Musa, Dr. H., University of Technology of Malaysia, Kuala Lumpur, Malaysia
Blom, Dr. C.E., Justus-Liebig Universität, Giessen, F.R. Germany
Bohle, Dr. W., Inst. für Angewandte Physik, Bonn, F.R. Germany
Brandão, J.C.C.P., University of Coimbra, Portugal
Braun, Dr. M., Universität Tübingen, F.R. Germany
Brüggemann, R., Kronstädter Str. 26, Bonn, F.R. Germany
Caballero, Dr. J.F., University of California, Davis, U.S.A.
Calvo Hernandez, Dr. A., University of Salamanca, Spain
Cefalas, Dr. A.C., National Hellenic Research Foundation, Athens, Greece
Darwin, D.C., University of California, Berkeley, U.S.A.
Davidsson, J., Chalmers University of Technology, Göteborg, Sweden
Del Olmo Pintado, A., Inst. de Estructura de la Materia, Madrid, Spain
Demuynck, Dr. C., Université de Lille, France
Dinelli, Dr. B.M., Ist. di Spettroscopia Molecolare, Bologna, Italy
Docker, M.P., University of Nottingham, U.K.
Domenech Martinez, J.L., Inst. de Estructura de la Materia, Madrid, Spain
Domingo Maroto, Dr. C., Inst. de Estructura de la Materia, Madrid, Spain
Donovan, Prof. R., University of Edinburgh, U.K.
Drysdale, S.L.T., University of Glasgow, U.K.
Escribano, Dr. J.R., Cadiz, Spain
Evenson, Dr. K.M., National Bureau of Standards, Boulder, U.S.A.
Falęcki, W., Instytut Fizyki UJ, Kraków, Poland
Ferreira, Prof. M.A.A., Universidade de Lisboa, Portugal
Gaspar, R.P.P.S., Universidade de Coimbra, Portugal
Gonzalez Valdenebro, A., Inst. de Estructura de la Materia, Madrid, Spain
Hanley, L., State University of New York at Stony Brook, U.S.A.
Hamad, K.I., University of Strathclyde, U.K.
Heber, K.-D. Freie Universität Berlin, F.R. Germany
Herrero, Dr. V.J., Inst. de Estructura de la Materia, Madrid, Spain
Hertzler, Ch., Universität Kaiserslautern, F.R. Germany
Hippler, Dr. H., Universität Göttingen, F.R. Germany
Hotokka, Dr. M., Åbo Akademi, Finland
Huet, Th., Université Libre de Bruxelles, Belgium
Jardim, M.E., Universidade de Lisboa, Portugal
Jones, Dr. H., Universität Ulm, F.R. Germany
Jorge, M.E.M., Universidade de Lisboa, Portugal
Kennedy, Dr. R.A., University of Birmingham, U.K.
Korppi-Tomola, Dr. J., University of Jyväskylä, Finland
Liu, D.-J., University of Chicago, U.S.A.
Lopes, J.C.F.N., Universidade de Aveiro, Portugal
McCoustra, M.R.S., Heriot-Watt University, Edinburgh, U.K.
Martin, P.A., University of Cambridge, U.K.

Mikropoulos, T., National Hellenic Res. Foundation, Athens, Greece
Milkman, I., University of Oregon, Eugene, U.S.A.
Miller, Dr. S., University College London, U.K.
Mohebati, A., University of Manchester, U.K.
Monteiro, Prof. L.F., Universidade Nova de Lisboa, Portugal
Monteiro, Prof. M.L., Faculdade Ciências, Universidade de Lisboa,
  Portugal
Natalis, Prof. P., Université de Liège, Belgium
Pais, A.C.C., Universidade de Coimbra, Portugal
Pape, D.A., University of Leicester, U.K.
Patsilinakou, E., University of Crete, Iraklion, Crete, Greece
Persch, G., Universität Kaiserslautern, F.R. Germany
Pfab, Dr. J., Heriot-Watt University, Edinburgh, U.K.
Pinto-Coelho, Prof. Dr. F., Universidade de Coimbra, Portugal
Plimmer, M.D., University of Oxford, U.K.
Ribeau-Teixera, Dr. M., LNETI, Salavem, Portugal
Ribeiro-Claro, P.J.A., Universidade de Coimbra, Portugal
Roman, Dr. P., U.S. Office of Naval Research, London, U.K.
Römheld, Dr. M., Siemens AG, Erlangen, F.R. Germany
Santos Gomez, J., Inst. deEstructura de la Materia, Madrid, Spain
Senekowitsch, J., Universität Frankfurt, Austria
Siebbeles, L., FOM Inst. for Atomic and Mol. Physics, Amsterdam,
  The Netherlands
Smith, Lt.Col. La R.K., Department of the U.S. Air Force, London, U.K.
Softley, Dr. T.P., University of Cambridge, U.K.
Spyrou, Dr. S.M., National Hellenic Res. Foundation, Athens, Greece
Stewart, Dr. R.S., University of Strathclyde, Glasgow, U.K.
Stickland, R.J., University of Bristol, U.K.
Stiefvater, Dr. O.L., University College of Wales, Bangor, U.K.
Tate, Dr. D.A., University of Oxford, U.K.
Teixera, Prof. M.E.F.A., Universidade de Aveiro, Portugal
Teixera-Dias, Prof. J.J.C., Universidade de Coimbra, Portugal
Valle, Dr. R.G.D., Università degli Studi-Bologna, Italy
Weitzel, K.-M., Universität Göttingen, F.R. Germany
Western, Dr. C., University of Bristol, U.K.
Williams, Dr. J.H., University of Cambridge, U.K.
Wiorkowski, Dr. P., Physikalische Institut, Tübingen, F.R. Germany
Zink, Dr. L.R., University of Oxford, U.K.

CONTRIBUTORS

P. Andresen
Max Planck Institut
   für Strömungsforschung
Bunsenstrasse 10
D-3400 Göttingen
F.R. Germany

P.E.G. Baird
Clarendon Laboratory
University of Oxford
Parks Road
Oxford OX1 3PU
U.K.

L.D. Barron
Chemistry Department
University of Glasgow
Glasgow G12 8QQ
U.K.

B. Cagnac
Laboratoire Spectroscopie Hertzienne
   de l'E.N.S.
Université Pierre et Marie Curie
75252 Paris Cedex 05
France

A Carrington
Department of Physical Chemistry
University of Oxford
South Parks Road
Oxford OX1 3QZ
U.K.

S.D. Colson
Sterling Chemistry Laboratory
Yale University
New Haven
Connecticut 06511
U.S.A.

R.N. Dixon
School of Chemistry
University of Bristol
Cantock's Close
Bristol BS8 1TS
U.K.

K.M. Evenson
Time and Frequency Division
National Bureau of Standards
Boulder
Colorado 80303
U.S.A.

S.C. Foster
Laser Spectroscopy Facility
Department of Chemistry
The Ohio State University
120 West 18th Avenue
Columbus
Ohio 43210
U.S.A.

J. Grotemeyer
Institut für Physikalische und
   Theoretische Chemie
Technische Universität München
Lichtenbergstr. 4
D-8046 Garching
F.R. Germany

D.A. Jennings
Time and Frequency Division
National Bureau of Standards
Boulder
Colorado 80303
U.S.A.

R.A. Kennedy
Department of Chemistry
University of Birmingham
P.O. Box 363
Birmingham B15 2TT
U.K.

A. Kiermeier
Institut für Physikalische und
   Theoretische Chemie
Technische Universität München
Lichtenbergstr. 4
D-8046 Garching
F.R. Germany

H. Kühlewind
Institut für Physikalische und
   Theoretische Chemie
Technische Universität München
Lichtenbergstr. 4
D-8046 Garching
F.R. Germany

S. Leach
Laboratoire de Photophysique
   Moléculaire du C.N.R.S.
Bâtiment 213
Université Paris-Sud
91405 Orsay
France

I.R. McNab
Department of Physical Chemistry
University of Oxford
South Parks Road
Oxford OX1 3QZ
U.K.

T.A. Miller
Laser Spectroscopy Facility
Department of Chemistry
The Ohio State University
120 West 18th Avenue
Columbus
Ohio 43210
U.S.A.

I.M. Mills
Department of Chemistry
University of Reading
Reading RG6 2AD
U.K.

C.A. Montgomerie
Department of Physical Chemistry
University of Oxford
South Parks Road
Oxford OX1 3QZ
U.K.

H.J. Neusser
Institut für Physikalische und
   Theoretische Chemie
Technische Universität München
Lichtenbergstr. 4
D-8046 Garching
F.R. Germany

T. Oka
Department of Chemistry and
   Department of Astronomy and
   Astrophysics
The University of Chicago
Chicago
Illinois 60637
U.S.A.

D.A. Ramsay
Herzberg Institute of Astrophysics
National Research Council of Canada
100 Sussex Drive
Ottawa
Ontario
Canada K1A OR6

E.W. Schlag
Institut für Physikalische und
   Theoretische Chemie
Technische Universität München
Lichtenbergstr. 4
D-8046 Garching
F.R. Germany

B.P. Stoicheff
Department of Physics
University of Toronto
Toronto
Ontario
Canada M5S 1A7

W. Urban
Institut für Angewandte Physik
Universität Bonn
Wegelerstr. 8
D-5300 Bonn 1
F.R. Germany

M.D. Vanek
Time and Frequency Division
National Bureau of Standards
Boulder
Colorado 80303
U.S.A.

R. Wallenstein
Institut für Quantenoptik
Universität Hannover
3000 Hannover
F.R. Germany

G. Winnewisser
I. Physikalisches Institut
Universität zu Köln
5000 Köln
F.R. Germany

SPECTROSCOPY WITHOUT LASERS

D. A. Ramsay
Herzberg Institute of Astrophysics
National Research Council of Canada
100 Sussex Drive
Ottawa, Ontario
Canada K1A 0R6

ABSTRACT. A discussion is given of the relative merits of grating spectroscopy, interferometry, and laser spectroscopy. The contributions of classical spectroscopy, particularly to our knowledge of the spectra of free radicals and molecular ions, are briefly summarized. Finally some philosophical remarks are made concerning the accuracy with which molecular constants can be determined for polyatomic molecules in excited electronic states.

1.   INTRODUCTION

1.1 Mandate

I was originally invited to give two lectures on Spectroscopy **before** Lasers but considered it more appropriate to give these introductory talks on Spectroscopy **without** Lasers. My personal view is that lasers provide a very powerful adjunct to the battery of techniques which spectroscopists employ to tackle various problems but do not completely supersede what now might be called the earlier classical techniques. My assignment is to point out the advantages and disadvantages of these techniques vis-à-vis laser techniques.

1.2 Classical Instruments

Historically the main instruments at the disposal of the optical spectroscopist have been the prism and grating spectrograph or spectrometer, and the interferometer. With such instruments one can cover the spectral region from the vacuum ultraviolet (100–2000 Å) through the near ultraviolet, visible and infrared to the far infrared (100–1000 µm). Usually a given instrument covers only a limited region of the spectrum but recently interferometers have become available which can cover the complete region from 2000 Å (50,000 cm^{-1}) to 1000 µm (10 cm^{-1}) with appropriate choice of beam splitter and detector.

1

A. C. P. Alves et al. (eds.), Frontiers of Laser Spectroscopy of Gases, 1–8.
© 1988 by Kluwer Academic Publishers.

1.3 Doppler Widths

The limitation on high resolution spectroscopy until the advent of the laser was provided by the Doppler widths of the lines. The frequency width $\Delta\nu$ at half maximum intensity (FWHM) is given by the formula

$$\Delta\nu = \frac{2\nu}{c}\sqrt{\frac{2RT\ln 2}{M}} = 7.162 \times 10^{-7}\,\nu\,\sqrt{\frac{T}{M}}.$$

where R is the gas constant, M the molecular weight and T the absolute temperature. Some typical Doppler widths are given in Table I.

TABLE I Doppler Widths of Lines (assuming T=300K and M=50) for Different Spectral Regions

λ	$\nu(cm^{-1})$	Doppler Width $\Delta\nu(cm^{-1})$
1200 Å	83 333	0.15
2000 Å	50 000	0.09
3000 Å	33 333	0.06
5000 Å	20 000	0.035
10000 Å	10 000	0.018
3 μm	3 333	0.006
5 μm	2 000	0.004
10 μm	1 000	0.002

1.4 Grating Instruments

In the region of the spectrum from 2000 Å to 12000 Å large Ebert or Czerny-Turner spectrographs (3.4m - 10m) with resolving powers of ~600,000 are in current use. These instruments can resolve molecular spectra down to the Doppler widths of the lines for molecules with molecular weights up to 50-100. In the vacuum ultraviolet the practical resolving power is less, although resolving powers of the order of 300,000 have been achieved in the region of 1500 Å. In the infrared the resolving powers of grating instruments rarely exceeded 200,000 and these instruments have been almost completely replaced by interferometers.

1.5 Interferometers

The resolving power of an interferometer depends on the maximum path length difference between two interfering beams. In high resolution instruments which are commercially available the moving mirror covers 1.25m and the resolution is $1/(2\times125) = 0.004$ cm^{-1}. This resolution is more than adequate for studies in the visible and ultraviolet and is usually sufficient for studies in the infrared as

far as 5 μm. In the far infrared higher resolving power is required
and resolutions of ~0.001 cm^{-1} have been achieved.

2.   ELECTRONIC SPECTRA

2.1 Classical Studies

I will now discuss the work which has been carried out by
classical spectroscopy on the electronic spectra of diatomic and
polyatomic molecules including unstable chemical species which can be
classified as free radicals or molecular ions. More than 1000
diatomic (1) and 100 polyatomic species (2) have been found and
studied with Doppler-limited resolution.

2.2 Free Radicals and Molecular Ions

These species are of particular interest since special techniques
are required for the production of these species. Most of the
diatomic species and some of the polyatomic species have been found
using some form of electrical discharge. Identification of the
molecular species responsible for a given spectrum is fraught with
danger if based on chemical intuition alone. Isotopic substitution
provides a valuable aid in determining which atoms are present and how
many. Rotational analysis provides information on the type of
electronic transition involved and on the moment(s) of inertia of the
molecule. Vibrational analysis gives useful, but by no means
unambiguous, information.
Several other techniques have been employed including high
temperature furnaces, flames and shock tubes. Chemiluminescence
sources have been used and will be discussed in more detail later.
Vacuum ultraviolet excited fluorescence and photochemical modulation
have also given interesting results.

2.3 Flash Photolysis Studies

One of the most powerful techniques for studying the absorption
spectra of free radicals, particularly polyatomic species, is that of
flash photolysis. A summary of the polyatomic species which have been
studied spectroscopically (3) is given in Table II; approximately
one-half of these have been found by flash photolysis. In this
technique a parent molecule is decomposed by an intense light flash

$$AB + h\nu \rightarrow A+B$$

and absorption spectra of the fragments are photographed using a
second flash lamp which provides a source of continuum and has a
duration of a few microseconds. It is interesting to reflect that a
2" x 18" photographic plate capable of resolving 40 lines/mm is
equivalent to a diode array containing approximately one million
elements. Since spectra can be recorded in a few microseconds the

rate of acquisition of data is quite phenomenal, ~10^{12} bits/sec. The photographic plate also has the merit that it can record spectra over a wide range, 1000 - 12000 Å, although the sensitivity is low for λ>9000Å. Quite a battery of expensive lasers would be required to give the equivalent wavelength coverage!

Table II Polyatomic Free Radicals and Molecular Ions Studied
by High Resolution Electronic Spectroscopy

H_3,	BH_2,	$A\ell H_2$,	CH_2,	SiH_2,	NH_2,	PH_2,	AsH_2,	SbH_2,
H_2O^+,	H_2S^+,	C_2H,	HCO,	HNO,	HPO,	HCF,	$HCCl$,	$HSiCl$,
$HSiBr$,	$HSiI$,	HNF,	HO_2,	HSO,	HS_2,			
C_3,	SiC_2,	CCN,	CNC,	NCN,	CCO,	NCO,	CNO,	NCS,
BO_2,	N_3,	CO_2^+,	COS^+,	CS_2^+,	N_2O^+,	NO_2,	PO_2,	CF_2,
SiF_2,	S_2O,	NSF,	ClO_2,					
CH_3,	CH_2Si,	$HNCN$,	$HCCS$,	NO_3,	F_2BO,	F_2CN,	NH_4,	CH_3O,
CH_2CH,	FSO_3,	CH_3NO,	CF_3NO,	CH_2CHO,	$C_4H_2^+$,	C_5H_5,	$C_6H_5CH_2$	

2.4 Fourier Transform Studies

I would now like to discuss some results obtained with interferometers. Absorption and emission spectra of numerous species have been obtained in the region 2000Å - 1000μ which covers both electronic and vibration-rotation transitions. As examples, I will show you spectra taken at Ottawa with a Bomem DA3.002 interferometer which has a maximum resolution of 0.004 cm^{-1}. These spectra include:

a) the $2^2_0 4^1_0$ band of formaldehyde in absorption near 3260Å (4),

b) bands of the A $^3\Sigma_u^+$ - X $^3\Sigma_g^-$ and c $^1\Sigma_u^-$ - X $^3\Sigma_g^-$ systems of oxygen in the near ultraviolet taken with a path length of 660m and pressures ranging from 0.5 to 2 atm (5),

c) vibration-rotation bands of H_2O in the visible region taken with a path length of 1200 m and a pressure of 17 Torr (6),

d) emission spectra of NH and NH_2 in the region of 3μ, including both electronic and vibration-rotation transitions (7),

e) the ν_4 band of diacetylene in absorption near 3μ (4),

f) the ν_{14} band of benzene in absorption near 10μm (8),

g) an absorption band of the Van der Waals complex, Ar - H_2, near 4500 cm^{-1} (9).

Recently, in collaboration with Professor E.H. Fink (Wuppertal, W.Germany), we have observed some weak emission spectra in the near infrared in chemiluminescent systems. The apparatus is shown in Fig.1. Metastable oxygen molecules (a $^1\Delta_g$) are produced in a

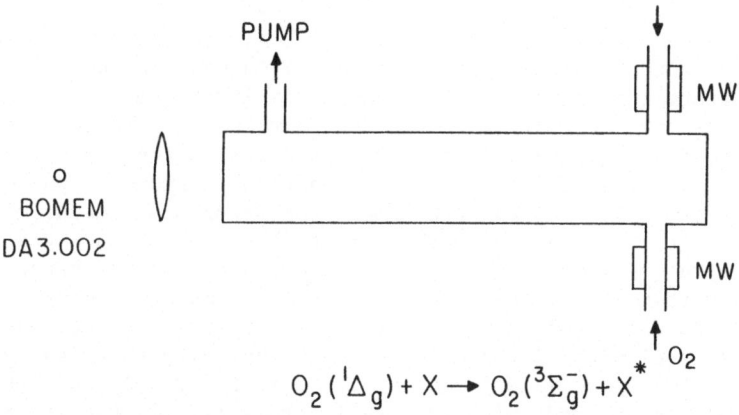

$$O_2(^1\Delta_g) + X \rightarrow O_2(^3\Sigma_g^-) + X^*$$

Figure 1. Chemiluminescence apparatus. The second molecule is introduced through the upper sidearm. In some cases it may be necessary to produce this molecule by means of a microwave discharge through an appropriate parent molecule.

microwave discharge and react with other molecules in their ground states. By energy exchange the latter are excited and give rise to the weak emission spectra. In this way high resolution spectra have been obtained for the b $^1\Sigma_g^+$ – X $^3\Sigma_g^-$ system of S_2 (10) as well as for the b O^+ – X_1 O^+ and b O^+ – X_2 1 subsystems of SO (11) and SeS (12). Bands of the $\tilde{A}$ $^2A'$ – $\tilde{X}$ $^2A''$ system of HO_2 and DO_2 have also been recorded (13). With oxygen alone, the b $^1\Sigma_g^+$ – a $^1\Delta_g$ electric **quadrupole** transition of O_2 (14) has been observed, indicating the sensitivity of the Fourier transform system in the near infrared.

3.   COMPARISON OF CLASSICAL AND LASER TECHNIQUES

3.1 Grating Instruments

Grating spectroscopy is still a viable activity in the region from approximately 1000 to 10000Å. In the infrared it has been almost completely superseded by interferometry. With large grating instruments, resolution is generally limited by the Doppler widths of the lines rather than by the optical system used, except for heavier molecular species. The use of the photographic plate permits a region of the spectrum to be studied at one time. A particular advantage is obtained with pulsed systems, e.g. flash photolysis, flash discharges, since the photographic plate readily integrates a series of optical pulses. A disadvantage is that the measuring process is rather tedious and furthermore it is not easy to obtain good intensity measurements. For the latter a photoelectric system is to be preferred.

3.2 Interferometers

Interferometers are particularly useful in the infrared where the performance of grating instruments is limited by the sensitivities and noise characteristics of the detectors used. The interferometer has the advantage that the complete spectral region under study is sampled all the time in contrast to a scanning spectrometer which samples only a small region at any one time. The interferometer normally has a greater aperture and hence greater throughput than a grating instrument (the Jacquinot advantage), and also the detector noise contributes only to the signal for the complete spectral region and not to each individual spectral element (the Fellgett advantage). Frequencies and intensities are automatically provided by the Fourier transform process and are usually more internally consistent than measurements made with grating instruments. In the visible and ultraviolet the interferometer gives essentially the same performance as a grating instrument. In principle it should be possible to detect weaker signals with an interferometer than with a photographic instrument, but in practice the latter has the advantage that very long exposure times (e.g. days) or very long path lengths (up to 10km) can be employed if necessary to detect weak emissions or absorptions. The most serious disadvantage of the interferometer is that it requires a reasonably steady light source, and hence cannot be easily used with pulsed systems, e.g. flash photolysis or flash discharge.

3.3 Lasers

The chief virtues of the laser are its monochromaticity and its high intensity. Spectra are recorded by scanning the laser if it is tunable or by tuning the molecule with an electric or magnetic field if the laser frequency is fixed. Line widths of the order of 1 MHz or better are readily obtainable and sub-Doppler Spectra can be recorded with a resolution ~1000 times higher than can be obtained with classical techniques. Hyperfine structure is frequently resolved. Another advantage of the laser is that two-photon and multi-photon spectra can be studied whereas classical techniques are limited to one-photon spectra. The chief disadvantage of the laser at the present time is the limited frequency coverage of a given laser system and its relatively high cost. The laser has high sensitivity and this is best demonstrated by work in the infrared where spectra of many free radicals and molecular ions have been found which have not been produced by classical techniques. In the visible, high sensitivities have also been claimed using intracavity techniques. However, while the 5-0 and 6-0 bands of HD have been found in absorption by classical techniques (15), only the 5-0 band has so far been reported in laser studies. Furthermore, in flash photolysis studies the $(0,11,0)-(0,0,0)$ band of $H^{13}CO$ has been observed in natural abundance (16) but has not been detected in laser studies.

4. PHILOSOPHICAL REMARKS

Finally I should like to make some philosophical remarks about the spectroscopy of polyatomic molecules in excited electronic states. Recently Dr. Fung and I carried out some intermodulated fluorescence experiments on the 4^1_0 band of the $\tilde{A}\ ^1A_2 - \tilde{X}\ ^1A_1$ system of H_2CS (17). Numerous perturbations were found in the excited state energy levels ranging up to a few hundred MHz. These perturbations are caused by interactions with the high rovibronic levels of the ground state. Further work using microwave-optical double resonance studies of the excited state of this and other molecules (18-20) has confirmed our suggestion "that these results may be a manifestation of a more widespread phenomenon, i.e. that in the spectra of polyatomic molecules numerous perturbations exist between the higher rovibronic levels of the ground state and the corresponding levels of electronically excited states." These perturbations limit the accuracy with which molecular constants can be determined for molecules in excited electronic states. Indeed it is possible that the constants derived from the analysis of classical spectra covering a wide range of J and K, are close to the best which can be determined.

5. REFERENCES

1. K. P. Huber and G. Herzberg, 'Molecular Spectra and Molecular Structure, IV Constants of Diatomic Molecules', <u>Van Nostrand Reinhold</u>, New York, U.S.A., 1979.

2. G. Herzberg, 'Molecular Spectra and Molecular Structure, III Electronic Spectra of Polyatomic Molecules', <u>Van Nostrand</u>, Princeton, N.J., U.S.A., 1967.

3. D. A. Ramsay, 'Electronic Spectra of Polyatomic Free Radicals, ch2 in Vibrational Spectra and Structure', <u>ed.James R. Durig</u>, **14,** 69-124 (1985), Elsevier.

4. D. A. Ramsay, unpublished.

5. W. S. Neil, D. A. Ramsay and M. Vervloet, to be published.

6. C.Camy-Peyret, J.-M. Flaud, J.-Y. Mandin, J.-P. Chevillard, J. Brault, D. A. Ramsay, M. Vervloet and J. Chauville, <u>J.Mol Spect</u>., **113,** 208-228 (1985).

7. M. Vervloet, private conversation.

8. J. Pliva and J.W.C. Johns, <u>J. Mol. Spect</u>. **107,** 318-323 (1984).

9. A. R. W. McKellar, <u>Proceedings, NATO Advanced Research Workshop on Structure and Dynamics of Weakly Bound Molecular</u>

8

Complexes, (Acquafredda di Maratea, Sept. 1986) Reidel, Dordrecht (1987).

10. E. H. Fink, H. Kruse and D.A. Ramsay, J. Mol. Spect. 119, 377-387 (1986).

11. E. H. Fink, H. Kruse and D.A. Ramsay, to be published.

12. E. H. Fink, H. Kruse, D.A. Ramsay and Ding-Chang Wang, Molecular Physics 60, 277-290 (1987).

13. E. H. Fink, H. Kruse and D.A. Ramsay, to be published.

14. E. H. Fink, H. Kruse, D.A.Ramsay and M. Vervloet, Can.J.Phys. 64, 242-245 (1986).

15. A. R. W. McKellar, W. Goetz and D.A. Ramsay, Ap. J. 207, 663-670 (1976).

16. J. M. Brown and D. A. Ramsay, Can. J. Phys. 53, 2232-2241 (1975).

17. K. H. Fung and D. A. Ramsay, J. Phys.Chem. 88, 395-397 (1984).

18. J. C. Petersen, D. A. Ramsay and T. Amano, Chem.Phys.Lett. 103, 266-270 (1984).

19. J. C. Petersen, S. Saito, T. Amano and D. A. Ramsay, Can.J. Phys. 62, 1731-1737 (1984).

20. J. C. Petersen, T. Amano and D. A. Ramsay, J. Chem.Phys. 81, 5449-5452 (1984).

INFRARED LASERS FOR SPECTROSCOPY

Wolfgang Urban
Institut für Angewandte Physik
Universität Bonn
Wegelerstr. 8
D-5300 Bonn 1

ABSTRACT. A survey of various IR-lasers is contained in the first chapter. Short explanations of the underlying mechanisms for both continuously tunable and stepwise tunable lasers suitable for high resolution spectroscopy are given.

After this introduction, we discuss in detail one example for each of the continuously and stepwise tunable IR-laser systems. In chapter 2, a model for the most common colour centre laser (CCL) is explained and linked to other phenomena. The essential points of a computer controlled CCL-spectrometer are also incorporated.

In the final chapter 3 the VV-pumping mechanism of the CO-laser is explained. Both the physics and technology of a widely tunable CO-laser are discussed in detail.

1. SURVEY OF LASERS, SUITABLE FOR SPECTROSCOPY IN THE MEDIUM INFRARED

The medium infrared spectral region contains typically vibrational transitions of molecules and their rotational substructure. Therefore it is obvious, that one can use vibration rotation transitions in a laser medium itself, provided there is an inversion mechanism available. However, in the gas phase such transitions are fairly narrow and therefore will not be the ideal source for spectroscopy, where one would like to have a continuously tunable laser source in order to scan across a series of vibration-rotation transitions of the molecular gas to be investigated. Although we can make use of it for very special situations e.g.for the spectroscopy of paramagnetic molecules, where Zeeman-tuning of the molecular transition can be achieved, we must use other types of gain media for a tunable infrared laser.

There are various approaches. Either one has a gain me-

A. C. P. Alves et al. (eds.), Frontiers of Laser Spectroscopy of Gases, 9–42.

dium with a very broad linewidth, and use coarse and fine tuning elements to scan the output wavelength across the gain bandwidth. This is very similar to dye lasers. Gain media can be either electronic transitions in ions distributed in a crystal lattice, such as transition metal ions, or electronic transitions of electrons trapped in a lattice vacancy, which is generally known as a colour centre. Both these types will be explained in the survey sections 1.1 and 1.2 resp. A detailed model for a colour centre laser will be extensively discussed in chapter 2.

A different type of a tunable infrared gain medium is found in the tunable diode laser. The gain is produced by carrier jumps across the band gap of a semiconductor. This band gap is temperature dependent and so is the corresponding gain wavelength. This way the coarse tuning is achieved. Further details will be explained in the subsequent chapter.

There are other methods to generate IR-radiation, starting off with a tunable dye laser by down conversion from the visible into the infrared. In the difference frequency laser, the dye laser radiation is mixed with a fixed frequency laser inside a birefringent crystal. The nonlinear behaviour of the refractive index under very intense radiation is utilized for this mixing. Stimulated Raman scattering produces sidebands that are shifted by an internal frequency of the Raman medium. This internal frequency can vary with an external parameter e.g. in the case of the spin-flip Raman laser (SFRL) thus producing tunable radiation from a fixed frequency pump laser. Alternatively the pump laser itself is tunable and the Raman shift occurs at a fixed internal frequency, such as the vibration of a diatomic molecule, preferably H_2. The shifting can occur not only by one quantum but also in subsequent processes by a series of quanta. However, one needs fairly high laser power in the first place, which can only be generated in a pulsed mode.

Finally, there is a way to use fixed frequency lasers for spectroscopy if one can achieve the tuning on the side of the molecules. Species with a permanent magnetic or electric dipole moment can be tuned into resonance by the Zeeman – or Stark-effect respectively. Tunability is very limited and therefore a densely distributed series of fixed frequency laser transitions is necessary for complete coverage of the spectrum.

The rotational structure of a molecular gas provides a suitable frequency ladder. The very high spectral quality of some of the molecular gas lasers forms an excellent source for spectroscopy with very high sensitivity. Therefore two molecular gas lasers, the CO_2-laser and the CO-laser will also be outlined in this survey in sections 1.2.1 and 1.2.3. The pumping mechanism and various interesting features of the CO-laser will be discussed in detail in chapter 3. A more complete compilation of laser systems is given in [1,1] and most of the tunable lasers are described in more detail in [1,2].

1.1. Broadband Gain Profile Lasers

There is a variety of laser systems, the gain media of which consist of centres incorporated in a solid. The nature of these centres can be different, e.g. transition metal ions or electrons trapped in a lattice vacancy. Anyway the gain medium is associated with some specific lattice defects randomly distributed in a periodic structure. Since the wavelength of the radiation involved in the laser process is large compared to the lattice unit cell or the defect diameter, the periodicity of the crystal structure is not relevant.

The tunability is due to the lasing transition linewidth, and the latter has different causes. In the transition metal ion lasers phonon- coupling changes the electronic transition frequency and in the colour centre lasers, very fast relaxation is producing the line broadening.

1.1.1 Transition Metal Ion Lasers

The very first laser, the ruby laser, belongs to the family of the transition metal ion lasers. However, its wavelength is fixed.Its lower lasing state is the ground state, the orbital momentum of which is quenched by the crystal field. There is no direct coupling of the lattice vibrations to this ground state, i.e. no phonon sideband can occur.

In the case of alexandrite, the model example of a tunable transition metal ion laser, the crystal field does not quench the orbital momentum of the Cr^{3+} and lattice vibrations can strongly couple to the electronic ground state. The lasing energy, determined by the electronic transition energy, is varied by the phonon energy that chops off a fraction of the electronic energy difference by transformation into the lattice vibration sink. Crystalline solids with high Debye-energies provide a means to vary their mean phonon energy with temperature. In alexandrite, the gain maximum is tuning by 5 cm^{-1} per degree temperature change at room temperature (Fig.1.1). The chromium in alexandrite is operating in the visible region, close to the ruby laser. However, there are other systems in the near to medium infrared, such as Ni^{2+} and Co^{2+} in MgF_2. The corresponding tuning ranges are 1.61 to 1.74 μm (Ni) and 1.63 to 2.11 μm (Co) respectively. Since these lasing materials have been developed quite recently, we can expect further progress in the field of transition

12

metal ion lasers. A qualitative energy level scheme for various laser materials is given in Fig.1.2. For further details see [1,4/1,1].

Fig.1.1 Gain curves for alexandrite at different temperatures (after Shand and Jensen [1,3])

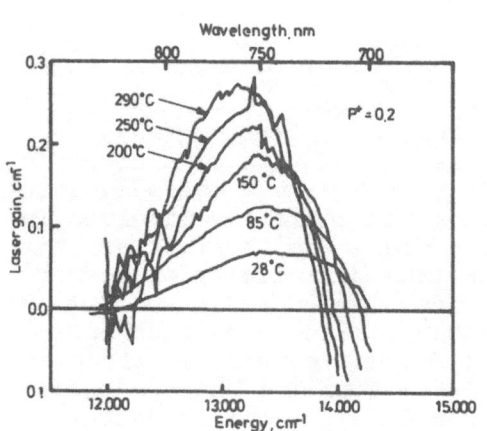

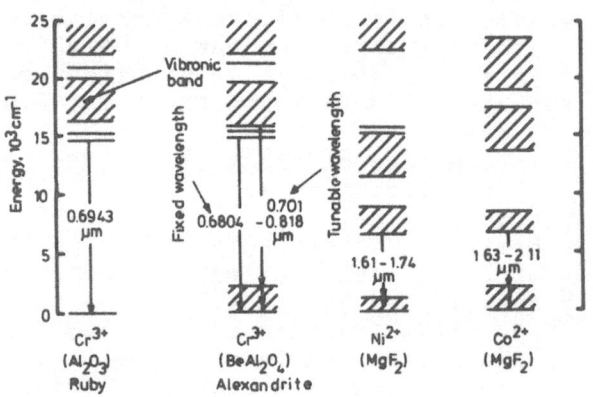

Fig.1.2 Qualitative energy level scheme for various transition metal ion systems [1,1]

1.1.2 Colour Centre Lasers

A very interesting laser system is the colour centre laser. Although it will be discussed in detail later, we will briefly outline its main features in this survey. In contrast to the transition metal ion laser, where the electronic energy is mainly determined by the ion trapped

at a lattice site and its structure is only modified by
the lattice interaction, the colour centre is an electron
trapped in a lattice vacancy. The energy level scheme is
therefore akin to that of an electron in a box. (Fig.1.3).

For laser action, rather special types of colour centre
conditions have to be met and we will postpone their
discussion. In short, the electron distribution in different states of excitation causes a local rearrangement
of the cage formed by the ions of the lattice. Therefore
in the ground state and in the excited state, the box, in
which the electron is sitting, has different effective
diameters. The energy difference between ground state and
excited state is smaller after atomic relaxation in the
excited configuration. Thus an excellent four level laser
scheme is defined (Fig.1.4). The local rearrangement of
the lattice is much faster than the electronic lifetime in
the upper state. The same is true for the lasing end
state: after electronic deexcitation, relaxation is fast
into the original ground state configuration. The local
rearrangement takes up quite a bit of energy which is then
distributed to the phonon system. The short lifetime of
the corresponding end states (#4 and #2 in the four level
scheme) causes the bandwidth for both the absorption and
the fluorescence band in the colour centre system. The
gain curve is homogeneously broadened, a very favorable
situation for a tunable laser.

A large variety of colour centres is available in the
medium and near infrared region [1,5]. Although in principle the spectral region starting from 3.4 μm down to 0.8
μm is completely covered (Fig.1.5), not all systems are
suitable for high resolution spectroscopy. In some cases
tunability is not smooth enough and some of the centres
are difficult to maintain in operation. The most reliable
F_{AII}-System in Na or Li doped KCl and Li doped RbCl have
still to be operated at liquid nitrogen temperature in
order to obtain a good quantum efficiency. The excited
electronic state of the F-centre is close to the conduction band, so that the electron can escape at room temperature thermal energies.

The redshift in the colour centre lasers is substantial,
particularly in the crystals operating in the very long
wavelength region which have to be pumped in the visible
with a rare gas ion laser [1,6]. The specific gain is very
high, thus the cavity and tuning design is very similar to
the dye laser, and we need not go into detail here.

Fig.1.3 Energy level scheme and eigenfunctions for an electron in a box of infinite boundary well.

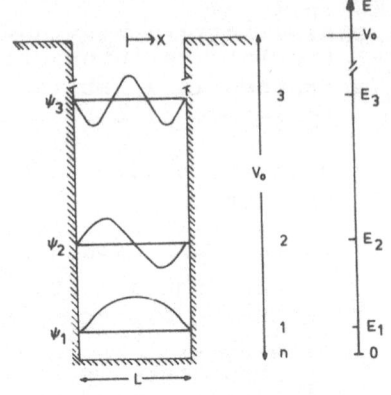

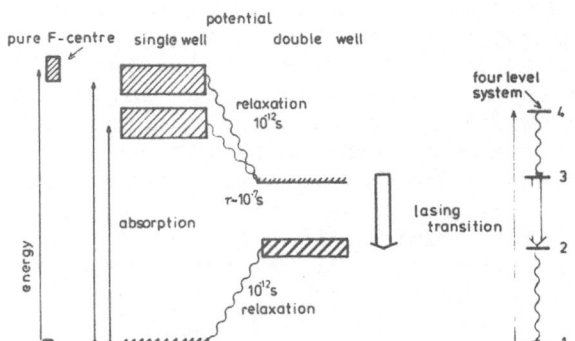

Fig.1.4 Four level laser scheme as can be defined by a colour centre in the starting (1 and 4) and in the relaxed (2 and 3) configurations

Fig.1.5 Fluorescence intensity distribution for various colour centre systems, suitable for laser operation

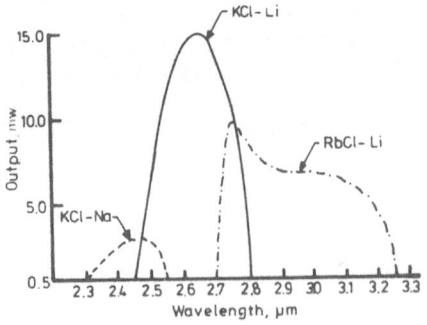

1.2. Narrowband Gain Profile Lasers

1.2.1 Semiconductor Lasers

In a so called "direct" band gap semiconductor it is possible under certain conditions to establish population
inversion across the gap. I do not intend to go into the
details of the inversion process, as only some essential
facts need concern here. Normally, at low enough temperatures, the valence band is filled completely and the conduction band is empty. In the region of a pn-junction, one
can fill the upper band with electrons and produce holes
in the lower band by a current across it and thus provide
inversion of states in this band gap. The distribution of
inverted states in the band is changing with temperature
(Fig.1.6). The band gap is determined by the particular
type of semiconductor (Fig.1.7), it also varies with temperature for a given compound (Fig.1.8) [1,7/1,8].

Thus the lasing frequency is determined by the semiconductor compound in the first place and it is possible to
tune the lasing emission with temperature. Since the population inversion is smeared out in the order of $k_B T$, it is
necessary to keep the system at low enough temperatures,
typical values being 8-80 K. The gain bandwidth is fairly
narrow and therefore no coarse tuning element is needed.
Tuning sensitivity varies with the type of semiconductor,
it can be in the order of 50 GHz/ degree Kelvin, and might
vary by one order of magnitude.

The effective temperature of the active region is not
only determined by the cryogenic system of the laser, but
varies also with the current across the diode, therefore
the change in current will tune the position of the maximum of the gain curve simultaneously.

The laser cavity is formed by the length of the semiconductor crystal and the active region, which is the contact
area between p and n type semiconductors (Fig.1.9). The
cavity length is defined by the other dimension of the
crystal and the refractive index. The latter varies also
with the electric current across the diode. Since the tuning rate of the refractive index differs from the tuning
rate of the gain curve, mode switching will normally occur
when tuning the laser by the current (Fig.1.10).

The tuning characteristics of a diode laser are not very
satisfying for spectroscopic applications. The gain curve
comprises mostly one to three longitudinal modes of the
short cavity and it is tuned across them by the current,
which modifies both temperature and the refractive index.
One of these modes can be selected externally by a grating
spectrograph of high transmission.

Recent experiments, both in Houston and in Bonn, have
shown, that temperature tuning of the diode laser can produce much better behaviour. Continuous scans of up to 5

wavenumbers have been recorded without modejumps
[1,9/1,10].

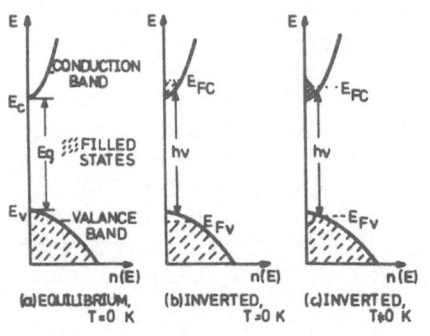

Fig.1.6 Valence band and
conduction band distribu-
tion for various conditions
of a pn-junction

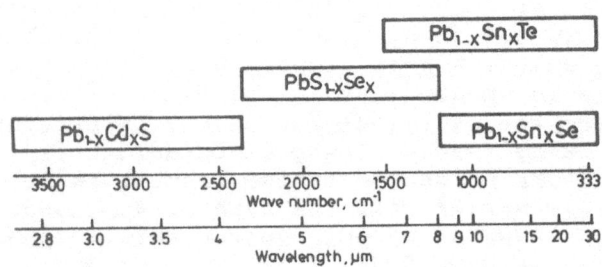

Fig.1.7 Variation of the band gap with semiconductor
compound [1,11]

Fig.1.8 Variation of the
band gap with temperature
in three semiconductors

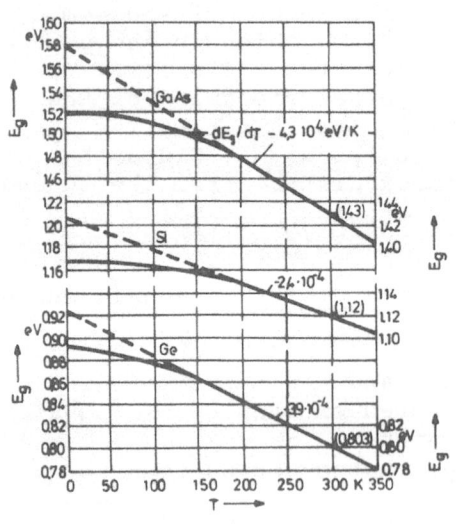

The intrinsic linewidth of the diode laser should be in the 10 KHz range. This differs drastically from experimental observations, where a jitter of 20-200 MHz can occur. The reason for this seems to be optical feedback into the laser cavity by any of the windows between the crystal and the spectroscopic system. Thus in most of the available systems, the periodic vibrations caused by the closed cycle refrigerator, can be seen directly in the jitter of the laser [1,12], (Fig.1.11).

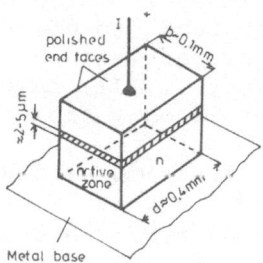

Fig.1.9 Geometry of a pn-junction laser

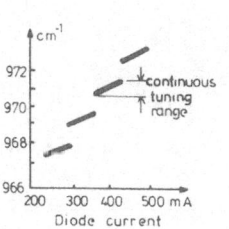

Fig.1.10 Tuning behaviour of a semiconductor diode laser with current

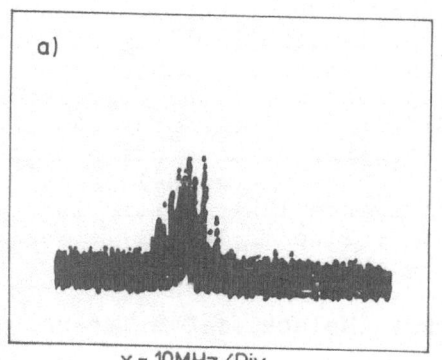

x = 10MHz/Div

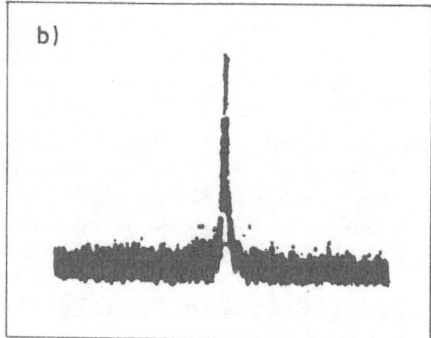

Fig.1.11 Heterodyne frequency between a diode laser and a fixed frequency CO-laser
a) with the close cycle cooler operation
 b) with the cooler switched off

1.2.2 Raman Shifted Lasers

As a further class of continuously tunable IR-lasers, we will discuss methods to change a given laser frequency by the stimulated Raman effect.

In the spin-flip Raman laser, the Zeeman energy of an electron in a magnetic field is taken away from a constant frequency CO-laser. This Zeeman energy can be varied with the magnetic field applied to the sample containing the electrons. In the direct semiconductor InSb the g-factor of the carriers is huge (g=50) and therefore the Zeeman splitting can vary by 2.3 cm $^{-1}$ per kilogauss, which gives a substantial tuning rate. Close to the band gap of InSb this tuning can give very reliable laser radiation suitable for high resolution spectroscopy [1,13], however the total tuning range for cw operation is only from 1800 to 1900 cm $^{-1}$ [1,14].

There is a different approach using the stimulated Raman effect by Raman shifting the radiation of a tunable laser by a fixed frequency. The laser source is a dye laser and the Raman shift is produced by the H_2-vibration. There are Raman shifted lines of the 1st, 2nd and even higher orders thus shifting the visible output of a dye laser into the near and medium infrared. Also antistokes lines can be observed [1,15].

1.3. Difference Frequency Mixing of Two Laser Outputs in a Nonlinear Crystal

As a final example to generate continuously tunable IR-laser radiation, I want to mention mixing the outputs of a fixed frequency laser and a tunable dye laser in a nonlinear process. Since different wavelengths normally have different refractive indices in transparent matter, an anisotropic medium is needed to phasematch the conversion of two propagating waves of different wavelengths. Another limitation of this technique is internal absorption of the IR-radiation by the nonlinear crystal. Nevertheless excellent performance of the difference laser system is possible, as has been verified by Pine [1,16/1,17]. For further details we refer to the article by T. Oka.

1.4. Stepwise Tunable IR-Lasers (Molecular Gas Lasers)

1.4.1 The CO_2-Laser

The most commonly used molecular gas laser is certainly the CO_2-laser. It is a powerful infrared laser operating near 10 μm. As the pumping mechanism is rather straight-forwardly analogous to the He-Ne laser it should be called "N_2-CO_2 laser".

The CO_2 molecule is linear and has three normal modes (ν_1, ν_2, ν_3) at different frequencies (Fig.1.12); the symme-

tric stretching ν_1 = 1388 cm⁻¹, the bending ν_2 = 667 cm⁻¹
and the asymmetric stretching ν_3 = 2349 cm⁻¹. The highest
frequency coincides with the vibration of the N_2-molecule.
A discharge in a gas mixture of CO_2, N_2 and He generates
strong population of the ν_3 level, via vibrational (VV-)
transfer from the N_2- into the CO_2-molecule. Thus population inversion between the (0 0 1) and both (1 0 0) and
(0 2 0) is readily achieved.

There are two different lasing bands according to the
two types of transitions, one near 9.6 μm and one near
10.6 μm. Both types consist of P- and R-branch transitions. For one isotope combination, e.g. the most abundant
¹²C¹⁶O₂, there are about 100 laser lines available. In
moderate systems 50 watts are readily obtained in the
strongest transitions. Further details can be found in the
literature [1,1/1,18/1,19]. Although the linewidth of the
gain in a gas laser is intrinsically determined by the
Doppler width, one can easily stabilize the CO_2-laser to a
lambdip in the corresponding molecular transition. Thus
the CO_2-laser can be used a secondary standard in the IR
[1,20]. On the other hand, it is also possible to increase
substantially the gain bandwidth by pressure broadening.
That way, considerable tunability of the CO_2-laser radiation is achievable [1,21/1,22].

Fig.1.12 Energy level
diagram for the three
normal modes in CO_2

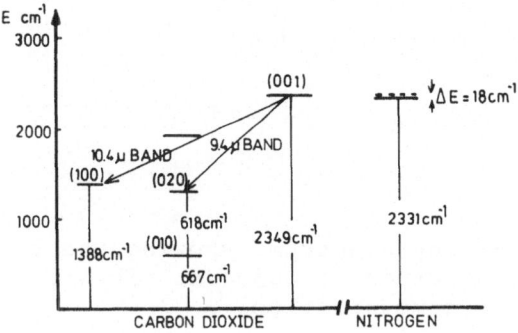

1.4.2 The CO-Laser

The carbon monoxide laser normally operates at roughly twice the frequency of the CO_2-laser. However, it can oscillate on several hundred lines between 4.8 and 8.4 μm [1,23]. This is due to the fact that many different vibrational levels in the deep, anharmonic potential curve of CO can be the starting point for laser emission. Under CW-condition, $v=3{\rightleftharpoons}2$ up to $v=37{\rightleftharpoons}36$-bands can be obtained and the frequency shift between adjacent bands is due to the anharmonicity of the CO-potential. The inversion mechanism is not as straightforward as in the CO_2-laser. Usually no complete inversion for adjacent vibrational levels can be obtained and therefore only P-branch transitions do occur. Nevertheless, the wavelength region is completely covered with the rotational distribution of the vibrational bands (Fig.1.13).

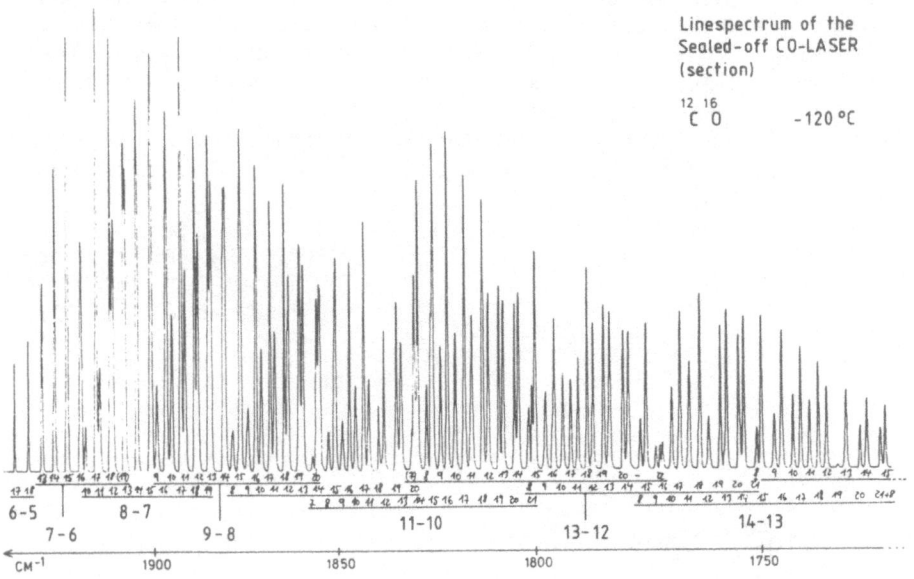

<u>Fig.1.13</u> Part of a CO-laser spectrum, showing the rotational distribution of a series of adjacent vibrational bands. (W.Bohle, private communication)

The essential pumping mechanism is the VV-transfer between vibrationally excited CO molecules, the so called Treanor mechanism [1,24]. The explanation of its physical background will not be given here, but will be postponed to chapter 3.

The most interesting feature of the CO-laser is its wide span of lasing frequencies, covering 1200-2070 cm^{-1} with more than 400 lines per isotopomeric species [1,23]. Within its wavelength region many vibrational transitions of

open-shell molecules occur, making the CO-laser an ideal
source for laser magnetic resonance (LMR) [1,25/1,26].
Recently, lambdip stabilization of a large number of CO-
transitions has been made possible [1,27]. Pressure broadening of the CO-transitions in order to get substantial
tunability has not yet been achieved.

2. THE COLOUR CENTRE LASER

The first colour centre laser has been set into operation by Fritz and Menke in 1964 [2,1]. At that time, however, this type of laser material was found to be of no
great value, since its lasing wavelength was not well defined. Only after the development of tuning techniques for
dye lasers, the intrinsic potential of a broadband gain
medium was recognized for the colour centres by Mollenauer
et al [2,2] and also by the group of Welling [2,3].

2.1. Physics of Colour Centres

A pure alkali halide crystal is transparent in the whole
wavelength range from the far infrared down to the ultraviolet, however, it can be easily coloured if point defects are generated in the crystalline structure. The most
elementary and most important point defect is the anion
vacancy. Since a negative charge is missing, an electron
will get trapped in such a vacancy, thus neutralizing the
charge distribution of the lattice. The spectroscopic properties of this defect can be described by an electron in
a three dimensional potential well - an electron in a box
- as has been mentioned in the previous survey (paragraph
1.1.2). This simple system, an electron trapped in the
anion vacancy of an alkali halide crystal, is well-known
as the "F-centre" (german: Farbzentrum). Although its
dynamic properties are not suitable for laser operation,
we will discuss its spectroscopic properties in some detail, since it is an excellent model case for the relevant
properties of any of the big family of F-centres.

2.1.1 The F-Centre
The box with infinite potential walls is certainly a
nonapplicable model for the cage defined by the cations
surrounding the anion vacancy (Fig.2.1), however, the simple structure of its energy levels give us a hint of some
essential features, particularly the role of the box diameter L.

$$E^n = ((n\pi)/L)^2 \quad \text{with } n=1,2,3,\ldots \quad (2-1)$$

The corresponding eigenfunctions are for

n=even	$\psi_n = \sin((n\pi)/L)X)$	(2-2)
n=odd	$\psi_n = \cos((n\pi)/L)X)$	(2-3)

This is plotted in Fig. 1.3 of the previous chapter.

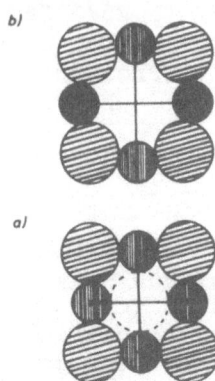

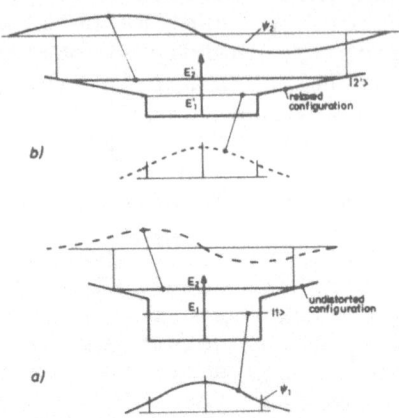

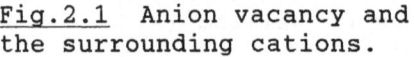

Fig.2.1 Anion vacancy and the surrounding cations.
 a) undistorted configuration of the cation cage
 b) relaxed configuration enlarged due to the lack of Coulomb attraction from the missing anion

Fig.2.2 Equivalent potential functions and the corresponding electron function describing the F-centre in the two configurations represented in Fig.1a and 1b respectively.

A more realistic approach to a potential is shown in Fig.2.2. Here, only the ground state is lying in the square part and only one excited level is present, extending into the border region of the "soup-plate" potential shape. The electron distribution in the ground- and excited state is quite different, and this is an essential point for the dynamics of the spectroscopic properties.

The position of the neighbouring cations is not fixed but will be different with the vacancy being empty or with the electron trapped in its ground state, where the charge distribution is very much localized in the centre of the vacancy. Due to the cubic point symmetry of the configuration, the collective motion of the neighbours defining the cage, can be described by one parameter, directly correlated to the cage diameter L (Fig.2.1). With this configuration coordinate the total energy varies and there is a minimum, corresponding to the equilibrium position of the F-centre ground state (Fig.2.3).

Fig.2.3 Dynamic behaviour of the energy level position when varying the cage diameter represented by the collective coordinate ρ.

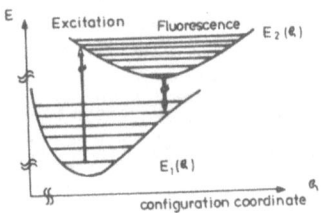

After excitation into the higher electronic state, the charge distribution changes drastically for two reasons. Firstly, we have to generate a nodal plane in the central region, where the maximum charge density is in the ground state and secondly, due to the shallow potential well the electron function extends much further beyond the boundary of the vacancy geometry. Thus we get a highly delocalized charge distribution in the excited state with respect to the electronic ground state. This produces a new equilibrium position of the cation cage, which is much closer to the "empty" vacancy configuration due to the highly reduced negative charge inside the vacancy volume (Fig.2.1b). There is also a widening of the equivalent box diameter L, which occurs on a very short timescale (10^{-13} sec). Widening of L causes a reduction of the energy levels, as shown by a formula (2-1). Therefore we will find a very low transition moment for the electronic transition back to the ground state, due to the highly delocalized charge distribution in state n=2 with respect to n=1, which means, that the F-centre in its relaxed excited state is not a strong induced emitter, particularly since there are competing processes, e.g. thermal excitation from the excited localized state into the nearby conduction band of the solid. The excitation and deexcitation process of the F-centre and its relaxation processes are indicated in Fig.2.3. The series of processes could, in principle form, a nice 4-level lasing system, however its efficiency is too low to drive a laser.

2.1.2 The $F_{A(II)}$-Centre.

As mentioned above, there is a large variety of colour centres, all correlated to the simple F-centre, but with a more complicated fine structure. Out of this family, we want to discuss the so called $F_{A(II)}$-centre in more detail, since it represents one of the most efficient types for spectroscopic application.

The $F_{A(II)}$-centre is an F-centre, where one of the nearest cations of the lattice is replaced by a different alkali ion. There is the combination of two point defects, which means that its point symmetry is axial rather than cubic, and so is the potential well, which is a single well with a diameter close to the pure F-centre in the x and y directions and slightly expanded in the z-direction, provided the disturbing cation is smaller than that of the host lattice (Fig.2.4a and 2.5a). This situation will hardly affect the ground state of the electron wave function, since it will remain mainly spherically symmetric, (close to a s-like electron function), however, after excitation, the n = 2 electron distribution can orient its nodal plane to coincide either with the xy-plane, which is favorable, or with the two perpendicular planes leading

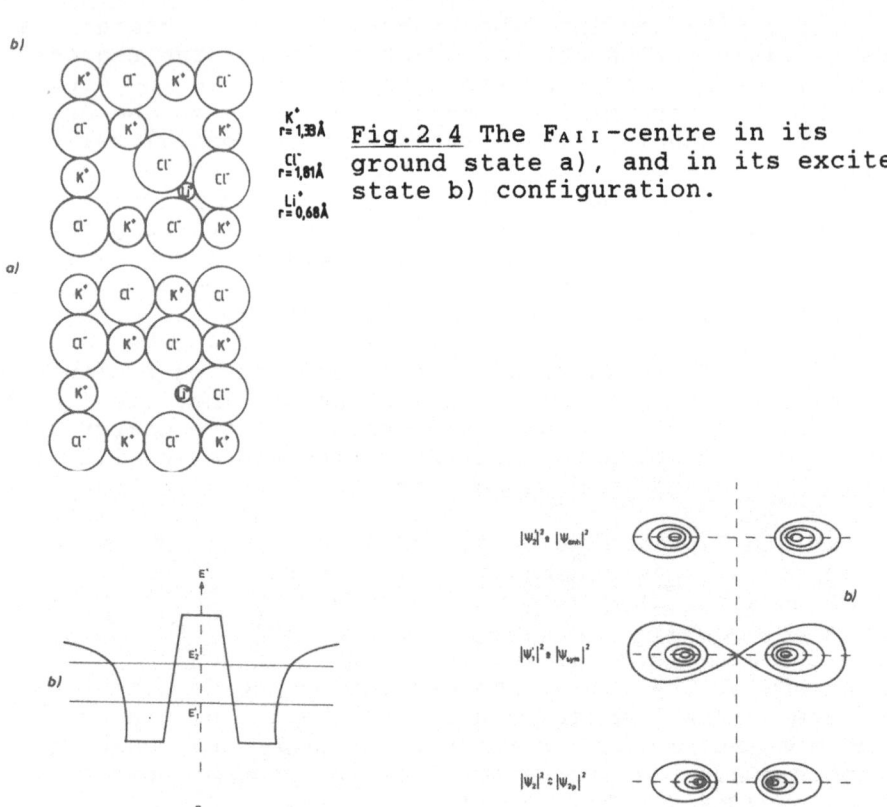

Fig.2.4 The F_{AII}-centre in its ground state a), and in its excited state b) configuration.

K⁺
r=1,33Å
Cl⁻
r=1,81Å
Li⁺
r=0,68Å

Fig.2.5 The equivalent potential well functions for the single well a) and the double well b) configuration. Cross sections taken perpendicular to the axis or plane of symmetry.(Fig.2.4)

Fig.2.6 The electron density distribution in the F_{AII} single well configuration a) and in the double well configuration b)

to slightly higher energies. Since these first electron states have equal symmetry to p-type electrons, they are sometimes referred to as p_x, p_y and p_z-functions. It follows, that two absorption bands will occur for the $F_{A(II)}$-centre (Fig.2.8).

As in the case of the F-centre, the n=2 state causes a redistribution of the cation cage, however, since one of the cations is much smaller, not only the size of the potential well will change, but also the geometry. For Li both in KCl and RbCl one of the Cl⁻ -ions flips into a diagonal position forming a line with K⁺Cl-Li⁺ and provid-

ing two wells instead of one. (Fig.2.4b and 2.5b) This
double well potential will accept a p-type electron distribution with its nodal plane coinciding with the line of
the $K^+Cl^-Li^+$. Since the boundary is formed by a double
well potential, the eigenstates have to be adapted to it,
which means, that they are solutions of a two centre problem. The corresponding eigenstates can be approximated by
linear combination of one-centre solution of a diatomic
molecule. Since the double well situation evolves from a
single well potential, we would expect to find corresponding eigenstates according to both the "separated atoms
approximation" (double well) and the united atoms approximation (single well) also in the $F_{A(II)}$-centre. Fig.2.6
shows the corresponding electron density distributions.

The p-type excited state will transform into the antisymmetric Ψ_{anti} two centre solution, and can go back into
the symmetric ground state Ψ_{sym} of the double well configuration. This latter, however, will interfere with the
ion distribution, since there is electron density where
the Cl^- is situated. Therefore the Cl^- will flip back into
its corner and we have the single well situation from
which the process began. Necessarily, the Ψ_{sym} will transform into the $\Psi_{(n=1)}$ of the single well eigenfunction. The
flipping process of Cl^- and Li^+ is connected to the relaxation process, triggered by the change of charge distributions in the lattice vacancy. The coupling to the lattice
vibrations is very strong and the relaxation processes
occur on a very short timescale (10^{-13} s), thus providing
extremely shortlived energy levels for $\Psi_{(n=2)}$ and Ψ_{sym}.

We can define coordinates describing the collective motion of the ions defining the $F_{A(II)}$-centre, however, they
are quite different in character for the single well and
for the double well configuration. This has to be kept in
mind in Fig.2.7.

Fig.2.7 Pumping and fluorescence scheme for the F_{AII}-
centre. The collective coordinates ρ and ρ' in the
single- and the double well
potential are different in
character. The minimum in
the ground state occurs for
ρ and the minimum in the
excited state for ρ'. Absorption and emission bands
are indicated.

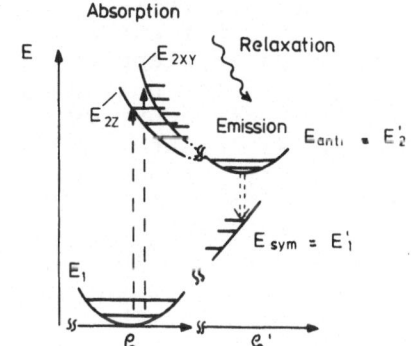

Since in the double well potential, the electron distribution is well localised, for both Ψ_{anti} and Ψ_{sym}, the
electronic transition moment is reasonably high, and
therefore the $F_{A(II)}$-centre will have considerable gain.
Nevertheless, the excited level is still close to the conduction band and therefore the crystal has to be kept at
low temperature for a high quantum efficiency. At liquid
nitrogen temperature the efficiency is 35 %, it increases
to 50 % at liquid helium temperature, however this increase does not justify the much more sophisticated cryogenic equipment that would be necessary.

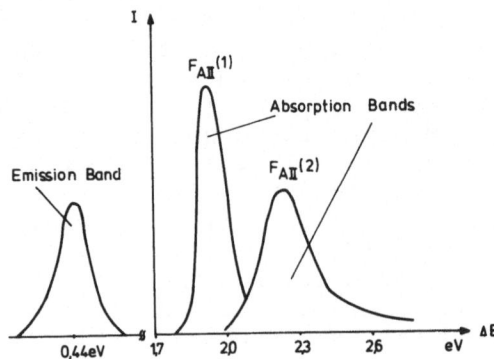

Fig.2.8 Photon energy
distribution for the
two absorption bands
and the emission band
of KCl:Li F_{AII}-centres.

2.2. The F-Centre Laser

The sequence of states that occurs after excitation of a
$F_{A(II)}$-centre forms an excellent four level laser scheme,
as has already been mentioned in Chapter 1.1.2. (see Fig.
1.3). The absorption and emission bands of the $F_{A(II)}$-
centre are plotted in Fig.2.8. The simple F-centre absorption is partly overlapping with the short wavelength
band of the $F_{A(II)}$-centre. The redshift of the fluorescence is substantial. The emission centres for Li-doped
KCl is at 2.6µm and for Li doped RbCl is at 2.9µm with
both bands overlapping. The $F_{A(II)}$-centres can be flash-
lamp-pumped for pulsed operation, however, for high resolution spectroscopy cw operation is achieved by pumping
with an Ar⁺ or Kr⁺ ion laser. Sufficient gain can only be
produced by strong focussing of the pump laser into a
small crystal volume. The F-centre crystals behave very
much like a laser dye and thus the resonator configuration
is also very similar. One difference, however, is the
large redshift of the fluorescence with respect to the
pump wavelength. Therefore it is possible to use dichroic
coatings, that are highly transparent at the sharp pump
wavelength and highly reflecting in a wide IR-region, or
vice versa, depending on the type of laser setup
[2,2/2,3].

As already explained in Chapter 1.1.2, the broad band
gain profile needs both coarse tuning and fine tuning elements. Coarse tuning is most readily achieved by a reflection grating. Fine tuning is mostly done with an intracavity etalon. If single mode, continuous tuning is needed, both grating and etalon have to be tracked synchro-
nously with the laser resonator.

2.3. Colour Centre Laser Spectrometer

In some applications, the tunability of a laser is used
to set the frequency to a particular region and only a
small wavelength region is scanned in the particular experiment. The narrow band tuning can be achieved by changing just one parameter of the laser cavity. If, however,
we want a wide single mode scan across let us say 100cm^{-1},
we need a rather sophisticated tracking system. The first
fully computer controlled F-centre laser spectrometer has
been set up in R. F. Curl's group [2,4/2,5]. We have set
up a slightly modified tracking procedure [2,6] in our
lab.

A complete spectrometer setup is presented in Fig. 2.9.
The stepping motor puts the grating into the wavelength
region required. The intracavity etalon then selects the
resonator mode and suppresses holeburning modes. It is
scanned by a PZT. The resonator mode, finally is determining the laser frequency and it is tuned by another PZT.

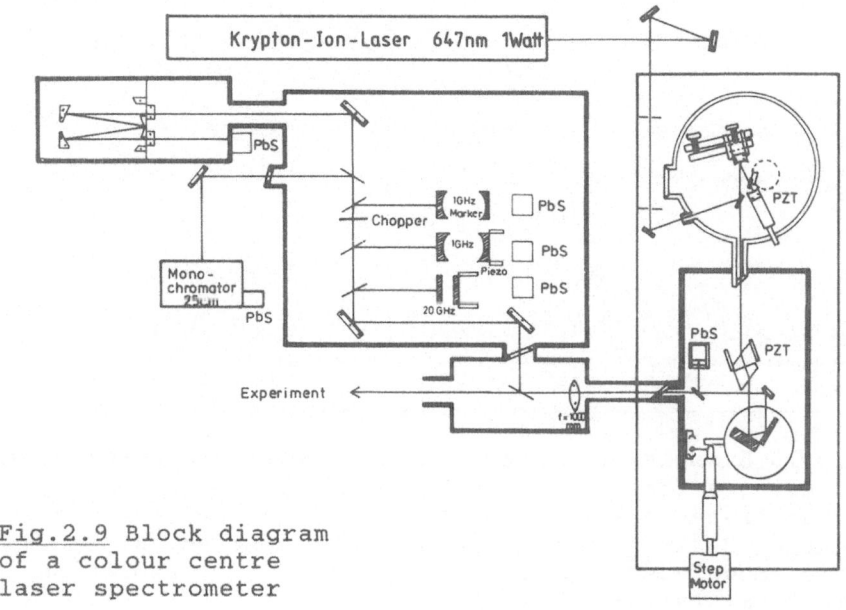

Fig.2.9 Block diagram
of a colour centre
laser spectrometer

The grating position and also the etalon-PZT-voltage can be calibrated and the calibration is stored in the computer memory. In our system the cavity length is tracked by an analog tracking loop within the tuning range of the intracavity etalon. The PZT systems can be set back in a well defined way to achieve continuous scans over large frequency regions [2,7].

In order to ensure single mode operation, a diagnostics system is added to the spectrometer, consisting of two scanning etalons of a different FSR (1 GHz/20 GHz). The wavelength calibration is achieved by a reference absorption White-cell filled with a gas of well known absorption wavelength. For interpolation purposes, the peaks of a marker etalon are also registered. A substantial part of the laser power is used for these controls and for regulation purposes. The rest is then used for spectroscopy of molecules, free radicals or ions. The computer displays traces of calibration gases, the signals to be investigated in various forms (e.g. first and second derivatives), the marker etalon fringes. An internal evaluation programme then produces a print out with a strictly linear frequency scale. Fig. 2.10 gives an example of a spectrum

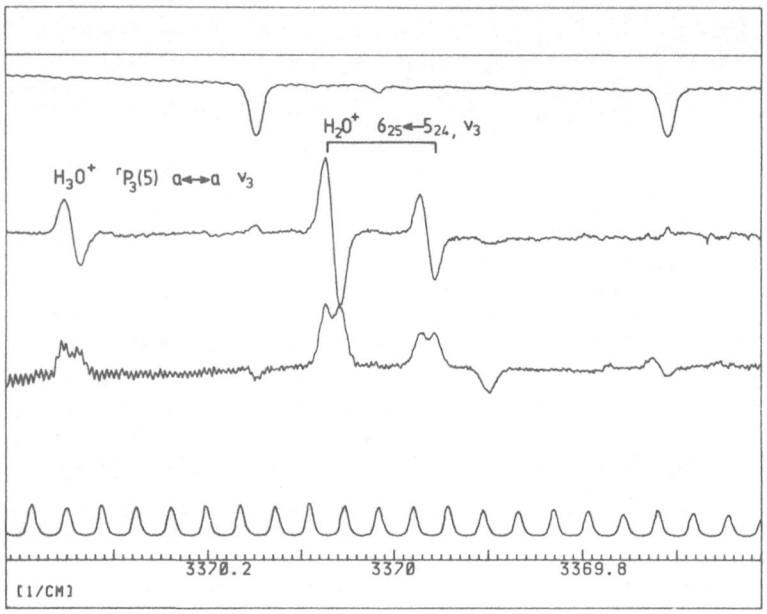

Fig.2.10 Example of a spectrum taken with a CCL-spectrometer, using a computer controlled mode of operation. The various traces are from top to bottom. 1) reference gas spectrum. 2) first derivative of ionic species. 3) concentration modulated absorption signals. 4) Interpolation marks from etalon.

of H_3O^+ and H_2O^+, that has been recorded with our spectrometer using Saykally's velocity modulation scheme for
ionic species [2,8]. From this we can tell that the F-
centre laser spectrometer shows both high resolution and
high sensitivity [2,9].

3. THE CARBON MONOXIDE LASER

Molecular gas lasers in the medium infrared, such as the
CO_2-laser and the CO-laser are very powerful, have a high
efficiency and can be operated with very high spectral
purity on single vibration-rotation transitions. The CO_2-
laser operates on two bands near 9.6 and 10.6 μm and the
CO-laser is generally known to work near 5.3 μm, however
its spectral range can be extended to cover 4.8 to 8.4 μm,
thus it is a most interesting source for spectroscopic
applications, as already mentioned in paragraph 1.4.2.
Since the pumping mechanism for CO is not at all straight
forward, and since some explanations given in the literature are slightly misleading, I want to go into some
detail here. We will start off with what might be taken
as a detour, the description of an experiment, where the
NO-molecule is excited by a CO-laser in order to have well
defined conditions. Only after that we will turn to the
excitation process of CO in an electric discharge.

3.1 The Treanor Pumping Mechanism

3.1.1 UV-Fluorescence from IR-Laser Excited
NO, Ar Gas Mixture
In Fig.3.1 we see the potential curves of a diatomic
molecule, NO in this case. The electronic groundstate $X^2\pi$
is fairly deep and comprises more than 30 vibrational
levels. There are various excited electronic states with
minima below the dissociation limit of the ground state
and there are well known UV-transitions from the $A^2\Sigma$ and
the $B^2\pi$ into the $X^2\pi$, the so called γ- and β-bands.
There are excellent coincidences between vibration rotation transitions of the NO fundamental in the $X^2\pi$ and some
strong CO-laser lines near 5.2 μm. If we shine a few watts
of this CO-laser light into an absorption cell containing
NO and Ar, those γ- and β-bands occur at IR-laser power
densities of less than 1 kW/cm². This means that there
must be a way for the energy that is put into the fundamental vibration of the diatomic molecule to get up the
ladder of vibrational states to the level of electronic
excitation or to the dissociation limit [3,2/3,3]. For a
diatomic molecule, particularly at these low power densities, multiphoton excitation is not possible.
The phenomenon described above can be understood in terms
of the very same model that has been developed to explain

the inversion mechanism in the CO-laser plasma. The IR-
laser pumping of NO, however, gives much neater boundary
conditions for demonstrating the mechanism. In this case
there is definitely only one vibration-rotation transition
by which the energy is fed into the molecule and all the
other processes have to be energy transfer via collisions
[3,2/3,3].

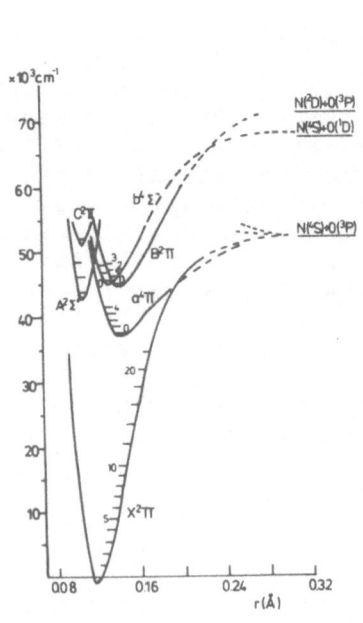

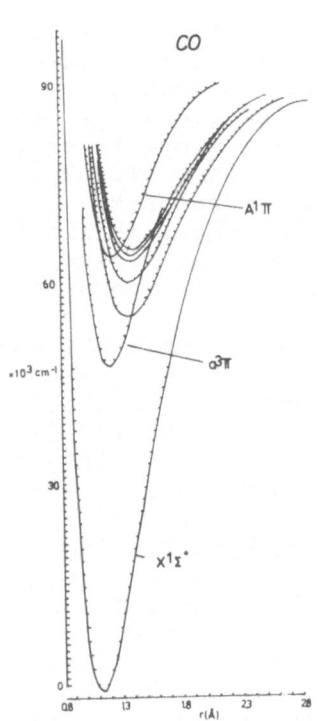

Fig.3.1 Potential curves
for the lower states of NO
(after [3.1]).

Fig.3.2 Potential curves
for the CO molecule.

3.1.2 Vibrational Pumping of the CO-Molecule

The CO-potential of the $X^1\Sigma$ ground state is even deeper
than that of NO (Fig.3.2). In a CO-laser plasma the main
excitation process occurs in the vibrational degree of
freedom by electron collision. At electron energies of a
few eV the excitation cross section for v=0 to v=1 has a
maximum, also further excitations up the ladder of vibrational states (always with Δv=1) are in resonance at that
energy range [3,4]. Therefore we can be sure that, in a
discharge at low electron energies, the CO molecules that
are present will become vibrationally excited, just as in
the case of the IR-laser excitation of NO described in the
previous paragraph. Now we have to ask, what happens if

two vibrationally excited CO-molecules collide? To begin
with, let us discuss two CO molecules with v=1:(Fig 3,3a)
CO(v=1) + CO(v=1) -> CO(v=0) + CO(v=2) + ΔE   (3-1)
The energy difference ΔE is the anharmonicity shift between v=1⇒0 and v=2⇒1, it is positive and enters into the
translational energy of the colliding particles. Such a
vibrational energy transfer between two CO-molecules takes
place after roughly 50 collisions. This is much slower
than the rotational equilibration which takes place after
almost every collision, but very much faster than the
direct transfer of a vibrational quantum into the translational energy of the colliding partners, which is in the
order of 10^5 gas kinetic collisions for CO. A recent study
of the VV-pumping in CO has shown that the corresponding
rate constants increase strongly with v [3,5].

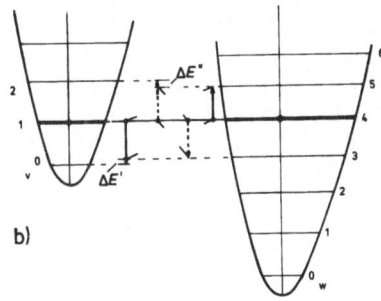

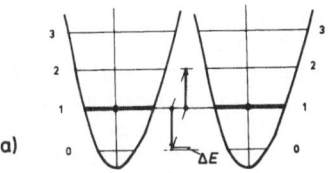

Fig.3.3 a) Two CO molecules
in v = 1 can exchange one
vibrational quantum during
collision. Both ways are
equally probable.

b) Two CO molecules,
one in v = 1 the other in
w = 4 can also exchange one
vibrational quantum during
collision. One direction is
more probable than the
other.

Considering these figures, we can estimate that once a
CO-molecule has reached a higher vibrational state, it
will bring the rotational distribution of this state into
equilibrium due to the rotation-translation relaxation
(RT) and will be ready for further vibration-vibration
(VV-) transfer, e.g. going from v=2 to v=3 etc. In each
step, the anharmonicity difference is contributed to the
translational energy. This process can go both ways since
in the case of two molecules in different vibrational
states, an exothermic and endothermic path is possible
(Fig.3.3b):

$$CO(v=1) + CO(v=w) \begin{cases} \xrightarrow{k_{v-1}^{w+1}} CO(v=0) + CO(v=w+1) + \Delta E' \\ \xrightarrow{k_{v+1}^{w-1}} CO(v=2) + CO(v=w-1) - \Delta E'' \end{cases}$$ (3-2)

The endothermic path has to take the energy from the translational reservoir and this path will have low probability if the energy difference $\Delta E''$ is bigger than the thermal energy $k_B T$ of the gas. We can take advantage of this by cooling down the translational temperature of the gas discharge. There will then be a substantial flow of excitations up the vibrational ladder.

A considerable amount of energy is stored in the vibrational degree of freedom, whereas the translational and rotational excitations are in thermal equilibrium. This is due to the fact that the electrons in the discharge have a high cross section for vibrational excitations and that the VV-transfer in the CO-plasma is close to resonance, and therefore absorbs energy much faster than do the other excitations.

This kind of process has first been described by Treanor, Rich, and Rehm [1,24] and has been used successfully to explain the vibrational distribution in diatomic gases. It is called "anharmonic VV-pumping" and is now very often referred to as "Treanor pumping". For quantitative treatment we refer to the literature [3,6]. However, we have to add a few more points to the discussion of the CO-plasma.

In formula (3-2) the endothermic path will die out according to an exponential containing $\Delta E''/k_B T$ and therefore the equilibrium can be strongly influenced by the translational temperature. At low temperatures the onset of VV-pumping will be earlier than at high thermal energies. Thus we have to cool the CO-laser plasma, and therefore use helium as a good thermal conductor and run the discharge in a tube cooled with liquid nitrogen or another coolant. The laser gas contains further components, the role of which we will briefly mention. At temperatures higher than liquid nitrogen, Xe is added in order to keep the electron temperature down in the optimum region for the vibrational excitation of CO. In a very cold plasma Xe has to be replaced by oxygen, which has a similar effect on the electrons. Oxygen also serves a second purpose. In the discharge, CO tends to dissociate

$$CO \rightleftharpoons C + O \qquad (3-3)$$

The reaction is pushed to the left side if excess oxygen is present. Finally, one often adds nitrogen to the gas. The nitrogen has a similar vibrational energy and can lose this only by VV-transfer from N_2 to CO. To be precise its vibrational quantum energy is slightly higher than CO and thus it will excite preferably lower vibrational states in CO. It helps to bring the lasing range down to short wavelengths [3,7]. Finally we should mention how in a single step highly excited vibrational states can be generated, thus initiating the VV-pumping. Very slow electrons pre-

sent in the discharge can form negative ions of CO. CO⁻ is
an unstable compound and dissociates into CO + e⁻ leaving
the CO in a vibrational state close to v=8. Not all processes are known to every detail, and from the facts collected so far we can see, that the excitation process of
CO in the laser plasma is far from being simple.

If we observe the fluorescence from a plasma, consisting
of CO, He, N_2 and O_2, and which is known to form a good
amplifier when run inside a liquid nitrogen-cooled jacket,
we can deduce a vibrational distribution of the form shown
in Fig. 3.4 [3,8/3,9/3,10]. This means that, in contrast
to CO_2, CO has no population inversion for vibrational
states. The vibrational "temperature" defined by the populations of neighbouring vibrational states is positive
over the whole vibrational ladder. However, there is a
region, called the "plateau" where the corresponding distribution would correspond to a very high vibrational
temperature T_{vib} ($T_{vib} > 36000$ K). In spite of the absence
of vibrational population inversion lasing is possible. We
have to look closer into this problem in order to understand the phenomenon called "partial inversion".

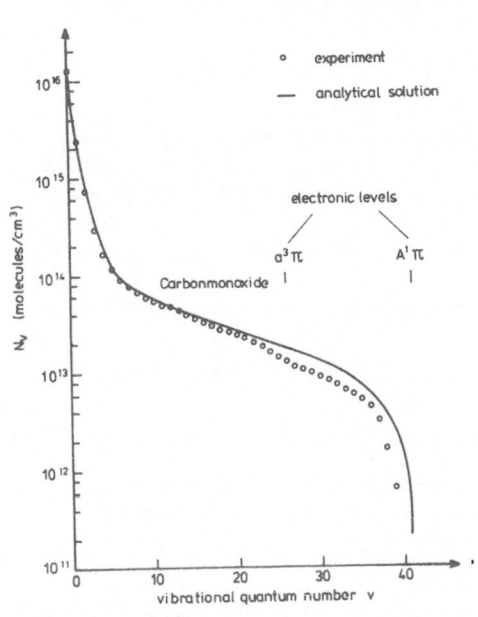

Fig.3.4 Vibrational distribution in a liquid N_2 cooled
CO plasma tube(after
[3,10]).

3.2   Partial Inversion in a CO-Plasma

It follows from the above discussion that a distribution
of internal excitations in the CO molecule is established
by VV-transfer that is far off a unified equilibrium. This

34

strong internal disequilibrium can lead to gain under
special conditions.

At a given thermal energy, we can calculate the population distribution for a vibrational state of a molecule
into the various rotational sublevels. In most cases vibrational energies are large compared to the thermal energy and low rotational levels have a difference which is
small compared to the thermal energy. Since the molecules
in a gas can be taken to be statistically independent, we
have a Boltzmann-distribution of the rotational energies,
defined by the translational temperature leading to a
population as plotted in Fig.3.5a. However, this Boltzmann
decay of higher levels has to be modified in the final
case by the degeneracy (Fig.3.5b) of the J levels leading
to the well known total distribution shown in Fig.3.5c.
Since the rotational redistribution of all vibrational
levels is very fast (RT-relaxation) we can assume that the
rotational temperature for each of the populated v-levels
is equal to the translational temperature of the gas. However, no unified vibrational temperature can be assumed
for a CO-laser plasma, therefore we just take the total
number of molecules in a given vibrational state explicitly in our final formula according to

$$N_v = \sum_J N(v,J), \qquad J=0,1,2,\ldots \qquad (3-4)$$

and
$$N(v,J,T) = \frac{N_v\,(2J+1)}{Z(B_v,T)} \exp[-B_v\,J(J+1)/k_b\,T] \qquad (3-5)$$

where $Z(B_v,T)$ is the distribution function B_v is the
rotational constant of the molecule and T is the translational temperature. The distribution for conditions close
to the situation in the CO-discharge (T=125K, B_v=1.9 cm^{-1})
corresponds to the case plotted in Fig.3.5c.

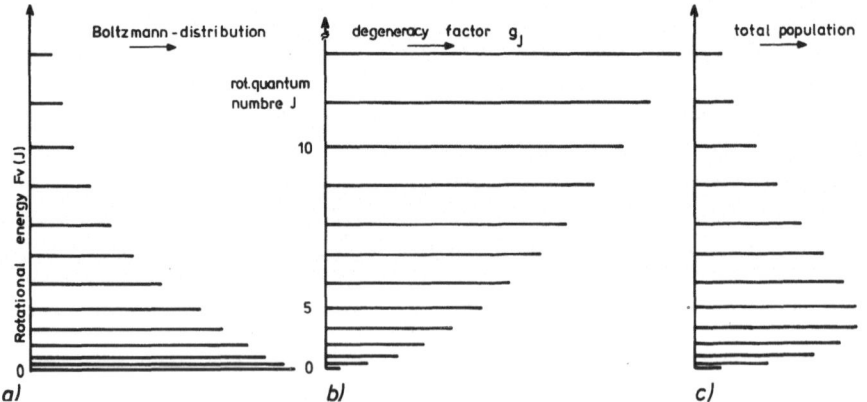

a)   b)   c)

Fig.3.5 Rotational distribution of a given vibrational
manifold a) pure Boltzmann distribution. b) Rotational
degeneracy factor. c) Total population distribution.
The relation is plotted for a molecule with $B_v = 1.9$ cm^{-1}
at a temperature T = 125K.

A very thorough analysis of the intensity distribution
of the fluorescence from our CO-laser tube cooled by
liquid nitrogen gives T~125K and in the plateau-region
N_{v+1} / N_v ~0.95, which would be equivalent to 35 · 10^3 K
[3,10]. It is generally accepted that inversion is only
possible if we can attribute "negative" temperatures to a
given population distribution, but there can also be gain,
if two internal degrees of freedom are so much out of
equilibrium, as for the CO plasma where T_{rot} = T_{trans} ~
125K and T_{vib} ~ 35 · 10^3 K. However, we have to look into
this in detail and be on our guard because very misleading
arguments can be encountered in the literature.

In the Einstein relation for induced radiation processes
the ratio of absorption and emission is proportional to
the population density in the two energy levels involved
however, we have to account for the degeneracy factor g.
This is of particular importance in vibration-rotation
transitions with the selection rules $\Delta J=-1$ and $\Delta J=+1$ for
P-branch and R-branch transitions, since there the J-
values and hence the degeneracy factors of the two levels
involved are different. The relation is

$$g_m \; B_{mn} = g_n \; B_{nm} \qquad\qquad (3-6)$$

where g_m is the degeneracy factor of the level m, and B_{mn}
is the corresponding Einstein factor for induced processes
m ⇒ n, etc. If m and n are the corresponding J-values, the
degeneracy factors are 2m + 1 and 2n + 1.

The result of this correction for the degeneracies is
the fact, that in considering the gain or loss of a vib-
rot-transition, the total population density does not
count, but this quantity divided by the degeneracy of the
corresponding level. This means that the Boltzmann-distribution enters into the expression of the gain factor
(Fig.3.6). Finally, however, the degeneracy factor that is
in common for both levels, will play a role of intensify-
ing the gain of higher J-values as compared to low J-
states.

For positive vibrational temperatures, gain can only
occur for P-branch transitions. In the evaluation given by
C.K.N.Patel [3,11] the Doppler-width of the spectrum is
also incorporated. The distribution function leading to a
term mainly containing B_v/$k_B T$ is combined with the Doppler
terms. Here is the Patel formula:

$$\alpha_0 = \frac{8\pi^3}{3h\epsilon_0} \cdot \frac{Bhc}{k_B T} \cdot (\frac{M}{2\pi RT})^{\frac{1}{2}} \cdot !R_{N''}^{N'}!^2 \cdot !m! \quad *$$

$$* \; F_n * (N,m)\frac{2J'+1}{2J''+1} \; [N_{v'} \, e^{\frac{-hcF_{N'}(J')}{k_B T}} \quad -N_{v''} e^{\frac{-hcF_{N''}(J'')}{k_B T}} \quad] \qquad (3-7)$$

Herein the following notations have been used:

36

B = $(B_{v'}+B_{v''})/2$ is the averaged rotational constant
$|R^{N'}_{N''}|$ is the electric dipole moment [3,12], m=-J" for
the P-branch and m=J"+1 for the R-branch transitions, F_n^*
is the Herman-Wallis factor and $F_{Nv}(J)$ is the rotational
term energy.

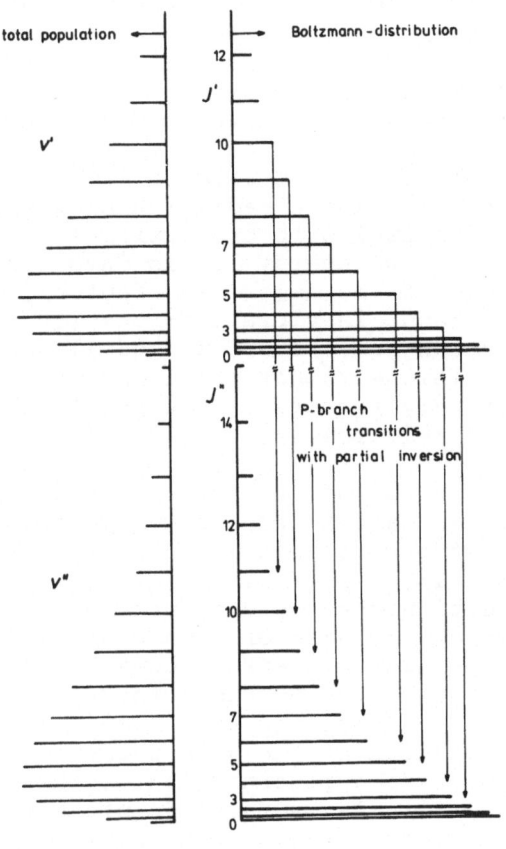

Fig.3.6 Not the total distribution (left side) but the Boltzmann-distribution of two adjacent vibrational states v'and v" is responsible for building up the partial inversion.

Fig. 3.7 shows plots of the gain coefficients evaluated
from Formula (3-7). In order to get a feeling for the
influence of the essential parameters T_{rot} and N_{n+1}/N_n the
vibrational population ratio is varied in Fig.3.7a and the
rotational temperature is varied in Fig.3.7b.

3.2.2 Experimental Observations

If we compare the results of the distributions evaluated
by Patel's formula and the actual intensity distribution
of the corresponding laser transitions, there is a discrepancy in the intensity relations of the rotational components within a vibrational band. The peak intensity
occurs at a higher J-value than calculated. This discrepancy is due to saturation effects. Patel's formula gives
the intensity for "small signal gain", whereas the experimental peak intensity is limited by saturation effects.

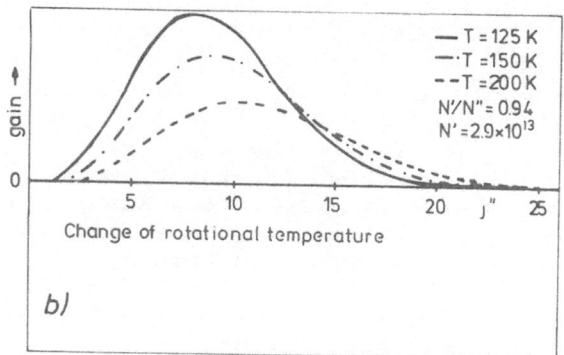

b)

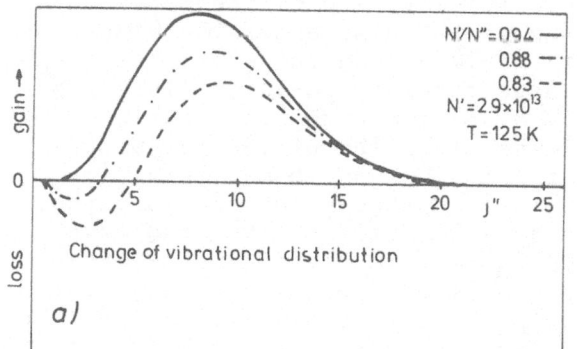

a)

Fig.3.7 J-dependent gain coefficients for a given vibrational band using Patel's formula for evaluation. In part a) the rotational temperature is fixed, and the vibrational population ratio is varied. In part b) the rotational temperature varies at a fixed vibrational population ratio. Note that for higher T higher J-values get a higher gain.

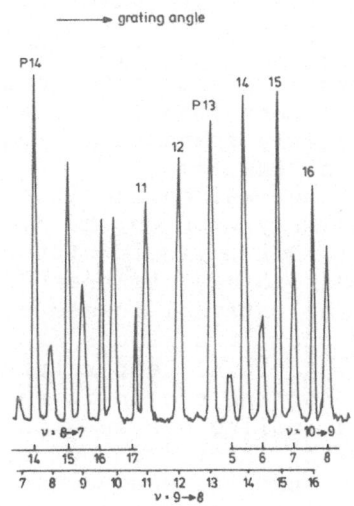

Fig.3.8 Part of a CO-laser spectrum obtained by tuning the grating angle (see Fig.3.9). The appearing linewidth does not indicate the laser linewidth but the angular region where the gain is above threshold.

For low J-values saturation plays a more important role than for high J-values. Therefore a different ratio for the onset and the maximum power within one particular line can be observed for low J- and high J-lasing transitions. This is in accordance with the apparent "linewidth" when the CO-laser is scanned by turning the angle of the re- flection grating (Fig.3.9). The angular range by which the grating position can be off the centre of a laser line is not determined by the actual linewidth but by the addi- tional losses the gain medium can cope with beyond threshold.

3.3 A CO-Laser for Spectroscopic Applications

The purpose of a CO-laser in spectroscopy is to provide a large manifold of laser lines in a wide frequency range. The lower the temperature of the discharge, the wider the extent of the plateau in Fig.3.4, and the wider the tuning range of the laser. We have observed CO-laser lines run- ning CW up to the $v = 37 \Rightarrow 36$ band with frequencies down to $\tilde{\nu} = 1205$ cm^{-1} [1.23] The other end of the frequency scale is $\tilde{\nu} = 2070$ cm^{-1} for a $v = 2 \Rightarrow 1$ band transition. Both values occur in a liquid nitrogen cooled discharge, but with different discharge conditions. The long wave- length end needs high CO-content in the plasma, whereas short wavelengths require a minimum of CO and very low current. The gas parameters of a flow system, used in the high liquid nitrogen cooled CO-laser, have to be optimised individually. A good starting mixture is defined in [1,23]. The experimental setup is outlined in Fig.3.8.

A very convenient way to extract the laser power is the Littow-mounted grating combined with a mirror at right angle to form a "corner cube reflector" (Fig.3.8). The zeroth order reflection of the grating is used to couple the laser power out of the cavity. This loss is present anyway, and thus we can avoid a partly transmitting end mirror and use a gold coated mirror instead.

The rotational distribution of the lasing transitions within a vibrational band depends very much on the dis- charge temperature (see Fig.3.6b). For a liquid nitrogen cooled system, the rotational manifold may start at J=5 and may go up to J=17, provided the internal losses are kept low [1,23]. If higher rotational transitions are re- quired, we need to increase the rotational temperature and this means that a less effective cooling system is re- quired. One can achieve this by a cooling system with pentane as a coolant, combined with a heat exchange device [1,26]. The increase in rotational and the equivalent translational temperature leads necessarily to a reduced span of vibrational lasing transitions. At these elevated temperatures, however, a "sealed off" gas mixture may be used, which provides the possibility to use rare isotopes in the gas mixtures in order to increase the manifold of available frequencies. As mentioned in paragraph 1.4.2, it is of particular importance for LMR-spectroscopy to make a dense distribution of fixed frequency-laser lines available.

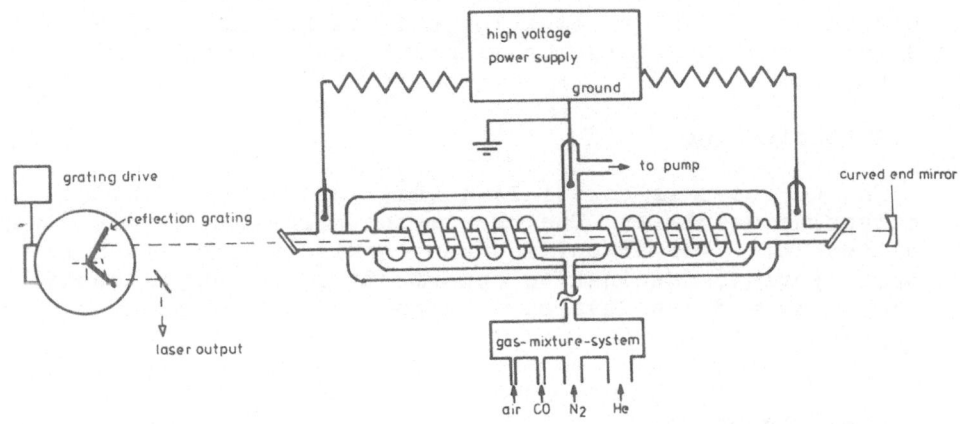

Fig.3.9 Experimental setup for a line tunable liquid nitrogen cooled CO-laser. The reflection grating acts as wavelength selective end mirror and as coupling device at the same time (Littrow mount).

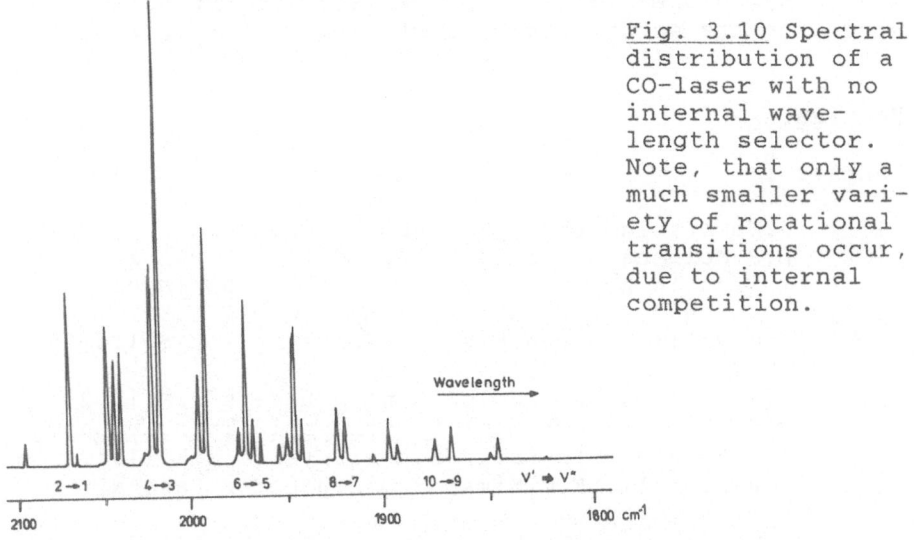

Fig. 3.10 Spectral distribution of a CO-laser with no internal wavelength selector. Note, that only a much smaller variety of rotational transitions occur, due to internal competition.

There are some cases, where one operates a CO-laser without intracavity line selection. Although this mode of laser operation is called "all line", the spectral analysis of the output shows a wavelength distribution, that is much less rich in rotational lines compared to the internal selection mode. The output spectrum of a CO-laser specially optimized for low vibrational bands, is plotted in Fig.3.10. The reduced number of rotational lines per lasing vibrational band is due to competition effects. The strongest lines are starting to lase before enough partial inversion can be built up in the wings of the rotational distribution. These competing effects have also been observed under CW operation with internal selection of laser lines [3,13].

3.4 Conclusions

The CO-laser is certainly a very interesting type of molecular gas laser. The underlying processes involve many different fields of physics and chemistry. The laser itself provides fascinating opportunities for experimental ingenuity and its development opens up a wide range of applications.

Acknowledgements

I wish to express my thanks towards Prof.Dr.D.J.Newman for his constructive comments which improved the quality of the manuscript substantially. I am grateful to Mrs.H.Hartmann for her careful preparation of the drawings and to Mrs.H.Baessler for her patience in typing the manuscript on the wordprocessor installed by K.Cloppenburg.

References

Chapter 1

[1,1] J.Hecht: The Laser Guidebook, McGrawHill (1987)
[1,2] L.F.Mollenauer, J.C.White, Editors: Tunable Lasers, Topics in Applied Physics, Vol 59, Springer (1987)
[1,3] M.L.Shand, H.P.Jenssen: IEEE J.QE-19, 480 (1983)
[1,4] J.C.Welling: "Tunable Paramagnetic-Ion Solid-State Lasers", in [1,2]
[1,5] L.F.Mollenauer: "Color Center Lasers" in [1,2]
[1,6] H.Welling, D.Frölich; "Progress in Tunable Lasers" in: Advances in Solid State Physics XIX (1979)
[1,7] R.W.Campell, F.M.Mims: Semiconductor Lasers, H.W.Sams (1972)
[1,8] A.Mooradian: "High Resolution Tunable Diode Lasers" in: Very High Resolution Spectroscopy, R.A.Smith Editor, Acad.Press (1976)
[1,9] C.B.Dane, R.Brüggemann, R.F.Curl, J.V.V.Kasper, F.K.Tittel: Appl.Opt.26, 95 (1987)

[1,10] R.Brüggemann, private communication (1987)

[1,11] W.Demtröder: <u>Laser Spectroscopy</u>, Springer 1982

[1,12] H.G.Wells, A.Hinz: private communication (1985)

[1,13] W.Urban, W.Herrmann: <u>Appl.Phys. **17**</u>, 325 (1978)

[1,14] S.R.J.Brueck, A.Mooradian, F.A.Blum: <u>Phys.Rev B**7**</u>, 5253 (1973)

[1,15] J.C.White: <u>"Stimulated Raman Scattering"</u> in [1,2]

[1,16] A.S.Pine : <u>J.Opt.Soc.Am.**64**</u>, 1683 (1974)

[1,17] A.S.Pine: <u>J.Opt.Soc.Am.**66**</u>, 98 (1976)

[1,18] W.W.Dudley: <u>CO_2 Lasers</u>, Acad.Press (1976)

[1,19] R.L.Abrams: in <u>Laser Handbook Vol.3</u>, M.L.Stich Edid., North Holland (1979)

[1,20] K.M.Evenson, J.S.Wells, F.R.Peterson, B.L.Danielson, G.W.Day: <u>Appl.Phys.Lett.**22**</u>, 192 (1972)

[1,21] T.Jaeger, G.Wang: <u>"Tunable High-Pressure Infrared Lasers"</u> in [1,2]

[1,22] J.J.Degnan: <u>Appl.Phys.**11**</u>, 1 (1976)

[1,23] J.X.Lin, W.Rohrbeck, W.Urban: <u>Appl.Phys. B**26**</u>, 73 (1981)

[1,24] C.E.Treanor, J.W.Rich, R.G.Rehm: <u>J.Chem.Phys.**48**</u>, 1789 (1968)

[1,25] W.Rohrbeck, A.Hinz, P.Nelle, M.A.Gondal, W.Urban: <u>Appl.Phys. B **31**</u>, 139 (1983)

[1,26] A.Hinz,D.Zeitz,W.Bohle,W.Urban: <u>Appl.Phys. B**36**</u>, 1 (1985)

[1,27] M.Schneider, A.Hinz, A.Groh, K.M.Evenson, W.Urban: <u>Appl.Phys.</u> (in press) (1987)

Chapter 2

[2,1] B.Fritz, E.Menke: <u>Solid State Commun. **3**</u>, 61 (1964)

[2,2] L.F.Mollenauer, D.H.Olson: <u>Appl.Phys.Lett. **24**</u>, 386 (1974)

[2,3] G.Litfin, R.Beigang, H.Welling: <u>Appl.Phys.Lett. **31**</u>, 381 (1977)

[2,4] G.Liftin, R.F.Curl, D.R.Pollock, F.K.Kittel: <u>J.Chem.Phys. **72**</u>, 12 (1980)

[2,5] J.V.V.Kasper, C.R.Pollock, R.F.Curl, F.K.Kittel: <u>Appl.Opt. **21**</u>, 23b (1982)

[2,6] H.Adams, R.Brüggemann, P.Dietrich, D.Kirstem, H.Solka, W.Urban: <u>J.Opt.Soc.Am. B**2**</u>, 815 (1985)

[2,7] D.Reinert, H.Solka, W.Urban: to be published. (D.Reinert, Thesis 1987)

[2,8] C.S.Gudemann, M.H.Begemann, R.J.Saykally: <u>Phys.Rev.Lett. **50**</u>, 727 (1983)

[2,9] A.Stahn, H.Solka, H.Adams, W.Urban: <u>Molec.Phys. **60**</u>, 121 (1987)

[2,10] H.Solka: private communication

42

Chapter 3

[3,1]   I.M.Campbell, R.S.Mason:
        J.Photochem.**11**, 53 (1979)
[3,2]   J.Kosanetzky, H.Vormann, H.Dünnwald, W.Rohrbeck,
        W.Urban: Chem.Phys.Letters **70**, 60 (1980)
[3,3]   H.Dünnwald, E.Siegel, W.Urban, J.W.Rich, G.F.Hmicz,
        M.J.Williams: Chem.Phys.**94**, 195 (1985)
[3,4]   R.E.Center: in Laser Handbook, Vol.3,
        M.Stich, Editor, North Holland (1979)
[3,5]   R.L.Delon, J.W.Rich: Chem.Phys.**107**, 283 (1986)
[3,6]   P.Brechignac, J.P.Martin, G.Taieb:
        IEEE Qe-10, 797, (1974)
[3,7]   N.Djeu: Appl.Phys.Letters **23**, 309 (1973)
[3,8]   G.Guelachvili, D.De Villeneuve, R.Farrenq, W.Urban,
        J.Verges: J.Mol.Spectrosc.**98**, 64 (1983)
[3,9]   R.Farrenq, C.Rossetti, G.Guelachivili, W.Urban:
        Chem.Phys.**92**, 389 (1985)
[3,10] R.Farenq, C.Rosetti: Chem.Phys.**92**, 401 (1985)
[3,11] C.K.N.Patel: Phys.Rev.**141**, 71 (1966)
[3,12] C.Chackerian, R.Farenq, G.Guelachvili,
        C.Rossetti, W.Urban: Can.J.Phys.**62**,1597
        (1984)
[3,13] X.Luo, M.Koch, W.Urban, C.L.Sung, Q.X.Yü, J.X.Lin:
        Appl.Phys. **B 44**, (in Press) (1987)

TUNABLE FAR INFRARED LASER SPECTROSCOPY*

K.M. Evenson, D.A. Jennings, and M.D. Vanek
Time and Frequency Division
National Bureau of Standards
Boulder, Colorado, 80303, USA

ABSTRACT. Tunable far-infrared (FIR) radiation has been generated using CO_2 laser difference generation in metal-insulator-metal diodes either from the difference between a fixed frequency CO_2 laser and a tunable waveguide laser, or from the difference between two fixed frequency CO_2 lasers plus microwave sidebands. Our tunable FIR source is being used to make highly accurate FIR frequency measurements of stable species to serve as frequency and wavelength calibration standards; to measure frequencies of transient species (including molecular ions) for astronomical searches; and to study line broadening and line shape parameters especially for atmospheric spectroscopy applications.

INTRODUCTION

Tunable far-infrared (FIR) radiation has been generated with four different techniques: harmonics of microwave oscillators (1), CO_2 laser difference frequency generation in GaAs (2), FIR laser plus microwave sidebands (3), and the CO_2 laser difference generation in a metal-insulator-metal diode (4,5) described in this paper. We are using our tunable FIR source to make highly accurate FIR frequency measurements of stable species to serve as frequency and wavelength calibration standards (6), to measure frequencies of transient species (including molecular ions) for astronomical searches (7-11), and to study line broadening and line shape parameters (12) especially for atmospheric spectroscopy applications.

We are using two related techniques of synthesizing tunable far infrared radiation: either from the difference between a fixed

*Contribution of the U.S. Government, not subject to copyright.

Work supported by NASA Contract W-15,047

An earlier version of this paper was submitted to the 11th International Conference on Infrared and Millimeter Waves, and was available only to delegates at the conference.

A. C. P. Alves et al. (eds.), Frontiers of Laser Spectroscopy of Gases, 43–51.
© 1988 by Kluwer Academic Publishers.

frequency CO_2 laser and a tunable waveguide laser (i.e., with second
order generation), or from the difference between two fixed frequency
CO_2 lasers plus microwave sidebands (i.e., with third order
generation). In the first of these, a tungsten-nickel diode is used as
the nonlinear element and in the second, a tungsten-cobalt diode is
used. The tungsten-nickel diode generates very little third order
radiation, and the tungsten-cobalt, very little second order. In
second order, tunability is achieved by using a waveguide CO_2 laser
with about ± 120 MHz of tunability; and in 3rd order, tunable microwave
sidebands are added to the fixed FIR frequency difference from two CO_2
lasers. The best 3rd order cobalt diodes produce about 1/3 as much FIR
radiation as the best 2nd order nickel diodes. The current-voltage
curves of these two different diodes predict their corresponding 2nd or
3rd order behaviors. Typical FIR powers of a few tenths of a microwatt
are obtained from 200 mW of CO_2 power.

SECOND ORDER SPECTROMETER

In the 2nd order spectrometer, shown in figure 1, the common
isotope of CO_2 is used in the waveguide laser and one of four isotopic
species are used in the fixed frequency CO_2 laser (I). Eighty-percent
of all frequencies from 0.3 to 4.5 THz can be synthesized, and the
coverage then decreases to a few percent at 6 THz. Ninety megahertz
opto-acoustic modulators are used to isolate the output beams of the
CO_2 lasers from the MIM diode and also to increase the tunability by an
additional 180 MHz. Feedback reduction decreases the amplitude noise
in the FIR radiation by an order of magnitude; hence, the spectrometer
sensitivity increases by this amount. Two fixed frequency CO_2 lasers
(I and II) stabilized to saturated fluorescence signals in CO_2 (13) are
used: the radiation from one is focused on the diode, and the other
serves as a frequency reference for the waveguide laser. The CO_2
radiation is coupled to the whisker by a 25 mm focal length lens using
the conical sharpened tip of the 25 μm diameter tungsten whisker as an
antenna (14). The 3 to 10 mm long whiskers (from the tip to a right
angle bend) radiates the FIR radiation in a long wire antenna pattern
(14). The FIR radiation is collimated using a 10 mm focal length off-
axis segment of a parabolic mirror.

THIRD ORDER SPECTROMETER

Figure 2 illustrates the 3rd order spectrometer. CO_2 lasers I and
II are stabilized to CO_2 sub-Doppler saturated fluorescence features
using separate (not shown) low pressure cells (13). Their radiation is
focused on the diode along with that of a frequency-synthesized,
microwave (2-20 GHz) source used to add the tunable sidebands to the

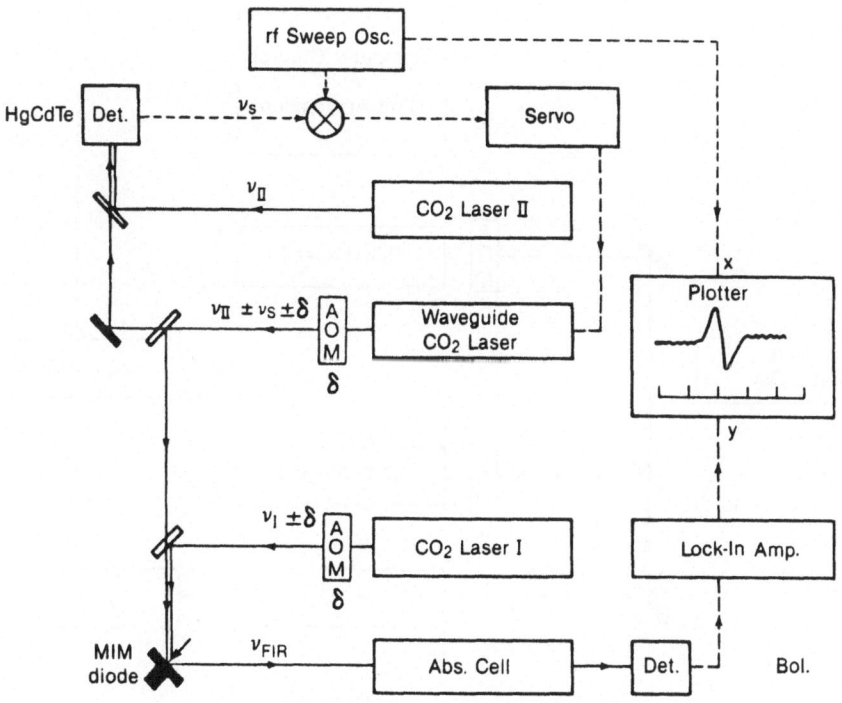

TuFIR SPECTROMETER

Figure 1 : The Second Order Tunable Far Infrared Spectrometer:
$\nu_{FIR} = (\nu_{II} \pm \nu_S \pm \delta) - (\nu_1 \pm \delta)$.

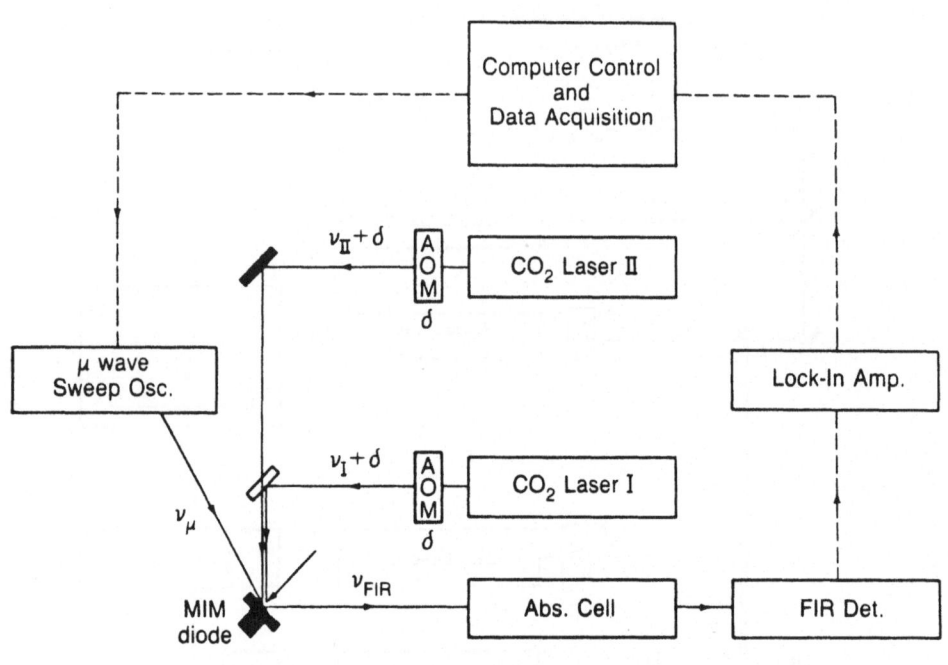

3rd ORDER
TuFIR SPECTROMETER
$$\nu_{FIR} = (\nu_I - \nu_{II}) \pm \nu_\mu$$

Figure 2 : The Third Order Tunable Far Infrared Spectrometer.

CO_2 difference frequency. Its accuracy is a few hertz, and it does not contribute to the uncertainty of the source.

EXPERIMENTAL TECHNIQUE

In each spectrometer both CO_2 reference lasers are frequency modulated at a 1 kHz rate using piezoelectric drivers on the end mirrors and are then servoed to the line center of the saturated fluorescence signals obtained from the external low-pressure CO_2 cells. The FIR detectors and lock-in amplifier detect at the modulation rate; hence, the derivatives of the absorptions are recorded. The depth of the frequency modulation of CO_2 laser I can be increased up to 7 MHz to enhance the FIR frequency modulation, thereby increasing the sensitivity for the broader lines.

Our lasers are locked with an uncertainty of 25 kHz; thus, the overall uncertainty in the FIR frequency is: $\sqrt{2}$ x 25 or 35 kHz. The frequencies of these CO_2 absorptions are known to about ±3 kHz (14,15); however, without special baselines slope correction and special locking techniques described in references 13 and 15, the overall uncertainties increase to the aforementioned 25 kHz level. The largest contributor to the overall uncertainties in the measurement of a transition frequency comes from finding the line centers of these Doppler broadened lines, about 0.1 of the line width (0.05 to 1 MHz).

Absorption cells from 0.5 to 3.5 meters in length with diameters ranging from 19 to 30 mm have been used in each spectrometer. The cells have either glass or copper walls and polyethylene or polypropylene windows.

Four different detectors have been used in the spectrometers: 1) an InSb 4K, liquid ^{4}He cooled, hot-electron bolometer, operating from 0.3 to 0.6 THz with an NEP of about 10^{-13}W/$\sqrt{Hz}$; 2) a gallium doped germanium bolometer, cooled to the lambda point of ^{4}He, operating from 0.6 to 6.5 THz with an NEP of about 10^{-13}W/$\sqrt{Hz}$; 3) a similar, but ^{3}He cooled bolometer, with a two-orders-of-magnitude smaller NEP; and 4) a Ge:Ga photoconductor, cooled to 4K, with an NEP of 10^{-14}W/$\sqrt{Hz}$, operating from 2.5 to 6.0 THz.

RESULTS

FIR spectra of a series of rotational transitions have been measured in CO, HCl, and HF and will be published elsewhere (6). These lines, about ten times more accurately measured than what is needed in the present state-of-the-art Fourier transform spectrometers, serve as excellent absolute calibration lines and are the most accurate available.

Radio astronomy has been extended to the terahertz region and requires accurate frequencies of transitions in order to identify the species and to determine the Doppler shifts of the sources. Several astronomically interesting molecules have been studied: OH (8), NaH (9), and MgH (10). The OH had previously been studied by FIR laser magnetic resonance, but the new measurements decrease the uncertainties of the transitions by about an order of magnitude. Rotational spectra of NaH were studied in several vibrational levels and accurate rotational constants were obtained. The measurements in MgH are accurate to better than one megahertz, and the frequencies were used for a preliminary astronomical search for MgH; further searches are necessary for a positive identification.

The long absorption cells naturally lend themselves to a hollow-cathode discharge configuration for the study of molecular ions. A preliminary experiment revealed the HCO^+ line at 1 THz with a signal-to-noise ratio (100:1 with a 1 s time constant) equivalent to that obtained using the laser sideband technique (16). Possible transitions in H_2D^+ and OH^- have also been observed; however they are weak and only tentatively identified, and further work is underway.

The instrumental resolution of the spectrometers is limited by the combined frequency fluctuation from each CO_2 laser (about 15 kilohertz). This, of course, is less than any Doppler limited linewidth and, therefore, does not limit our resolution except for possible sub-Doppler work. This high resolution provides an excellent way of studying pressure shifts and line shape studies of spectral lines. The measurement of OH concentrations in our atmosphere as a function of altitude using absorption and emission measurements requires an accurate knowledge of its linewidth in the atmosphere.

Figure 3 shows a spectrum of methanol taken with our 3rd order spectrometer. Spectra from both sidebands are observed simultaneously and are superimposed; however, they are easily distinguishable because the phases of the derivatives are opposite. That is: the absorptions between 3 499 588.4 and 3 504 588.4 MHz start out positive (from left to right in the figure); those on the other sideband (3 489 5884.4 and 3 484 588.4 MHz) start out negative. A corresponding scan taken with a high resolution FTIR show a jumble of unresolved lines; in contrast, this scan shows resolved lines even though two sidebands are superimposed.

CONCLUSION

Improvements in these different techniques may come from either improved diodes or detection schemes. The nonlinearities measured in the current-voltage curves of our MIM diodes are extremely small and conversion efficiencies could be 100 times larger. We are optimistic that better materials which will result in larger FIR powers may be found. Differential detection schemes would also significantly improve the sensitivity and permit the detection of weaker lines. The sensitivity, however, is still only about 1% of that of laser magnetic resonance. Laser magnetic resonance is useful only for paramagnetic

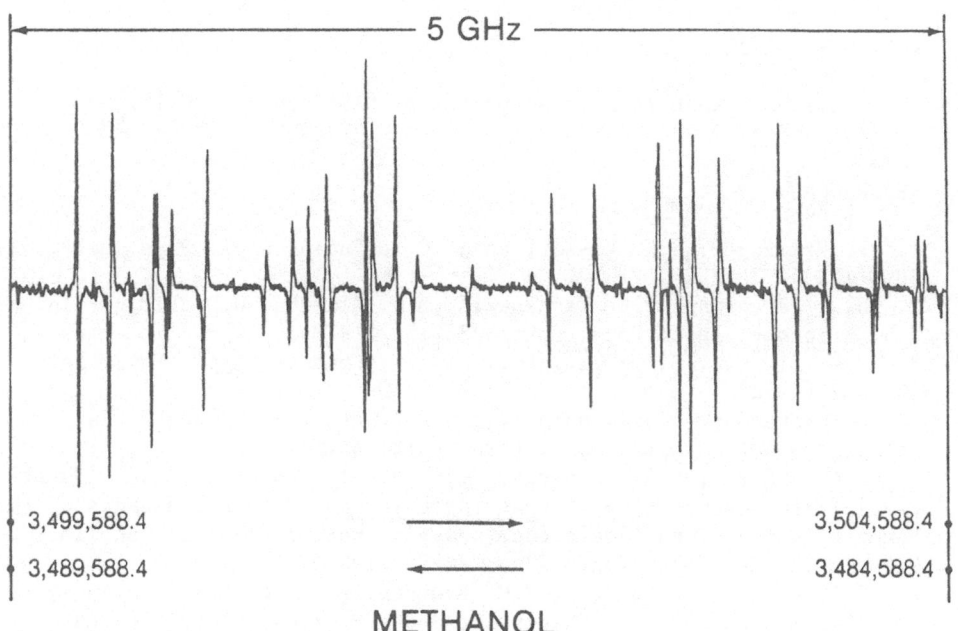

5 GHz

3,499,588.4   ⟶   3,504,588.4

3,489,588.4   ⟵   3,484,588.4

METHANOL

Figure 3 : A far infrared scan of methanol: a pressure of about 10 Pa
of methanol and a frequency modulation of 2.5 MHz were used to record
the spectrum. Both sidebands are shown: the lines which are positive
on the left are those in the increasing frequency sideband and vice
versa for those which are negative on the left. The frequencies are in
megahertz.

species and is less accurate than the TuFIR technique by about an order of magnitude. We believe that this technique of laser difference spectroscopy in the FIR is now well established and we are looking forward to many exciting discoveries.

ACKNOWLEDGMENTS

This work is partially supported by NASA contract W-15, 047.

REFERENCES

1. P. Helminger, J.K. Messer, and F.C. DeLucia, Appl. Phys. Lett. **42**, 309 (1983).

2. H.R. Fetterman, P.E. Tannenwald, B.J. Clifton, W.D. Fitzgerald, and N.R. Erickson, Appl. Phys. Lett. **33**, 151 (1978).

3. D.D. Bicanic, B.F.J. Zuidberg, and A. Dymanus, Appl. Phys. Lett. **32**, 367 (1978).

4. K.M. Evenson, D.A. Jennings, and F.R. Petersen, 'Tunable Far Infrared Spectroscopy,' Appl. Phys. Lett. **44**, 576 (1984).

5. K.M. Evenson, D.A. Jennings, K.R. Leopold, and L.R. Zink, 'Tunable Far Infrared Spectroscopy,' Laser Spectroscopy VII, Proceedings of the Seventh International Conference, Hawaii, Springer-Verlag, Berlin, Heidelberg, New York, Tokyo, June 24-28 (1985).

6. I.G. Nolt, G. DiLonardo, K.M. Evenson, A. Hinz, D.A. Jennings, K.R. Leopold, M.D. Vanek, J.V. Radostitz and L.R. Zink, 'Far-Infrared Frequency Measurements of CO, HCl, and HF: Frequency Standards for the 0.3 to 6 THz (10 to 200 cm^{-1}) Region,' J. Mol. Spectrosc. **125**, accepted (1987).

7. K.R. Leopold, L.R. Zink, K.M. Evenson, D.A. Jennings and M. Mizushima, 'The Far Infrared Spectrum of Magnesium Hydride,' J. Chem. Phys. **84**(3), 1935-1937 (1985).

8. J.M. Brown, L.R. Zink, D.A. Jennings, K.M. Evenson, A. Hinz, and I.G. Nolt, 'Laboratory Measurement of the Rotational Spectrum of the OH Radical with Tunable Far-Infrared Radiation,' Ap. J. **307**, 210 (1986).

9. K.R. Leopold, L.R. Zink, K.M. Evenson, and D.A. Jennings, 'The Far Infrared Spectrum of Sodium Hydride,' J. Mol. Spectrosc. **122**, 150-156 (1987).

10. K.R. Leopold, L.R. Zink, K.M. Evenson, and D.A. Jennings, 'The Far Infrared Spectrum of Magnesium Hydride,' J. Chem. Phys, **84**, 1935-1937 (1986).

11. D.A. Jennings, K.M. Evenson, L.R. Zink, C. Demuynck, J.L. Destombes, B. Lemoine, and J.W. Johns, 'High Resolution Spectroscopy of HF from 40 to 1100 Wavenumbers: Highly Accurate Rotational Constants,' J. Mol. Spectrosc. **122**, 477-480 (1987).

12. D.A. Jennings, K.M. Evenson and M.D. Vanek, I.G. Nolt, J.V. Radostitz and K.V. Chance, 'Air- and Oxygen-Broadening Coefficients for

the O_2 Rotational Line at 60.46 cm^{-1},' Geophys. Research Letters, **14**, 722-725 (1987).

13. F.R. Petersen, E.C. Beaty and C.R. Pollock, 'Improved Rovibrational Constants and Frequency Tables for the Normal Laser Bands of $^{12}C^{16}O_2$,' J. Mol. Spectrosc. **102**, 112-122 (1983).

14. K.M. Evenson, M. Inguscio, and D.A. Jennings, 'Point Contact Diode at Laser Frequencies,' J. Appl. Phys. **57**(3), 956-960 (1985).

15. L.C. Bradley, K.L. Soohoo, and C. Freed, 'Absolute Frequencies of Lasing Transitions in Nine CO_2 Isotopic Species,' IEEE J. Quantum Electronics, **QE-22**, 234- (1986).

16. F.C. Van Den Heuvel and A. Dymanus, 'Observation of Far-Infrared Transitions of HCO^+, CO^+ and HN_2^+,' Chem. Phys. Lett., **92**, 219-222 (1982).

Generation of Coherent Tunable Radiation

R. Wallenstein
Institut für Quantenoptik
Universität Hannover
3000 Hannover FR Germany

INTRODUCTION

One of the most important tools which are used today for the spectroscopy of atoms, molecules and solid states is the laser. During the
past 27 years - since its invention - a large number of very different
lasers have been developed. The shortest wavelength of laser emission
is presently about 16 nm, the longest wavelengths are in the mm-range.
Powers of more than 10^{13} W have been generated. The spectral bandwidth
ranges from a few Hz $(\Delta\nu/\nu\sim10^{-15})$ up to more than 10^{14} Hz. Laser action
in thin layers of 10^{-4} cm have been achieved and lasers of more than
100 m have been built. The smallest laser has a volume of a fraction
of mm^3 while large laser systems fill a building of several stories.
The duration of laser emission varies over several orders of magnitude.
The shortest laser pulses generated so far have a duration of about
$8\cdot10^{-15}$sec, which corresponds to about 4 wavelengths of the electromagnetic field.

These few numbers illustrate the very different properties of
laser systems as they exist today. The laser is,in principle, a light
source with fixed and stable frequency. The emission frequency is
determined by the optical transition of the laser medium and the frequencies of the modes of the laser resonator. In fact the monochromaticity and the high stability of the frequency is the basis of many
spectroscopic experiments with lasers. For this reason systems with
high frequency stability have been developed. In addition the spectroscopy requires, however, light sources with tunable frequency. The broad
application of lasers for spectroscopy is thus closely related to the
development of dye lasers, since this laser provided for the first
time coherent light of broadly tunable wavelength.

DYE LASERS

The dye laser is the most efficient tunable laser. Its qualities of
continuous tunability, efficient narrowband operation, wide spectral
range and simplicity have made it an indispensible tool in many areas
of laser research and applications. Since the discovery of the dye laser
this field has progressed rapidly and dye laser technology is now at

A. C. P. Alves et al. (eds.), Frontiers of Laser Spectroscopy of Gases, 53–61.
© 1988 by Kluwer Academic Publishers.

a rather advanced level of development. This is evident from the commercial availability of a number of sophisticated dye laser systems, both pulsed and continuous wave (CW). Detailed theoretical models for pulsed and CW dye lasers have also led to a better understanding of the working of this unique laser. Both the physics and the present technology of dye lasers have been discussed in the excellent review given by L.G. Nair[1] and shall thus not be outlined here in any further detail.

GENERATION OF TUNABLE UV AND IR

With presently available laser dyes, dye laser operation is limited to the wavelength range from about 340 to 1200 nm. The wavelength range can be extended towards shorter and longer wavelengths by frequency mixing of dye laser radiation in nonlinear optical materials[2]. In this way, tunable ultraviolet radiation down to 189 nm has been obtained by second harmonic and second-order sum frequency generation in nonlinear crystals. In the long wavelength range the generation of tunable near and middle infrared as the difference frequency of two laser oscillators has been very successful[2].

Frequency doubling of visible laser light in nonlinear optical materials is a technique which was well established with fixed frequency lasers. The physics of this process has been discussed in several excellent reviews[3,4,5]. If the peak power of the laser radiation is sufficiently high (in the kW to MW range), the doubling crystal can be placed outside the laser cavity. At smaller power levels, as supplied by CW dye lasers, low-loss laser cavities with an intracavity doubling crystal provide the best efficiency for frequency conversion. Because the phase matching condition requires different crystal orientations for different wavelengths, doubling of the frequency-tuned output of a dye laser necessitates the rotation of the crystal, synchronuously with dye laser tuning. At low laser intensities the energy conversion of the visible laser radiation into coherent ultraviolet light is proportional to the input intensity and the UV output increases as the square of the input. At sufficiently high intensities, the conversion efficiency approaches a constant value, which is usually of the order of 10 to 40 percent. To obtain an optimum in conversion, the dye laser light is focused in general into the doubling crystal. If the focusing is too tight the efficiency may be reduced, however, because the angular spread of the beam is too large to achieve simultaneous phase matching for the different angular directions. Highest conversion efficiencies are obtained with KDP and ADP. Although these crystals are transparent down to 200 nm, they do not permit phase matching for second harmonic generation at UV wavelengths below 261 nm (ADP) and 259 nm (KDP), if the crystals are kept at room temperature. Temperature tuning of a 90° phase matched ADP crystal between -116°C and 52°C extends the generation of coherent radiation to the spectral range of 246 to 265 nm. It should be mentioned that UV radiation at somewhat shorter wavelengths can be obtained in these crystals as the sum frequency of two laser waves of different frequencies[2].

Besides ADP and KDP a new efficient nonlinear material became available more recently [6-11]. This material, barium-β-borate(BBO), has a nonlinear coefficient four times that of KDP, a high damage threshold of 13 GW/cm^2, a wide transparency range from about 190 nm to 2.25 μm and it can be phase-matched through the entire transmission region. Efficient second harmonic generation of dye laser radiation at 410 to 700 nm thus generates intense UV light in the wavelength range of 205 to 350 nm[12]. BBO can efficiently double, triple and quintuple the output of a Nd-YAG laser[9]. Optical parametric oscillators made from BBO[13] can cover the tuning range from the UV across the visible and into the near infrared.

TUNABLE VUV AND XUV RADIATION GENERATED BY FREQUENCY MIXING IN GASES

Nonlinear frequency mixing in gases is a well established method for the generation of optical radiation in the spectral region of the vacuum ultraviolet (VUV) at wavelengths λ_{VUV} = 100 - 200 nm and in the extreme ultraviolet (XUV) at λ_{XUV} <100 nm.

The principles of frequency mixing and the experimental progress towards the generation of coherent VUV radiation have been reviewed by W. Jamroz and B.P. Stoicheff[14], by C.R. Vidal[15] and by R. Hilbig et al[16].

In this contribution we discuss a few examples of more recent investigations of third- and fifth order sum- and difference frequency mixing of dye laser radiation in rare gases. These frequency conversions are simple, reliable methods for the production of tunable coherent VUV and XUV radiation of narrow spectral width and high spectral intensity. The results of current investigations are very promising for a further, considerable increase of the pulse power and for an extension of the tuning range to even shorter wavelengths.

THIRD-ORDER FREQUENCY CONVERSION

Nonresonant third-order sum- and difference frequency mixing (ω_{VUV} = $2\omega_1 \pm \omega_2$) in the rare gases Xe and Kr and in Hg vapor generated broadly tunable VUV radiation in the wavelength range λ_{VUV}=110-200 nm[17-]. Frequency tripling in Ar and Ne produced XUV light in spectral regions between 72 and 105 nm[25,26].

In these experiments laser pulse powers of 1 to 5 MW provided conversion efficiencies of typically 10^{-5} to 10^{-6}. The power of the generated VUV light pulses was in the range of 1 to 20 W (0.3 - 6·10^{10} photons/pulse). Main limitations on the efficiency are caused by dielectric gas-breakdown in the focus of the laser light and by nonlinear intensity dependent changes of the refractive index[15,27] which destroy the phase-matching.

The VUV intensity generated by nonresonant conversion methods is sufficient for most investigations in linear (absorption or fluorescence) spectroscopy. Other applications (like multiphoton excitation and ionization or photodissociation) require more powerful light pulses.

The pulse power can be increased by several orders of magnitude using resonant conversion methods[28].

Tuning the laser frequency, for example, to a two-photon resonance the induced polarization is resonantly enhanced. This enhancement provides conversion efficiencies of $\eta > 10^{-4}$ even at input powers of only a few kilowatts.

The two-photon resonant conversion is usually of the type ω_{VUV} = $2\omega_1 \pm \omega_2$ where ω_1 is tuned to a two-photon transition, and ω_2 is a variable frequency.

In the past this type of frequency conversion has been investigated in detail in metal vapors (like Sr[29,30], Mg[31-33], Cs[34], Ba[35], Hg[36-38] and Zn[39]) and in the rare gases Xe and Kr[23,40-43].

For the experimental realization of the frequency mixing rare gases are very advantageous. Enclosed in a simple glass or metal cell. (equipped with appropriate windows) these gases provide a nonlinear medium of homogeneous, easily variable density. These gases are thus an appropriate medium for the construction of a reliable VUV light source.

Very promising for this purpose is the resonant frequency conversion in Kr[44]. Because of the high excitation energy of the lowest two-photon resonance $4p-5p|5/2,2|$ the sum- and difference frequency ($\omega_{VUV}=2\omega_R \pm \omega_T$) of the two-photon resonant radiation ($\lambda_R=216.6$ nm) and of tunable dye laser light ($\lambda_T=217-900$ nm) generates radiation in the region of the XUV ($\lambda_{XUV}=72.3-96.7$ nm) and of the VUV ($\lambda_{VUV}=123-216$ nm), respectively. Tuning ω_T for example in the range $\lambda_T=219-364$ nm the sum frequency generated light at $\lambda_{XUV}=72.5-83.5$ nm. In agreement with theoretical predictions the conversion efficiency η is almost constant within this spectral range. At input powers $P_R=14$ kW and $P_T=400$ kW the pulse power of the XUV exceeded $P_{XUV}=20$ W. Absorptions in the Kr gas reduced, however, the power of the detected XUV light to about 5 W (effective efficiency $\eta=1.2\cdot10^{-5}$). With laser light at $\lambda_T=272-737$ nm the difference frequency generates continously tunable radiation at $\lambda_{VUV}=127-180$ nm. In this range the conversion efficiency increases with wavelength by more than one order of magnitude. At $\lambda_{VUV}=135$ nm, for example, input powers $P_R=0.2$ MW and $P_T=1.2$ MW generate VUV light with $P_{VUV}=250$ W ($\eta=1.8\cdot10^{-4}$). At $\lambda_{VUV}=175$ nm, a lower input ($P_R=80$ kW, $P_T=560$ kW) produced VUV light pulses of $P_{VUV}=1.8$ kW($\eta=2.8\cdot10^{-3}$). This spectral variation of η is in agreement with the calculated wavelength dependence of the nonlinear susceptibility and of the gas pressure required for optimum VUV output.

In these experiments the UV radiation of the wavelength $\lambda_R=216.6$ nm (resonant with the two-photon transition $4p-5p|5/2,2|$) was generated by sum frequency mixing in KDP ($\omega_R=\omega_{UV} +\omega_{IR}$) of frequency doubled dye laser radiation ($\omega_{UV}=2\omega_L$ with $\lambda_L=544$ nm) and of the fundamental (ω_{IR}) of the Nd:YAG laser. In principle, radiation at $\lambda_R=216.6$ nm can be generated also by doubling the frequency of a blue laser ($\lambda_L=433$ nm) In the past the only crystal suited for the second harmonic generation was the deuterated KB5 crystal[45]. Because of the low conversion efficiency of 2 to $4\cdot10^{-2}$ the generated pulse powers are typically 60 to 120 kW.

Considerably higher efficiences (of about 15 percent) are now obtainable with the new nonlinear optical material barium-β-borate (BBO). Besides the high conversion efficiency this material generates the second harmonic of radiation at wavelengths as short as 410 nm. Therefore not only the $4p-5p|5/2,2|$ transition can be used for the two-photon resonant enhancement of $\chi^{(3)}$ but also the resonances $4p-5p$

$|3/2,2|$ and $4p-5p|1/2,0|$ which require UV radiation at $\lambda_{UV}=214.7$ nm and 212.5 nm, respectively. The use of the transition $4p-5p|1/2,0|$ is of particular advantage since in this case the resonant enhancement of $\chi^{(3)}$ is considerably larger than the one obtained with the transition $4p-5p|5/2,2|$ which has been used in previous investigations.

If ω_R is tuned to the different two-photon transitions the calculated ratios[28] of the output P_{XUV} of the resonant third harmonic $\omega_{XUV}=3\omega_R$ are, for example, $R_1=8.7$ and $R_2=10.6$ where $R_1=P_{XUV}(5p|1/2,0|)/P_{XUV}(5p|5/2,2|)$ and $R_2=P_{XUV}(5p|5/2,2|)/P_{XUV}(5p|3/2,2|)$.

In the experiment the frequency of a Stilbene 3 dye laser was doubled in a suitable BBO-crystal. The frequency tripling was achieved by focusing the UV laser light into a pulsed Kr jet (lens: f= 200 mm). The measurements - performed with the same (low) UV power level of about 40 kW - provide values of $R_1=6.0$ and $R_2=7.7$. These values are in good agreement with the theoretical estimations if saturation is taken into account.

The resonant frequency conversion with UV laser radiation ω_R - generated by the efficient frequency doubling in BBO - offers several advantages. First, the frequency ω_R can be tuned to the two photon transition, which provides the largest resonant enhancement of $\chi^{(3)}$. Second, the spectral width of the UV light is determined by the line width of the dye laser radiation and thus narrowband radiation is easily obtained (the spectral width of the radiation at $\omega_R= \omega_{UV}+\omega_{IR}$ is usually limited by the rather large line width of the Nd:YAG laser light). Finally, the wavelength ($\lambda_L=425$ nm) of the radiation resonant with the transition $4p-5p|1/2,0|$ is located in the center of the tuning range of Stilbene 3 which is efficiently excited by the third harmonic of a Nd:YAG laser as well as by the radiation of excimer lasers.

Considering these new advantageous possibilities (together with the results of a general theoretical treatment of the frequency mixing in gases)[46] the two-photon resonant conversion in Kr might become a standard method for the generation of broadly tunable coherent radiation in the spectral region of the VUV and of the XUV.

FIFTH-ORDER FREQUENCY MIXING

In the experiments decribed so far third order frequency mixing of dye laser radiation ($\lambda_L=216-800$ nm) generated continuously tunable VUV in the wavelength range of 72 to 200 nm. In principle an extension of this tuning range to shorter wavelengths is possible by conversion processes of higher order[41,47]. Sum-frequency mixing of fifth order, for example, should produce coherent VUV at wavelengths $\lambda_{XUV}=42-72$ nm.

In the past fifth harmonic generation has been investigated with powerful fixed frequency solid state (Nd-YAG or Nd-Glass) and gas (XeCl, KrF) lasers[42,47-49]. In one of these experiments[49] input powers of more than 300 MW (mode-locked Nd-YAG fourth harmonic, $\lambda=266.1$ nm) provide, for example, conversion efficiencies of 10^{-5} to 10^{-6}. Since the pulse power of most dye laser systems is lower by one or two orders of magnitude nonresonant fifth-order frequency mixing of this radiation would produce intensities below a useful level.

A considerable increase of the generated VUV power is expected, however, from resonant six-wave mixing. In fact resonant fifth-order conversion has been demonstrated in Ar[16]. In these investigations the UV radiation (λ_{UV}=318.9 nm) of a frequency doubled dye laser (λ_L= 637.8 nm) is resonant with the four-photon Ar-transition 3p-9p$|1/2,0|$. Simultaneously $3\omega_{UV}$ is close to the transition frequency ω_{res} of the first Ar resonance ($3p^1S_o$-$4s|3/2,1|$). The energy difference $\Delta E=3\omega_1-\omega_{res}$ of 91 cm^{-1} is sufficiently small so that the fifth-order conversions $\omega_{XUV}=5\omega_{UV}$ and $\omega_{XUV}=4\omega_{UV}+\omega_L$ are not only four-photon but also (almost) three-photon resonant. The conversion $\omega_{XUV}=3\omega_{UV}+2\omega_L$ is five- and near-ly three-photon resonant.

Tunable radiation is generated by $\omega_{XUV}=4\omega_{UV}+\omega_V$. With the radiation of a second dye laser (λ_V=216-800 nm) the six-wave mixing $\omega_{XUV}=4\omega_{UV}+\omega_V$ should cover the whole range of 58 to 72 nm.

In the experiment λ_V was tuned, for example, in the range of 550-580 nm (dye: Rhodamine 6G) and of 275-290 nm (the frequency doubled dye laser radiation). In this way the generated XUV was tunable in the wavelength regions λ_{XUV}=69.85-70.1 nm and λ_{XUV}=62.0-62.6 nm, respectively.

Of special interest is of course the XUV pulse power obtainable by this method. Measurements of the dependence of the power P_F of the fifth harmonic on the laser power P_L (P_L=0.2-1.5 MW) confirmed that P_F is proportional to P_L^5. The value of P_F could be estimated by comparison with the known pulse power of the VUV generated by fourwave mixing processes. The measured ratio of P_F and of the tripled power P_T is about one hundred. Since for the laser pulse power P_{UV}=1.5 MW (used in the present investigation) P_T is on the order of 10 W the power P_F should be about 0.1 W or $3\cdot10^8$ photons/pulse.

Analogous to the results in Ar, six-wave mixing in Ne should generate VUV at even shorter wavelengths. In the case of Ne the UV radiation of a frequency doubled dye laser (λ_{UV}=282.1 nm) and of the sum frequency $\omega'_{UV}=\omega_{UV}+\omega_{IR}$ (λ'_{UV}=223.0 nm) is four-photon resonant with the 2p-6p$|1/2,0$ transition ($\omega(2p$-$6p)=3\omega'_{UV}+\omega_{UV}$). The frequency $3\omega'_{UV}$ is close to ω_{res} of the resonance transition $2p^1S_o$$\rightarrow$$3s|3/2,1|$. The difference $\Delta E=3\omega_1-\omega_{res}$ is only 69.8 cm^{-1}. The (almost) three- and four-photon resonant conversion $\omega_{XUV}=3\omega'_{UV}+\omega_{UV}+\omega_V$ (with λ_V=216-800 nm) should thus be well suited for the production of tunable XUV in the range λ_{XUV}=46.2-54.8 nm.

At present the resonant six wave mixing in Ar and Ne is subject of detailed investigations. The results obtained so far indicate that the mixing schemes described above are appropriate for the desired extension of the tuning range of the VUV light generated by nonlinear frequency up-conversion.

APPLICATIONS

The discussed experimental results demonstrate that nonlinear optical frequency conversion produces widely tunable VUV radiation. Because of the narrow spectral width and the high intensity the VUV light is a powerful tool for high resolution spectroscopy of atoms and molecules. This has been demonstrated, for example, by absorption spectroscopy (carried out on $CO^{50,51}$, $H_2^{52,53}$ and N_2^{54}, by excitation spectroscopy (in $CO^{21,55}$, $NO^{56,57}$, $H_2^{53,58}$, $Xe_2^{59,60}$, Kr_2 and Ar_2^{61} or by state selective resonant - excitation-ionization spectroscopy -

first demonstrated in CO and NO[62] - which made possible state-selective photoionization and photodissociation spectroscopy of the H_2 molecule from excited states[63].

With narrow band L_α-radiation it was possible to measure the spatial velocity distribution of atomic hydrogen generated by the photo-dissociation of HI[64]. Other applications of L_α radiation included the resonant photodissociation of H and D[65], the investigation of the H atom in an external electric[66] or strong magnetic field[67] and the hydro-gen-atom photofragment spectroscopy[68].

The powerful VUV generated by resonant frequency mixing is suited for applications which require very intense laser radiation like the multiphoton excitation of atoms[38] and molecules, photodissociation studies of molecules[68] or plasma diagnostics.

Today the number of spectroscopic applications of coherent laser-generated VUV light is still small. Because of the simple way of genera-tion and the excellent spectral properties there is no doubt that in the future this radiation will be very useful for a large variety of spectroscopic applications.

REFERENCES

1 L.G. Nair, Progress Quantum Electronics 7, 153 (1982)
2 see for example: R.Wallenstein, Laser Handbook Vol.III, M.L. Stitch ed. North Holland Publ.Comp. (1979) and references therein.
3 N. Bloembergen, Nonlinear Optics, Benjamin, New York (1965)
4 R.W. Terhune and P.D. Maker, Lasers Vol. 2, A. K. Levine Ed. Dekker, New York (1968).
5 F. Zernike and J.E. Midwinter, Applied Nonlinear Optics, J. Wiley, New York (1973).
6 C. Chen, B. Wu, A. Jiang und G. You, Sci. Sinica (Ser.B) 28, 235 (1985)
7 J. Liebertz und S. Stähler, Z. Kristallogr. 165, 91 (1983)
8 K. Kato, IEEE J. Quantum Electron. QE-22, 1013 (1986)
9 C. Chen, Y.X. Fan, R.C. Eckardt and R.L. Byer, in Digest of the Conference on Lasers and Electrooptics(Opt. Soc. of America, Washington, D.C. 1986) Seite 322 ff.
10 K. Miyazaki, H. Sakai und T. Sato, Opt. Lett. 11, 797 (1986)
11 H. Schmidt and R. Wallenstein, Laser u. Optoelectron. 19, 302 (1987)
12 P. Lokai, B. Burghardt, D. Basting und W. Mückenheim, Laser u. Optoelectron. 19, 296 (1987)
13 Y.X. Fan, R.C. Eckardt and R.L. Byer; C. Chen and A.D. Jiang, Postdeadline paper, ThT4, CLEO 1986.
14 W. Jamroz and B.P. Stoicheff, in Progress in Optics, E. Wolf ed., North Holland, Amsterdam 1983, vol. 20, pp. 326-380.
15 C.R. Vidal, in Tunable Lasers (I.F. Mollenauer and J.C. White, eds., Springer Verlag, Heidelberg 1984)
16 R. Hilbig, G. Hilber, A. Lago, B. Wolff, and R. Wallenstein, Comments At. Mol. Phys., part D18, 157 (1986)
17 G.C. Bjorklund, IEEE J. Quantum Electron. QE-11, 287 (1975).
18 R. Mahon, T.J. McIlrath and D.W. Koopman; Appl. Phys. Lett. 33, 305 (1978)

19 D. Cotter, Optics Commun. 31, 397 (1979).
20 R. Wallenstein, Optics Commun. 33,119 (1980)
21 R. Hilbig and R. Wallenstein, IEEE J. Quantum Electron. QE-17, 1566 (1981)
22 R. Hilbig, PhD thesis, 1984(to be published in Appl. Phys.).
23 R. Hilbig, G. Hilber, A. Timmermann, and R. Wallenstein; "Laser Techniques in the Extreme Ultraviolet", in Proc AIP Conf. vol.119, p.1 (1984)
24 R. Hilbig and R. Wallenstein; Appl. Optics 21, 913 (1982).
25 R. Hilbig and R. Wallenstein; Optics Commun. 44, 283 (1983).
26 R. Hilbig, A. Lago and R. Wallenstein, Optics Commun. 49, 297 (1984)
27 H. Langer, H. Puell, and H. Röhr; Opt. Commun. 34, 137 (1980)
28 D.C. Hanna, M.A. Yuratich, and D. Cotter, Nonlinear Optics of Free Atoms and Molecules; Berlin: Springer Verlag, 1979.
29 R.T. Hodgson, P.P. Sorokin, and J.J. Wynne; Phys. Rev. Lett. 32, 343 (1974)
30 H. Scheingraber, H. Puell, and C.R. Vidal; Phys. Rev. A18, 2585 (1978)
31 D.M. Bloom, J.T. Yardley, J.F. Young, and S.E. Harris; Appl. Phys. Lett. 427 (1974).
32 S.C. Wallace and G. Zdaziuk; Appl. Phys. Lett. 28, 449 (1976).
33 H. Junginger, H.B. Puell, H. Scheingraber, and G.R. Vidal, IEEE J. Quantum Electron. QE-16, 1132 (1980).
34 K.M. Lang, J.F. Ward, and B.J. Orr, Phys. Rev. A9, 2440 (1974).
35 J. Heinrich and W. Behmenburg; Appl. Phys. 23, 333 (1980).
36 F.S. Tomkins and R. Mahon; Opt. Lett.6, 179 (1981), IEEE J. Quantum Electron. QE-18, 913 (1982).
37 J. Bokor, R.R. Freeman, R.L. Panock, and J.C. White, Opt. Lett.6, 182 (1981);
M. Jopson, R.R. Freeman, and J. Bokor; Proc. XIIth IQEC, App. Phys. B 28, 203 (1982); see also "Laser Techniques in the Extreme Ultra-violet" in Proc. AIP Conf., vol.90.
38 R. Hilbig and R. Wallenstein, IEEE J. Quantum Electron. QE-19, 1759 (1983).
39 W. Jamroz, P.E. LaRocque, and B.P. Stoicheff; Opt. Lett.7, 617 (1982)
40 R. Hilbig and R. Wallenstein, IEEE J. Quantum Electron. QE-19, 194 (1983).
41 J. Bokor, P.H. Bucksbaum and R.R. Freeman; Opt. Lett.8, 217 (1983).
42 K.D. Bonin and T.J. McIlrath, JOSA B2, 527 (1984).
43 A. Lago, Ph.D. Thesis, Bielefeld 1987.
44 G. Hilber, A. Lago, and R. Wallenstein, JOSA B (in press).
45 J.A. Paisner, M.L. Spaeth, D.C. Gerstenberger, and I.W. Rudermann, Appl. Phys. Lett. 476 (1978).
46 A. Lago, G. Hilber, and R. Wallenstein, Phys. Rev. A (in press).
47 J. Reintjes, Appl. Opt. 19, 3889 (1980) and references therein.
48 J. Reintjes, L.L. Tankersley and R. Christensen; Opt. Commun. 39, 355 (1981).
49 J. Reintjes, R.C. Eckardt, C.Y. Shen, N.E. Karangelen, R.C. Elton, and R.A. Andrews; Phys. Rev. Lett. 37, 1540 (1976).
50 A.C. Provorov, B.P. Stoicheff and S.C. Wallace, J. Chem. Phys. 67, 5393 (1977).
51 J.C. Miller, R.N. Compton and R.W. Cooper, J.Chem.Phys.76, 3967 (1982)

52 M. Rothschild, H. Egger, R.T. Hawkins, J. Bokor, H. Pummer and C.K. Rhodes, Phys. Rev. A23, 2=6 (1981).

53 E.E. Marinero, C.T. Rettner and R.N. Zare, Chem. Phys. Lett. 95, 486 (1983).

54 P.R. Herman and B.P. Stoicheff, Opt. Lett. 10, 502 (1985).

55 P. Klopotek and C.R. Vidal, Can. J. Phys. 62, 1426 (1984).

56 J.R. Banic, R.H. Lipson, T. Efthimiopoulos, and B.P. Stoicheff, Opt. Lett. 6, 461 (1981).

57 H. Scheingraber and C.R. Vidal, JOSA B2, 343 (1985).

58 F.J. Northrup, J.C. Polanyi, S.C. Wallace and J.M. Williamson, Chem. Phys. Lett. 105, 34 (1984).

59 R.H. Lipson, P.E. LaRoque, and B.P. Stoicheff, Opt. Lett. 9, 402 (1984)

60 R.H. Lipson, P.E. LaRoque, and B.P. Stoicheff, J. Chem. Phys. 82, 4470 (1985).

61 B.P. Stoicheff and A.A. Madej (EICOLS'87, ARE (Schweden))

62 H. Zacharias, H. Rottke, and K.H. Welge, Opt. Commun.35, 185 (1980).

63 H. Rottke and K.H. Welge, Chem. Phys. Lett. 99, 456 (1983); J. Physique 46, (1-127) (1985).

64 R. Schmiedl, H. Dugan, W. Meier, and K.H. Welge, Z.Phys. A304, 137 (1982).

65 H. Zacharias, H. Rottke, J. Danon and K.H. Welge, Opt. Commun. 37, 15 (1981).

66 H. Rottke and K.H. Welge, Phys. Rev. A 33, 301 (1986).

67 A. Holle, J. Main, G. Wiebusch, and K.H. Welge (EICOLS'87, ARE (Schweden))

68 H.J. Krautwald, L. Schnieder, K.H. Welge, and M.N.R. Ashfold, Faraday Discuss. Chem. Soc. 82, 7 (1986).

TUNABLE COHERENT SOURCES FOR VACUUM ULTRAVIOLET SPECTROSCOPY

B. P. Stoicheff
Department of Physics
University of Toronto
Toronto, Ontario
Canada   M5S 1A7

ABSTRACT. Tunable coherent radiation in the ultraviolet and vacuum
ultraviolet has been generated by stimulated Raman scattering, by
anti-Stokes Raman lasers, and by frequency mixing processes in nonlinear media. The theory and experimental progress in the development
of these laser-driven sources is reviewed, and examples of available
systems and their characteristics are discussed. Various applications
in spectroscopy of radiation tunable in the wavelength region 200-90 nm
are presented.

1. INTRODUCTION

The availability of tunable lasers in the visible and infrared
wavelength regions has made possible significant advances in atomic
and molecular spectroscopy. At the present time, however, there is a
lack of lasers and especially of tunable lasers in the ultraviolet (UV),
vacuum ultraviolet (VUV, from 200 to 100 nm), and extreme ultraviolet
(XUV, from 100 to $\sim$20 nm) regions of the spectrum. In fact, only a
few lasers have been made to operate at these short wavelengths, in
spite of considerable efforts being made in the past decade. The
excimer lasers, such as XeF (315 nm), XeCl (308 nm), KrF (248 nm), ArF
(193 nm), Xe_2 ($\sim$170 nm), and Ar_2 ($\sim$120 nm), and the H_2 laser ($\sim$110 nm)
have been available for some time now, but these emit at discrete
wavelengths or are tunable only over their relatively narrow bandwidths.

The difficulty in producing stimulated emission in the ultraviolet regions is well known, and arises from the basic relation that
the probability for spontaneous emission, A, varies as $\nu^3 B$, where B is
the probability for stimulated emission. Thus, losses in excited
state population due to spontaneous emission increase rapidly at short
wavelengths, and put severe demands on pumping sources in order to
achieve inverted population. Various techniques are being explored to
overcome these difficulties including recombination processes and
excitation of ions; and in principle, the free-electron laser could
operate at these short wavelengths. Much effort in this direction is

A. C. P. Alves et al. (eds.), Frontiers of Laser Spectroscopy of Gases, 63–88.
© 1988 by Kluwer Academic Publishers.

being made in many laboratories, and we hope that new VUV and XUV lasers will soon be reported.

In the meantime, there has been substantial progress in producing VUV and XUV radiation by Raman scattering processes, by harmonic generation, and by frequency mixing of laser radiation in nonlinear media. The resulting radiation is coherent, monochromatic, directional, and tunable over broad regions, and thus has all the characteristics of laser radiation except high intensity. Nevertheless, the intensities achieved to date are sufficient for many applications in absorption and fluorescence spectroscopy. In fact, recent progress has reached a stage where VUV and XUV sources will soon be found in many spectroscopic laboratories.

It is my purpose to review these accomplishments, to sketch the basic theoretical concepts, describe the experimental methods used in generating tunable, coherent, VUV and XUV radiation, and to discuss some recent spectroscopic applications, mainly based on work in my laboratory.

2. STIMULATED RAMAN SCATTERING

Since the first observations of stimulated Raman scattering [1], it was clear that this process could be used to generate coherent radiation at prescribed wavelengths [2] by selection of the pump laser and Raman medium. In the intervening years, laser-stimulated Raman Stokes and anti-Stokes radiation has been produced at specific wavelengths in the infrared, visible, and extending into the vacuum ultraviolet region. It is of interest to note that some of the earliest experiments were carried out with the liquids N_2, O_2, and H_2, and with the gases H_2, D_2, and CH_4 [3]. These are the most efficient Raman scattering media, and today, the gases H_2, CH_4, and N_2 are the most generally useful for Raman shifting into the ultraviolet regions.

There are several comprehensive reviews [4] of theories of stimulated Raman processes and of salient experiments. For our purpose, we need only consider that when intense laser radiation (at frequency ω_0) is incident on a Raman-active medium, there is exponential growth of the spontaneous Stokes wave (at frequency ω_s) given by

$$I_s(1) = I_s(0)e^{g1}$$
$$\text{and the gain } g = (N_a-N_b)(\frac{d\alpha}{dq}) \frac{4\pi^2\omega_s}{n_0 n_s \mu \omega_0 \Gamma} \tag{1}$$

Here, N_a and N_b are populations in the lower and upper states of the Raman transition, $d\alpha/dq$ is the rate of change of polarizability with normal coordinate, n_0 and n_s are the refractive indices at ω_0 and ω_s, Γ is the Raman linewidth and μ the reduced mass. Eq. (1) describes a Stokes Raman laser producing coherent radiation at ω_s. In this brief and simplified description, we may consider that the strong pump and Stokes waves generate a coherent material excitation at ω_r. This oscillation causes variations in the refractive index which then modulate and scatter the incident laser radiation (ω_0) thus producing sidebands or many orders of coherent Stokes and anti-Stokes radiation at frequencies $\omega_0 \pm n\omega_r$.

From Eq. (1) it is apparent that in molecular media, totally symmetric modes of vibration would lead to the highest Raman gain. This follows from our knowledge that in spontaneous Raman scattering these vibrational modes exhibit the largest rates of change of polarizability and the sharpest Raman lines. Also, since usually $N_a - N_b > 0$, an inverted population is not required for a Stokes Raman laser or for coherent anti-Stokes scattering. However, if $N_a < N_b$, then one has the possibility of an anti-Stokes Raman laser, with exponential gain at a higher frequency ($\omega_a = \omega_0 + \omega_r$) than that of the pump. Both anti-Stokes coherent Raman scattering and anti-Stokes Raman lasers now provide UV and VUV radiation. Recent progress will be reviewed below.

2.1 Anti-Stokes Raman Scattering

While the earliest anti-Stokes (AS) Raman shifting experiments with H_2, CH_4, and N_2 were carried out with ruby laser excitation any significant progress in reaching the far UV and VUV regions was only achieved with the availability of excimer laser radiation in the late 70's. Initially several orders of stimulated Stokes emission and one or two orders of anti-Stokes emission were generated with KrF (248 nm) and ArF (193 nm) excitation [5]. However, up to 5 AS lines were observed [6] with pressurized H_2 ($\Delta\nu = 4155$ cm^{-1}) and ArF excitation, with the shortest wavelength being 138 nm. In another experiment tunable dye laser radiation [7] (Rh 101 dye and 9 ns pulses) of 30 mJ energy provided 9 AS components of H_2 to reach 185 nm. This was extended to the 13th AS line to generate VUV radiation at 138.5 nm (Fig. 1) using the experimental arrangement [8] shown in Fig. 2. A Nd:YAG laser was used to generate $\sim$300 mJ of second harmonic radiation at 532 nm, and this in turn pumped a dye oscillator-amplifier system to emit $\sim$100 mJ at 545 nm. This radiation was focussed in a one meter Raman cell of H_2 at a pressure of 2-3 atm to yield up to 13 AS lines. The emitted energy was detected and measured with an evacuated prism spectrometer. In a more recent modification of this experiment, Döbele et al. [9] amplified the radiation of the 8th H_2 AS component at 193 nm in three ArF amplifiers and used the output ($\sim$40 mJ) to excite up to the 6th H_2 AS component at 130 nm. Under these circumstances high power pulses ($\sim$0.8 kW) of a few nanosecond duration were obtained at 130 nm.

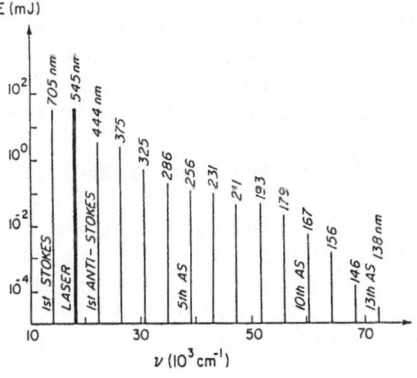

Figure 1. Graph of measured energy of first-Stokes and up to 13th anti-Stokes emission in H_2 gas (at 5 atm pressure) excited by laser radiation at 545 nm using the experimental arrangement shown in Figure 2 [8].

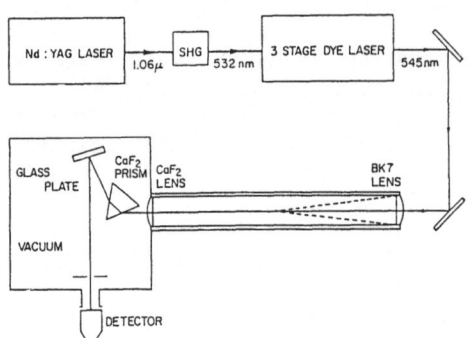

Figure 2. Experimental arrangement for generating many orders of anti-Stokes coherent emission in H_2 gas [8].

Recently, a XeCl laser (308 nm) with high spectral brightness has been used by Maeda and Takahashi [10] to excite up to the 11th AS component at 128.3 nm in H_2 gas at 10 atm. This source and therefore the AS Raman emission is tunable over the bandwidth of the XeCl laser ($\sim$10Å). The potential for emission at much shorter wavelengths now exists with excitation of stimulated Raman scattering with an E-beam pumped Ar_2 excimer laser operating at 124 to 127.5 nm with $\sim$2 MW output power. In preliminary experiments, Sasaki et al. [11] have generated $\sim$100 kW power in the first and second order Stokes emission (at 134 and 141 nm) of H_2 at 8 atm pressure.

2.2 Anti-Stokes Raman Lasers

As already mentioned above, when a population inversion exists between the upper and lower Raman levels, that is when $N_b > N_a$ in Eq. (1), stimulated anti-Stokes emission may be induced at a wavelength shorter than that of the pump laser. Such an anti-Stokes Raman laser would in principle yield high output energy since the upper Raman level is usually metastable, and moreover would be tunable by tuning the pump laser.

In the earliest experiments [12] gain was measured in the anti-Stokes signal when the excited $^2P_{1/2}$ state of I* was inverted relative to the $^2P_{3/2}$ ground state by flash photolysis of CF_3I, and then pumped by 1.06 μm radiation from a Nd:YAG laser. The first anti-Stokes Raman laser was reported in 1982 by White and Henderson [13]. Thallium vapor was used as the Raman medium with inversion of the $6p^2P_{3/2}$ level relative to the $6p^2P_{1/2}$ ground level produced by selective photodissociation of TlCl by 193 nm radiation from an ArF laser. With the second and third harmonics (at 532 and 355 nm, respectively) of a Nd:YAG laser as pump sources, anti-Stokes laser emission was observed at 376 and 278 nm, respectively. The experimental arrangement and relevant energy levels are shown in Figs. 3 and 4.

In a series of papers on various anti-Stokes laser media, White and Henderson [13] demonstrated such laser emission at 178 nm from I, at 149 nm from Br, and 410 nm from In. White [14] also proposed that anti-Stokes lasers with emission from 100 to 206 nm could be produced based on metastable population inversions in the group VI elements O, S, Se from selective photodissociation of N_2O, OCS, and OCSe by VUV radiation. In recent experiments, Ludewigt et al. [15] have achieved anti-Stokes

Raman-laser emission in Se at 158.7 and 167.5 nm using pump-laser radiation at 199.5 and 254.8 nm, respectively. Anti-Stokes Raman lasers using
Sn and Pb have also been reported [16].

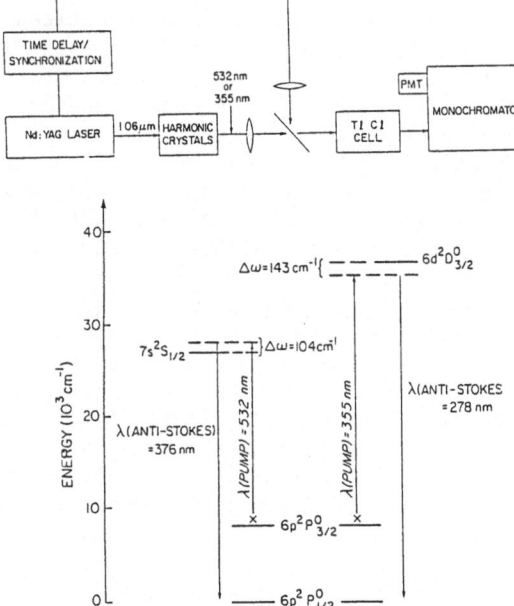

Figure 3. Experimental
arrangement used to produce
a Thallium anti-Stokes Raman
laser [13].

Figure 4. Energy levels involved in the Thallium anti-
Stokes Raman laser with inverted population in the 6p
$^2P_{3/2}$ level and pump sources
at 532 and 355 nm [13].

3. HARMONIC GENERATION AND FREQUENCY MIXING

3.1 Theory

Laser-driven VUV sources are based on third harmonic generation
(THG) or 4-wave sum mixing (4-WSM) in nonlinear media. These processes
are usually described by the induced macroscopic polarization of the
medium when irradiated by intense laser light (Armstrong et al [17]). It
is well known that the polarization of a medium in the presence of a
monochromatic field $\bar{E}(r,t) = \Sigma_i E(\omega_i)$ can be written as

$$\bar{P}(\omega_i) = \chi^{(1)}(\omega_i) \cdot \bar{E}(\omega_i) + \Sigma \chi^{(2)}(\omega_i = \omega_j + \omega_k) \cdot \bar{E}(\omega_j) \cdot \bar{E}(\omega_k) +$$

$$+ \Sigma \chi^{(3)}(\omega_i = \omega_j + \omega_k + \omega_\ell) \cdot \bar{E}(\omega_j) \cdot \bar{E}(\omega_k) \cdot \bar{E}(\omega_\ell) + \ldots$$

where $\chi^{(n)}$ are the susceptibility tensors of nth order. The lowest
order term producing nonlinear effects is $\chi^{(2)}$. However, this tensor
has nonzero components only in noncentro-symmetric systems: isotropic
media such as cubic crystals, liquids, and gases do not exhibit quadratic
nonlinearities. For third order processes such as THG and 4-WSM we need
be concerned only with $\chi^{(3)}$, whose principal term may be written

$$\chi^{(3)} (\omega_0 = \omega_1 + \omega_2 + \omega_3)$$

$$= \frac{3e^4}{4\hbar^3} \frac{<g|\mu|a><a|\mu|b><b|\mu|c><c|\mu|g>}{(\Omega_{cg} - \omega_1 - \omega_2 - \omega_3)(\Omega_{bg} - \omega_1 - \omega_2)(\Omega_{ag} - \omega_1)} \tag{2}$$

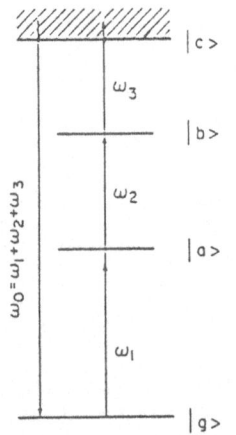

Figure 5. Schematic diagram of 4-WSM process $\omega_0 = \omega_1 + \omega_2 + \omega_3$ with a 2-photon resonance $\Omega_{bg} = \omega_1 + \omega_2$.

Here, $<g|\mu|a>$ is the electric dipole matrix element between the ground state $|g>$ and an excited state $|a>$, having a lifetime Γ_a, and $\Omega_{ag} = \omega_{ag} - i\Gamma_a/2$ is the energy difference (Fig. 5) between states $|a>$ and $|g>$, e is the electronic charge and $\hbar = h/2\pi$, with h being Planck's constant.

Equation (2) shows that $\chi^{(3)}$ will be resonantly enhanced whenever the applied frequencies, ω_1, ω_2, ω_3, are such that the real part of the resonance denominator vanishes, namely when $(\Omega_{ag} - \omega_1) = 0$, or $(\Omega_{bg} - \omega_1 - \omega_2) = 0$, or $(\Omega_{cg} - \omega_1 - \omega_2 - \omega_3) = 0$, corresponding to one, two, or three photon resonance, respectively. If any of ω_1, ω_2, ω_3 is set equal to a resonance frequency (Ω_{ag} etc.) $\chi^{(3)}$ will be enhanced but the incident radiation will be strongly absorbed. Similarly if $\omega_0 = \omega_1 + \omega_2 + \omega_3$ equals a resonance frequency, the generated radiation will be absorbed. If however, $\omega_1 + \omega_2$ is equal to a 2-photon resonance (Ω_{bg}), the incident radiation at $\omega_1 + \omega_2$ is expected to be only weakly absorbed by the 2-photon transition, while the resonance enhancement of $\chi^{(3)}$ could be just as strong as for the 1-photon resonances.

For third harmonic generation (THG), $\chi^{(3)}$ simplifies to

$$\chi^{(3)} (\omega_0 = 3\omega) = \frac{3e^4}{4\hbar^3} \frac{<g|\mu|a> \text{ etc.}}{(\Omega_{cg} - 3\omega)(\Omega_{bg} - 2\omega)(\Omega_{ag} - \omega)} \tag{3}$$

When 2ω approaches resonance, $\chi^{(3)}$ undergoes strong ($>10^4$) enhancement. For efficient THG, collinear phase-matching is necessary, that is, the refractive index $n(3\omega) = n(\omega)$ in order to yield a maximum effective interaction length. With focused incident radiation, THG can be observed only in negatively dispersive media. Tunability is achieved by varying the incident frequency ω.

For generating tunable radiation by 4-WSM, the process $\omega_0 = 2\omega_1 + \omega_2$ is of interest. $\chi^{(3)}$ then becomes

$$\chi^{(3)} (\omega_0 = 2\omega_1 + \omega_2) = \frac{3e^4}{4\hbar^3} \frac{<g|\mu|a> \text{ etc.}}{(\Omega_{cg} - 2\omega_1 - \omega_2)(\Omega_{bg} - 2\omega_1)(\Omega_{ag} - \omega_1)} \tag{4}$$

Strong enhancement is again achieved by tuning $2\omega_1$ to a parity-allowed 2-photon resonance, Ω_{bg}. Tunability (and further enhancement) is then obtained by selecting ω_2 so that $2\omega_1 + \omega_2$ corresponds to the ionization continuum or to broad autoionizing levels above the ionization limit (Hodgson et al [18]). More detailed treatments of the relevant theory including phase-matching, saturation effects, and converson efficiencies have been given by Vidal [19] and Jamroz and Stoicheff [20].

3.2 Experimental Techniques

Obviously, any medium used for generating radiation at wavelengths below 200 nm must be transparent to such radiation. This condition is generally met by the rare gases and some metal vapors. New and Ward [21] were the first to demonstrate THG in gases. Subsequently, Harris and Miles [22] showed that relatively high conversion efficiency of THG and 4-WSM could be obtained by using phase-matched metal vapors as non-linear media, and that efficiency could be improved further by resonance enhancement.

Frequency conversion into the VUV and XUV regions has been achieved by a variety of laser systems. Powerful pulsed lasers such as ruby, Nd:YAG, Nd in glass, flashlamp pumped dye (FPD), rare gas excimer, and rare gas halide exciplex lasers provide the primary coherent radiation. In some systems, tunable radiation from dye lasers (>320 nm) is used directly, and in others laser radiation (>400 nm) is doubled once or twice in nonlinear crystals to produce coherent radiation in the UV to about 200 nm. Subsequently, the coherent UV radiation is converted to coherent VUV and XUV by THG or frequency mixing in rare gases or metal vapors. These methods all use a cell or heat pipe to contain the non-linear gas. Windows of LiF are used to the limit of transmittance at ∿104 nm; for generation to shorter wavelengths, pinholes [23] and capillary arrays [24] with differential pumping are used. Recently, pulsed supersonic jets have been introduced for harmonic generation with rare gases [25]. Specific atomic (and molecular) systems are selected because of their large third order nonlinear susceptibility, suitability of energy levels for resonance enhancement, and low absorption at the desired VUV or XUV wavelength.

In our laboratory, we use essentially the same 4-WSM technique as that described by Hodgson, Sorokin, and Wynne [18] who generated VUV radiation tunable from 200 to 177 nm in Sr vapor. By using Mg, Zn, or Hg [27] vapor as the nonlinear medium, we have extended the tuning range to the transmission limit of LiF windows at ∿104 nm, and to 87.5 nm by replacement of the LiF window with a capillary array [24]. The experimental arrangement is shown in Fig. 6. A N_2 laser (Molectron-UV1000) or excimer KrF/XeCl laser (Lumonics TE-861M) pumps two dye lasers at frequencies ν_1 and ν_2. With the N_2 laser pump, each dye laser produces ∿10 kW pulses of ∿7 ns duration with linewidths ∿0.1 cm^{-1}. Resonance enhancement of $\chi^{(3)}$ is achieved by setting $2\nu_1$ to a 2-photon resonance in the atomic system, and the second laser (ν_2) is scanned over the breadth of the ionization continuum or autoionizing level shown in Fig. 7 for Mg, to obtain a broad range of tunability. A schematic diagram of the relevant 2-photon resonances and continua for Mg, Zn, and Hg which provide this broad tunability are shown in Fig. 8. The two laser beams are spatially overlapped in a Glan-Thompson prism, and focused near the exit end of a heat-pipe containing Mg vapor at ∿200 torr. The resulting coherent radiation is continuously tunable from 174 to 135 nm with a flux >10^8 photons per pulse. For generating shorter wavelength radiation, the excimer laser is used to pump two dye laser oscillator-amplifier

systems with outputs of ∿50 kW. One of the beams is frequency-doubled
in a KDP crystal to produce the 2-photon resonance enhancement, $2\nu_1$, of
a suitable level in Mg, Zn or Hg vapor. With Zn vapor, a flux of ∿10^7
photons per pulse is generated from 140 to 106 nm, and with Hg vapor an
increase of 10^2 to 10^3 in flux is achieved over the tuning range of 120
to 104 nm, but decreases rapidly on going to 87.5 nm [24].

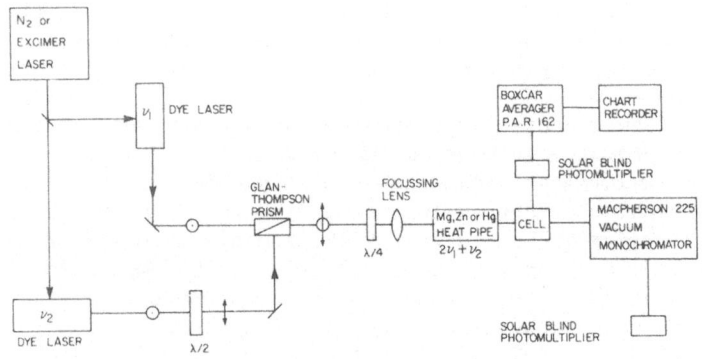

Figure 6. Method of exciting two tunable dye lasers and of
combining radiation at ν_1 and ν_2 to generate tunable coherent
VUV radiation at $2\nu_1+\nu_2$ by 4-WSM in a metal vapor [26].

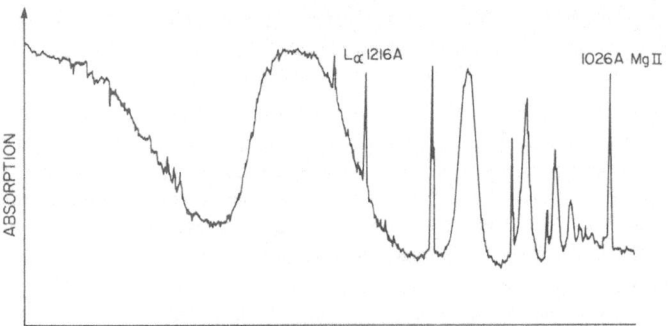

Figure 7. Mg absorption spectrum in the region 145 to 100 nm,
somewhat beyond the ionization limit at 162.1 nm.

For work with the Mg and Zn vapors, uniform vapor densities resulting in stable VUV emissions are obtained over periods of several hours
with the use of double heat-pipe ovens while the heat-pipe oven for Hg
vapor (Fig. 9) is a simple cell of pyrex glass. Liquid Hg and its vapor
are confined to the central heated section by water-cooled jackets at
each end which are tapered to return condensed Hg back to the hot zone.
For generation of VUV radiation from 126 to 104.5 nm, a LiF plate ∿0.5 mm
thick forms the exit window, and a vapor pressure of up to 95 torr is
used with an equal pressure of He buffer gas. For XUV generation, $\lambda <$

104.5 nm, the exit window is a glass disc 2 mm thick with a central section ∿2 mm diam containing an array of 50 μm diam capillaries. These pores reduce the gas conductance by a factor of ∿750 over that of a single aperture of equivalent area, at a He pressure of 35 torr, and provide an efficient XUV window with ∿50% transmission. [24].

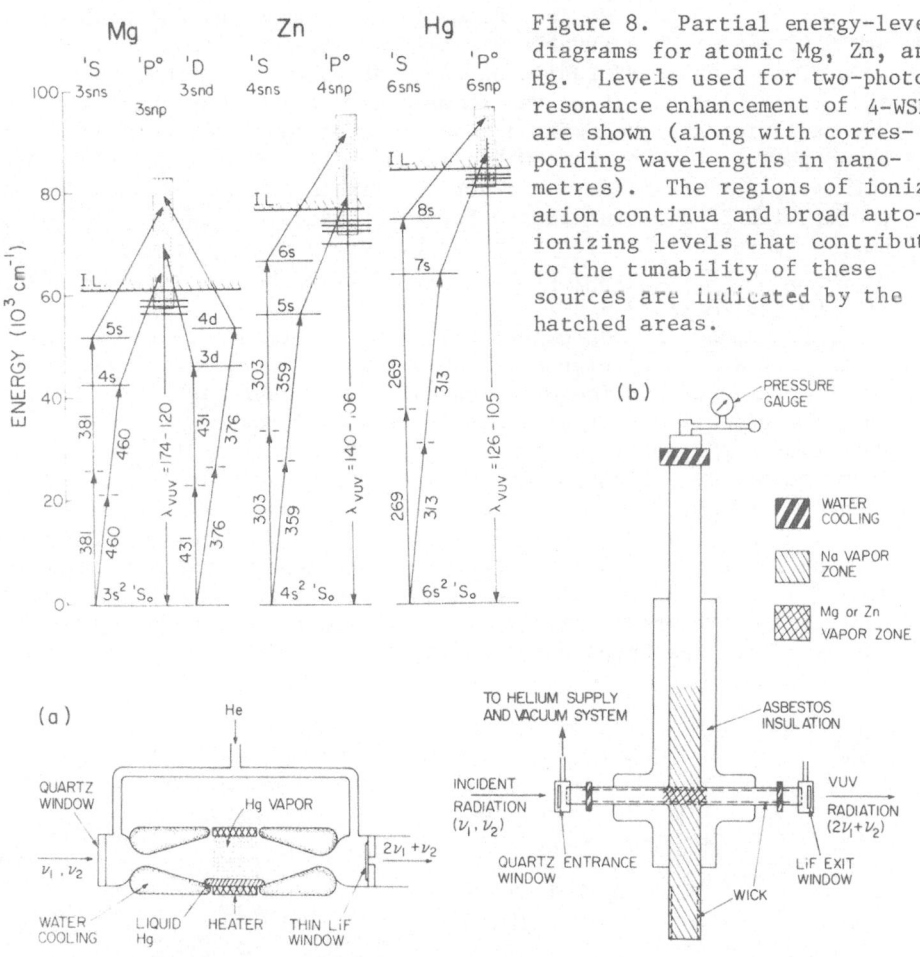

Figure 8. Partial energy-level diagrams for atomic Mg, Zn, and Hg. Levels used for two-photon resonance enhancement of 4-WSM are shown (along with corresponding wavelengths in nanometres). The regions of ionization continua and broad autoionizing levels that contribute to the tunability of these sources are indicated by the hatched areas.

Figure 9. Heat pipes used to produce stable densities of (a) Hg vapor, and (b) Mg and Zn vapors.

The early work of Harris and coworkers [27] on nonlinear mixing in Xe and Ar was followed by work on Kr and Xe. Extensive wavelength tunability with rare gases was reported by Hilbig and Wallenstein [28]. They used sum-frequency mixing ($\omega_{VUV} = 2\omega_{UV} + \omega_D$) in Kr and Xe to generate VUV radiation of ∿20 W power, tunable over most of the range 110 to 130 nm where these gases are negatively dispersive. They also generated VUV radiation of ∿50 W power at longer wavelengths in Xe by difference-frequency mixing. The process $\omega_{VUV} = 2\omega_{UV} - \omega_D$ resulted in radiation from 185 to 207 nm, and $\omega_{VUV} = 2\omega_{UV} - \omega_L$ at shorter wavelengths, from 160 to 190 nm. (The frequencies ω_L, ω_D, and ω_{UV} refer to the output of a Nd:YAG laser, a dye laser and harmonic of the dye laser, respectively.) For those experiments, Hilbig and Wallenstein used the second harmonic radiation of a Nd:YAG laser (λ_L) to excite a dye-laser oscillator-amplifier system. This system operated at λ_D = 550 to 650 nm, with output powers of 3 to 5 MW in pulses of ∿6 ns duration, and with a bandwidth of ∿0.02 cm^{-1}. This tunable visible radiation was then doubled in KDP to produce tunable UV radiation (λ_{UV}) at powers of ∿1 MW. Both visible and ultraviolet radiation was focused in the rare gas, and the resulting VUV radiation analyzed with a monochromator and detected by a solar-blind photomultiplier and NO ionization cell. This multi-laser system, with further flexibility in order to fill in the few gaps in tuning range, will be a useful source for spectroscopy in the region 100 to 200 nm. More recently Wallenstein and his group have generated XUV radiation by nonresonant frequency tripling in Ar (from 97.4 - 104.8 nm) and in Ne (in the region 72-74.4 nm).

Rhodes and his colleagues [29] have developed KrF and ArF laser systems of extremely high brightness for use in THG and 4-WSM below 100 nm. A schematic outline of a typical multi-laser system is shown in Fig. 10. The tunable output of a single-frequency, cw dye laser at ∿580 nm was pulse-amplified in a three-stage, XeF-pumped, amplifier. The ∿10 ns, 20 mJ pulses were focused into Sr vapor to generate third harmonic radiation at ∿193 nm, in 5 ns pulses of 200 mW peak power. These pulses were then amplified in two ArF laser amplifiers to produce ∿5 ns pulses of 6 MW peak power and ∿0.01 cm^{-1} bandwidth, tunable within the ArF gain profile. This radiation was focused and tripled to 64 nm in Kr, Ar, and H$_2$. Another similar system with a final KrF amplifier generated radiation at 248 nm. This was tripled in flowing Xe to produce radiation at ∿83 nm.

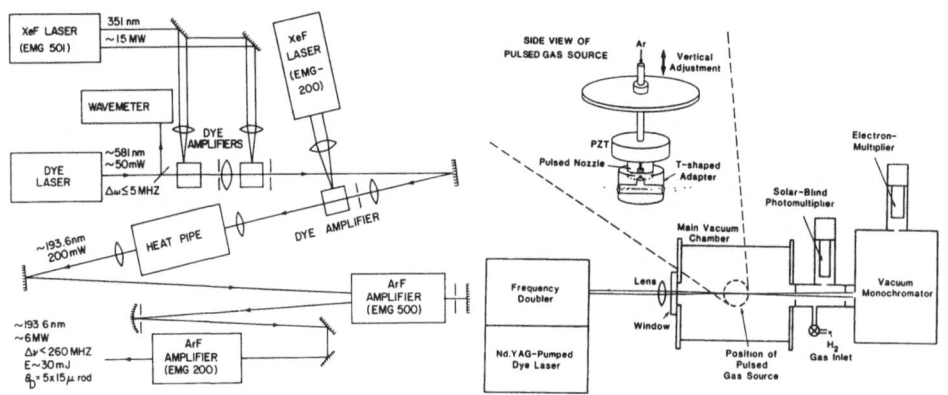

Figure 10. A high spectral brightness laser system for use in generating VUV radiation [29].

Figure 11. Laser and pulsed supersonic jet with rare gases used to produce VUV and XUV radiation by third-harmonic generation [30].

In Fig. 11 is shown the system developed by Kung and his colleagues [25, 30] for use with gases at high densities, emitted from a pulsed supersonic nozzle, as nonlinear media. The pump source of Nd:YAG - dye laser - and frequency doubler provided tunable radiation of ∿10 MW at ∿300 nm, for frequency tripling in Xe, Kr, Ar, Ne, H_2 or CO. Ar was the most efficient, giving ∿10^8 photons per 3 ns pulse (and 10W peak power) over the range 102.3 to 97.3 nm.

Higher order frequency conversion has been used to generate radiation at fixed frequencies to wavelengths as short as 38 and 35.5 nm. Reintjes et al [31] have used the fundamental, second, and fourth harmonics of a mode-locked Nd:YAG laser to generate XUV radiation in rare gases in third through fifth and seventh harmonic conversion. Fifth harmonic of 266.1 nm radiation produced radiation at 53.2 nm in He, Ne, Ar and Kr, with peak pulse powers of ∿1 kW in He. Seventh harmonic at 38 nm was observed in He with peak power of ∿100 W. Recently, Bokor et al [25] have used a supersonic He gas jet to produce radiation at 35.5 nm by seventh-harmonic conversion of 248 nm radiation from a KrF excimer laser.

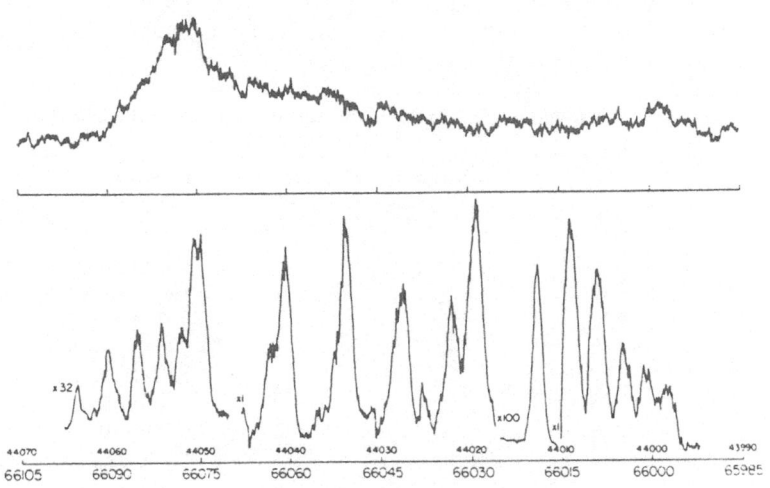

Figure 12. Coherent, tunable, third harmonic radiation at 151 nm from NO at pressures of 50 Torr (bottom) and 10 atm (top). The structure at low pressure is that of the 0-0 band of the $A^2\Sigma+ - X^2\pi_{\frac{1}{2}}$ system used in the resonant enhancement [32].

As an alternative to metal vapors and rare gases, Innes et al [32] explored the use of molecular systems as nonlinear media for VUV generation, and found that nitric oxide, NO, was admirably suited for this purpose. A strong 2-photon transition was available for resonant enhancement: thus one dye laser was tuned so that $2\nu_1$ was in resonance with the $A^2\Sigma+\leftarrow X^2\pi$ transition and a second dye laser was tuned to the $C^2\pi$ manifold, yielding a rich rotational structure (Fig. 12). VUV radiation was generated in the γ bands at 151, 143, 136, and 130 nm (each of breadth ∿600 cm^{-1}). At an NO pressure of ∿90 Torr, a photon yield of ∿10^7 photons per pulse was obtained for incident laser powers of 20 kW. Significant pressure broadening occurred at 10 atm, and the rotational structure of the 2-photon transition was essentially eliminated. This provided continuously tunable VUV radiation by simple THG using a single laser. In similar experiments with CO at a few Torr pressure, resonantly enhanced 4-WSM was used to generate radiation in the region of 115 nm [33] and resonantly enhanced THG in the region of 147 nm [34].

The nonlinear media and laser systems used at the present time to generate tunable, coherent, VUV and XUV radiation are summarized in Tables I and II, along with their regions of tunability. Only systems with demonstrated broad regions of tunability are included here. (Many other systems which make use of rare gases and molecular gases have been reported, but have limited tunability [20].) The resulting tunable radiation retains all of the properties of the generating laser radiation including coherence, directionality, narrow linewidth and short-pulse duration, with the exception of high intensity. Nevertheless the generated VUV radiation has peak spectral brightness several orders of magnitude greater than that available from incoherent sources, including synchrotrons. Moreover, the monochromaticity achieved to date provides resolving powers $\lambda/\Delta\lambda > 10^5$, and thus these sources are ideally suited for high-resolution spectroscopy.

TABLE I. Tunable Generation in Metal Vapors

λ(nm)	Nonlinear Medium (Ioniz. Limit in nm)	Primary Laser	Reference
195.7-177.8	Sr(217.8)	N_2-Dye	18
174 -130	Mg(162.1)	N_2-Dye	26
129 -120	Mg	KrF-Dye	26
140 -106	Zn(132.0)	XeCl/KrF-Dye	26
125.1-117.4	Hg(118.0)	Nd:YAG-Dye	35
115.0- 93.0	Hg	Nd:YAG-Dye	36
196 -109	Hg	Nd:YAG-Dye	28
126 - 87.5	Hg	XeCl-Dye	24, 26

TABLE II. Tunable Generation in Rare Gases

λ(nm)	Nonlinear Medium	Primary Laser	Reference
206 -16C	Xe	Nd:YAG-Dye	28
195 -163	Xe	Nd:YAG and PO	27
147 -118	Xe	Nd:YAG and PO	27
147 -140	Xe:Kr	Nd:YAG-Dye	28
130 -11C	Kr	Nd:YAG-Dye	28
123.6-120.3	Kr	KrF-Dye	37
123.5-120	Kr:Ar	Nd:YAG-Dye	28
106.8-105.8	Xe	Nd:YAG-Dye	38
102.3- 97.3	Ar	Nd:YAG-Dye	30, 28
73.6- 72.0	Ne	Nd:YAG-Dye	28

4. APPLICATIONS IN SPECTROSCOPY

To date, the laser-driven sources based on harmonic generation and frequency mixing have been used in a number of experiments to measure radiative lifetimes and to record absorption and laser-induced fluorescence spectra. While some of the anti-Stokes Raman sources have demonstrated high intensity and are being applied to specific problems of impurity detection in plasmas, their general utility in spectroscopy awaits the development of systems with broad tunability. Here, several typical spectroscopic applications of the above VUV and XUV sources are given to illustrate their usefulness.

First, are shown absorption and fluorescence spectra in Fig. 13 of a portion of the 7-0 Lyman band of H_2 obtained by Marinero et al. [30] using the experimental arrangement given in Fig. 11. As shown by the lower trace taken at 1/10 the pressure of that for the absorption spectrum (upper trace), detection by laser-induced fluorescence is clearly a far more sensitive technique.

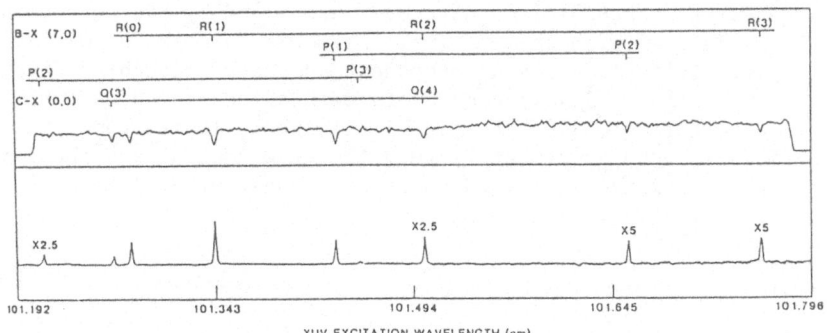

Figure 13. H_2 absorption spectrum (at top) taken with a pressure of 5×10^{-5} Torr, and fluorescence excitation spectrum (at bottom) with a pressure of 4×10^{-6} Torr, in the region 101.2 - 101.8 nm [30].

The N_2 absorption spectrum shown in Fig. 14, is part of the $b'^1\Sigma_u^+$ (12) - $X^1\Sigma_g^+$ (0) band system near 89 nm. This spectrum was obtained with a 20 cm path of N_2 at 50 m Torr, at a resolving power of $\sim2 \times 10^5$. It represents the shortest wavelength spectrum obtained to date using the frequency mixing techniques described above [24].

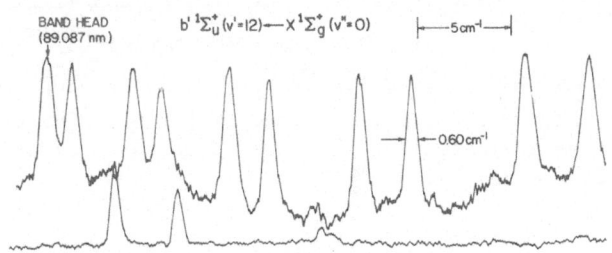

Figure 14. Absorption spectrum of N_2 near 89 nm [24].

4.1   Radiative Lifetimes of Perturbed Rovibronic Levels of CO

The monochromaticity ($\sim$0.2 cm^{-1}) and short pulse duration ($\sim$1 to 5 ns) of these sources makes possible the measurement of radiative life-times of specific rovibronic levels.   As an example, recent results [39] with CO are shown in Fig. 15.   Initially, the fluorescence excitation spectrum of the (0,0) band of the transition $A^1\Pi \to X^1\Sigma$ was recorded (154 to 155.5 nm) with an effective resolution of $\sim$0.3 cm^{-1}.   The observed lines were identified; then the exciting radiation was tuned to each rovibronic line and the decay of fluorescence intensity with time was measured.   In this way, lifetimes for levels $J' = 1$ to 29 of the $v' = 0$ level were obtained from the decay rates for transitions in the P, Q, and R branches.   The results indicate strong perturbations at $J' = 9$, 16, and 27, as shown by the P and R transitions (Fig. 15) and at $J' = 12$ and 27 as determined from Q transitions, with lifetimes almost double those of unperturbed levels ($\tau_u \sim$10 ns).   The $J' = 27$ perturbation has been attri-buted to interaction with the $d^3\Delta_1$ (v=4) state, with all three triplet components affecting the $A^1\Pi$ levels of high J'.   At lower J'-values, the observed perturbations have been explained by the selection rules for interaction with the $e^3\Sigma^-$ (v=1) state which provide for the F_2 (J=N) com-ponent affecting the Π^- component and thus perturbing the Q branch, while the F_1 and F_3 (J=N±1) components each affect the Π^+ components, resulting in perturbations of the P and R branches.   It is seen (Fig. 15) that there is very good agreement between the experimental and theoretical [40] values for the perturbed as well as unperturbed lifetimes over the range of J' levels from 1 to 29.

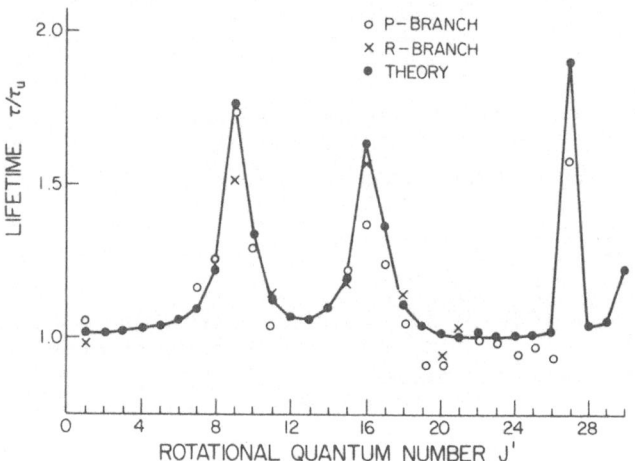

Figure 15.   A graph of lifetimes as a function of rotational quantum numbers J', obtained from P and R branch lines.   Life-times are normalized by $\tau_u = 10$ ns, and compared with theoret-ical values [40].

4.2   Spectra of Xe$_2$, Kr$_2$, and Ar$_2$

The electronic spectra of rare gas dimers have been a subject of interest for many years, mainly because these dimers are model systems for studying van der Waals interactions, and because of their potential as media for VUV and XUV lasers.   Yet very little is known about the excited states of these dimers.   Two experimental techniques were combined in our laboratory for this investigation: four-wave summixing (4-WSM) and a pulsed supersonic jet to produce rotationally and vibrationally cold dimer molecules.   In this way it was possible to resolve rovibronic structures in several isotopic band systems of Xe$_2$, Kr$_2$ and Ar$_2$, in the region 150 to 104 nm, to determine the relevant molecular constants, and to calculate potential energy curves for the ground states and the three lowest (stable) excited states, for all three dimers [41,42,43].

The potential energy curves shown in Fig. 16 for Xe$_2$ are representative of the potentials for all three dimers, including the transitions investigated in recent studies.   In Fig. 17 is a spectrum of band system II recorded with VUV beam excitation as close as possible to the jet nozzle.   The observed broad bands are similar to those reported earlier by Freeman et al [44] and Castex and co-workers [45] who carried out experiments at moderately high pressures ($\sim$150 Torr) in cells cooled to $\sim$150K.   For comparison, the same band system is shown in Fig. 18 when examined $\sim$15 mm from the nozzle, where the vibrational and rotational temperatures are $\sim$1-10K.   The observed spectrum shows ten highly structured vibrational bands, each of $\sim$30 cm^{-1} width, and separated from each other by $\sim$50 cm^{-1}.   Each band has a similar structure consisting of ten or more narrow features of $\sim$1.5 cm^{-1} width degraded to the blue and separated from one another by $\sim$3.5 cm^{-1}.   These resolved features are vibronic bands of isotopic Xe$_2$ and their resolution and identification (Fig. 21) has led to the unambiguous vibrational numbering of the observed bands.   This same structure is evident in spectra of band systems I and III given in Figs. 19 and 20.

At the low temperatures obtained with the supersonic jet it was assumed that only the v'' = 0 vibrational level was populated.   Thus measured frequencies could be described by the term expression (in cm^{-1}):

$$\nu_i = T_e' + \rho_i \omega_e'(v'+1/2) - \rho_i^2 \omega_e' x_e'(v'+1/2)^2 -$$

$$- \frac{21.12}{2}\rho_i + \frac{0.65}{4}\rho_i^2 - \frac{0.003}{8}\rho_i^2.$$

T_e' is the electronic energy of the upper state, ω_e' and $\omega_e' x_e'$ are the vibrational frequency and anharmonic constant, respectively, and ρi is the ratio $[\mu(^{129,132}Xe_2)/\mu_i]^{\frac{1}{2}}$ where μ_i is the reduced mass of a particular dimer i.   The ground state constants, $\omega_e'' = 21.12$ cm^{-1}, $\omega_e x_e'' = 0.65$ cm^{-1}, and $\omega_e y_e'' = 0.003$ cm^{-1} were derived experimentally by Freeman et al. [44] for $^{129,132}Xe_2$.   Similar analyses for the three band systems yielded values of the constants for all three excited states of Xe$_2$ for the first time.   Potential energy curves were calculated from the derived constants and these are shown in Fig. 22.

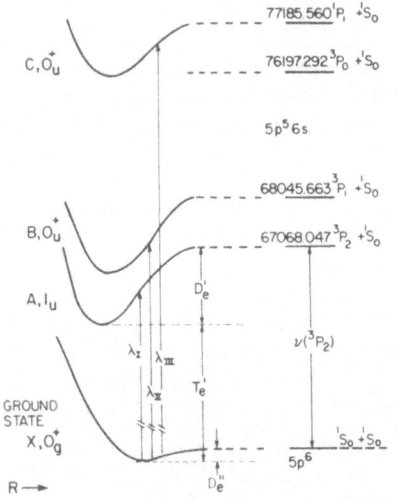

Figure 16. Schematic diagram of potential energy curves of the lowest-lying stable states of Xe₂.

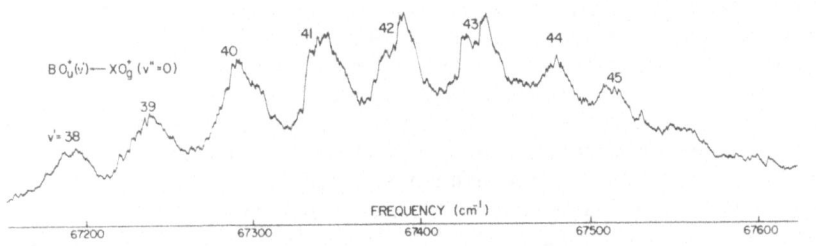

Figure 17. Fluorescence excitation spectrum of band system II of Xe₂ probed ∿2 mm from nozzle.

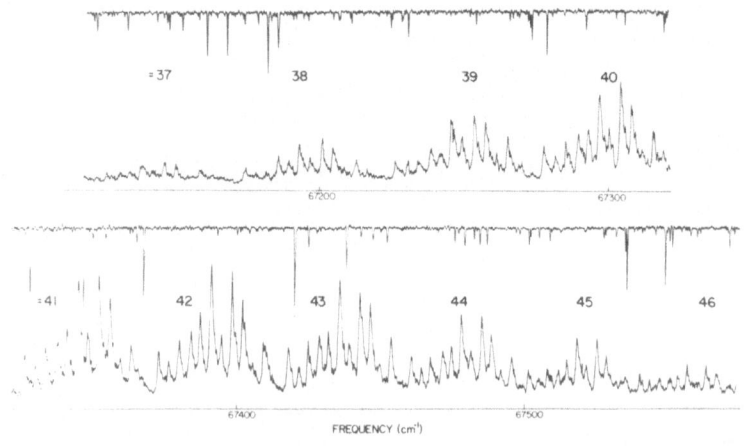

Figure 18. Fluorescence excitation spectrum of band system II, $BO_u^+(v^1)$ ←$XO_g^+(v''=0)$ at ∿148.5 nm, probed 15 mm from nozzle [41].

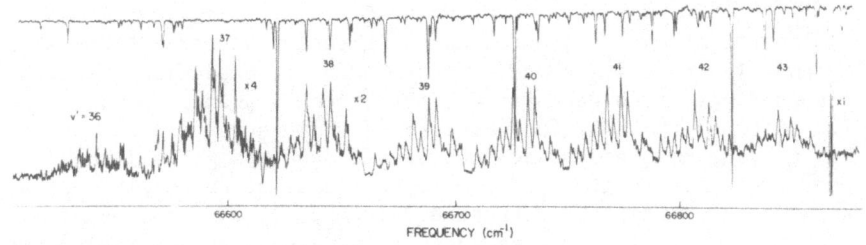

Figure 19. Fluorescence excitation spectrum of band system I, $A1_u(v')\leftarrow X0_g^+(v''=0)$ of Xe_2 at $\sim$150 nm [41].

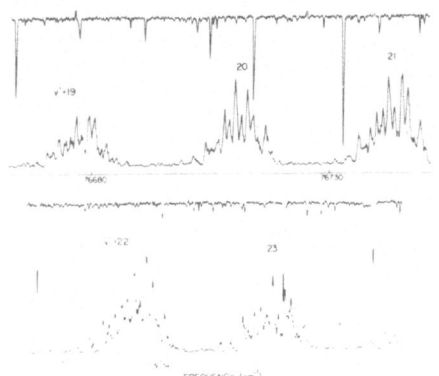

Figure 20. Fluorescence excitation spectrum of band system III, $C0_u^+(v')\leftarrow X0_g^+(v''=0)$ at $\sim$130 nm [41].

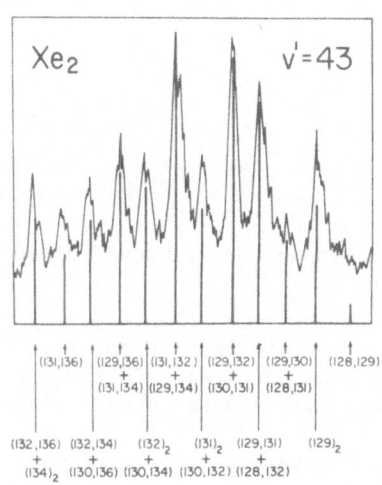

Figure 21. Observed and calculated spectra of isotopic vibronic bands 43-0 of system II [41].

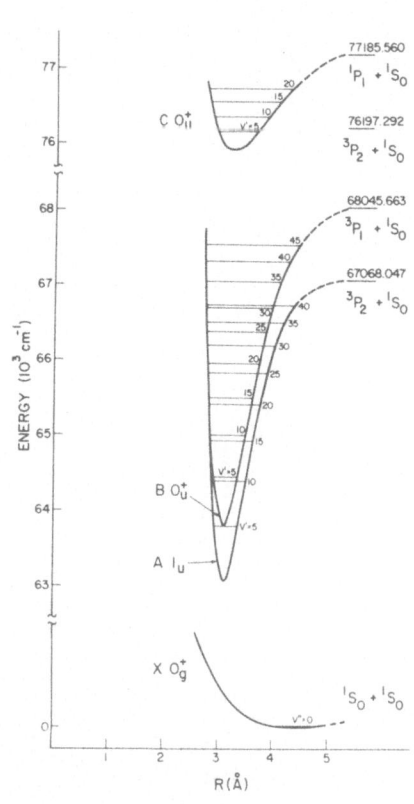

Figure 22. Calculated potential energy curves for excited states $A1_u$, $B0_u^+$, $C0_u^+$, based on analysis of band systems I, II, III of Xe_2 [41].

Investigations of fluorescence excitation spectra of Kr_2 at 125, 124, and 117 nm, also formed by supersonic jet expansion, yielded correspondingly rich data for all three band systems. Since Kr has fewer isotopes than Xe, the observed vibronic structure is more open and exhibits four main peaks and four less intense peaks, all shaded to higher frequency (Fig. 23). The lighter mass of Kr permitted the resolution of rotational structure (Fig. 24) in each band system, and rotational constants and internuclear distances for Kr_2 were determined for the first time. These constants along with vibronic data were used to calculate potential energy curves for the ground and lowest three excited states of $^{84}Kr_2$.

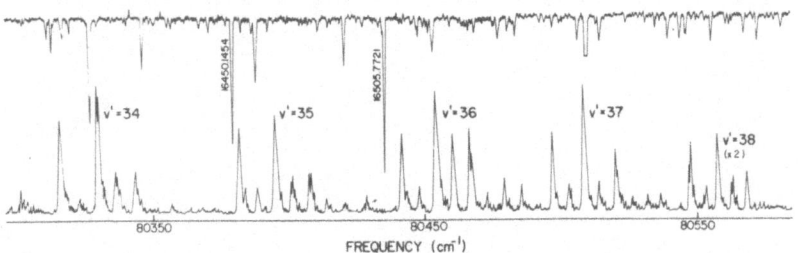

Figure 23. A portion of the fluorescence excitation spectrum of Kr_2 near 124 nm showing 5 of the observed vibronic bands (v' = 30 - 38) with structure due to isotopic molecules [42].

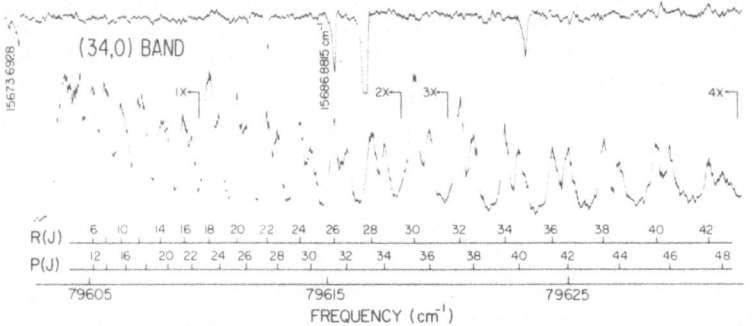

Figure 24. Rovibronic structure in the 34-0 band of system I for $^{86}Kr_2$ [42].

For Ar_2, with its single isotope, ^{40}Ar, it was not possible to determine the vibrational numbering of the observed band systems at 108, 107, and 105 nm. Thus an enriched sample of ^{36}Ar was used, providing spectra of $^{40}Ar_2$ and $^{36}Ar^{40}Ar$ with a large isotope separation of vibronic bands, as shown in Fig. 25. The very much lighter mass of Ar resulted in clear resolution of rotational structure even in spectra of low dispersion (Fig. 25). Analyses of the observed spectra yielded the molecular constants for calculating potential energy curves for $^{40}Ar_2$, as was done for Xe_2 and Kr_2.

The rovibronic spectrum presented in Fig. 26 was recorded at the highest power, of 5×10^5, obtained to date with the VUV laser-driven sources described here. This represents a factor of 3 to 5 improvement over spectroscopic resolution obtained with grating instruments in this wavelength region. The 25-0 band of Fig. 26 is one of 12 vibronic bands originating from $v'' = 0$ or 1 levels of the ground state to levels $v' = 23$ to 31 of the Al_u excited state. For each of the 24-0, 25-0, 26-0, 26-1, and 27-1 bands, three rotational branches were clearly resolved. Three branches were less conspicuous in other bands or unobservable because of blending of lines, but all bands could be analysed in terms of P, Q, and R branches.

Prior to the recent study by Herman et al [43], both of Hund's coupling cases (b) and (c) were used in the literature to describe the rare gas dimers. The rotational levels, symmetries of electronic states,

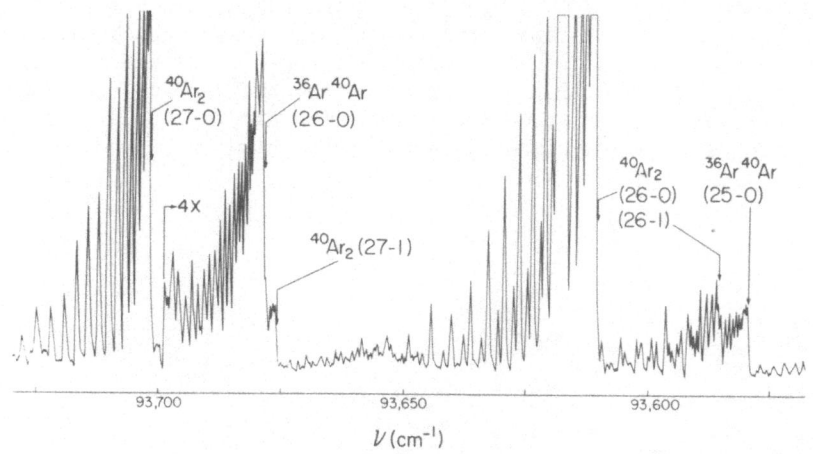

Figure 25. Rovibronic structure in several bands of system II for $^{40}Ar_2$ and $^{36,40}Ar_2$. Note the large rotational spacing for homonuclear $^{40}Ar_2$ (with every second level missing) which is almost twice that of heteronuclear $^{36,40}Ar_2$.

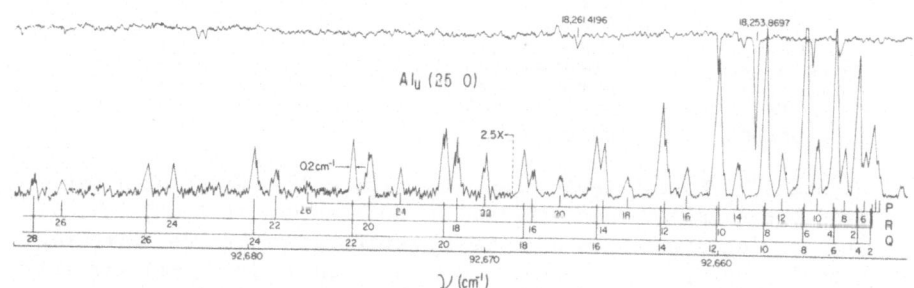

Figure 26. Rovibronic structure in the 25-0 band of the $Al_u \leftarrow XO_g^+$ transition in $^{40}Ar_2$ (at 108 nm) observed at a resolving power of 5×10^5. Three rotational branches are resolved and identified, thus establishing Hund's coupling case (c) for Ar_2[43].

allowed transitions, and correlations between the two coupling cases are shown in Fig. 27. Four branches, P, ^PQ, ^RQ, and R are allowed for $A^3\Sigma_u^+ \leftarrow X^1\Sigma_g^+$ transitions in case (b) coupling. In going over to case (c) coupling, the ^PQ branch correlates to forbidden $0_u^- \leftarrow 0_g^+$ transitions, and consequently disappears. This leaves A $1_u \leftarrow X0_g^+$ transitions for which three rotational branches, P, Q, and R, are expected. In case (b) coupling, the ^PQ branch nearly overlaps the P branch and the ^RQ nearly overlaps the R branch. A small splitting between the overlapping branches should be linear in K. In case (c) coupling, the P and R transitions terminate in the lower levels of each doublet, and Q transitions in the upper levels, and the splittings between doublets should vary as J(J+1). The observation of P, Q, and R branches (Fig. 26) and the dependence of the Ω-type doublet splitting on J(J+1) shown in Fig. 28, establishes that Hund's case (c) is the dominant coupling for the high vibrational levels (v'>17) of the $A1_u$ state of Ar_2. The observed band system is thus properly designated as $A1_u - X0_g^+$. Most of the other electronic states of Ar_2 should have symmetries compatible with case (c) coupling when internuclear separations are near to or greater than the equilibrium separation for the ground state. Similar behaviour would be expected for the heavier dimers, but even higher resolution will be required to demonstrate this for Kr_2 and Xe_2.

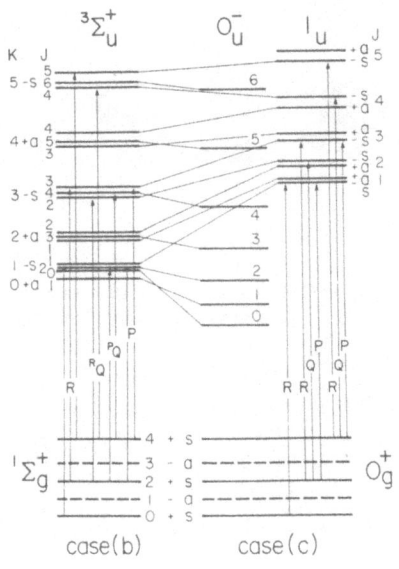 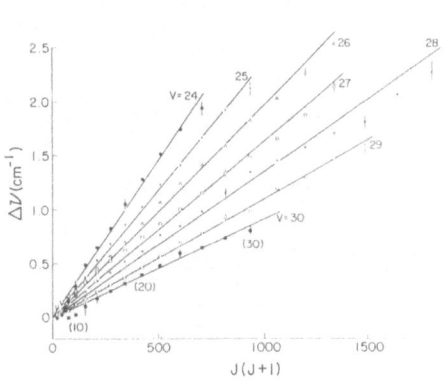

Figure 27. Energy level diagram for coupling cases (b) and (c) showing symmetries of rotational levels, their correlations for the two coupling cases, and the allowed transitions. Transitions between the A 0_u^- and X 0_g^+ states are forbidden [43].

Figure 28. Graph of Ω-type doublet splittings in the $A1_u$ state of Ar_2. The splittings are linear in J(J+1) for all observed J (in parentheses) in all vibronic bands from v' = 24 to 30 [43].

4.3 Radiative Lifetimes of Xe_2, Kr_2, and Ar_2 Excimers and Dependence on Internuclear Distance

Time-resolved fluorescence studies [46] of the rare gas excimers have been used to measure radiative lifetimes of the excimer states $(A1_u)$ involved in the operation of electron-beam pumped excimer lasers. As described above for the studies of the spectra of Xe_2, Kr_2, and Ar_2, dimers were formed in their weakly bound van der Waals ground states by expansion of high pressure gas through a pulsed supersonic nozzle into a region of 1 - 10 m Torr pressure. Excitation to selected vibronic levels of the $A1_u$ excimer states (Fig. 29) was achieved by monochromatic and tunable vacuum ultraviolet radiation of ~3 ns pulse duration. Fluorescence radiation was detected by a solar-blind photomultiplier with signal averaging and storage provided by a programmable transient digitizer. The time response of this detection system was ~2.5 ns.

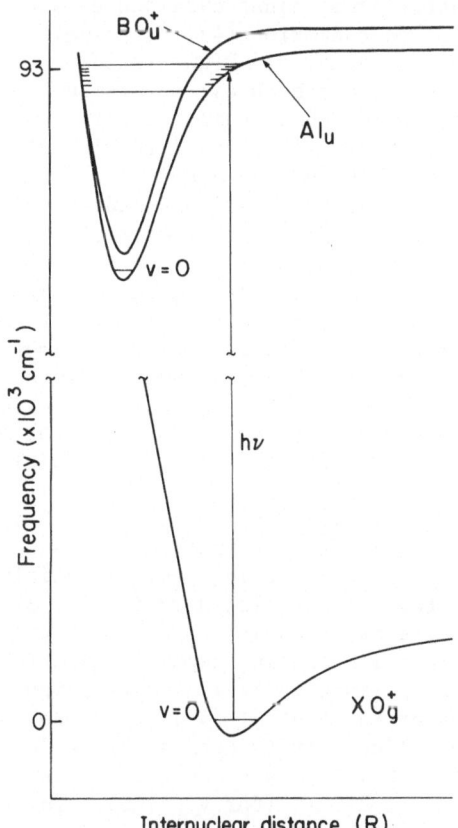

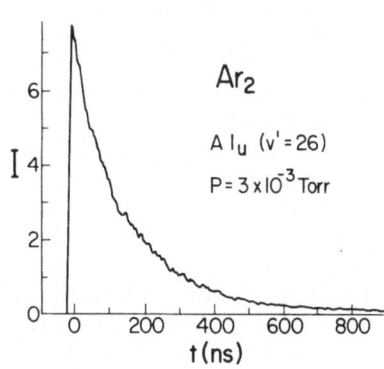

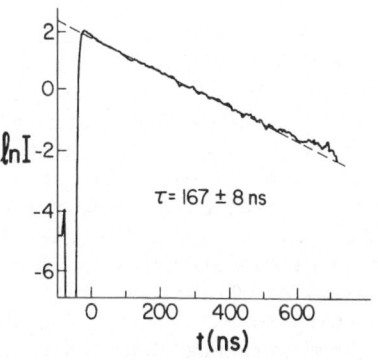

Figure 29. Schematic diagram of potential energy curves for the lowest stable states of Ar_2 showing transitions from v = 0 of the ground state to high vibrational levels of the $A1_u$ state.

Figure 30. Typical observed fluorescence intensity vs. time for v' = 26 level, including a semilogarithmic plot (below) illustrating single exponential decay [46].

The Al_u - $X0_g^+$ band systems of Ar_2, Kr_2, and Xe_2 occur at 107, 126 and 150 nm, respectively. Because of the relative positions of the potential energy curves for the strongly bound excimer states and the shallow ground states, (Fig. 29), only high vibronic levels of the Al_u states are accessible from the ground states. Thus only the levels $v' = 24$-30 for Ar_2, $v' = 32$-38 for Kr_2, and $v' = 36$-43 for Xe_2 could be investigated. A typical fluorescence decay curve of intensity versus time is given in Fig. 30 for Ar_2. Such curves clearly revealed single-exponential decays with time for all three excimers, yielding the radiative lifetimes shown in Fig. 31 for each level investigated.

It is seen that for Ar_2 and Xe_2 the lifetimes are essentially constant for the high vibrational levels studied leading to average values of $\tau = 160\pm10$ ns and 47 ± 5 ns, respectively. For Kr_2 there is a small but noticeable increase in lifetime of $\tau = 50$ ns at $v' = 32$ to $\tau = 68$ ns at $v' = 38$, with an average of $\tau = 55\pm5$ ns. These results for Ar_2, Kr_2, and Xe_2 differ significantly from values obtained earlier by use of charged particles and synchroton radiation [47] for fluorescence excitation of these excimers formed at relatively high pressures (>100 Torr) in cell experiments. At these pressures, collisional frequencies are sufficiently high that rapid vibrational relaxation occurs, resulting in fluorescence emission from low vibrational levels ($v' \sim 0$) of the excited states. The values of τ for $v' \sim 0$ are 2.9 s for Ar_2, 265 ns for Kr_2, and 100 ns for Xe_2. These results imply reductions in lifetimes by factors of ~20 for Ar_2, 5 for Kr_2, and 2 for Xe_2, in going from $v' \sim 0$ to $v' \sim 20$-40 in the Al_u states. While differences of a factor of two in radiative lifetimes for vibrational levels of the same electronic states are not uncommon in molecular spectroscopy, a factor of 20 is unique. Nevertheless such a large difference in lifetimes is not totally unexpected since Schneider and Cohen [48] had predicted a decrease by a factor of ~10 from $v' = 0$ to $v' = 10$ for the Al_u state of Ne_2. The present results for Kr_2 and particularly Ar_2 provide the only confirmation for such large variations in lifetime with vibrational levels.

These large differences in radiative lifetimes arise from changes in the electronic transition moment $\mu(R)$ with internuclear distance R. The calculated dependence of $\mu(R)$ on R for electric dipole transitions involving the Al_u and ground (0_g^+) states of Ar_2 is shown in Fig. 32. In the united-atom and separated-atom limits, transitions are forbidden, but at intermediate distances, spin-orbit coupling causes transitions to be weakly allowed. The values of $\mu(R)$ given in Fig. 32 were computed on the basis of ab initio calculations by Yates et al [49], with the addition of the effect of spin-orbit coupling. It is seen that the derived values of $\mu(R)$ increase rapidly from ~1.0 x 10^{-2} a.u. at 4.5 a_0 (corresponding to $v' \sim 0$) to ~8 x 10^{-2} a.u. at 6.8 a_0 (for $v' \sim 30$). These values were then used in calculations of spontaneous emission probabilities and lifetimes from the well-known relation (in atomic units):

$$A_{v'} = \tau^{-1} = \frac{4}{3c^3} \sum_{v''} (\Delta E)^3 |<\phi_2,v'|\mu_{12}(R)|\phi_1,v''>|^2$$

Here ΔE is the energy between two vibronic states (1,v'' and 2,v'), $\phi_{1,v''}$ and $\phi_{2,v'}$ are vibrational wave functions for the ground and excited state

states, and $\mu_{12}(R)$ is the electronic transition moment connecting the
states 1 and 2.

The resulting values of radiative lifetimes for Ar_2 are compared
with measured values in Fig. 33. There is excellent agreement for high
vibrational levels and an apparent discrepancy of a factor of 3 for
$v' = 0$. However, calculations have shown that this difference can arise
from a 5% increase in the equilibrium internuclear distance found for
the Al_u state from our recent spectroscopic data. This is well within
the quoted error of ~10% in R_e caused by the necessarily long extrapolation of constants from $v' = 24-30$ to $v' = 0$ for Ar_2. Similar calculations of lifetimes for Kr_2 and Xe_2 have yielded good agreement with
measured values at high vibrational quantum numbers. This investigation
has revealed surprisingly large changes in radiative lifetimes between
low ($v' = 0$) and high ($v' \sim 30$) levels of the Al_u states in Ar_2, Kr_2, and
Xe_2. Calculations have confirmed the respective lifetime ratios of ~20,
5, and 2, and shown that they arise from large variations of the electronic transition moments with internuclear distance.

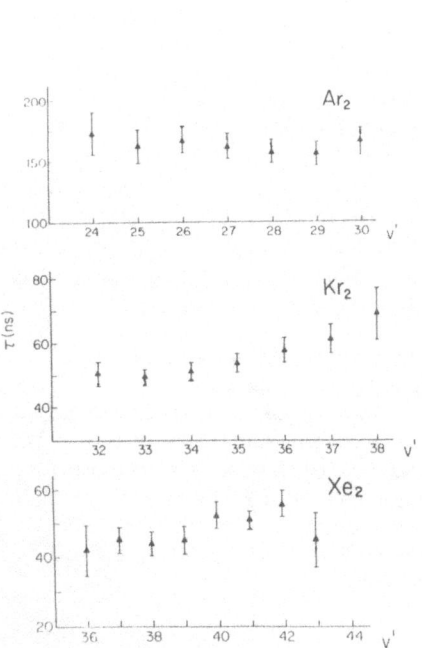

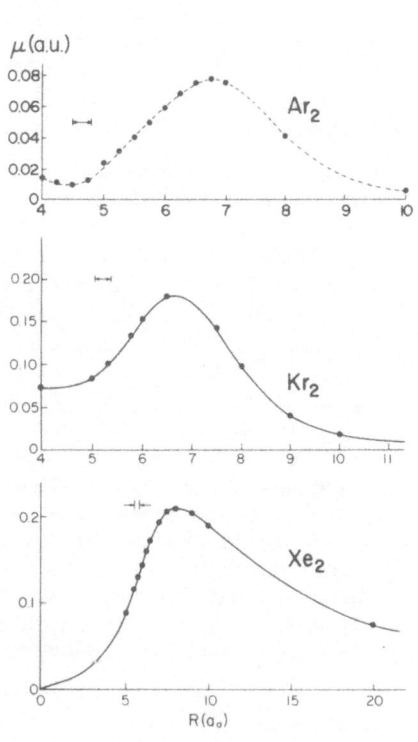

Figure 31. Measured radiative
lifetimes for various vibronic
levels of Al_u states of Ar_2, Kr_2,
Xe_2.

Figure 32. Calculated values of
electronic transition moments for
Al_u-$X0_g^+$ transitions of Ar_2, Kr_2,
Xe_2 including effects of spin-
orbit coupling.

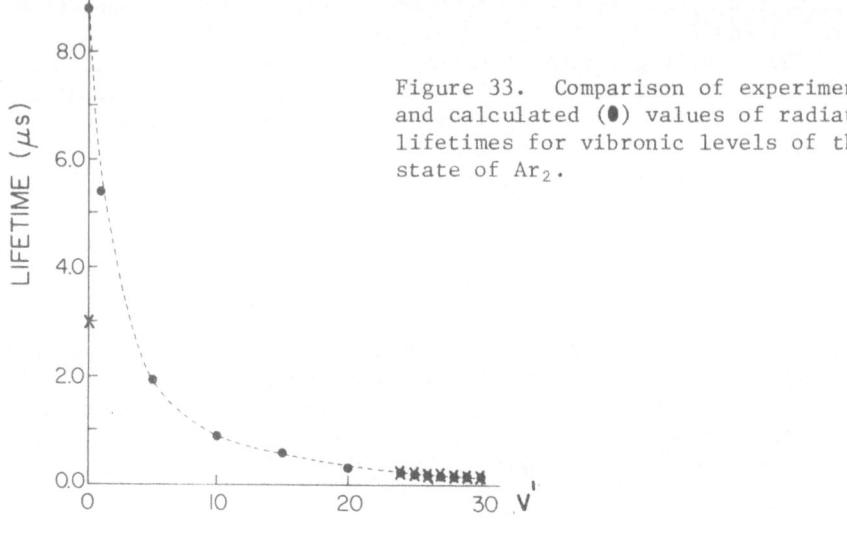

Figure 33. Comparison of experimental (X) and calculated (●) values of radiative lifetimes for vibronic levels of the $A1_u$ state of Ar_2.

5. CONCLUSIONS

Several different laser techniques for generating VUV and XUV radiation are now available, and these have been discussed briefly, along with the relevant theory. Of these, harmonic generation and four-wave frequency mixing have been shown to provide coherent and monochromatic radiation which is tunable over broad regions of the spectrum, from 200 nm to ~70 nm, and with limited tunability to ~50 nm. While these laser-driven sources are not presently available commercially, they have been developed in several laboratories around the world, and have been used in various fields of research.

In the present paper, their application to molecular spectroscopy has been described, with several examples of spectra which could not have been obtained by other means. These results have depended primarily on the tunability and monochromaticity of this radiation. Its directionality has also permitted the use of small gas volumes, and the short duration (of a few nanoseconds) has led to its importance in studies of lifetimes. Expected improvements in intensity, tunability, linewidth, and pulse duration of these laser-driven sources will further enhance their use in spectroscopy.

6. ACKNOWLEDGEMENTS

The development of some of the sources described here and their use in molecular spectroscopy was carried out by many collaborators and graduate students over the past decade. I am indebted to them for their valuable contributions. During this time, the research was supported by the Natural Sciences and Engineering Research Council of Canada, and by the University of Toronto.

REFERENCES

1. E.J. Woodbury and W.K. Ng, Proc. IRE, 50, 2347 (1962): G. Eckhardt, R.W. Hellwarth, F.J. McClung, S.E. Schwarz, D. Weiner, and E.J. Woodbury, Phys. Rev. Lett., 9, 455 (1962).
2. B.P. Stoicheff, in Proc. International School of Physics "Enrico Fermi", Course XXXI, 1963, C.H. Townes and P.A. Miles, eds., Academic Press, New York, pp. 306-325 (1964): B.P. Stoicheff, Phys. Lett. 7, 186 (1963).
3. R.W. Minck, R.W. Terhune, and W.G. Rado, Ap. Phys. Lett., 3, 181 (1963).
4. N. Bloembergen, Am. J. Phys., 35, 989 (1967): A. Penzkofer, A. Laubereau, and W. Kaiser, Prog. Quant. Electr., 6, 55 (1979). Y.R. Shen The Principles of Nonlinear Optics, Wiley-Interscience, New York (1984).
5. T.R. Loree, R.C. Sze, and D.L. Barker, Appl. Phys. Lett., 31, 37 (1977); T.R. Loree, R.C. Sze, D.L. Barker, and P.B. Scott, IEEE J. Quantum Electron., QE-15, 337 (1979).
6. R.S. Hargrove and J.A. Paisner, in Topical Meeting on Excimer Lasers Opt. Soc. Am. Washington, D.C. (1979), Paper ThA6.
7. V.Wilke and W. Schmidt, Appl. Phys., 16, 151 (1978); 18, 177 (1979).
8. H. Schomburg, H.F. Döbele, and B. Rückle, Appl. Phys., B30, 131 (198
9. H.F. Döbele, M. Röwekamp, and B. Rückle, IEEE J. Quantum Electron., QE-20, 1284 (1984).
10. M. Maeda and A. Takahashi, private communication (1986); A. Takahash M. Sumi, and M. Maeda, Opt. Commun., 55, 193 (1985).
11. W. Sasaki, Y. Ulhara, K. Kurosawa, E. Fugiwara, Y. Kato, and M. Yamanaka, Rev. Laser Engineering, 14, 370 (1986); Opt. Lett., 10, 487 (1985).
12. R.L. Carmen and W.H. Lowdermilk, Phys. Rev. Lett., 33, 190 (1974).
13. J.C. White and D. Henderson, Phys. Rev., A25, 1226 (1982): Opt. Lett 7, 204 (1982): Opt. Lett., 8, 520 (1983): IEEE J. Quantum Electron QE-20, 462 (1984).
14. J.C. White, Opt. Lett., 9, 38 (1984).
15. K. Ludewigt, H. Schmidt, R. Dierking, and B. Wellegehausen, Opt. Lett., 10, 606 (1985).
16. K. Ludewigt, K. Birman, and B. Wellegehausen, Appl. Phys., B33 (1984 B. Wellegehausen, K. Ludewigt, and H. Welling, Proc. Soc. Photo-Opt. Instrum. Eng., 492, 10 (1985).
17. J.A. Armstrong, N. Bloembergen, J. Ducuing, and P.S. Pershan, Phys. Rev., 127, 1918 (1962).
18. R.T. Hodgson, P.P. Sorokin, and J.J. Wynne, Phys. Rev. Lett., 32, 343 (1974).
19. C.R. Vidal, Appl. Opt. 19, 3897 (1980).
20. W. Jamroz and B.P. Stoicheff, in Progress in Optics XX, E. Wolf ed., North-Holland, Amsterdam, 325 (1983).
21. G.H.C. New and J.F. Ward, Phys. Rev. Lett., 19, 556 (1967).
22. R.B. Miles and S.E. Harris, Appl. Phys. Lett., 19, 385 (1971).
23. K.D. Bonin and T.J. McIlrath, J. Opt. Soc. Am., B2, 527 (1985).
24. P.R. Herman and B.P. Stoicheff, Opt. Lett., 10, 502 (1985).
25. A.H. Kung, Opt. Lett., 8, 24 (1983): J. Bokor, P.H. Bucksbaum, and

R.R. Freeman, Opt. Lett., $\underline{8}$, 217 (1983).

26. P.R. Herman, P.E. LaRocque, R.H. Lipson, W. Jamroz, and B.P. Stoichef
    Can. J. Phys., $\underline{63}$, 1581 (1985).

27. A.H. Kung, J.F. Young, G.C. Bjorklund, and S.E. Harris, Phys. Rev.
    Lett., $\underline{29}$, 985 (1972): A.H. Kung, Appl. Phys. Lett., $\underline{25}$, 653 (1974).

28. R. Hilbig and R. Wallenstein, Appl. Opt., $\underline{21}$, 913 (1982); IEEE J.
    Quantum Electron., $\underline{QE-17}$, 1566 (1981).

29. H. Egger, T. Srinivasan, K. Hohla, H. Scheingraber, C.R. Vidal, H.
    Pummer, and C.K. Rhodes, Appl. Phys. Lett., $\underline{39}$, 37 (1981); Opt. Lett.
    $\underline{5}$, 282 (1980).

30. E.E. Marinero, C.T. Rettner, R.N. Zare, and A.H. Kung, Chem. Phys.
    Lett., $\underline{95}$, 486 (1983).

31. J. Reintjes, C.Y. She, and R.C. Eckardt, IEEE J. Quantum Electron,
    $\underline{QE-14}$, 581 (1978).

32. K.K. Innes, B.P. Stoicheff, and S.C. Wallace, Appl. Phys. Lett., $\underline{29}$,
    715 (1976); S.C. Wallace and K.K. Innes, J. Chem. Phys., $\underline{72}$, 4805
    (1980).

33. J. Lukasik, S.C. Wallace, W.R. Green, and F. Vallée, in Proc. Laser
    Techniques for Extreme Ultraviolet Spectroscopy, R.R. Freeman and
    T.J. McIlrath, eds., Amer. Inst. Phys., New York (1982).

34. J. H. Glownia and R.K. Sander, Appl. Phys. Lett., $\underline{40}$, 648 (1982).

35. R. Mahon and F.S. Tomkins, IEEE J. Quantum Electron., $\underline{QE-18}$, 913
    (1982).

36. R.R. Freeman, R.M. Jopson, and J. Bokor, in Laser Techniques for
    Extreme Ultraviolet Spectroscopy, T.J. McIlrath and R.R. Freeman,
    eds., Amer. Inst. Phys., New York, 422 (1982).

37. D. Cotter, Opt. Commun., $\underline{31}$, 397 (1979).

38. F.J. Northrup, J.C. Polanyi, S.C. Wallace, and J.M. Williamson, Chem.
    Phys. Lett., $\underline{105}$, 34 (1984).

39. M. Maeda and B.P. Stoicheff, in Laser Techniques for Extreme Ultra-
    violet Spectroscopy, R.R. Freeman and T.J. McIlrath, eds., Amer. Inst
    Phys., New York, 162 (1982); A.C. Provorov, B.P. Stoicheff, and S.C.
    Wallace, J. Chem. Phys., $\underline{67}$, 5393 (1977).

40. R.W. Field, Ph.D. Thesis, Harvard University (1972).

41. R.H. Lipson, P.E. LaRocque, and B.P. Stoicheff, J. Chem. Phys., $\underline{82}$,
    4470 (1985).

42. P.E. LaRocque, R.H. Lipson, P.R. Herman, and B.P. Stoicheff, J. Chem.
    Phys., $\underline{84}$, 6627 (1986).

43. P.R. Herman, A. A. Madej, and B.P. Stoicheff, Chem. Phys. Lett., $\underline{134}$,
    209 (1987).

44. D.E. Freeman, K. Yoshino, and Y. Tanaka, J. Chem. Phys., $\underline{61}$, 4880
    (1974).

45. M.C. Castex, Chem. Phys. $\underline{5}$, 448 (1974); J. Chem. Phys., $\underline{74}$, 759
    (1974).

46. A.A. Madej, P.R. Herman, B.P.Stoicheff, Phys. Rev. Lett., $\underline{52}$, 1574
    (1986).

47. T.D. Bonifield, F.H.K. Rambow, G.K. Walters, M.V. McCusker,
    D.C. Lorents, and R.A. Gutcheck, J. Chem. Phys., $\underline{72}$, 2914 (1980).

48. B. Schneider and J.S. Cohen, J. Chem. Phys., $\underline{61}$, 3240 (1974).

49. J.H. Yates, W.C. Ermler, and N.W. Winter, Lawrence Livermore National
    Laboratory, Informal Report No. UCID-2-224 (1984), unpublished.

THE FREE ELECTRON LASER, OFFSPRING OF SYNCHROTRON RADIATION

Sydney Leach

Laboratoire de Photophysique Moléculaire du C.N.R.S.

Bâtiment 213, Université Paris-Sud

91405 ORSAY, FRANCE

and

DAMAP, Observatoire de Paris-Meudon

92190 MEUDON, FRANCE

ABSTRACT : Free electron lasers (F.E.L.) have been developed in the last ten years based on the technology of various devices which produce synchrotron radiation through free-free transitions of electrons. The properties of synchrotron radiation sources are described, leading to a discussion of wiggler, undulator and optical klystron systems in which photon and electron beams interact and result in the production of spontaneous and coherent electromagnetic radiation to be amplified in the optical resonator of a FEL. Harmonic generation in the vacuum ultraviolet is treated, including the case of external sources. Results obtained with the Orsay storage ring FEL are discussed. Present and projected FELs of the Compton scattering (low electron density) and Raman scattering (high electron density) types are presented. Prospects for scientific and technical use of FELs are explored.

I - INTRODUCTION

In conventional lasers, the emitted radiation occurs between energy states involving <u>bound</u> electrons in atoms or molecules. Energy state population is modified by pumping the initial state to higher levels by means of an external source. The electron modes of motion are discrete so that absorption and emission of photons involves two (or more) quantized levels (except where the final state in the laser emission process is above the dissociation limit e.g. in some excimer lasers). On the other hand, free electron lasers (FEL) operate on free-free transitions i.e. by energy changes of <u>unbound</u> electrons, so that any state of electron motion

89

A. C. P. Alves et al. (eds.), Frontiers of Laser Spectroscopy of Gases, 89–152.
© *1988 by Kluwer Academic Publishers.*

or energy value may be involved. Photon absorption or emission is achieved by constraining the "free " electron to accelerate or decelerate its motion (1).

The basic elements of a FEL are a high velocity (relativistic) electron beam, a device enabling the electrons to undergo oscillatory motion so as to generate - and interact with - photons, and a resonant optical cavity, the latter having the same feedback function as in a conventional laser (fig. 1). The electron oscillations are usually produced by a suitably periodic magnetic field such as that existing in an undulator device. Some recent FELs have been conceived in which electron oscillation occurs by electrostatic means, or even by interaction with a high intensity light wave. Characteristics of the undulator, discussed later, enable coupling to occur between the electrons traversing the undulator and the electromagnetic waves which they emit as a result of their oscillatory motion. Undulators can also be used to generate harmonics of a fundamental electromagnetic frequency and thus have important scientific and technical applications independent of their value in FEL devices. The discussion of undulator devices will therefore include their harmonic generation aspects.

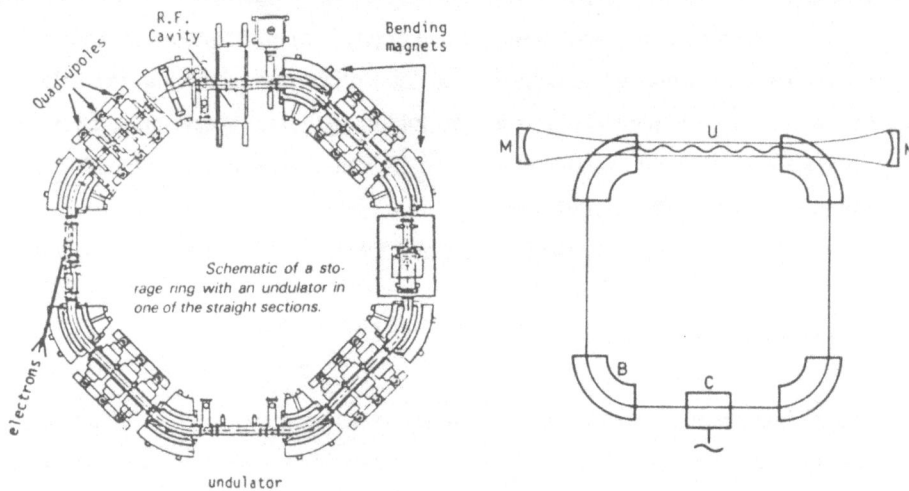

Fig. 1. Schematics of a storage ring Free Electron Laser
U = Undulator ; M = Optical Cavity mirror : B = Bending
magnets and focusing elements : C = Radiofrequency cavity.

Electron-photon coupling can be obtained, in other types of FEL, without the use of undulators. In these cases the light wave is slowed down so as to enhance interaction with the travelling electrons (2). In one type this is achieved through the Cerenkov effect in which an electron moving through a dielectric (gaseous) medium acts as a radiating superluminous charge. Another type of conceivable non-undulator FEL is the Orotron, which uses the Smith-Purcell effect in which the electron radiates in passing over a grating. The electromagnetic wave suffers a modification of the phase velocity so as to enhance coupling with the electron beam. Yet another technique, related to the cyclotron maser, or gyrotron, involves spiralling electrons. The physics and relative advantages and disadvantages of FEL-Cyclotron hybrids, Cerenkov FELs and Orotrons have been discussed in detail by Marshall (2). Our discussion will mainly be involved with FELs based on undulators and their variants.

There are two main types of FEL : (i) the low electron density FEL in which electrons individually experience the undulator magnetic field and the photon field, and (ii) the high electron density FEL in which electrons interact collectively with the undulator and photon fields. These types are distinguished by their electron scattering regimes, Compton for the low and Raman for the high electron density versions. The first is mainly useful from the XUV to the I.R. while the Raman FEL is of value for the cm to mm wavelength region.

The generation of electromagnetic radiation in a FEL is intimately related to synchrotron radiation devices since the basic physics of both involves effects of high velocity on the radiation patterns of electrons in accelerating fields. This will be described in section 2 and some basic relations between instrumental parameters and the emitted synchrotron radiation will be given. The development of a number of insertion devices and their characteristics will be presented in section 3 and a discussion given of optical klystrons, which are modified undulators and are used to improve the coherence and other characteristics of the emitted radiation. Section 4 is devoted to harmonic generation via insertion devices. It is theoretically a short step from the insertion device to the free electron laser as is clear from Fig.1. FEL analogues of the characteristics of conventional lasers will be discussed in section 5 so as to bring out the similarities and differences in the basic physics and properties of the two classes of lasers.

A description will be given of the FEL devices constructed at ORSAY and the results to date will be presented. Various electron accelerators (section 6) and FEL devices operating or under construction in the world (section 7) will then be considered, and some comments made as to future expectations. Possible physico-chemical and technical applications of undulator and FEL radiation are briefly reviewed in the final section.

The general references (1) through (5) provide some introductory material (1), a text book giving the situation of FEL devices in 1984, with some emphasis on Raman type FELs (2), a recent review article (3), some of the ORSAY results (4) and a picture of world-wide FEL development and devices in recent conference proceedings(5).

2 - SYNCHROTRON RADIATION

An electron in an accelerating field will lose energy by radiation. This is the basis of synchrotron radiation and FEL devices, in which positive or negative electrons undergo radial acceleration, usually by magnetic constraint. In this case the radiation emitted can be considered as magnetobrehmsstrahlung. A low velocity electron radiates with a pattern close to that of a normal Hertzian electric dipole (Fig. 2a) but this pattern is severely distorted at high electron velocity, due to relativistic effects. The zeros of the radiation pattern then occur at angles $\theta = (1-v^2/c^2)^{1/2}$ from the direction of motion. The radiation pattern as seen by a stationary observer is indicated in fig. 2b. The relativistic (Lorentz) transformation causes the power radiated to be projected into a very small foward cone of the order of a milliradian in angle.

It is useful to distinguish between electron synchrotron and electron storage rings as synchrotron radiation sources. An electron synchrotron is modulated by three superimposed frequencies, the electron injection frequency (typical value $\approx$ 50 Hz), the electron orbital frequency ($\approx$ 1 MHz) and the cavity radiofrequency used to accelerate and restore energy lost by radiation ($\approx$ 400 MHz). The electrons are injected periodically ($\approx$ every 20 ms), and are accelerated with a periodically varying magnetic field. There is consequently a continuous time variation in the intensity, frequency and angular distribution of the synchrotron radiation. An electron storage ring, (fig. 1) contains a number of dipole "bending" magnets, which force the electrons to undergo a circular trajectory where they radiate. Between the bending magnets are a number of straight sections where reside quadrupole magnets to focus the electron beam, and insertions devices such as undulators. There may exist also the possibility of beam correction with additional, sextupole, magnet coils. Electrons, once injected into the magnet lattice, can have their energy modified or restored by a radiofrequency cavity. Electron storage rings are thus particularly interesting as providers of electrons for certain types of low electron density FEL devices.

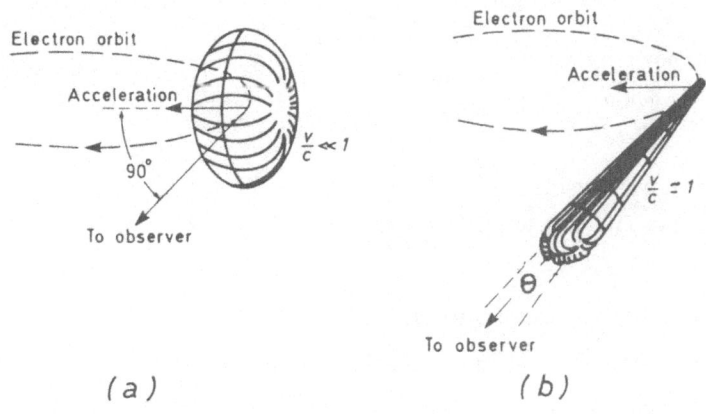

Fig. 2. Schematic radiation patterns for electrons in circular
orbit (a) at low velocities ; b) at relativistic velocities,
$\theta = (1-v^2/c^2)^{1/2}$.

The basic relations between instrumental parameters and synchrotron radiation characteristics are as follows (6,7) (where the electron energy E is in GeV, the magnetic radius of curvature R is in meters, the electron current I in amps and the magnetic field B in Teslas) :

(i) total power radiated by a relativistic electron, for $(v/c) \approx 1$

$$P = \quad I(\lambda,\Psi) \; d\lambda d\Psi = (2/3)(e^2c/R^2)(E/(m_0c^2))^4........(1)$$

(ii) energy loss per revolution, per electron

$$\delta E(keV) = 4\pi e^2 \aleph^4/3R = 88.47 \; E^4/R = 26.5 \; E^3 B..........(2)$$

where the relativistic factor $\aleph = E/m_0c^2 = 1957 \; E$

(iii) total power radiated
$$P_{tot}(kW) = 0.2654 \; E^3 .B.I.................................(3)$$

(iv) critical wavelength
$$\lambda_c \; (A) = 4\pi R/3\aleph^3 = 5.59 \; R/E^3 = 18.64/BE^2...............(4)$$

(v) characteristic (so-called critical) energy
$$\epsilon_c(eV) = 2218 \; E^3/R = 2.96 \; x \; 10^{-7}\aleph^3/R$$
$$= 665.1 \; BE^2 ....(5)$$

(vi) mean emisssion angle $\theta \approx \aleph^{-1}$ ....(6)

Because of the Lorentz transformation, an observer at a particular angle in the orbit plane will detect synchrotron radiation emitted over an arc length $R/\aleph$ in the bending magnet. The radiation pulse will have a duration $\Delta t \approx R/c\aleph^3$. The power spectrum comprises the Fourier components of the synchrotron radiation pulse and will contain harmonics of the orbit frequency ν up to $\omega = \nu\aleph^3$. The fundamental frequency is Doppler shifted from the MHz region into higher frequencies ; the harmonics blur into a continuum, so that the observed spectrum extends from the far infra-red to the ultraviolet and x-ray regions approximately up to the critical wavelength λ_c. The divergence of the resulting light, in the vertical plane is comparable to that of a laser.

Fig. 3 gives the radiation spectrum of an electron moving in a curved trajectory, per GeV. This is expressed in terms of the photon flux, which is of practical interest, as a function of the radiation wavelength divided by the critical wavelength. Its peak occurs at $\lambda/\lambda_c \approx 4$. Although the flux falls off rapidly below $4\,\lambda_c$, it is still of useful intensity down to $\approx 0.1\,\lambda_c$.

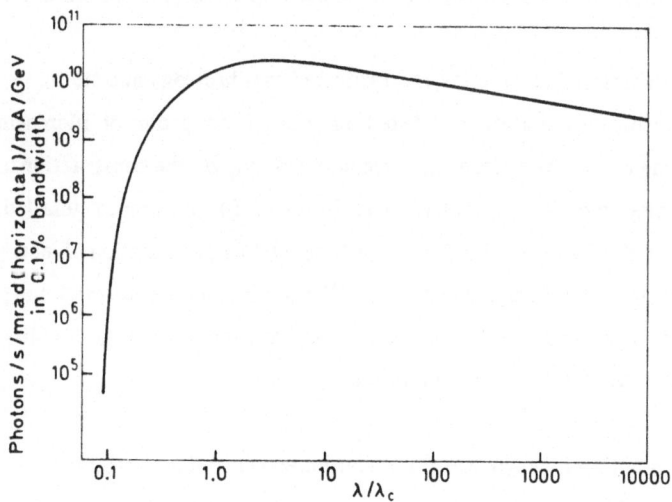

Fig. 3. Normalized radiation spectrum of an electron moving in a
curved trajectory, per GeV. λ_c = critical wavelength.

The principal properties of synchrotron radiation of interest as a spectroscopic source or excitation source in physicochemistry are as follows (8) :

1) Tunability over a very broad energy range from below 0.1eV to above 1000 eV.

2) Pulsed time structure (ps-ns range) enabling time resolved experiments to be carried out.

3) High intensity (brightness).

4) High degree of spatial collimation (1 mrad).

5) Quasi-complete linear polarisation ; possibility of other polarisations.

6) Temporal and spatial stability.

Many of these properties carry over to, or are improved upon in, insertion devices and FELs.

Characteristics of present-day and projected synchrotrons and storage rings used or useful for synchrotron radiation are listed in table I, in order of increasing critical energy (decreasing critical wavelength). A recent workshop on the construction and commissioning of dedicated synchrotron radiation facilities (100) gives more detailed information on a number of the facilities listed in table I, as well as presentation of new projects in several countries, including Brazil and Taiwan. The candid manner in which the participants of this workshop described the positive and negative aspects of their facility's development is particularly admirable and very instructive.

It is worth noting that synchrotron radiation was considered originally, in practice, simply as a detrimental by-product of high energy particle accelerators and storage rings i.e. as a kind of energy dissipative "friction" of the particles destined for high energy collision experiments. The modern recognition of synchrotron radiation as a desirable tool in itself dates from about 1970. The first experiments using synchrotron radiation were carried out in parasitic mode on some of the first generation high energy electron collision physics machines e.g. ACO at Orsay (9). It is only since about 1975 that machines designed specifically for synchrotron radiation production and studies have been designed and built. These newer machines have taken advantage of progress in accelerator and storage ring design and are tailored to enhance the intensity and quality of synchrotron radiation output. However, many of them were designed almost exclusively for photon beam lines issuing from simple bending magnets. It is only in the last few years that the use and advantages of insertion devices as means of enhancing and controlling synchrotron radiation qualities, as well as for the construction of FELS, has come to the forefront (1). Some existing storage rings are being fitted with insertion devices but generally in less than optimal spatial dimensions. Storage rings recently designed have included insertion devices ab initio.

TABLE I. - Synchrotron Radiation Sources

These are storage rings unless specified otherwise.
D = Dedicated ; P.D = Partially dedicated ; UC = Under Construction ;
S = Synchrotron ; PAR = Parasitic mode ; POSS = Possible use for SR.

MACHINE NAME	LOCATION	E/(GeV)	I/mA	R/m	λ_c/A	Remarks
N-100	Kharkov, USSR	0.10	50	0.5	2800	D
SURF II	Washington, USA	0.25	25	0.84	300	D
TANTALUS I	Madison, USA	0.24	200	0.64	260	D
SOR RING	Tokyo, Japan	0.38	250	1.1	112	D
TERAS	Tsukuba, Japan	0.6	150		52	D
SIBERIA I	Moscow, USSR	0.45				UC ; D
MAX	Lund, Sweden	0.55	370	1.2	40.3	
ACO	Orsay, France	0.54	150	1.1	39	D
FIAN C-60	Moscow, USSR	0.67	10	1.6	30	S
UVSOR	Okazaki, Japan	0.75	500	2.2	29	D
NSLS I	Brookhaven, USA	0.75	500	1.9	25	D
HESYRL	Hefei, China	0.8	300	2.22	24	UC ; D
VEPP-2M	Novosibirsk, USSR	0.67	100	1.22	23	P.D
BESSY	Berlin, FRG	0.8	500	1.83	20	D
SUPER ACO	Orsay, France	0.8	500	1.75	19.1	UC ; D
SRL	Stanford, USA	1.0	500	2.1	11.7	UC ; P.D
IN-ES	Tokyo, Japan	1.3	30	4.0	10.2	S
PAKHRA	Moscow, USSR	1.3	300	4.0	10.2	S
COSY	Berlin, FRG	0.63		0.44	9.8	UC ; D
SIRIUS	Tomsk, USSR	1.36	15	4.23	9.4	S
ADONE	Frascati, Italy	1.5	100	5.0	8.3	P.D
BONN	Bonn, FRG	2.0	50	7.6	5.3	P.D
ALADIN	Madison, USA	1.3	500	2.08	5.3	D
SRS	Daresbury, U.K.	2.0	370	5.56	3.9	D
DCI	Orsay, France	1.85	300	3.82	3.4	D
PHOTON FAC-TORY	Tsukuba, Japan	2.5	250	8.33	3.0	D
VEPP-3	Novosibirsk, USSR	2.25	100	6.15	3.0	D
NSL II	Brookhaven, USA	2.5	500	8.17	2.9	D
BEPC	Beijing, China	2.8	150	10.35	2.6	P.D
SIBERIA II	Moscow, USSR	2.5	300	5.0	1.8	UC ; D
ARUS	Erevan, USSR	4.5	1.5	24.6	1.5	S
ELSA	Bonn, FRG	3.5	50	10.1	1.3	P.D
SPEAR	Stanford, USA	4.0	100	12.7	1.1	P.D
ESRF	Grenoble, France	5.0	200		0.9	UC ; D
DORIS II	Hamburg, FRG	5.0	50	12.1	0.54	P.D
TRISTAN ACC.	Tsukuba, Japan	6-8				UC ; PAR
CESR	Ithaca, USA	8.0	100	32.5	0.35	PAR
VEPP-4	Novosibirsk, USSR	7.0	10	16.5	0.27	PAR
PETRA	Hamburg, FRG	18.0	18	192.0	0.18	POSS
PEP	Stanford, USA	18.0	10	165.5	0.16	PAR
TRISTAN	Tsukuba, Japan	30.0				UC ; PAR

Among the machines listed in table I, special mention should be made of COSY which has been designed as a "table-top" synchrotron radiation source. This compact ring was conceived as a source for X-ray lithography for sub-micron structures and would have application in semiconductor fabrication. If successfully commercialized, it is conceivable that COSY could be the forerunner of "University Sized" Synchrotron Radiation and FEL sources.

3 - INSERTION DEVICES

Insertion devices have two important characteristics which enhance the value of simple bending magnet synchrotron radiation sources i.e. tunability and increased intensity. Many of these devices pertain to storage rings where the insertion is in a field-free straight section of the ring lattice. Ideally, it is possible to thus insert a magnetic field of zero field integral, with respect to the orbit coordinate (for example by an arrangement of fields of alternate polarity), without any consequent alteration of the closed orbit of the ring electrons. The local magnetic field simply produces a transverse acceleration of the electrons with no overall deflection or displacement. Any value of the local magnetic field can be used, quite independent of that in the normal bending magnets of the storage ring. Thus, in principle, insertion devices can operate independently of the rest of the storage ring without affecting electron beam properties or other experiments. Some insertion devices are found, however, to reduce the $1/e$ duration of the beam, due to increased possibilities of collisions with residual gases in the confined regions of the device.

3.1. Wavelength shifter

The simplest of the insertion devices is a <u>wavelength shifter</u> (fig.4) which consists of adding an additional bending magnet, of smaller radius of curvature, into the storage ring magnet lattice. The local reduction in radius results in a modification of the spectral distribution of the synchrotron radiation emitted at that site, with a shift of the spectral maximum (fig. 3) to shorter wavelengths (cf. eqn.4). In principle it is also possible to use a lower local magnetic field (greater value of R) and so produce a "softer" synchrotron radiation spectrum without altering the electron energy.

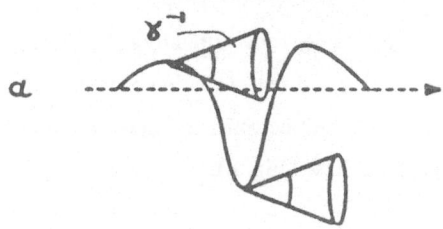

WAVELENGTH SHIFTER

a

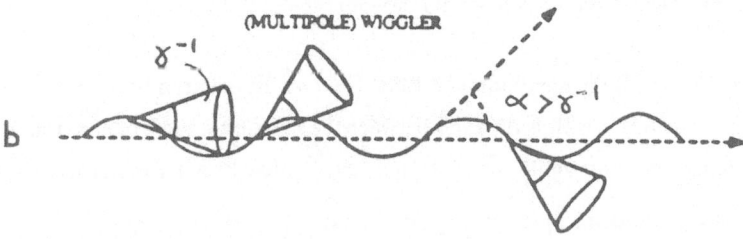

(MULTIPOLE) WIGGLER

b

$\alpha > \gamma^{-1}$

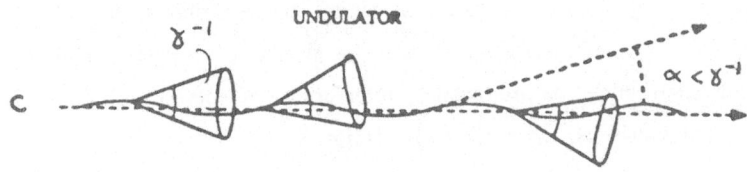

UNDULATOR

c

$\alpha < \gamma^{-1}$

Fig. 4. Representative schematic electron trajectories and photon emission cones for insertion devices of decreasing field index K and maximum deflection angle α.
(a) wavelength shifter, K ≫ 1, (b) wiggler, K > 1, (c) undulator, K < 1. γ^{-1} = mean emission angle.

3.2 Wiggler

A transverse wiggler is a multipole device constituted by a sequence of three or more magnets with the field alternating in polarity and in a direction perpendicular to the electron orbit. The maximum deflection angle α of the electrons, and therefore the mean emission angle, is less than in a single wavelength shifter (Fig. 4).

The spectral characteristics of insertion devices are largely determined by the dimensionless deflection parameter, or field index :

$$K = \alpha\gamma = eB\lambda_w/2\pi \, m_0c = 0.0934 \, B_0\lambda_w \, \dots\dots\dots\dots \, (7)$$

where B_0 (Teslas) is the maximum magnetic field on the electron trajectory and λ_w (mm) is the wiggler period. K is thus the ratio between the maximum angle of electron deflection and the radiation emission angle $\theta \approx \gamma^{-1}$, and has a value $K \gg 1$ for normal wigglers.

The modification of the emitted spectrum as a function of K will be discussed in detail in the next section. At this point it suffices to say that wigglers are often designed so that the oscillation of the electron beam is comparable to the transverse beam dimensions. Radiation from all poles is then superimposed <u>incoherently</u>. The photon emission is greatly enhanced in intensity with respect to a single bending magnet ; the total photon flux is 2N times that of a single period (3 pole device), where N is the number of wiggler magnetic field periods. Coherent radiation can be produced by interference effects in multipole wigglers having adapted characteristics, as described in section 3.3.

Wiggler magnets can be oriented so as to deflect the electrons in the plane of the ring bending magnets or in the vertical plane. The synchrotron radiation emitted in the electron trajectory direction is linearly polarised, with the E vector in the direction of electron acceleration i.e. polarised in a plane perpendicular to the field. Thus a vertically deflecting wiggler gives rise to a vertical ribbon of synchrotron radiation which is polarised vertically, whereas the ring bending magnets create horizontal polarisation.

Two important wiggler characteristics are the emission angle, which is of the order of $10\gamma^{-1}$ to $50 \, \gamma^{-1}$, and the linear dependence of the differential angular distribution of the power $dP/d\Omega$ on N. The emission spectrum of standard wigglers producing one or a small number of transverse oscillations of the electron beam is reasonably continuous in spectral distribution. This is the $K \gg 1$ condition of fig.8 and is the useful condition for producing high energy radiation, e.g. X-rays, in the high frequency tail of the synchrotron radiation spectrum.

Fields in the ring magnets are generally smaller than 1.2 Teslas but wiggler fields of 2 Teslas or more are possible using iron-core magnets and 5 Teslas or more with superconducting magnets (10). Samarium-cobalt permanent magnets, which have high remanent fields, are now considered to be among the best materials for wigglers and undulators (11). Wigglers have been constructed or are being planned at most storage rings and at a few synchrotrons throughout the world.

3.3 - Undulators

Undulators are multipole wiggler devices having a smaller deflection angle α than in standard wigglers, and a field index $K<1$. The small value of α means that radiation can be collected from practically the whole electron trajectory in the undulator (Fig. 4). These multipole wigglers have relatively modest magnetic fields and usually a greater number of periods than in a standard wiggler. Undulators with values of N greater than 100 have been built. The possibility of using undulators to produce relatively narrow spectral bands ($\Delta\lambda/\lambda \approx 1\%$), via interference effects, make them of considerable interest as photon sources for various uses, including FEL construction. Undulator configurations will first be treated and a discussion given of the basic resonance condition which determines the spectrum and the radiation time structure.

Two undulator configurations are usually considered: (i) a transverse sinusoidal magnetic field with period λ_w and maximum field amplitude B_0, and (ii) a helical magnetic field in which the field amplitude remains constant but the direction of the field vector rotates about the axis of the undulator as a function of the distance along the axis. A permanent magnet undulator, first proposed by Halbach (12), operates with magnets rotated 90° from one site to the next along the undulator. Fig. 5 shows two periods of such an undulator; the magnetic field along the electron beam axis z is given by $B(z)=B_0 \sin (2\pi z/\lambda_w)$. The undulator period

λ_w is usually in the cm range (typically 2-3 cm) but so-called micro-undulators are being developed, having λ_w values in the mm range, to take advantage of the fact that one can then use smaller electron generators (see later).

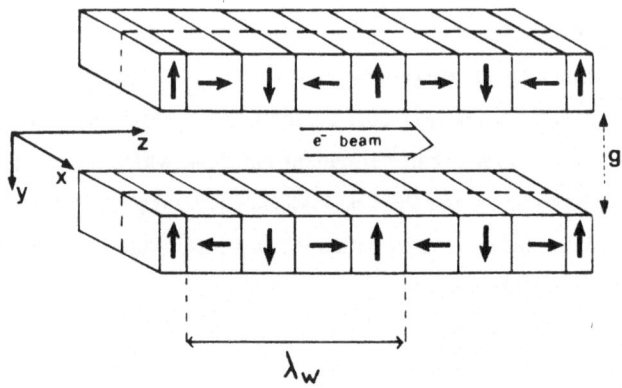

Fig. 5. Halbach-type permanent magnet undulator.

Two periods are shown. The field integral and double integral are compensated by adjustable half-poles at the extremities, giving zero net deviation of the electron.

g = undulator gap, λ_w = undulator period.

The wiggling electron in the undulator becomes equivalent to a dipole oscillator in the electron frame of reference. Since these are high velocity electrons, the electron sees the undulator relativistically contracted, so that it is forced to oscillate and emit dipole radiation of wavelength $\approx \lambda_w / \gamma$. The radiation emitted undergoes a Lorentz transformation in the laboratory frame, just as described in section 2, (see also fig.15) the electromagnetic radiation being Doppler shifted to wavelength $\approx \lambda_w / 2\gamma^2$ and is projected forward in a small cone whose angular dimension is $\approx \gamma^{-1}$. The motion in the electron frame becomes more complex with increasing K, giving rise to the emission of harmonics. The fundamental (n = 1) or harmonic (n > 1) wavelength of this spontaneous emission is given by the <u>resonance condition</u> (13-15) :

$$\lambda_n = (\lambda_w / 2\gamma^2)(1 + cK^2 + \gamma^2\theta^2)/n \ldots\ldots\ldots\ldots (8)$$

where θ = angle of observation with respect to mean electron trajectory. The constant $c = 1/2$ for a planar undulator, $c = 1$ for helical field undulators. The fraction of the power emitted in the fundamental is given by $(1 + K^2/2)^{-2}$, e.g. about 44 % for $K = 1$, but only about 10 % for $K = 2$.

Another perception of the properties of undulator radiation can be given in terms of interference between the radiation by the same electron at different points on its trajectory through the undulator. Each electron traversing the undulator will undergo radial forces and so emit synchrotron radiation over a continuous spectrum. The resonance condition holds when the electron, which moves slower than the velocity of light, is overtaken by one period of the photon field as it passes through one period of the undulator (Fig. 6). There will be a lag distance D between electron and photon. Destructive interference will develop after a number of periods since phase matching will not occur for most frequencies of the synchrotron radiation. However phase matching will exist for a particular ratio D/n (n=1,2,3...) of the lag distance. Contributions from different electrons will be incoherent unless the electrons are structured in microbunches as will now be discussed.

The electrons enter the wiggler/undulator in unstructured packets (e.g. storage ring case) or in a quasi-continuous stream. As stated above, the electromagnetic waves, produced by oscillation of the various electrons traversing the device, add incoherently. There will be partial destruction since positive and negative amplitudes add algebraically. The emitted amplitude is proportional to the square root of the number of electrons in the packet, i.e. to $N_e^{1/2}$, so that the radiation intensity will be proportional to N_e.

However, if the electrons are grouped together at regular intervals within the electron packet, the emitted radiation can have a coherent component whose relative importance will depend on the degree and the quality of the microbunches. The existence of energy exchange between photons and the electrons gives rise to modulation of the density of the electrons as they progress down the insertion device. This process can be explained in terms of the ponderomotive force description of the interaction between the electromagnetic field and the charged particles (16).

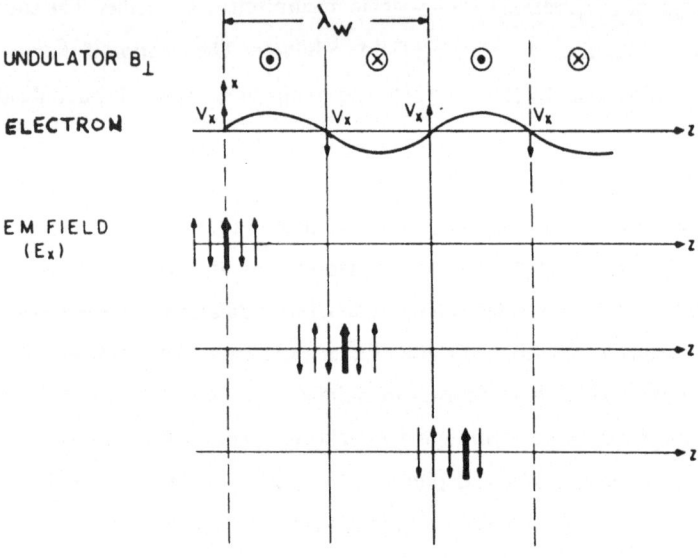

Fig. 6. Photon field relative to electron trajectory in an undulator magnetic field $B_\perp$, illustrating the resonance condition (cf eqn. 8). The boldface arrow in the electromagnetic field is a reference.

The ponderomotive energy corresponds to a travelling wave whose phase velocity $v = (c\lambda)/(\lambda_w^{-1} + \lambda_r^{-1})$. If the electron motion is considered in a frame moving with the ponderomotive potential, the electron dynamics reduces to that of a simple pendulum (17, 18). The energy exchange, calculated from the Lorentz force, is given by $\delta\gamma\, mc^2 = -e\int v E dt$, where $\delta\gamma mc^2$ is the variation of energy of the electron traversing the insertion device, v its velocity and E the electric field of the light wave. The integral is taken over the electron trajectory. Energy conservation implies that any energy lost or gained by the electron is respectively given to or received from the radiation. The interaction creates an electron acceleration when $\delta\gamma mc^2 > 0$ and electron deceleration accompanied by light amplification when $\delta\gamma mc^2 < 0$. The value of $\delta\gamma mc^2$ will be zero at resonance, and close to zero elsewhere except when electron and light wave are nearly in resonance (cf.eqn.8). Thus this process leads to absorption or to amplification of the electromagnetic wave.

Since the electrons have varying velocities along their trajectory in the insertion device, those that are accelerated by the light wave interaction will catch up with those injected half a wavelength earlier that were decelerated. Thus the electrons will have a tendency to group together in microbunches. The evolution of the longitudinal electronic density is illustrated in fig.7. It should be noted that the modulation of the electron density is relatively small with respect to the mean density. This is because electron repulsion and a number of other effects prevent the microbunching of more than a small fraction of the electron in a packet ($N_e \approx 10^{10}$). However, it is these microbunched electrons which create the coherent component of the radiation. As the coherent component increases, the tendency will be for the emitted amplitude to be proportional to N_e, i.e. for the intensity to vary as N_e^2.

The gain is a relatively small but since the total gain is proportional to L^3 where L is the length of the undulator, the gain per unit length can be quite large at the further end of the undulator and so be useful for laser devices. In principle, the longer the insertion device, the much greater is the gain. However, the length L is limited not only by the beam design e.g. the magnet lattice in a storage ring, but also because it is difficult to maintain a good quality electron beam in a long undulator, especially due to energy and angular dispersion of the electrons.

Due to the relativistic Doppler effect the radiation is concentrated within horizontal and vertical half-angles $\theta_h \simeq (1+K^2/2)\gamma^{-1}$ and $\theta_v \simeq \gamma^{-1}$ respectively. The gain in power emitted in a solid angle $2\gamma^{-2} \times 2\gamma^{-2}$, as compared with normal bending magnets, is of the order of N, the number of undulations. An increase of the order of N^2 in spectral brilliance can be obtained for a well collimated beam. The spectral width W will depend on a number of factors : a) finite number of poles, since W is proportional to $(nN)^{-1}$; this can be considered as homogeneous broadening ; b) angular spread; c) electron beam aperture ; d) acceptance angle θ, since W is a function of $\gamma^2\theta^2$.

The fundamental wavelength can be changed either by modifying K, i.e. by varying the undulator gap, of by observing off axis, i.e. θ variation. The range of tunability can be calculated from eqn.8 by putting in suitable practical values of the parameters.

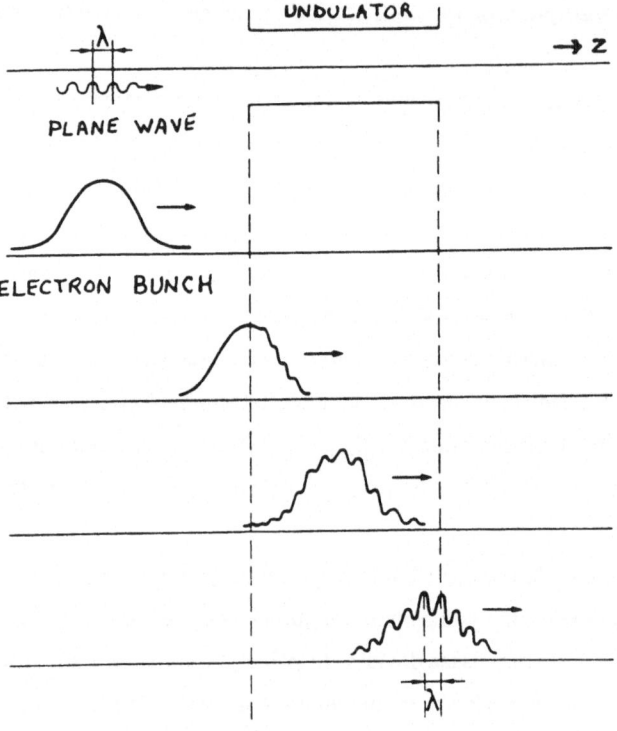

Fig. 7. Evolution of longitudinal density of electrons

through electron-photon (plane wave) interaction,

illustrating micro-bunching of electrons.

The harmonic content of undulator emission increases with magnetic field amplitude. The form of the spectrum depends on K, which is independent of the electron energy(19) and on the parameter Σ, which takes into account the angular spread of the electron (σ') and the acceptance angle of the detector (Ω) :

$$\Sigma = [(\gamma\sigma')^2 + \gamma^2 - \Omega/\pi]^{1/2} \ldots\ldots\ldots\ldots(9).$$

As discussed earlier, for K « 1 only one spectral harmonic is emitted, at $\lambda=\lambda_w/2\gamma^2$ (Fig.8); its wavelength is independent of B_0 and the power emitted is proportional to B_0^2 ; the wavelength spread is small if the radiation is observed through a pinhole on the axis of the undulator (Fig.8a). Fig.8c illustrates that the spectrum includes many harmonics when K>1; under certain conditions, their envelope ressembles a normal synchrotron radiation spectrum (fig.8d).

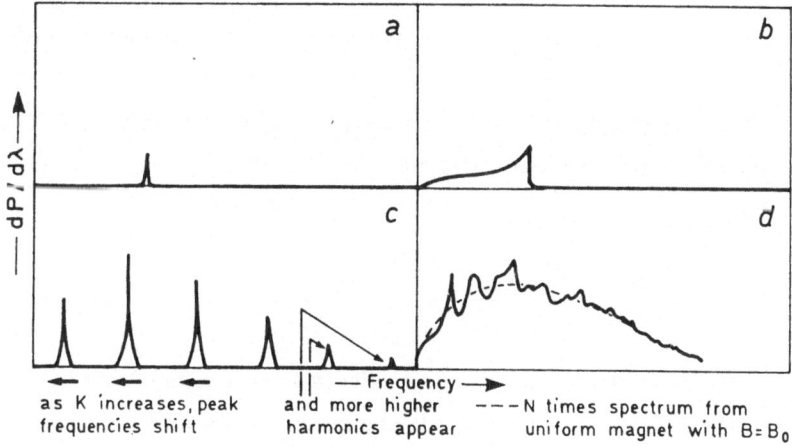

Fig. 8. Schematic representation of spectral characteristics
   of undulators see text for definition of parameters
   K and Σ.
   a) K « 1, Σ « 1 : weak field, well collimated beam
   b) K « 1, Σ ⩾ 1 : weak field, σ' large or no pinhole
   c) K > 1, Σ « 1 : strong field, well collimated beam
   d) K > 1, Σ ⩾ 1 : strong field, σ' large or no pinhole

The phase relation between the electric field components perpendicular and parallel to the insertion device's magnetic field will determine the ellipticity of the polarisation of the radiation. This phase relation depends on the particular configuration of the magnetic field. The phase difference is π/2 for a bending magnet, giving rise to circular polarisation, whereas for standard wigglers and undulators the phase is the same for both components so that the polarisation will be linear in a direction which is a function of the direction of observation. Most undulators have a vertical magnetic field since the vertical gap is smaller than the horizontal gap, so that a much higher field can be achieved for a given period, or a much shorter period can be used for a given magnetic field. The polarisation of the radiation will then be in the ring orbit plane. Circular polarisation can be obtained by using crossed undulators, asymmetric wigglers, or helicoidal undulators (20). Indeed, if the magnetic field configuration is helical instead of transverse, the emission dependence on θ and K are very similar to the latter case (see eqn.8) except that at $\theta = 0$ only the fundamental is observed and polarisation is circular ; at $\theta > 0$, harmonics with elliptical polarisation are emitted.

The time structure of insertion devices depends on the value of K. If T_1 is the time for an electron to span a wiggler period, then this is equal to the time between successive radiation peaks :

$$T_1 = (\lambda_w/2\gamma^2 c) (1 + K^2/2) \dots\dots\dots\dots (10)$$

where c= velocity of light. For the wiggler case, the duration T_c of a light pulse is given by $T_c = (4\pi/3) (R_0/c\gamma^3)$, where R_0 is the radius of curvature of the electron trajectory at the centre of the wiggler. The time structure of an actual electron bunch is that corresponding to its length and its repetition rate. The spectrum of the radiation, schematised in Fig.8, is the Fourier transform of its time domain structure. The periodicity of the radiation field produces, in principle, radiation at wavelengths given by the resonance condition (eqn.8), whose observed spectrum will also be defined by parameters discussed above (cf. fig.8).

In a wiggler, K » 1 so that, because of the large value of the deviation angle α, the electrons emit in a given direction only during a sequence of short time intervals (Fig.9). The radiation will contain many harmonic components of the basic repetition frequency which here will be given by the magnetic periodicity rather than the electron orbit frequency. Because of the finite emittance of the electron beam, the short wavelength harmonics are not observed separately, so that blending into a quasi-continuum occurs (fig.8d). However in the undulator, where K is often less than unity and α is small, the observer can detect light over most of the electron trajectory within the device since the angular excursion of the electron is smaller than the radiation angle θ. The time structure of the radiation is then much closer to a modulated sinusoidal wave train (Fig.9), corresponding to radiation at a single frequency. In the case of K≃1, several harmonics are observed. Harmonic generation is discussed more fully in section 4.

The spectral and time structure properties of insertion devices carry over to FELs in which they play a part.

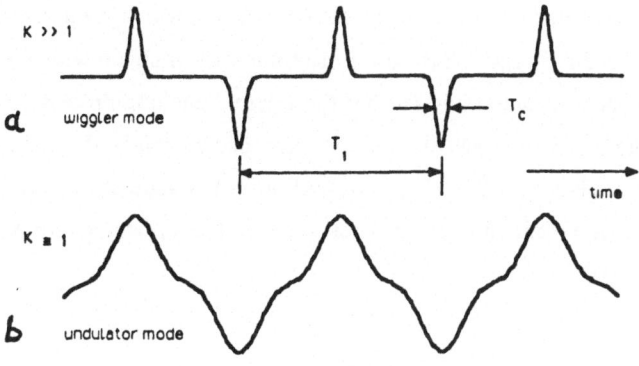

Fig. 9. Radiation time structure as a function of the field index K (a) K » 1, wiggler, (b) K ≈ 1, undulator.

3.4- Optical Klystrons

As discussed above, electron microbunching in undulator devices creates coherent spontaneous emission (and also leads to more efficient harmonic generation, see later). This is of importance in improving the gain of FEL oscillators. The optical klystron, proposed by Vinokurov and Skrinsky (21), enables electron bunching be further enhanced. The ordinary klystron, invented by the Varian brothers in 1939, (and named from the Greek term for waves beating on the shore, i.e. bunching of water "packets") is constituted by two coupled resonators whose corresponding electric fields are in phase. In the first resonator, the interaction between the oscillating electric field and a stream of electrons causes the latter to bunch up spatially. The electron packets can amplify the electric field in the second resonator, enabling one to extract radiation which in the case of the klystron is usually in the centimeter wavelength region. The optical klystron uses the same general principles, except that the electrons are coupled with transverse components of the electromagnetic field, instead of the longitudinal components in the ordinary klystron.

The optical klystron consists of two undulators, respectively the "modulator" and the "radiator", separated by a dispersive section (fig.10) in which the transit time of the electrons depends on their energy. Thus slower electrons will catch up with faster electrons in the dispersive section, increasing further the micro-bunching process for a given undulator length. The maximum increase of the gain occurs when the modulator and radiator undulators are identical (22). The optical klystron is a good device for optimizing the small signal gain. A recent theoretical description of the optical klystron has been given by Elleaume (23).

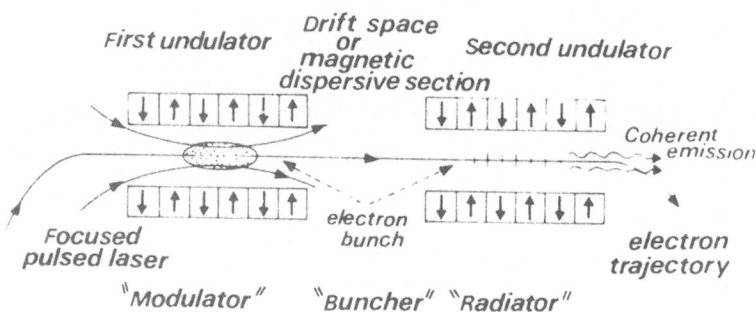

Fig. 10. Optical Klystron. Schematics of undulator and dispersive sections. The focused pulsed laser case is discussed in section 4.2.

A number of optical klystrons have been built or are projected. The Novosibirsk OK-3 optical klystron has a total length of 2m, and a total of 11 periods (24) ; the Orsay optical klystron is 1.3 m long and has 2 x 7 periods (25). The maximum gain is 12 and 6.4 times greater than that for an equivalent length undulator for the respective optical klystrons (23). Recently there has been developed a variable period undulator, known as the tapered undulator, which enhances the energy exchange between electrons and photons by compensating for changes in the energy of the propagating electrons(2). It is mainly useful with Linac sources, from which much energy can be extracted.

4 - HARMONIC GENERATION

Harmonic generation of radiation via insertion devices promises to have important applications in photophysics and photochemistry. The basic physical principles were discussed in section 3.3 and the conditions schematized in fig. 8. Spatial bunching of electrons, as described in previous sections, serves not only to create coherent radiation but also to enhance the higher Fourier components in the emitted radiation.

4.1. Intrinsic Harmonic generation

Experiments have been carried out with an undulator at Orsay at various electron energies between 150 and 536 MeV in which intrinsic harmonic generation (i.e. without an external source) was studied. For example, with E = 240 MeV, up to 23 harmonics were observed (26). In principle only odd harmonics should be seen in the forward direction ($\theta \simeq 0$) (cf. eqn. 8), but in practice harmonics of even rank are also observed, due to various imperfections of the set-up, e.g. misalignment, size of solid angle of the detector. The spectral widths of the harmonics are broadened by the angular dispersion of the electron beam ; this increases with electron energy, as does the relative intensity of the even harmonics. Although all electrons may leave the accelerator having the same energy, instrumental effects create a certain distribution of electron momenta between parallel and transverse components. It is this distribution that characterizes the so-called emittance (2). Among the storage rings listed in table I, those of low emittance include BESSY, Super-ACO and the ESRF. To obtain a low emittance electron beam it is necessary to use strongly focusing quadrupole magnets and the machine to provide the most appropriate beam for a particular insertion device.

A typical spectrum in the VUV measured at Orsay with the storage ring operating at 536 MeV is shown in fig. 11. This was achieved with an undulator having a gap = 32.5 mm, at K = 2.29. The fundamental λ_1 = 1280 A. Harmonics 2 - 9 are displayed in fig. 11. The even rank harmonics are quite evident even though the observations were made at $\theta \simeq 0$. The even harmonics are relatively weaker in mesurements carried out at a beam energy E = 240 MeV, mostly due to the smaller beam emittance. For example, the intensity at n = 2 is then only about 15 % of that at n = 1. The weakness of harmonics of rank higher than 7 is due to the fall off in the spectral response of the monochromator below 200 A.

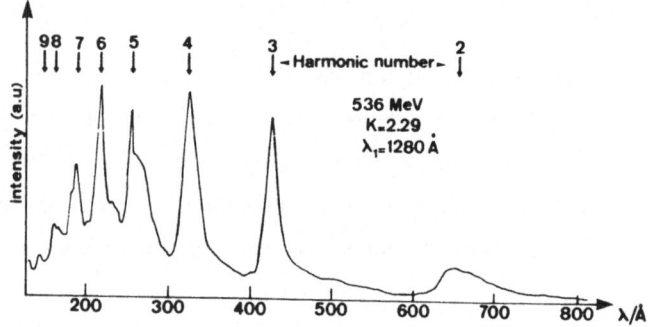

Fig. 11. Spectrum of radiation emitted between 100 and 800 A by electrons of energy E = 536 MeV in an undulator of gap = 32.5 mm, K = 2.29, at $\theta \approx 0$.
Uncorrected for spectral response of the monochromator which falls off sharply at λ < 200 A.

Equation 8 shows that the wavelength of the n th harmonic will vary with the observation angle θ. From this and from the dependence of spectral width on $(nN)^{-1}$ at $\theta = 0$, it can be shown that in order to prevent line broadening due the θ effect, the observation cone aperture has to be smaller than an angle θ_m, where :

$$\theta_m = \gamma^{-1} \left[(1 + K^2/2)\,(nN)^{-1} \right]^{1/2} \quad .... (11)$$

At ACO, for E = 536 MeV, γ = 1050, so that $\theta_m \approx 0.5$ mrad for the first harmonic with undulator operation at field index K = 2.29. Measurements made at E = 153 MeV gave a FWHM = $\Delta\lambda$ = 104 A for the n = 3 harmonic at λ = 5000 A ; for E = 169 MeV, the n = 5 peak at 2400 A had a FWHM = 44 A (4b). These FWHM values are somewhat larger than the theoretical values given by W $\simeq (nN)^{-1}$, and the main culprit is the emittance. Broadening due to this factor can be calculated (4b). Typical values of $\lambda_n/\Delta\lambda_n$ resulting from the emittance, for an undulator operating at K = 2.3 are as follows for rings at Orsay (table I) : $\lambda_n/\Delta\lambda_n$ = 37 ; 250 ; 2 ; 15 ; 200 respectively for ACO 536 MeV ; ACO 200 MeV ; DCI 1700 MeV ; DCI 600 MeV ; SUPER-ACO 800 MeV. These values give the ultimate bandwidths that could be possibly obtained, independent of the harmonic number. The brilliance of high order harmonics thus depends strongly on emittance dependent effects. Data on brilliance (expressed in number of photons /s/A/0.1 % band/(0.1 mrad)^2), gain and $\lambda_n/\Delta\lambda_n$ are given in table II for the NOEL permanent magnet undulator at Orsay. The characteristics of the undulator are listed in table III. The emittance reduces the expected brilliance by a factor of 1 to 10 between 1300 and 100 A. But the spectral brightness can be several orders of magnitude greater than for conventional bending magnet synchrotron sources as is illustrated in table II.

TABLE II. - Central brilliance of the NOEL undulator at K = 2.29 on the ACO storage ring at 536 MeV (brilliance is expressed in number of photons/s/0.1 A/0.1 % band/(0.1 mrad)^2. This brilliance is compared with that of a normal bending magnet of ACO at the same electron energy. The expected effect of the emittance of the beam on the undulator performance is also displayed. The total emitted power is 2.5 W for the undulator and 700 W for the ring.

Harmonic number	1	3	5	7	9
Wavelength (A)	1 280	426	256	183	142
Theoretical line-width ($\lambda_n/\Delta\lambda_n$)	18	54	90	126	162
Undulator theoretical brilliance/10^{12}	3.6	5.8	6.3	6.0	5.6
Calculated bending magnet brilliance/10^{10}	0.10	0.55	1.2	2.2	3.2
Gain over bending magnet radiation brillance	3 600	1 000	520	270	170
Gain over bending magnet including emittance	3 200	560	200	76	40

Table III - Characteristic of the permanent magnet
undulator NOEL.

Magnetic material	$SmCo_5(B_r \approx 0.85\ T)$
Full length	1.33 m
Number of periods	17
Period	78 mm
Transverse pole width	100 mm
Vacuum Chamber full gap	25 mm
Undulator minimum gap	33 mm
Undulator maximum field	0.31 T
Undulator K range	0-2.3
ACO electron energy range	150-536 MeV

Intense harmonics in the VUV are expected from a set of undulators in the Berkeley project for an Advanced Light Source (ALS) Synchrotron Radiation Facility operating with a low emittance beam at 1.5 GeV (27). The proposed ALS will have 12 long straight sections for setting up undulators optimized for use in various spectral ranges in the VUV and soft X-ray regions. The radiation will be tuneable over a broad spectral range and will be created in 20 ps pulses at a 500 MHz repetition rate. For comparison it should be noted that ACO pulses are $\approx$ 1 ns long at repitition rates of 13.6 or 27.2 MHz. SUPER-ACO will have pulses < 100 ps and operate at 100 or 500 MHz (see section 5.4.5).

4.2 - Harmonic generation with an external source or FEL source

As discussed above, harmonic generation is intimately linked to electron microbunching and interference effects in the insertion device. Increased enhancement of electron density modulation can be achieved by focussing an external laser source into an undulator or an optical klystron(28). In the latter case it is focussed into the modulator where it creates an energy modulation transformed into a spatial modulation of the electrons in the dispersive section. The ensuing microbunches are separated by the external laser wavelength λ_ℓ. The Fourier transform in space of the electron density $\rho(z,t)$ contains all harmonics of λ_ℓ, the intensity of each harmonic being dependent on the laser electric field. Light emitted by the electrons traversing the second undulator, or radiator, becomes coherent for these harmonics at wavelengths λ_ℓ /n. One of the important advantages of this technique is that an optical resonator is not required and that it is possible to produce radiation in the vacuum ultraviolet region using an external laser source in the visible or near ultraviolet. This is particularly helpful in avoiding the use of mirrors, which are necessary in FELs, whose degradation under VUV radiation can create severe radiation losses, as will be discussed in a later section.

This type of up-conversion is to be distinguished from laser harmonic generation obtained by non-linear optical devices. In the undulator case, the coherent output power is generated by loss of electron kinetic energy, i.e. an energy transfer process, rather than taken from the laser pump.

Some particular results at Orsay using external photon sources are as follows. In 1984, coherent VUV harmonics of a Nd-YAG laser emitting at 1.06 μm were obtained by focussing the laser beam into an optical klystron operating with an electron beam energy E = 166 MeV (28). In particular, the third-harmonic output at 355 nm was observed to have an intensity about 3000 times that of the spontaneous, incoherent, third-harmonic signal. Experimental values of R_3, the coherent/incoherent emission amplification ratio are shown in fig. 12, as a function of beam current and are compared with theoretical values (curve a). The modulation rate of the spontaneous emission measured with (curve b) and without (curve c) the laser are also shown ; the abscissa corresponds to the storage ring current. The measurements were made with a bandpass of 0.7 A and a solid angle of 0.2 mrad^2. The real spectral width of the coherent emission is certainly less than 0.1 A, so that the measured value of R_3 underestimated the real value by a factor of at least 7. The maximum number of coherent photons emitted per pulse was 6 x 10^5 as compared with a theoretical value of about 4 x 10^6.

There are many loss factor parameters (28). It was estimated, however, that this external laser technique should be able to create 10^{10}-10^{12} photons per pulse with modern storage rings operating at their nominal energy, which was not the case in the ACO experiment (E = 166 MeV, nominal working energy E = 536 eV). Intensities of this magnitude should have many applications in high energy photophysics and photochemistry (8). From these experimental results it was possible to foresee an extension of the spectral range to the VUV region. Indeed, very recently (29) coherent radiation at λ_3 = 1773 A and λ_5 = 1064 A was observed in an

116

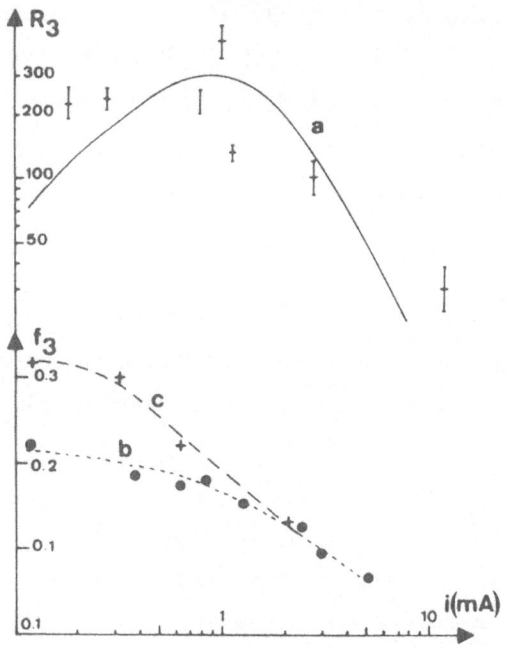

Fig. 12. Ratio R_3 of coherent/incoherent emission measured for
$\Delta\lambda = 0.7$ A and a solid angle of 0.2 mrad^2. Experimental
points are compared with theoretical curve a. Data in b
and c are of f_3, the modulation rate of spontaneous emission
measured respectively with and without the laser. Abscissa is
stored current. E = 166 MeV ; λ_1 = 1.06 μm, λ_3 = 0.355 μm.

experiment in which radiation of λ_ℓ = 5320 A (0.4 J in 6ns, $\Delta\lambda = 0.35$ A) from a frequency
doubled Nd-YAG laser was focussed into the modulator of an optical klystron. These
experiments require very precise superposition of external photon beam with the electron
beam. This was a source of difficulty in a similar experiment attempted at the NSLS in a Bell
Labs - Brookhaven collaboration (30). At ACO, 10^5 coherent photons per pulse were
obtained in the 10 eV region for an input power of 50 MW. With the new SUPER-ACO
storage ring under construction at Orsay, it is expected to obtain 10^{10} photons per pulse down
to about 500 A, with peak powers of the order of 1 KW. The results of some numerical
calculations are given in fig. 13 for harmonics n = 3, 5, 7 of an external laser operating at λ_1 =
3000 A (31).

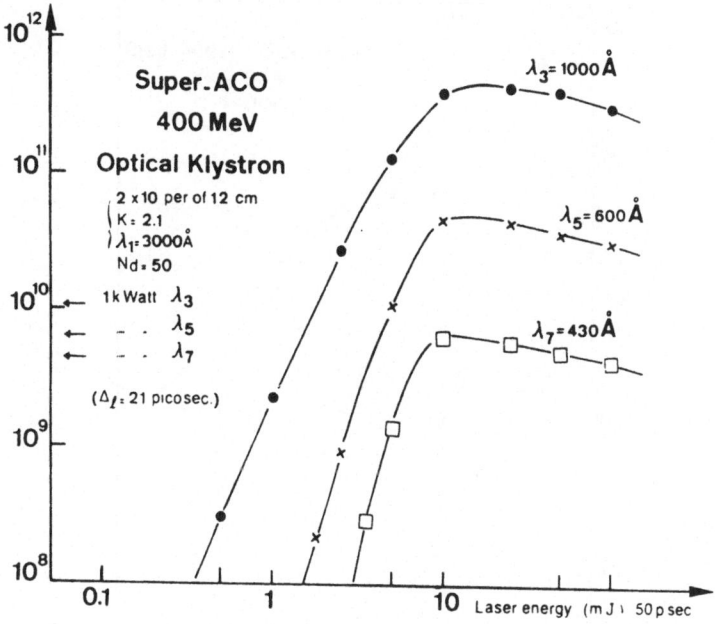

Fig. 13. Calculated number of coherent photons produced, as a function
of input laser energy, by an external laser λ_1 = 3000 A
focused into an optical klystron. E = 400 MeV at SUPER-ACO.

The optimum conditions for operation of the input laser are rather severe with storage rings
such as SUPER-ACO and would lead to rather complex laser systems. An alternative
procedure would be to use a FEL as the pump laser. Calculations of coherent harmonic
production expected with a FEL operating in a c.w. and also in a Q-switched mode at the
SUPER-ACO storage ring (see table IX) have been made by Ortega (31). He showed that the
c.w. mode is not of practical interest for generating harmonics. The operation of a FEL in the
Q-switched mode has been demonstrated at Orsay (32,33). It is expected to produce in
practice a smaller output of photons than by the use of an external pulsed dye laser. This is
illustrated in fig. 14, where the results of calculations are presented for two cases : (i) for 200
MW and (ii) for 20 MW of available tunable input laser power in the 3000 A region. However,
there are several potential advantages of using a Q-switched FEL as input. In particular
tuning the laser between 2000 and 4000 A should be relatively easy through the use of
broadband metallic mirrors. Furthermore, the usable spectral region can be extended since
the practical operation of FELs at λ < 2000 A should become possible within the next few
years.

118

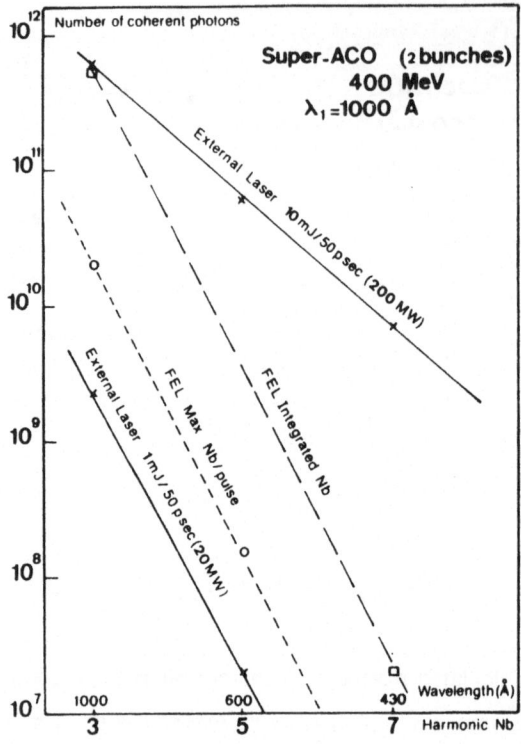

Fig. 14. Calculated number of coherent photons produced
by an external laser and the Q-switched FEL
operating under SUPER-ACO conditions. Results
for the external laser are for two values,
200 MW and 20 MW, of tunable laser power
at $\lambda \approx 3000$ A.

5 - FREE ELECTRON LASER

5.1 - Compton and Raman FELs

Theoretical analysis of free-free transitions of electrons moving in undulating magnetic
fields has much developed since the original work of Ginzburg (34) and Motz (13). Its
renewed importance is related to the development of Free-Electron Lasers using free-free

transitions of relativistic electrons. The present theoretical situation has been summarized by Fedorov (35), by Marshall (2), by Colson (36) and by Elleaume (36). As mentioned in the Introduction, two main types of FEL have been proposed : (i) the low electron density FEL in which electrons individually experience the undulator magnetic field and the photon field, and (ii) the high electron density FEL where electrons interact collectively with the undulator and photon fields. In the first type, or Compton type, there is negligible interaction between the electrons. Operation is with high energy, low current beams. All wavelengths are usefully created, but especially from the I.R. to the XUV. In the dense electron beam, or Raman type, there are strong interelectronic interactions i.e. between collective modes of the plasma. The electrons are in low energy, high current beams, and the useful wavelengths are mainly from in the cm to mm range.

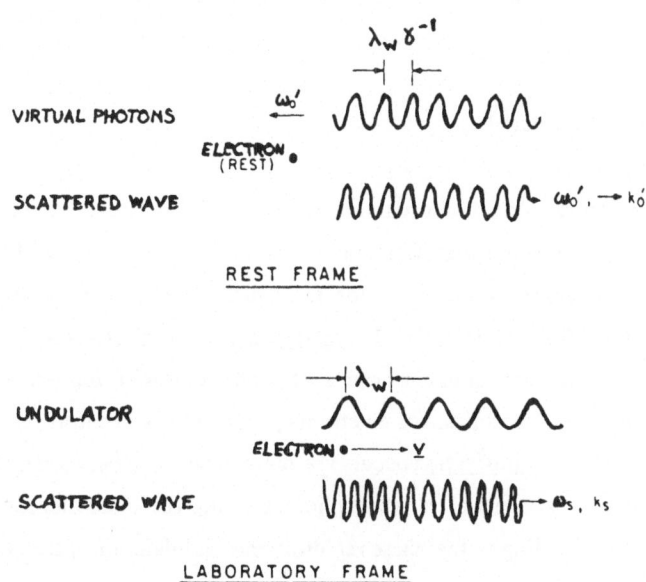

Fig. 15. Radiation emitted by an electron moving in an undulator.
Stimulated Compton scattering regime. The situation is shown
for the laboratory frame and the electron rest frame.

In the electron reference frame, the periodic undulator field appears as an electromagnetic field corresponding to virtual photons. It can thus amplify a real photon field travelling in the same direction as the relativistic electrons, the two fields being coupled by the electron charges (stimulated Compton scattering (37) (Fig. 15). The quantum mechanical description by Madey (38) led to his building the first FEL (in the Stanford linear accelerator) (39). In the Raman type FEL, the virtual photons corresponding to the undulator field act as the pump mode, the plasma wave is equivalent to an idler mode that absorbs the energy remaining from inelastic Raman scattering, the scattered light being the signal mode. The whole process corresponds to stimulated Raman scattering of photons from the electron beam.Dense electron beams, for which resonance effects appear at plasma frequencies (40,41), cannot be achieved in storage rings. Our discussion will mainly consider the low density electron beam case, using planar field undulators.

5.2 - Laser Characteristics

The three familar processes occurring in conventional lasers, i.e. spontaneous emission, stimulated emission, and amplification, take the following forms in FELs. The spontaneous emission wavelength is given by the resonance condition (eqn. 8). The "transition" is symbolised in fig. 16. Stimulated emission arises from the electron bunching process described earlier, where emission from a set of electrons with approximately the same phase gives rise to coherent addition of electromagnetic fields and constructive interferences. As mentioned earlier amplification occurs via energy exchange between electrons and photons, the interaction between electrons and photons giving rise to acceleration or deceleration of the electrons according to the phase region in the undulator or optical klystron. Net energy transfer to photons leads to a concentration of electrons in the deceleration phases.

Saturation mechanisms limit the power output of undulators. The energy loss from an electron increases with increasing radiated power. At high power, the electron behaviour is perturbed, e.g. increase of angular dispersion, so that the resonance condition given by eqn. 8 is only valid in part of the undulator. The saturated power decreases with increasing N, however there is a trade-off between maximizing the gain and maximizing the output power since the maximum gain is proportional to N^3 (see later). The power output of a FEL could be improved by the use of tapered wigglers (2) since they can be designed so that their magnet parameters vary to compensate for loss of electron energy resulting from the photon-electron interaction. Thus it is theoretically possible for them not to saturate until the radiation intensity at the electron beam attains hundreds of gigawatts per cm^2. However, this introduces new factors in the quality of the laser beam optics, including problems of mirror damage.

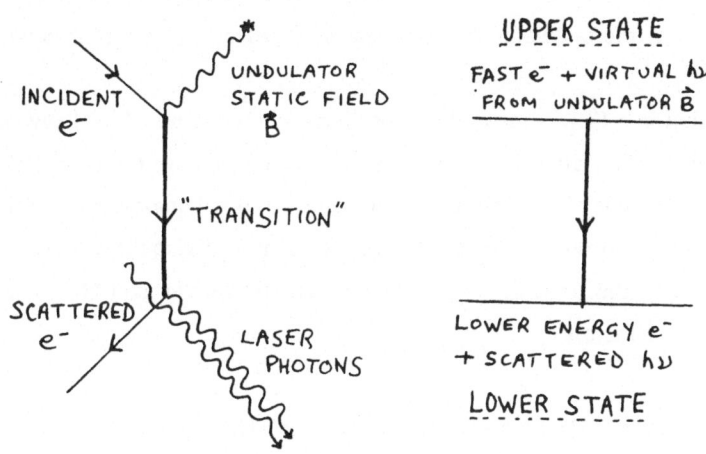

Fig. 16. Feynmann-type diagram of the Stimulated Compton scattering process and nature of upper and lower states of Free Electron Laser "transition".

The FEL beam has spatial and time structures that reflect the electron beam structure. The radiation pulse length is proportional to the electron bunch length ℓ_e, and the radiation frequency width is given by $\delta\omega = 2\pi c / \ell_e$. Individual laser modes have linewidths which are approximately equal to the inverse of the correlation time. The practical limit to the correlation time, and therefore to the widths of individual lines in the spectrum, is set by mirror microphonics. The laser power $P_L = G_{max} \times P_{RF}$ where P_{RF} is the power given to the electron beam by the RF cavity.

Complete coherence of emission requires spatial (transverse) coherence, related to source divergence, and temporal (longitudinal) coherence, related to source monochromaticity. The undulator has a high spatial coherence, of the order of γ^{-1}. For example for E = 500 MeV, $\gamma^{-1} \approx 1$ mrad. The parameters of the projected European Synchrotron Radiation Facility (ESRF) are E = 5 GeV, ϵ_x(horizontal emittance) = 6 x 10^{-9} mrad, ϵ_z = 6 x 10^{-10} mrad, giving a total spatial coherence for $\lambda > 19$ A. The optical beam may have poor temporal coherence at the exit of the undulator, but this can be improved by adding a monochromator.

The gain in a FEL is inversely proportional to the square of the linewidth of spontaneous radiation. The latter has two components (i) a homogeneous width due to the finite interaction time of electrons in the undulator, (ii) an inhomogeneous width due to the angular divergence and velocity spread of the electrons as well as to the variation with position of the strength of the periodic magnetic field. The maximum gain G_{max} is proportional to a number of factors :

$$G_{max} = k\rho_e B_0^2 \lambda_w^{3/2} \lambda^{3/2} N^2 L \dots\dots\dots (12)$$

where k is a proportionality constant, $L = N\lambda_w$ is the undulator length and ρ_e is the electron density (assuming that the electron energy distribution is Gaussian). Non linear effects similar to self-focusing can occur in the case of a long insertion device and high gain. This can counteract the diffractive spreading of the light beam.

5.3 - Gain measurements

Gain measurements were first made by Madey and his coworkers at Stanford (42). They used a superconducting transverse helical undulator, period λ_w = 3.2 cm, and electrons in a linear accelerator. Spontaneous gain of an external photon source was observed for 10.6 μm CO_2 laser radiation with 24 MeV electron energy. A gain of 7 % per pass was measured with an electron current of 70 mA. A 12 m long optical cavity was then added to the undulator so as to make a FEL ; with 43.5 MeV electrons, laser emission was observed at 3.417 μm with an average power of 0.36 W, peak power 7 KW (39). The role of the optical resonator, as in all laser systems, was to favour amplification of modes whose gain is maximum.

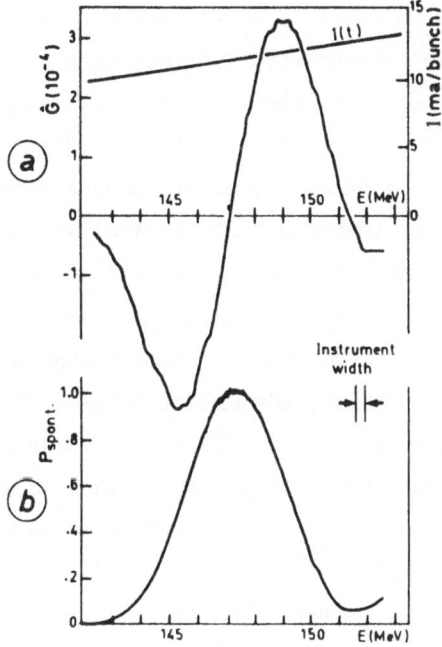

Fig. 17. a) Gain and b) Spontaneous spectrum
as a function of electron energy. I(t)
indicates the fall-off of Synchrotron
Radiation intensity as the electron energy
is lowered during the experiment.

Gain measurements on visible radiation have been made at Orsay using a 23 period
superconducting undulator (4 cm wavelength and 0.4 Tesla design field) inserted in a straight
section of the ACO storage ring, and an Ar^+ laser as an external pump. Spontaneous
emission was measured for electron energies of 240 MeV and 150 MeV (43,45). Fig. 17 shows
the gain and spontaneous emission curves as a function of electron energy at 4880 A. The
negative part of the gain curve corresponds to net absorption ; zero gain coincides with the
spontaneous emission maximum at the resonance energy. The curves behave as expected
from theory (see below). The largest measured peak gain, averaged over the laser mode was
$\hat{G}$ = 4.3 x 10^{-4} per pass. Higher gains have been obtained through the use of optical
klystrons. Fig. 18 compares the results for an undulator and an optical klystron concerning

the following : a) spontaneous emission as a function of the magnetic gap in mm ; the "sinusoidal" structure in a2 is due to interference of the synchrotron radiation emitted by the same electron in the modulator and the radiator ; b) the laser induced lengthening of the electron bunch as a function of the gap. As predicted from Madey's first theorem (46), which relates stimulated and spontaneous emission, bunch lengthening is proportional to the spontaneous spectrum ; this is clear from a comparison between the curves of a1 and b1 as well as between a2 and b2 in fig. 18 ; the lengthening effects are very similar for undulator and optical klystron ; c) the gain profiles versus magnetic gap measured with an external argon laser. It is found that these curves correspond approximately to the derivatives of the forward spontaneous emission curves, as predicted from Madey's second theorem (46), which expresses the spontaneous emission in terms of first order perturbation effects due to the electric field seen by the electrons. This derivative relation differs from that of a conventional laser, in which the gain follows the spontaneous emission line function.

The c1 experiments were carried out under the following conditions : E = 243 MeV, I = 30 mA, 2 electron bunches of 0.52 ns length ; the c2 experiments with E = 234 MeV, I = 30mA, single bunch of 0.67 ns length. The gain enhancement is up to a factor of 7, depending on electron energy and ring current, with respect to the original 17 period undulator. The observations show that the FEL should radiate more power with the optical klystron than the undulator, since the main saturation process is expected to result from bunch lengthening.

The good agreement with the predictions of Madey's theorem indicates that the spontaneous spectrum, relatively easy to measure, is a reliable diagnostic of the small signal gain which is much more difficult to determine directly. However, further detailed measurements of both spontaneous spectrum and the gain curve, carried out at Orsay (47), showed that Madey's theorem holds only at the low divergence limit. With divergent optical modes, which occur when the Rayleigh range of the optical wave is of the order of the length of the interaction zone, the gain spectrum exhibits shifts and distortions.

In order to convert the undulator or optical klystron set-up to a FEL (Fig. 1), cavity mirrors of exceptional quality are required in the visible and ultraviolet regions. The technical problems of producing adequate, radiation resistant, mirrors have not yet been solved as will be seen shortly.

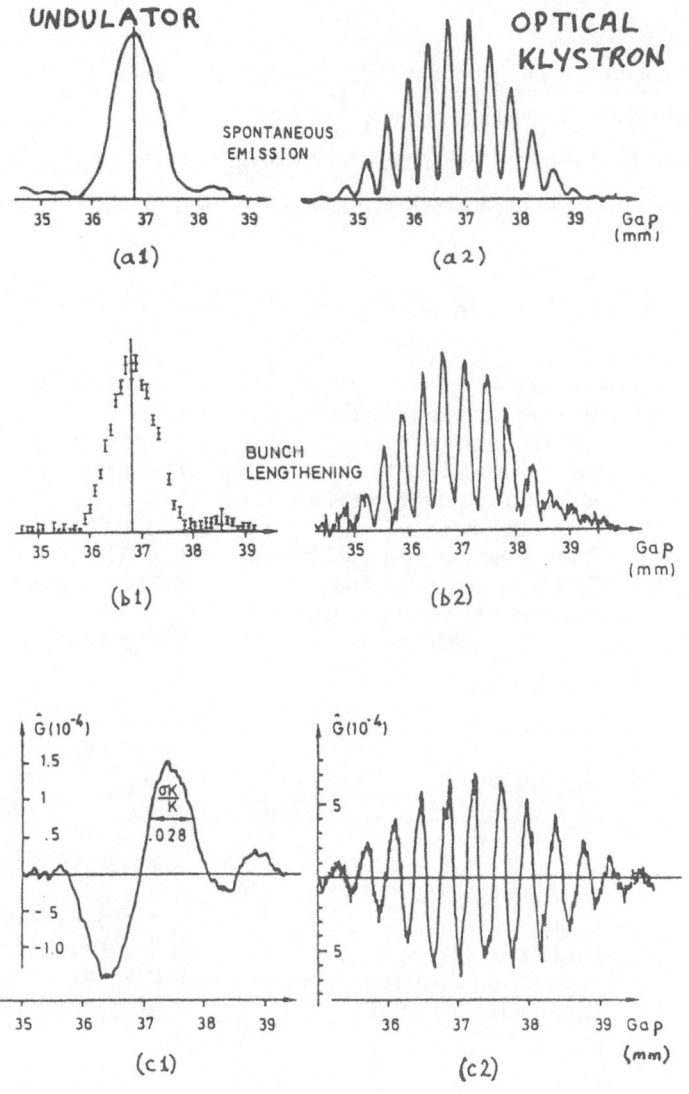

Fig. 18. Comparison of undulator and optical klystron : a1, a2 :
Spontaneous emission ; b1, b2 : laser induced electron
bunch lengthening ; c1, c2 : gain profiles at λ = 4880 A
and λ = 5145 A respectively. Measurements as a function of
undulator gap.

5.4 - FEL experiments at Orsay

FEL experiments carried out at Orsay used the ACO characteristics given in table IV, the optical klystron parameters of table V and the two sets of optical cavity characteristics of table VI. Time structure and Q-switching studies have also been carried out (4a, 48). Orsay FEL performances in three sets of experiments (4c) are listed in table VII.

TABLE IV
ACO characteristics in the FEL experiments

Energy	160-224 MeV
Bunch to bunch distance	11 m
RF frequency	27.2361 MHz
RMS bunch length	0.5 to 1 ns
RMS bunch transverse dimension	0.3 to 0.5 mm
RMS relative energy spread	0.9×10^{-3} to 1.3×10^{-3}
Electron beam lifetime	60 to 90 min
Electron beam current for oscillation	16 to 100 mA

TABLE V
Optical klystron characteristics

Overall length	1.3 m
Type	SmCo permanent magnets
Undulators	2 X 7 periods (λ_w= 78 mm)
K in the FEL experiments	1.1 to 2
Dispersive section	$70 < N_d < 100$

Laser operation necessitates excellent synchronism between the electron bunch revolution frequency in the storage ring and the light pulse round trip frequency in the optical resonator. It is better to fine tune by modification of the RF frequency rather than by mirror translation, so as to avoid backlash and misalignment of the cavity mirrors.

TABLE VI
Optical cavity characteristics

Length   5.5 m

Mirrors : radius of curvature   3m
  TiO_2 /SiO_2 multi-dielectric layers

First set of mirrors (most external layer : SiO_2)
  Mirror transmission : 3×10^{-5}
  Wavelength of maximum Q :   6200 to 6800 A
  Round trip cavity losses at 6500 A :
    First experiment : 7×10^{-4}
    After 40 runs : 17×10^{-4}

Second set of mirrors (most external layer : TiO_2)
  Mirror transmission : 2×10^{-4}
  Wavelength of maximum Q :   5400-6000 A
  Round trip cavity losses at 5800 A :
    First experiment : 16×10^{-4}

TABLE VII : Orsay FEL results

	Experiment 1 (32)	Experiment 2 (48)	Experiment 3 (4c)
Wave-length tunability (A)	6500 $\pm$ 50 (100)	5750-6440(690)	4630-4860(230)
Smallest spectral bandwidth observed (A)	3	1.7	2
Electron energy (MeV)	166-210	180-245	195-233
Laser lifetime after each injection of the ring (h)	1	3-4	2
Ring current (mA)	20-100	25-200	60-100
Max gain per pass for $I \sim 70$ mA (%)	0.18(200 MeV)	0.33(220 MeV)	0.35 (220 MeV)
Mirrors		SiO_2/TiO_2 dielectric layers	
Round trip cavity losses a) before irradiation (%)	0.02	0.04	0.13
b) after the first irradiation of about 2 h (%)	0.07	0.13	0.20
Mirror transmission	3.10^{-5}	1.10^{-4}	8.10^{-5}
Average laser power intracavity (W)	13	20-30	15
exit power per mirror (mW)	0.4	2-3	1.2
total extracted power (mW)	20	60	60
Peak power (27 MHz pulsed laser) intracavity (MW)	0.05	0.5	0.3
exit power per mirror (W)	1.5	20-50	25

128

5.4.1 - Temporal structure of the laser emission

The FEL emits micropulses at the storage ring frequency. The width of the laser micropulse is certainly smaller than the electron bunch length (0.5 - 1ns) but has not yet been measured. A pulsed macro-temporal emission structure has been described theoretically (33) and observed experimentally (4a, 25).In the ACO case, where the electron pulses are 1ns long and the repetition period 37ns, the FEL emission macro-temporal structure has a pseudo-period of 10-20ms and a pulse duration of about 2-3ms (Fig. 19a).The laser time structure is thus similar to the relaxation oscillation of a conventional laser, but here the excursions of the electron beam energy spread about a mean value play a role equivalent to that of the excursions of an inverted population density.

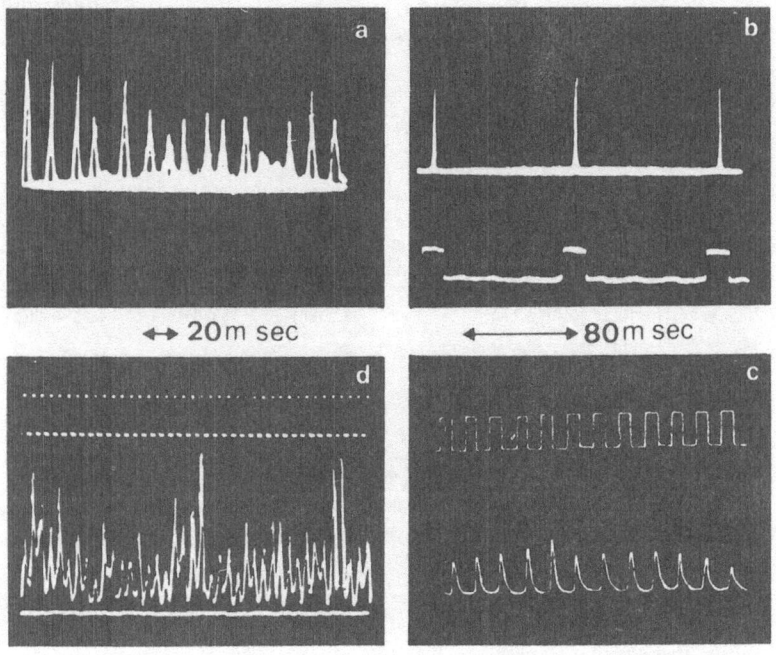

Fig. 19. FEL macro-temporal structure : (a) "natural" operation,
(b) Q-switched operation (f ≈ 12 Hz) ; lower trace = trigger
signal, (c) Q-switched operation (f = 600 Hz), (d) quasi-
continuous operation.

It is difficult to achieve a perfect periodic signal since this requires perfect synchronisation, as discussed above, so that "natural" operation, less than perfect, is somewhat noisy (Fig. 19a). The noise can be reduced considerably by Q-switching, using either of two techniques peculiar to FELs. One is a "detuning" technique in which the optical gain is cancelled by varying slightly the RF frequency f by $(\Delta f/f) \approx 10^{-6}$ (Fig. 19 c). The second method uses electric field modulation of the transverse position of the electron beam. The beam is moved periodically away from the optical resonator axis, thus modulating the gain so that Q-switching occurs. For the Orsay FEL, a 0.5mm displacement decreases the gain by a factor of 4. The gain could be modulated at frequencies 10 Hz - 1 KHz with no change in the average power. 12 Hz and 600 Hz operations are shown in Figs. 19b and 19c respectively. A quasi-continuous regime can be obtained by precise adjustment of the gain reduction or the duty cycle during "high" frequency operation. This is shown in Fig. 19d which corresponds to 770 Hz periodic translation of the electron beam ; in this case the gain was switched between g_{max} and $g_{max}/2$.

A good periodicity of the laser is necessary for physicochemical experiments ; this can be achieved by choosing appropriate FEL conditions as shown above. At low frequency, there is the added advantage of conservation of average power and simultaneous reduction in the temporal width of the pulse. The peak power of a macropulse is thus increased by a factor of about 100 with respect to the average power. For a micropulse emitted by the Orsay FEL, this peak power must be multiplied by a factor of at least 37 i.e. by the ratio between the 37 ns period and the 1ns width of the micropulse.

5.4.2 - Laser power

The FEL power output (table VII) is limited by the saturation mechanisms discussed in section 5.2. Saturation of the FEL gain occurs because the laser-induced electron energy spread creates inhomogenerous broadening effects and lengthening of the electron bunch. From measurements of bunch lengthening and energy spread (49), it was shown that for the Orsay FEL, bunch lengthening it is relatively negligible as a saturation inducer since it decreases the gain by only 1 %, whereas inhomogeneous broadening is a much more important factor, being responsible for a 12 % variation of the gain.

The low power obtained with the FEL in the ACO storage ring is due to several limitations of this ring. It has a small diameter and a small space for insertion of an undulator or optical klystron (L = 1.3 m ; recall that the gain G is proportional to L^3). Furthermore, at the low beam energies used for FEL operations the synchrotron power emitted is relatively small and the electron density is low ($\rho_e \simeq 2 \times 10^{10}$ electrons/cm^3). The beam quality could be improved by the use of positrons rather than electrons in the storage ring.

5.4.3 - Spectral structure

Two spectra are presented in Fig. 20 : (a) was recorded with the optical cavity detuned (zero amplification) and (b) with a tuned cavity. Spectrum (b) shows that the laser operated at three wavelengths. Each of these wavelengths corresponds to a maximum in the gain curve as a function of wavelength. The dominant laser wavelength is at 6476 A. The number of laser wavelengths depends on the gain to loss ratio and would decrease with a smaller ratio.

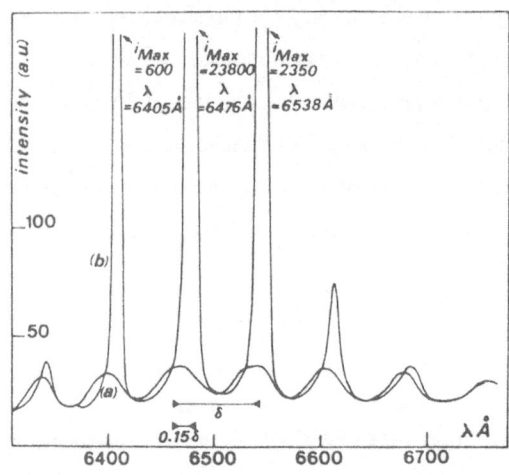

Fig. 20. Spectra of optical cavity output. (a) cavity
detuned, (b) cavity tuned, showing laser oscillation at 3 wavelengths.

The laser line spectral bandwidth has been determined by measurements averaged over 1 second and also by time-resolved spectroscopy with a photodiode array. For different macropulses the line shape is close to a Gaussian with FWHM $\approx$ 2A. The same result was obtained for all wavelengths in the 4630-4830 A region. The spectral width was observed to be related to the microtemporal construction of the laser wave rather than to the theoretically expected Fourier transform of the electron bunch length. Theoretical analysis of the spectral and microtemporal structure of a storage ring FEL (36) indicates that with the Orsay FEL it should be possible to obtain relative spectral widths as small as 10^{-6} by using an intracavity etalon, since the spectral round trip gain curve would then become much narrower.

5.4.4 - Cavity losses from mirror degradation

Because of the low gain intrinsic to the ACO storage ring FEL, it is necessary to use dielectric mirrors of very high reflectivity in the optical resonator. However the latter is subject to losses due to diffusion, absorption and transmission of the mirrors. These losses represent one of the most important parameters determining the FEL threshold and its power output. Furthermore they are time dependent in the sense that the mirror quality is degraded, sometimes rather rapidly, by high energy synchrotron radiation and/or spontaneous VUV and XUV radiation emitted by the undulator or optical klystron.

Several sets of mirrors, whose initial losses per mirror were of the order of 10^{-4}, were experimented with in the Orsay FEL. These dielectric mirrors were made up of 24 to 28 layers, with each layer having an optical thickness equal to $1/2$ or $1/4$ of the central laser wavelength. The layers were TiO_2 and SiO_2 or $Ta_2 O_5$ and SiO_2. Various techniques were used for measuring the mirror losses. They were based on cavity decay time measurements using either an external laser, the synchrotron radiation stored in the cavity, or the FEL itself. Loss variations as low as $10^7 \, s^{1/2}$ could be detected (50). Three main processes of radiation induced degradation of TiO_2/SiO_2 dielectric mirrors were identified, although to a certain extent each mirror appears to have a different history perhaps due to variation of the microstructure of the multilayer coatings. The effects of the radiation damage are complex, not understood in their physico-chemical detail, but their characteristics can be summarised as follows. The first degradation process occurs during initial exposure and is a rapid volume degradation due to low energy photons. The second process corresponds to absorption in the layer stack mainly at the upper SiO_2- air interface ; unlike the first process, it does not recover on annealing. Another reversible volume degradation process occurs at high photon energies and is responsible for considerable losses at the forward mirror. The general results are summarised in table VIII. It was found that SiO_2/TiO_2 dielectric mirrors having SiO_2 as an external layer are much less subject to radiation-induced degradation than SiO_2/TiO_2 mirrors with a TiO_2 external layer or to SiO_2/Ta_2O_5 mirrors. Furthermore, a fused silica Brewster window was observed to be much less radiation resistant than any of these dielectric mirrors. The data reported in table VII were obtained with the SiO_2/TiO_2 - SiO_2 outer layer mirrors.

TABLE VIII. - Orsay FEL : Three Processes of Degradation Observed on TiO_2/SiO_2 Dielectric Mirrors

N°.	Nature	Energy of incriminated photons	Effect of annealing	Final absorption losses		Losses vs wavelength
				Forward mirror $(x10^{-4})$	Rear mirror $(x10^{-4})$	
1	Volume	Several eV	Recover	10	6	Uniform in λ
2	Surface	$\gtrsim 30$ eV	Unchanged	40 a	25 a	Minimum at $\lambda = 6500$ A
3	Volume	$\gtrsim 160$ eV	Recover	230	<10	Uniform in λ

a Measured at $\lambda = 6328$ A.

It is clear that obtaining laser oscillation at short wavelengths (4630 A obtained at Orsay is the shortest FEL wavelength reported so far) will depend on overcoming mirror degradation. Most storage rings are designed to operate at E > 500 MeV where the VUV flux is high, with the consequent tendency to enhance radiation damage of FEL cavity mirrors. An extensive research programme on radiation damage of mirrors and other optical elements necessary for a FEL has been undertaken by Deacon. His recent analysis of the situation (51) includes a discussion of the severe problems that would arise with the otherwise very attractive high power tapered wiggler FEL systems. The emphasis is on finding ways of reducing the causes of energy deposition into the substrates and coatings of the optical elements as well as devising means of reducing the thermal effects of the absorbed energy.

5.4.5 - Super-ACO and its FEL

The intrinsic limitations of the ACO FEL have been mentioned earlier. Much improvement is expected with a FEL designed for operation with the Super-ACO storage ring under construction at Orsay. Super-ACO parameters are given in table IX which also includes the relevant parameters used for calculations of harmonic generation (31) discussed in section 4.2. It is a low emittance ring and although its nominal energy is 800 MeV it will be possible to operate it down to at least 400 MeV. The energy spread, proportional to the beam energy, can thus be minimized, which will lead to an increase in the optical gain of the FEL. The 500 MHz RF cavity frequency will enable electron bunches of less than 100 ps to be produced. Furthermore the ring can be injected with positrons instead of electrons which will prevent an increase in the transverse sizes of the beam with the storage ring current and improve the beam half-life.

TABLE IX

(i) Characteristics of the Super-ACO storage ring

Overall length	72 m
Nominal energy	800 MeV
Energy range	400-800 MeV
RF frequency	100 and 500 MHz
Injection	with e^+ or e^-
Energy spread (800 MeV)	5×10^{-4}
Damping time (800 MeV)	16 ms

(ii) Parameters used for Super-ACO in VUV harmonic generation calculations (31)

	400 MeV	800 MeV
Bunch sizes σ_x (mm)	0.25	0.32
σ_y (mm)	0.25	0.32
σ_z (mm)	2.7	7.6
Single bunch stored current (mA)	6	20
Energy spread	2.5×10^{-4}	5×10^{-4}
Touscheck lifetime	1 h	2.5 h
Emittance ε_x	2.8×10^{-8}	3.7×10^{-8}

Super-ACO has eight bending magnets, each of which has two beam lines ; each beam line is capable of having three exit ports. There are six straight sections for insertion devices. Undulators, optical klystrons and high K wigglers will be inserted, including an optical klystron/undulator for the FEL experiments. Typical characteristics for the optical klystron are as follows : overall length 3.3 m ; magnetic period $\lambda_w = 12$ cm; number of periods: 2×10; maximum field index K = 6 ; dispersive section strength $N_d = 0 - 300$. With such an optical klystron, and under the operating conditions of table IX, the expected wavelength range for the FEL will be 1200 - 7000 A, with beam energies varying from 400 to 800 MeV. As discussed in section 4, it will be possible to increase the photon energy by the generation of harmonics down to at least 300 A : pulses of the order of 10^8 photons have been estimated for $\lambda = 273$ A. Super-ACO is expected to be fully operating in 1988 ; current in the ring was obtained, and fully up to design expectations, in February 1987.

6 - ELECTRON ACCELERATORS

The electrons entering the FEL should have optimum characteristics for maximizing the gain (cf eqn.12) and avoiding effects leading to line broadening. The desirable electron parameters can be resumed as follows : low emittance and energy spread, high peak current, proper bunch length in the micropulse, stability during the macropulse. Thus the feasibility of FELs is intimately linked to the quality of electron accelerator technology. It is only relatively recently that attention has been paid to the development of higher current electron beams, rather than the higher energy beams useful for collision physics. The most common accelerator devices, their typical beam energies and useful wavelength ranges of FELs using these as electron sources, are as follows :

1) Electron storage rings, E > 200 MeV, $\lambda = 500$ A-1 μm

2) Radiofrequency LINAC : E > 20 MeV, $\lambda = 1$-20 μm

3) Induction Linac : E ≈ 0.1 - 50 MeV , $\lambda = $ mm-cm range

4) Van der Graaff electrostatic accelerator : E < 10 MeV, λ = 1μm-1mm

5) Classical Microtron : E < 40 MeV, λ = 10-500 μm

6) Racetrack Microtron : E = 10-500 MeV, λ = 10-500 μm

7) Pulse-Line : E $\approx$ 0.1-5 MeV, λ = 80-3000 μm

The induction linac consists of an injector and a series of acceleration modules each of which submits the electron beam to a relatively constant electrostatic accelerating voltage, whereas the basic principle of a radiofrequency linac is the operation on the electron of an electromagnetic wave that, on the average, presents a constant phase to the electron (52). The pulsed induction linac is a single shot device, capable of a pulse duration of about 2 μs, a duty cycle of 100 s^{-1} and currents in the kA range. Microtrons, which are related to RF Linacs, are compact electron-beam sources (53). The circular microtron has a single homogeneous magnetic field (0.1-0.2 T) and a single cavity resonator for the acceleration. The recently developed racetrack microtron uses two magnets and has a field-free region between them that modifies the resonance condition such that the energy per revolution, and therefore the magnetic field strength, can be increased at will ; acceleration takes place in a linear accelerator (54). The pulse-line accelerator uses a Marx generator to drive a pulse-forming transmission line, which provides a square pulse output to a matched load, the accelerator diode. Pulses that are flat over 100 ns can be achieved. It should be noted that typical electron pulse lengths are 3-50 ps for linacs and microtrons, 0.1-3ns for storage rings and of the order of 50 μs for Van der Graaff machines. The longer the electron pulse, the narrower will be laser line produced by a FEL. Further characteristics of these and other accelerators have either been mentioned earlier or will be noted in a presentation of FEL devices in the next section.

7 - FREE ELECTRON LASER FACILITIES

Table X lists Free Electron Laser Facilities and their typical parameter values.

TABLE X. - Actual and projected Free Electron Lasers, their typical parameters and beam technologies (EBT)

Laboratory	Beam Energy (MeV)	Beam Current a (amp)	Peak Power (Oscillator Mode)	Wavelength	Notes	Reference
EBT : RF Linac						
Stanford Univ.	24	0.1		10.6 μm		42
Stanford Univ.	43	1.3	130 KW	3.3 μm	b	39
Los Alamos	20	10		10.6 μm		55
Los Alamos	20	30-60	10 MW	9-35 μm		56,57
MSNW/Boeing	20	5		10.6 μm		58
TRW	25	10		10.6 μm		59
TRW/Stanford	66	0.5-2.5	1.2 MW	1.6 μm		60
NRL	35	5	17 MW	16 μm		5,61
Boeing/Spectra T.	120	100	0.6 GW	0.5 μm	c	62,63
Glasgow	40-160	2.5		2-20 μm		64,65
Orsay	50		1 MW	1-15(50)μm		66
Stanford Univ.	43	240	5 MW	2.6-3.1 μm		67,68
EBT : Storage ring						
Orsay	240	2	13 W	0.5 μm	d	4,69
Orsay	240	2	50 W	0.5 μm	e	32,48
Frascati	600	10		0.5 μm		61,70,71
Novosibirsk	340			6 μm		5,61
Brookhaven	500	108		0.35 μm	f	5,61
INP- Novosibirsk	370			0.6 μm		72
Stanford Univ.	1000	270	100 MW	< 1000 A		73
Livermore	50			mm-I.R:visible		5,61
Berkeley	500-750		150 MW	400-1000 A	g	74

TABLE X. - Continued

Laboratory	Beam Energy (keV)	Beam Current [a] (amp)	Peak Power (Oscillator Mode)	Wavelength	Notes	Reference
EBT : Microtron						
Bell Labs	10-20	5	100 KW	100-400 μm		75
Frascati	20-35	0.6		10-35 μm		76,77
NBS	20-200	1.0	0.1-1 KW	100 μm-1500 A	h	78,79
Darmouth/Frascati	5	2		10-100 μm	i	80
			10-200 KW	375 μm-1mm	i	81
EBT : Induction Linac						
Livermore	3.6	850	1.9 GW	8.7 mm		82,83
NRL	0.7	200		8 mm		5,61
Livermore	5	400	80 MW	3-6 mm		5,61
EBT : Pulse line						
NRL	2	30 000		0.4 mm		84
NRL	1.35	1 500	75 MW	4 mm		84,85
NRL/Columbia	1.2	25 000		0.4 mm		5,61
Columbia Univ.	0.86	5 000	4 MW	1.7-3.5 mm		86
Columbia Univ.	0.9	10 000		0.6 mm		5,61
MIT	1	5 000		3 mm		61
Ecole Polytechnique (Palaiseau)	1	2 000		2 mm		61
Osaka	0.6	40 000		4 mm		87

TABLE X. - Continued

EBT : Electrostatic accelerator

Laboratory	Beam Energy (MeV)	Beam Current[a] (amp)	Peak Power (Oscillator Mode)	Wavelength	Notes	Reference
UCSB	3	2	8 KW	0.36 mm	j	88,89
UCSB	6	2		0.1-1 mm		90
UCSB	6	10,20	2 MW		k	91
UCSB	1			703 1.08 μm sub-mm	l	92
UCSB	25			I.R., vis	l	92
KMS Fusion	3				m	93
Rehovot	6	1		200 μm	n	94

Notes :

a : Peak current except for Pulse-Line accelerators

b : First FEL operation (1977)

c : Tapered wiggler, design goal 5 % extraction of energy

d : First FEL in visible (1983)

e : Shortest wavelength operation 4630 Å

f : Projected wavelength 3500 Å. Not achieved because of inadequate electron beam characteristics.

g : By-pass FEL in storage ring

h : c.w. electron accelerator. Racetrack Microtron

i : Cerenkov laser

j : pulses up to 50 μs duration

k : 2-stage FEL.

l : Micro-undulator, gap ≈ 3 mm

m : Design for a 2-stage FEL with cylindrically symmetric quasi-optical cavity for long (100-100 μm) and short (1-10 μm) wavelengths.

n : Design to produce many harmonics of $\lambda_1 = 200$ μm

At the present time about a twenty of these FELs have operated successfully as lasers in the oscillator mode or as amplifiers of an external source. Others are in various stages of conception, design and achievement ; some of them may have gone into action before this text appears.

The FELs in Table X are grouped as a function of electron beam technology, which largely determines the spectral range for amplification and oscillation and whether they are repetitive or single pass devices. Storage rings inherently recirculate the electrons whereas the other beam sources operate in the pulsed mode, usually at a frequency of several Hz. Recirculation or beam recovery devices have been conceived and constructed for a number of the latter. Notes to table X give some additional information for particular FELs. Detailed accounts of each FEL can be obtained from the cited references. This compilation, although certainly incomplete, gives a picture of the FEL situation as of Easter 1987. Theoretical work on FELs, as well as other proposed FEL technologies or projects can be found in references 5a-5e.

Further remarks on some of the FEL projects in table X are in order.

A development programme at Stanford seeks to build a broadly tuneable, medium power infra-red FEL based on existing small-scale linear accelerator technology. The long-term objective is to be able to furnish a general purpose tuneable FEL to laboratories inexperienced in avanced accelerator techniques. The technical aims of the Mark III infra-red FEL built at Stanford (67), operating with the 20 year-old Mark III Linear Accelerator (E = 43 MeV, I = 240 mA) are as follows : wavelength range 2-10 μm (achieved 2.6-3.1 μm), 500 mJ/pulse (achieved 140 mJ/pulse), 60 Hz operation, macropulse 10 μs (achieved 3.5 μs), high peak power (achieved $\approx$ 5 MW), gain achieved per traversal 28 % with a 2.5 μs macropulse. The device is compact, with an optical cavity 178.5 cm long and an undulator of length 125 cm constructed from hybrid permanent magnets with 47 periods of λ_w = 2.3 cm. The peak field is B_0 = 0.7 T.

An analogous project (CLIO) is being studied at Orsay (66). The RF-Linac will be built specially for FEL operation, with an adjustable energy E = 30-65 MeV. A wavelength range 1-15 μm is expected to be covered in the first instance, with later extension to 50 μm. Some of the characteristics of CLIO are as follows : peak power 1MW, energy macro-pulse $\simeq$ 1J, energy/micro-pulse $\simeq$ 0.2 mJ, macro-pulse length $\approx$ 10 µs micro-pulse length 10 ps, micro-pulse interval 2 ns, repetition rate 50 Hz.

The Stanford XUV-FEL project (73) seeks to achieve oscillation at λ < 1000 A. Because the gain drops as wavelength decreases, and also in view of the mirror problems associated with an optical resonator operating in the VUV, it is necessary for the FEL to operate in the single pass gain mode, with a gain greater than 3, so that a long undulator and a high quality electron beam are required. The Stanford X-ray center (SXRC) 1 GeV storage ring is the first constructed specifically for a FEL. The hybrid undulator is constructed of rare earth cobalt/vanadium permendur. Its length is 27 m, with a minimum gap g = 30 mm, and periods λ_w = 6.4 - 11.4 cm are envisaged, with 422 to 237 periods respectively, and peak fields of 0.38 - 0.89 T. Oscillation should be feasible down to 220A with a 6.4 cm undulator period, but with only a small tuning range.

In the 2-stage FEL project at the University of California, Santa Barbara (91), the first stage produces radiation at 703 µm which then acts as equivalent to a short period undulator that interacts with the electron beam to generate the near I.R. at 1.08 µm. The quality of the electron beam -electron energy spread, emittance- is crucial to the success of this project.

The Electron Laser Facility (ELF) project at Livermore (82,83) uses a 3.6 MeV, 850 A electron beam produced by the Experimental Test Accelerator (ETA) which is a linear induction accelerator. It has operated as an amplifier at 34.6 GHz, but saturating at an undulator-length of 2.2 m (B_0 = 0.43 T). The gain up to saturation is $\approx$ 15.6 dB/m. Recent experiments give a peak microwave power of $\approx$ 1.9 GW at 34.6 GHz. In the future it is planned to use the 50 MeV Advanced Test Accelerator and so up-grade the photon energy to the infrared and visible, using a beam current of 3 kA in a single-pass FEL.

The advantage of the by-pass arrangement in the Berkeley high power XUV project (74), in which the FEL is in a by-pass in the storage ring, is to avoid degrading the electron beam as would happen with an undulator in a normal straight section of the ring. The undulator will have the following characteristics for FEL operation in the 400-1000 A region: λ_w = 1.85 - 2.91 cm, K = 2.50 - 5.26, peak magnetic field 1.45 - 1.93 T. At 400 A the peak power is calculated to be $\approx$ 150 MW, with an average power $\approx$ 0.3 W.

Boeing/Spectra Technology are building a FEL with the following characteristics (62,63) : RF linac, E = 120 MeV, peak current 100 A in 20 ps micropulses, macropulses 100 μs at 50 ms intervals. The hybrid (tapered) undulator will have L = 5 m, λ_w = 2.18 cm, g = 4.8 mm, B $\leqslant$ 1.02 T. The expected gain varies from 20 % to 400 % according to magnetic field conditions. The optical cavity of 55.4 m will operate as a FEL at 0.5 μm, with an average power of a few hundred watts, a peak power of 0.6 GW and a design goal of 5 % extraction.

A broad spectrum of FEL wavelengths, from λ = 100 μm to λ = 0.15 μm is expected from the NBS Racetrack Microtron FEL at NBS which is under construction (78,79). One of the touted applications is for medical research. The racetrack microtron operates at 185 MeV with a continuous wave electron accelerator. The electrons can be deflected at several points along their trajectory, so as to produce beams of various desired energies. The full 185 MeV corresponds to the complete maximum trajectory, and is suitable for the shortest wavelength to be achieved in the FEL. Undulator characteristics to cover the expected wavelength range are as follows : L = 5 - 8 m, λ_w = 20 - 2.4 cm, B_0 = 0.1 - 0.63 T. Average powers in the hundreds to thousands of watts are anticipated.

8 - SUMMARY OF FEL PROPERTIES AND APPLICATIONS

The properties of FELs which are considered to be possible with existing or foreseen improved technology can be summarized as follows. The operating wavelength is a continuous function of the electron energy and the magnetic field in the undulator; the range is from the mm region to the vacuum ultraviolet, with λ = 500 A being a lower wavelength limit beyond which it would be difficult to achieve sufficient gain using an optical resonator. It should be possible to tune the emitted radiation over about one octave in the I.R., visible and U.V. regions. The best electron source varies with wavelength region: storage rings in the visible and U.V., microtrons in the I.R., and Van de Graaf machines in the mm region. Average powers greater than 1 KW have been obtained and much higher values, MW and even GW, are reached or contemplated with improved technology ; overall net energy conversion efficiencies of the order 1-30 % are achieved or expected. In the case of a FEL operated with an electron beam from an RF accelerator the optical cavity length must be set to a multiple of the electron bunch spacing; the optical field in such a FEL is mode-locked by the periodic variation in electron current ; picosecond pulses can be obtained at high repetition rates (10 MHz - 1 GHz). All the usual laser techniques used to minimize linewidths can be applied to the FEL so that it should be possible to obtain individual lines with $\Delta \nu <$ 1MHz. An important advantage of the FEL is the possibility for it to operate simultaneously at different frequencies, or with a set of independent undulators. Furthermore, a storage ring FEL also generates significant amounts of broadband incoherent synchrotron radiation in addition to the emitted (laser) coherent radiation ; this SR could be of interest for independent experiments or for use as a secondary source in conjunction with the laser.

Some possible uses for free electron lasers are summarized as follows. 1) Far infrared : interesting wavelength range ; useful for saturation pumping, multiphonon processes, surface state spectroscopy, solid state spectroscopy and photophysics. 2) U.V. - V.U.V. : advantage can be taken of flexibility, power density, polarisation and wavelength range for studies on spectroscopy, multiphoton ionization, high energy-vibronic states, low density targets such a transients e.g. radicals, ions, muonium, positronium. 3) Pump and probe experiments could be carried out by FEL + S.R., FEL + separate laser, or FEL operating in multifrequency mode. 4) Study of laser assisted collisions and general photochemistry, in particular state and/or species selective photochemistry using the multifrequency resources. 5) Industrial applications of photophysical and photochemical processes. Some other scientific and technical applications are mentioned in ref. 5a.

Another suggested application of the FEL is as a driver for inertial confinement fusion (95). There are also prospective medical applications of FELs. These reflect the desire to achieve specific effects on particular types of biological tissue without affecting neighbouring structures. The promise is in the areas of laser surgery and in photoradiation medicine. The FEL qualities of great interest to these ends are the possibilities of varying the power, wavelength, spot size and pulse width (96). These are important qualities for situations where the skin or other tissue absorption depth is a critical factor.

Finally it is not inappropriate to mention the lunatic fringe (reaching for the moon ?) of FEL applications. This concerns the proposed use of the FEL as an offensive weapon in the context of the U.S. Strategic Defense Initiative (SDI) (97,98). The military application envisaged is as a ground-based directed energy weapon to destroy ballistic missiles during their boost and early midcourse trajectory, using relay mirrors in space. This is being investigated at the Lawrence Livermore National Laboratory (LLNL) and the Los Alamos National Laboratory (LANL), whose design concepts use the Induction Linear Accelerator (IND-LINAC) and Radiofrequency Linear Accelerator (RF-LINAC) techniques, respectively, for creation of the accelerated electron beam. A joint venture between Boeing Aerospace and Spectra Technology is also using the LINAC technique with same aim as LLNL and LANL i.e. to be chosen to build a FEL at the White Sands Missile Range in New Mexico. The objective of this US Army facility is research on the construction and operation of a network of ground-based centres, each containing about six FELs. Each FEL, designed to radiate a near infrared photon beam with several hundred MW average power, is estimated to cost about 10^9 US dollars. In order to avoid damage to optical components, it is planned to allow each IND-LINAC based FEL photon beam to expand in a long vacuum pipe, several km long. It has been estimated (99) that for strategic defence applications, a ground-based FEL should produce an average power of at least 1 gigawatt at 1 μm wavelength, corresponding to peak powers of 0.1 to 1 terawatts.

The critical propagation problems which are being addressed at the various research centres are : Stimulated Raman scattering by air molecules ; thermal blooming, due to laser heating of the air ; electrical breakdown of the air, which could generate plasmas that attenuate the beam ; atmospheric turbulence, which could perturb the focusing of the laser beam.

The idea of sending a FEL into space, along with its associated accelerator and undulator, and so avoiding some of the optical problems, has been proposed but is difficult to imagine as a practical solution.

A sober, and sobering, assessment of the possibility of military application of FELs (and other lasers, as well as other possible directed energy weapons) has been given very recently (99).

We end with a note of caution in stressing (i) that the output of theoretical work on free-electron lasers still exceeds the experimental activity ; (ii) that much technological improvement is required before the FEL properties discussed in this and previous sections become completely effective. Feasibility studies are in progress in many FEL areas and the direction of the field and its real possibilities have become more clear over the past three years (cf. ref. 1). Rapid progress is envisaged in FEL development under the impetus of scientific and publicity factors whose nature is evident from this summary section.

ACKNOWLEDGEMENTS : Thanks are due to J.M. Ortega and M. Velghe for useful comments on the manuscript.

REFERENCES

1 - S. Leach in "Laser Applications in Chemistry", eds.K.L. Kompa and J. Wanner, Plenum, N.Y. (1984) a) p.35 b) p.47.

2 - T.C. Marshall, "Free Electron Lasers", MacMillan, N.Y. (1985)

3 - W.B. Colson and A.M. Sessler, Ann. Rev. Nucl. Part. Sci. 35, 25 (1985)

4 - a) M. Billardon, P. Elleaume, J.M. Ortega, C. Bazin, M. Bergher, Y. Petroff and M. Velghe, I.E.E.E. J. Quant. Elec. QE-21, 805 (1985)
b) J.M. Ortega, M. Billardon, G. Jezequel, P. Thiry and Y. Petroff, J. Physique 45, 1883 (1984)
c) M. Billardon, P. Elleaume, J.M. Ortega, C. Bazin, M. Bergher, M.E. Couprie, Y.Lapierre, R. Prazeres, M. Velghe and Y. Petroff, Europhysics Lett. 3, 689 (1987)

5 - a) "Applications of Free Electron Lasers", Nucl. Inst. Meth. 237, 371-433 (1985)
b) I.E.E. J. Quant. Elec. QE-21, pp 804-1119 (1985)
c) Int. Q. Electronics Conference (Abstracts) J. Opt. Soc. Am. 3, n° 8 (August 1986)
d) Proc. 7th Int. Conf. Free Electron Lasers, Nucl. Inst. Meth. A250, 1-490 (1986)
e) Proc. 8th Int. Conf. Free Electron Lasers, Nucl. Instr. Meth. A259, 1-316 (1987)

6 - D. Ivanenko and I. Pomeranchuk, Phys. Rev. 65, 343 (1944)

7 - J. Schwinger, Phys. Rev. 70, 798 (1946) ; 75, 1912 (1949) ; Proc. Nat. Acad. Sci. USA 40, 132 (1954)

8 - J. Jortner and S. Leach, J. Chim. Phys. 77, 7 (1980)

9 - P. Dagneaux, C. Depautex, P. Dhez, J. Durup, Y. Farge, R. Fourme, P.M. Guyon, P. Jaegle, S. Leach, R. Lopez-Delgado, G. Morel, R. Pinchaux, P. Thiry, C. Vermeil and F. Wuilleumier, Ann. Phys. (Paris) 9, 9 (1975)

10 -J.E. Spencer and H. Winick in "Synchrotron Radiation Research", ed. H. Winick and S. Doniach, Plenum, N.Y. (1980) p. 663

11 -J.M. Ortega, C. Bazin, D.A.G. Deacon, C. Depautex and P. Elleaume, Nucl. Instr. Meth. 206, 281 (1983)

12 -K. Halbach, Nucl. Instr. Meth. A187, 109 (1981) ; J. Physique (Paris) 44, C1-211 (1983)

13 -H. Motz, J. Appl. Phys. 22, 527 (1951)

14 -D.F. Alferov, Yu A. Bashmakov and E.G. Bessonov, Sov. Phys.Tech. Phys. 18, 1336 (1974)

15 -A. Hofmann, Nucl. Inst. Meth. 152, 17 (1978)

16 -L.D. Landau and E.M. Lifshitz, Electrodynamics of Continuous Media, Pergamon, Oxford (1966) p.242

17 -A. Bambini, A. Renieri and S. Stenholm, Phys. Rev. A19, 2013 (1979)

18 -W. Colson, Phys. Lett. 64A, 190 (1977)

19 -European Synchrotron Radiation Facility, Supplement II : "The machine" eds. D.J. Thomson and M.W. Poole, European Science Foundation, Strasbourg (1979) p. 52 et seq.

20 -European Synchrotron Radiation Facility : Foundation Phase Report, ESRF, Grenoble (1987)

21 -N.A. Vinokurov and A.N. Skrinsky, Preprint INP77-59, Novosibirsk (1977) ; N.A. Vinokurov, Proc. 10th Int. Conf. High Energy Charged Particle Accelerators, Serpukhov, vol.2, 454 (1977)

22 -P. Elleaume, Physics of Quantum Electronics, vol 8, Addison-Wesley, N.Y. (1982), chap. 5

23 -P. Elleaume, Nucl. Instr. Meth. A250, 220 (1986)

24 -G.A. Kornyukhin, G.N. Kulipanov, V.N. Litvinenko, N.A. Mesentsev, A.N. Strinsky, N.A. Vinokurov and P.D. Voblyi, Nucl. Instr. Meth. A237, 281 (1985)

25 -M. Billardon, P. Elleaume, J.M. Ortega, C. Bazin, M. Bergher, Y. Petroff and M. Velghe, Nucl. Instr. Meth. A237, 244 (1985)

26 -M. Billardon, D.A.G. Deacon, P. Elleaume, J.M. Ortega, K.E. Robinson, C. Bazin, M. Bergher, J.M.J. Madey, Y. Petroff and M. Velghe, J. Physique Colloque 44, C1-29 (1983)

27 -Report of the Workshop on an Advanced Soft X-Ray and Ultraviolet Synchrotron Source : Applications to Science and Technology, November 13-15 (1985), Lawrence Berkeley Laboratory PUB - 5154 (1985)

28 -B. Girard, Y. Lapierre, J.M. Ortega, C. Bazin, M. Billardon, P. Elleaume, M. Bergher, M. Velghe and Y. Petroff, Phys. Rev. Lett. 53, 2405 (1984)

29 -R. Prazeres, J.M. Ortega, C. Bazin, M. Bergher, M. Billardon, M. E. Couprie, H. Fang, M. Velghe and Y. Petroff, Europhysics Letters 4, 817 (1987)

30 -Physics Today, (November 1984) p. 19

31 -J.M. Ortega, Nucl. Inst. Meth. A250, 203 (1986)

32 -M. Billardon, P. Elleaume, J.M. Ortega, C. Bazin, M. Bergher, M. Velghe, Y. Petroff, D.A.G. Deacon, K.E. Robinson and J.M.J. Madey, Phys. Rev. Lett. 51, 1652 (1983) ; P. Elleaume, J.M. Ortega, M. Billardon, C. Bazin, M. Bergher, M. Velghe, Y. Petroff, D. Deacon, K. Robinson and J. Madey, J. Physique (Paris) 45, 989 (1984)

33 -P. Elleaume, J. Physique (Paris) 45, 997 (1984)

34 -V.L. Ginzburg, Izv.Akad. Nauk (USSR), ser. Phys. 11, 165 (1947)

35 -M.V. Fedorov, Prog. Quant. Electr. 7, 73 (1981)

36 -P. Elleaume, Nucl. Instr. Meth. A237, 28 (1985) ; W.B. Colson, ibid p.1

37 -P.L. Kapitza and P.A.M. Dirac, Proc. Cambridge Phil. Soc. 29, 297 (1933)

38 -J.M.J. Madey J. Appl. Phys. 42, 1906 (1971)

39 -D.A.G. Deacon, L.R. Elias, J.M.J. Madey, G.J. Ramian, H.A. Schwettman and T.I. Smith, Phys. Rev. Lett. 38, 892 (1977)

40 -T. Kwan, J.M. Dawson and A.T. Lin, Phys. Fluids 20, 581 (1977)

41 -D.B. Mc Dermott, T.C. Marshall, S.P. Schlesinger, R.K. Parker and V.L. Granatstein, Phys. Rev. Lett. 41, 1368 (1978)

42 -L.R. Elias, W.M. Fairbank, J.M.J. Madey, M.A. Schwettman and T.I Smith Phys. Rev. Lett. 36, 717 (1976)

43 -C. Bazin, M. Billardon, D. Deacon, Y. Farge, J.M. Ortega, J. Pérot, Y. Petroff and M. Velghe, J. Physique Lett. 41, L 547 (1980)

148

44 -D.A.G. Deacon, J.M.J. Madey, K.E. Robinson, C. Bazin, M. Billardon, P. Elleaume, Y. Farge, J.M. Ortega, Y. Petroff and M. Velghe, I.E.E.E. Trans. Nucl. Sci. NS-28-3142 (1981)

45 -D.A.G. Deacon, K.E. Robinson, J.M.J. Madey, C. Bazin, M. Billardon, P. Elleaume, Y. Farge, J.M. Ortega, Y. Petroff and M. Velghe, Opt. Comm. 40, 373 (1982)

46 -J.M.J. Madey, Nuov. Cim. 50B, 64 (1979)

47 -D.A.G. Deacon and M. Xie, I.E.E.E. J. Quant. Electr. QE-21, 939 (1985)

48 -M. Billardon, P. Elleaume, J.M. Ortega, Y. Lapierre, Y. Petroff, M. Bergher, C. Bazin, and M. Marilleau, Nucl. Instr. Meth. A250, 26 (1986)

49 -J.M. Ortega, P. Elleaume, M. Billardon, D.A.G. Deacon, B. Girard and Y. Lapierre, Nucl. Instr. Meth. A237, 254 (1985)

50 -P. Elleaume, M. Velghe, M. Billardon and J.M. Ortega, Appl. Opt. 24, 2762 (1985)

51 -D.A.G. Deacon, Nucl. Instr. Meth. A250, 283 (1986)

52 -R.K. Cooper, P.L. Morton, P.B. Wilson, D. Keefe and A. Faltens, J. Physique (Paris) 44, C1-185 (1983)

53 -S. P. Kapitza and V.N. Melekhin, "The Microtron", (ed. E.M. Rowe) Harwood, London (1978)

54 -S. Rosander, J. Physique (Paris) 44, C1-233 (1983)

55 -R.W. Warren, B.E. Newnam, L.M. Young, W.E. Stein, J.G. Winston and C.A. Brau, I.E.E.E. J. Quant. Elect. QE-19, 391 (1983)

56 -B.E. Newnam, R.W. Warren, R.L. Sheffield, W.E. Stein, M.T. Lynch, J.S. Fraser, J.C. Goldstein, J.E. Sollid, T.A. Swann, J.M. Watson and C.A. Brau, I.E.E.E. J. Quant. Electr. QE-21, 867 (1985) ; J.M. Watson, Nucl. Instr. Meth. A250, 1 (1986)

57 -J.C. Goldstein, B.E. Newnam, R.W. Warren and R.L. Sheffield, Nucl. Instr. Meth. A250, 4 (1986)

58 -W.M. Grossman, J.M. Slater, D.C. Quimby, T.L. Churchill, J. Adamski, R.C. Kennedy and D.R. Shoffstall, Appl. Phys. Lett. 43, 745 (1983)

59 -H. Boemher, M.Z. Caponi, J. Edighoffer, S. Fornaca, J. Much, G.R. Neil, B. Saur and C. Shih, Phys. Rev. Lett. 48, 141 (1982)

60 -J.A. Edighoffer, G.R. Neil, C.E. Hess, T.I. Smith, S.W. Fornaca and H.A. Schwettman, Phys. Rev. Lett. 52, 344 (1984)

61 -P. Sprangle and T. Coffey, Physics Today, (March 1984) p.44

62 -J. Slater, T. Churchill, D. Quimby, K. Robinson, D. Shemwell, A. Valla, A.A. Vetter, J. Adamski, W. Gallagher, R. Kennedy, B. Robinson, D. Shoffstall, E. Tyson, A. Vetter and A. Yeremian, Nucl. Instr. Meth. A250, 228 (1986)

63 -K.E. Robinson, T.L. Churchill, D.C. Quimby, D.M. Shemwell, J.M. Slater, A.S. Valla, A.A. Vetter, J. Adamski, T. Doering, W. Gallagher, R. Kennedy, B. Robinson, D. Shoffstall, E. Tyson, A. Vetter and A. Yeremian, Nucl. Instr. Meth. A259, 49 (1987)

64 -W.A. Gillespie, P.F. Martin, M.W. Poole, G. Saxon, R.P. Walker, J.M. Reid, M.G. Kelliher, C.R. Pidgeon, S.D. Smith, W.J. Firth, D.A. Jaroszinski, D.M. Tratt, J.S. Mackay and M.F. Kimmitt, Nucl. Instr. Meth. A250, 233 (1986)

65 -C.R. Pidgeon, D.A. Jaroszynski, D.M. Tratt, S.D. Smith, W.J. Firth, M.F. Kimmitt, C.W. Cheng, M.W. Poole, G. Saxon, R.P. Walker, J.S. Mackay, J.M. Reid, M.G. Kelliher, E.W. Laing, D.V. Land and W.A. Gillespie, Nuc. Instr. Meth. A259, 31 (1987)

66 -J.M. Ortega, Y. Petroff, J.C. Bourdon, P. Brunet, J. Courau, J.L. Malglaive, P. Carlos and C. Hezard, LURE Report 1985-1987, Orsay, p. 479 (1987)

67 -S.V. Benson, J.M.J. Madey, J. Schultz, M. Marc, W. Wadensweiler, G.A. Westenskow and M. Velghe, Nucl. Instr. Meth. A250, 39 (1986)

68 -T.I. Smith, H.A. Schwettman, R. Rohatgi, Y. Lapierre and J. Edighoffer, Nucl. Instr. Meth. A259, 1 (1987)

69 -M. Billardon, P. Elleaume, J.M. Ortega, C. Bazin, M. Bergher, M.E. Couprie, Y. Lapierre, Y. Petroff, R. Prazeres and M. Velghe, Nucl. Instr. Meth. A259, 72 (1987)

70 -R. Barbini, G. Vignola, S. Trillo, S. De Simone, S. Faini, S. Guiducci, M. Preger, M. Serio, B. Spataro, S. Tazzari, F. Tazzioli, M. Vescovi, A. Cattoni, C. Sanelli, M. Castellano, N. Cavallo, F. Cevenini, M.R. Masulo, P. Patteri, R. Rinzivillo, A. Cutolo and S. Solimeno, J. Physique (Paris) 44, C1-1 (1983)

71 -M. Ambrosio, G.C. Barbarino, M. Castellano, N. Cavallo, F. Cevenini, M.R. Masulo, P. Pateri, M. Preger and A. Cutola, Nucl. Instr. Meth. A250, 239 (1986)

72 -See reference 24 and literature cited therein.

73 -J.E. La Sala, D.A.G. Deacon and J.M.J. Madey, Nucl. Instr. Meth. A250, 262 (1986)

74 -M. Cornacchia, J. Bisognano, S. Chattopadhyay, A. Garren, K. Halbach, A. Jackson, K.J. Kim, H. Lancaster, J. Peterson, M.S. Zisman, C. Pellegrini and G. Vignola, Nucl. Instr. Meth. A250, 57 (1986)

75 -E.D. Shaw, R.J. Chichester and S.C. Chen, Nucl. Instr. Meth. A250, 44 (1986)

76 -U. Bizzarri, F. Ciocci, G. Dattoli, A. de Angelis, M. Ercolani, E. Fiorentino, G.P. Gallerano, T. Letardi, A. Marino, G. Messina, A. Renieri, E. Sabia and A. Vignati, J. Physique (Paris) 44, C1-313 (1983)

77 -U. Bizzarri, F. Ciocci, G. Dattoli, A. de Angelis, G.P. Gallerano, I. Giabbai, G. Giordiano, T. Letardi, G. Messina, A. Mola, L. Piccardi, A. Renieri, E. Sabia, A.

Vignati, E. Fiorentino and A. Marino, Nucl. Instr. Meth. A250, 254 (1986)

78 -C.M. Tang, P. Sprangle, S. Penner, B.M. Kincaid and R.R. Freeman, Nucl. Instr. Meth. A250, 278 (1986)

79 -X.K. Mariyama, S. Penner, C.M. Tang and P. Sprangle, Nucl. Instr. Meth. A259, 259 (1987)

80 -J.E. Walsh, B. Jonhson, C. Shaughnessy, F. Ciocci, G. Dattoli, A. de Angelis, A. Dipace, E. Fiorentino, G.P. Gallerano, T. Letardi, A. Renieri, E. Sabio, I. Gabbai and G. Giordiano, Nucl. Instr. Meth. A250, 308 (1986)

81 -E.P. Garate, J. Walsh, C. Shaughnessy, B. Johnson and S. Moustaizis, Nucl. Instr. Meth. A259, 125 (1987)

82 -T.J. Orzechowski, B.R. Anderson, J.C. Clark, W.M. Fawley, A.C. Paul, D. Prosnitz, E.T. Scharlemann, S.M. Yarema, D.B. Hopkins, A.M. Sessler and J.S. Wurtele, Phys. Rev. Lett. 57, 2172 (1986)

83 -T.J. Orzechowski, B.R. Anderson, W.M. Fawley, D. Prosnitz, E.T. Sharlemann, S.M. Yarema, A.M. Sessler, D.B. Hopkins, A.C. Paul and J.S. Wurtele, Nucl. Instr. Meth. A250, 144 (1986)

84 -J.A. Pasour and S.H. Gold, I.E.E.E. J.Q. Electr. QE-21, 845 (1985)

85 -J.A. Pasour, J. Mathew and C. Kapetanakos, Nucl. Instr. Meth. A259, 94 (1987)

86 -J. Masud, Y.G. Yee, T.C. Marshall and S.P. Schlesinger, Nucl. Instr. Meth. A250, 342 (1986)

87 -N. Ohigashi, K. Mima, S. Miyamoto, K. Imasaki, Y. Kitagawa, H. Fujita, S.I. Kuruma, S. Nakai and C. Yamanaka, Nucl. Instr. Meth. A259, 111 (1987)

88 -A. Amir, L.R. Elias, D.J. Gregoire, J. Kotthaus, G.J. Ramian and A. Stern, Nucl. Instr. Meth. A250, 35 (1986)

89 -L.R. Elias, R.J. Hu and G.J. Ramian, Nucl. Instr. Meth. A237, 203 (1985)

90 -J. Gallardo and L. Elias, Nucl. Instr. Meth. A250, 438 (1986)

91 -I. Kimel, L.R. Elias and G. Ramian, Nucl. Instr. Meth. A250, 320 (1986)

92 -G. Ramian, L. Elias and I. Kimel, Nucl. Instr. Meth. A250, 125 (1986)

93 -S.B. Segull, M.S. Curtin and S.A. Von Laven, Nucl. Instr. Meth. A250, 316 (1986)

94 -E. Jerby, A. Gover, S. Ruschin, H. Kleinman, I. Ben-Zvi, J.S. Sokolowski, S. Eckhouse, Y. Gorren and Y. Shiloh Nucl. Instr. Meth. A259, 263 (1987)

95 -R.A. Jong and E.T. Scharlemann, Nucl. Instr. Meth. A259, 254 (1987)

96 -Photonics Spectra, July 1983 p. 40 et seq.

97 -Photonics Spectra, April 1987, p. 159 et seq.

98 -Aviation Week and Space Technology, Aug. 18, 1986, pp 40-79

99 -N. Bloembergen, C.K.N. Patel, P. Azivonis, R.G. Clem, A. Hertzberg, T.H. Johnson, T. Marshall, R.B. Miller, W.E. Morrow, E.E. Salpeter, A.M. Sessler, J.D. Sullivan, J.C. Wyant, A. Yariv, R.N. Zare, A.J. Glass and L.C. Hebel, Rev. Mod. Phys. 59, n° 3, Part II, July 1987, pp S1-S200

100 -"Construction and Commissioning of Dedicated Synchrotron Radiation Facilities", Proc. Workshop Brookhaven National Laboratory, Oct. 16-18, 1985 (ed. R.W. Klaffky), BNL 51959 (1986)

LASER DOPPLER-FREE TECHNIQUES IN SPECTROSCOPY

Prof. B. Cagnac
Laboratoire de Spectroscopie Hertzienne de l'E.N.S.
Université Pierre et Marie Curie
75252 PARIS CEDEX 05
FRANCE

ABSTRACT. We present a review of all techniques which permit in Laser spectroscopy to overcome the Doppler broadening due to the atomic or molecular motions.

1. IN RADIOFREQUENCY DOMAIN

1.1 Cancellation of Doppler shift with radiowaves

Fig. 1a recalls the principle of calculation of the Doppler shift of the apparent frequency seen by a moving receiver : the representation of the spatial variation of the electric field of the wave permits to count the number of wavelengths $\lambda = c/\nu$ between the emitter and the receiver. If, during the time interval Δt, this number decreases by the quantity $v\Delta t/\lambda$, the receiver counts as much additional periods ; and that permits to calculate the apparent frequency shift $\delta\nu_D = v/\lambda = v\nu/c$.

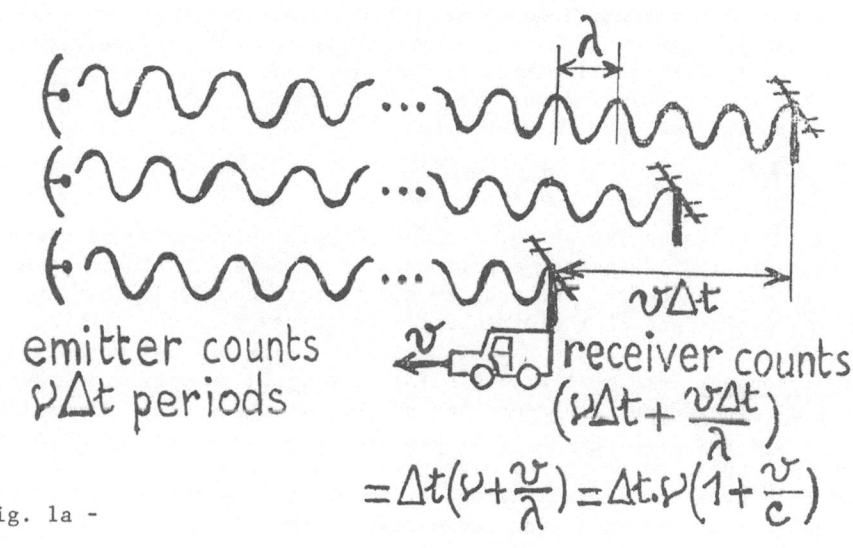

emitter counts
$\nu\Delta t$ periods

receiver counts
$\left(\nu\Delta t + \dfrac{v\Delta t}{\lambda}\right)$

$= \Delta t\left(\nu + \dfrac{v}{\lambda}\right) = \Delta t \cdot \nu\left(1 + \dfrac{v}{c}\right)$

Fig. 1a -

153

A. C. P. Alves et al. (eds.), Frontiers of Laser Spectroscopy of Gases, 153–185.
© 1988 by Kluwer Academic Publishers.

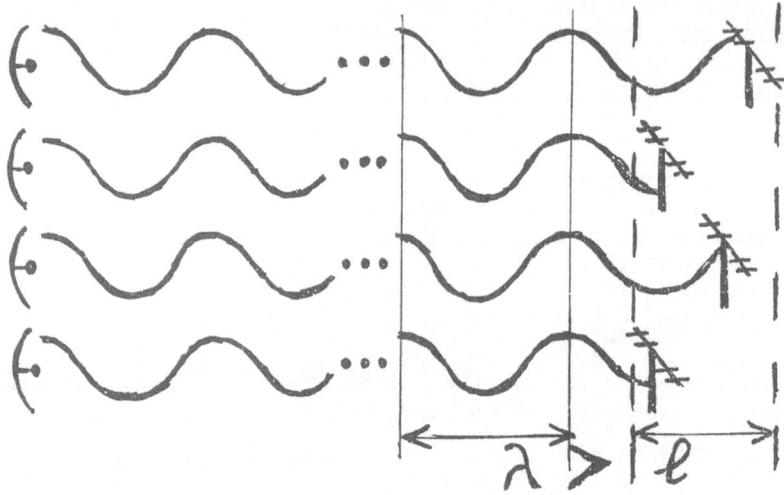

Fig. 1b -

By comparison we represent on Fig. 1b the case of a receiver in
random motion in a box of length $\ell < \lambda$ (ℓ can be the actual length of
the experimental vapour cell in radio-experiments, or can be the mean
free path due to collisions in optical experiments). In that case, the
number of wavelengths between the receiver and the emitter remains constant, and the integer numbers of periods counted by the emitter and the
receiver remains permanently equal ; only the decimal fraction of that
number fluctuates, and that does not produce really a Doppler shift.
An exact calculation has been done by Dicke |1| ; but the phenomenon
can also be understood as analogous to the motional narrowing in N.M.R.
|2| : the instantaneous "inhomogeneous" shift (i.e. different from an
atom to another one) would be $\delta\omega_D = \omega v/c$; but that shift fluctuates
randomly with a correlation time of the order of $\tau_c = \ell/\sqrt{\bar{v}^2}$. The theory
shows that the actual residual frequency broadening is $\Delta\omega_b \simeq \overline{\delta\omega_D^2} \cdot \tau_c$.
Compared to the ordinary Doppler broadening $\sqrt{\overline{\delta\omega_D^2}}$, it is reduced by the

$$\text{factor } \sqrt{\overline{\delta\omega_D^2}} \cdot \tau_c \simeq \omega\frac{\ell}{c} = 2\pi\frac{\ell}{\lambda}.$$

For that reason the actual residual Doppler broadening is always
negligible in radiofrequency experiments in vapours (in atomic beam, the
motion is not random and the problem can be more complex). That explains
the huge development of radiofrequency spectroscopy in the fifties
and sixties. Most of this work has been done using optical methods :
double resonance for excited levels |3| or optical pumping for ground
levels |4|. The advent of Lasers has not deeply changed the situation.
Nevertheless the light power available with tunable Lasers permits to
extend the double resonance to highly excited levels which can be attained only by stepwise excitation, or to extend the optical pumping
technique to very short living radioactive atoms.

1.2 Optical techniques without radiowave

We describe in this section methods which study small energy differences in the radiofrequency range, without using a radiowave, but only
wing optical measurements. These methods are not absolutely dependent
on Lasers, but some of them have been in fact largely developped only
with tunable Lasers.

1.2.1 <u>The quantum beats</u> are observed in situations such as the one
represented on Fig. 2 : fluorescence light can be emitted by an atom
or molecule from two close by excited levels E_1 and E_2, decaying to the
same final level E_f (this final level E_f can be different from the
ground level E_g, as represented in Fig. 2; or it can be the same as
E_g ; it does nõt matter). The interference of the two fluorescence
waves, with close by frequencies, $(E_2 - E_f)/\hbar$ and $(E_1 - E_f)/\hbar$, contains
the beat frequency :

$$\omega_{12} = \frac{E_2 - E_1}{\hbar} = \frac{\Delta E}{\hbar}$$

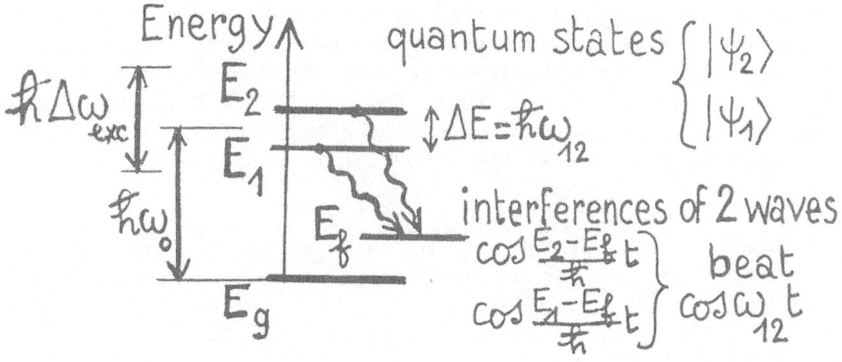

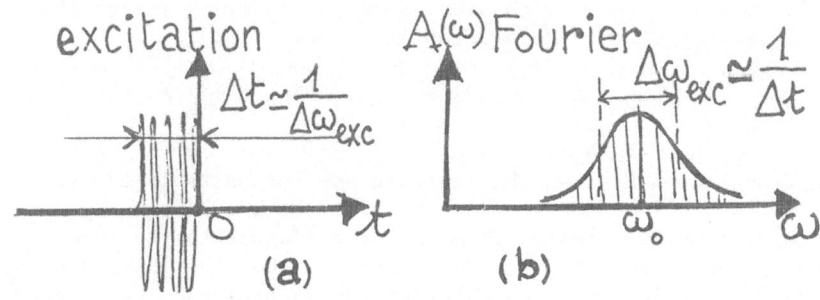

Fig. 2 - Energy diagram for quantum beats.
    In ordinary conditions, this interference cannot be observed be-
cause the two fluorescence waves are emitted from different atoms with
incoherent random phases. But the observation is possible in particular
conditions where the excitation of the atoms is quite sudden ; Fig. 3a

Fig. 3a - Temporal variation of the excitation for quantum beats
    3b - Fourier spectrum of the same excitation.

gives a typical representation of such an excitation (modulus of the interaction Hamiltonian versus the time), whose duration Δt is very short ; Fig. 3b represents the corresponding Fourier spectrum of that pulse interaction : its width $\Delta\omega_{exc}$ is of the order of the inverse of the pulse duration Δt. If Δt is short enough, the width $\Delta\omega_{exc}$ is larger than the beat frequency $\omega_{12} = \Delta E/\hbar$ (see Fig. 2). In these conditions, the atom can be excited simultaneously in the two levels E_1 and E_2 : just after the pulse Δt, at the time called zero, the atom is in that particular state called "coherent superposition" of the two states E_1 and E_2 and which can be written :

$$|\Psi(0)\rangle = \alpha_1|\Psi_1\rangle + \alpha_2|\Psi_2\rangle$$

where $|\Psi_1\rangle$ and $|\Psi_2\rangle$ represent the quantum states corresponding to the energies E_1 and E_2 ; α_1 and α_2 are two coefficients whose values depend on the exact terms of the interaction Hamiltonian but their real values do not matter. It is enough that they are simultaneously not equal to zero.

After the time zero, the wave functions $|\Psi_1\rangle$ and $|\Psi_2\rangle$ have the classical free evolution (in the absence of interaction) characterized by the phase factors : exponential of $- i(E/\hbar)t$. But we must add to the phase factors a real factor, exponential of $- (\Gamma/2)t$, representing the spontaneous decay to the final state with the decay rate Γ, inverse of the lifetime (the decay rate $\Gamma/2$ on the amplitudes gives the decay rate Γ on the intensities, when you square the amplitudes). So we obtain the wavefunction of the excited state to time t positive :

$$|\Psi(t)\rangle = \alpha_1 e^{-i\frac{E_1}{\hbar}t} e^{-\frac{\Gamma}{2}t}|\Psi_1\rangle + \alpha_2 e^{-i\frac{E_2}{\hbar}t} e^{-\frac{\Gamma}{2}t}|\Psi_2\rangle$$

As you know, the probability of any atomic transition is calculated from the square of the matrix element of the interaction Hamiltonian between the initial excited states $|\psi(t)\rangle$ and the final state $|f\rangle$. Hence the measured fluorescence signal is proportional to :

$$S(t) = |\langle f|D_\epsilon|\Psi(t)\rangle|^2$$

where $D_\epsilon = \vec{\epsilon}.\vec{D}$ is the projection of the electric dipole operator $\vec{D}$ on the particular polarization $\vec{\epsilon}$ choosen for the observation.
Using the above expression of $|\psi(t)\rangle$ we obtain :

$$S(t) = \Big|\underbrace{\langle f|D_\epsilon|\Psi_1\rangle\alpha_1 e^{-i\frac{E_1}{\hbar}t}e^{-\frac{\Gamma}{2}t}}_{A_1} + \underbrace{\langle f|D_\epsilon|\Psi_2\rangle\alpha_2 e^{-i\frac{E_2}{\hbar}t}e^{-\frac{\Gamma}{2}t}}_{A_2}\Big|^2$$

we call A_1 and A_2 the two constant factors appearing inside that expression ; that permits to develop it easier :

$$S(t) = e^{-\Gamma t}\Big[A_1A_1^* + A_2A_2^* + A_2^*A_1 e^{i\frac{E_2-E_1}{\hbar}t} + A_1^*A_2 e^{i\frac{E_1-E_2}{\hbar}t}\Big]$$

$$S(t) = e^{-\Gamma t}\Big[|A_1|^2 + |A_2|^2 + 2\mathcal{R}A_1^*A_2 e^{-i\omega_{12}t}\Big]$$

As represented on Fig. 4, the fluorescence intensity is always decaying with the rate Γ ; but an oscillation with the beat frequency ω_{12} is superimposed on that decay. It is quite analogous to the interference of two fluorescence waves with different frequencies $(E_2-E_f)/\hbar$ and $(E_1-E_f)/\hbar$; but in fact, the interference is produced inside each atom and is independent of all dephasing factors occuring from waves propagation or from different origins (different atoms). The modulation of

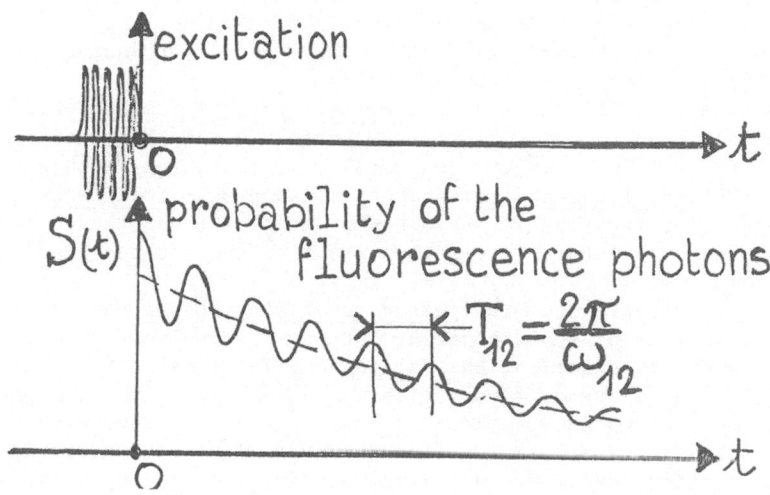

Fig. 4 - Quantum beats : theoretical response versus the time (bottom curve) after the pulse excitation (top curve).

S(t) is quite independent of the external parameters of the atoms (positions and velocities) and is not submitted to Doppler-broadening.

One can wonder about the energy of the fluorescence photons : it is $E_2 - E_f$ or $E_1 - E_f$? But this question has no definite answer, as we do not know if the atom passes through the level E_1 or the level E_2 : we know only that the atom passes from the ground state E_g to the final level E_f, with emission of one photon, through <u>two undistinguishable paths</u> (E_1 or E_2) and, following the quantum rules, we have summed the amplitudes corresponding to these two undistinguishable paths. If we wish to measure the energy of the photon, we can use a selective monochromator which transmits only the photons of a definite energy, E_1-E_f for example ; but in these new conditions, we see only atoms passing through the particular path E_1 and we have no more interference ; the fluorescence intensity is no more modulated.

That phenomenon is, in some respects, analogous to the classical phenomenon of Young and Fresnel fringes produced by two closely parallel slits. You observe on the screen the impact of photons which have passed through two undistinguishable paths (slit 1 or slit 2). If you wish to know the path of the photons, you can obscure one of the two slits ; but in these new conditions you do not observe fringes.

In the case where you have not only two, but many close by excited levels E_n, the short excitation pulse produces a coherent superposition of all these excited levels :

$$|\Psi(0)\rangle = \sum_n \alpha_n |\Psi_n\rangle$$

and the summation of amplitudes of the many undistinguishable paths (through E_1, E_2, ..., E_n) produces all beat frequencies :

$$\omega_{mn} = \frac{E_m - E_n}{\hbar}$$

On the contrary, if you have several final states f, f', f", etc..., the transitions to E_f, $E_{f'}$ or $E_{f''}$ are distinguishable (at least theore-

tically) and one must not sum the amplitudes of these distinguishable
processes, but only their intensities ; beats do not appear.

In practice, several methods can be used for producing the coherent superposition of excited states :

a- The irradiation with a <u>classical spectral lamp</u> through <u>a fast
shutter</u> has been used in the first experiments |5|. But the low intensi-
ty of the signal obliges to use complicated averaging technique.

b- <u>Pulsed electronic bombardment</u> |6| and <u>beam foil technique</u> with
highly accelerated ions |7| have been used also ; but these two methods
do not produce selective excitation and the fluorescence light spreads
out in many quite different wavelengths.

c- The irradiation with <u>pulsed tunable dye Laser</u> has permitted a
huge development of that technique because it combines the selectivity
and the high efficiency of the excitation. The first demonstrative ex-
periments |8| observed Zeeman splittings produced in small magnetic
fields. But the method has been used afterwards for real measurements
of hyperfine structure |9| or fine structure |10| intervals. The step-
wise excitation, using two different but synchronized dye Lasers, permits
to extend the studies towards highly excited levels. The extent of energy intervals which can be measured is limited by the rapidity of the
electronic equipment (currently hundred MHz, or some hundreds). One will
find a good review on quantum beats in reference |11|.

1.2.2 The <u>level-Crossing</u> technique is relatively old and has been widely
used for sixty years, a long time before lasers, together
with the double resonance technique |12|. We are interested here to ex-
plain its relation with quantum beats. The observation of level-crossing
arises when some energy levels of an atom are depending on an external
parameter as magnetic field B for example (see Fig. 5), if, for a particular value B_c of this parameter, the energies of two different atomic
states become equal. The observation of the level crossings uses the
old classical technique of producing fluorescence with a c.w. broad band
light source (at least the Doppler width of a traditional spectral lamp).
We call $\Delta\omega_{exc}$ the width of the light source ; you know that the coherence time of such a broad band light source is of the order of
$\Delta t = 1/\Delta\omega_{exc}$; that is to say : the phase of the sinusoïdal electric
field of the light wave remains almost stable on time intervals shorter

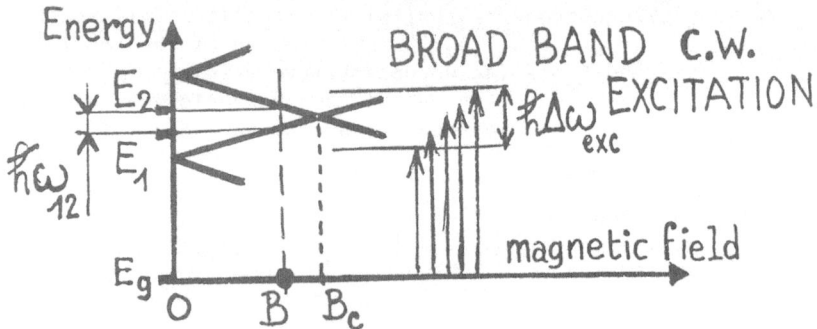

Fig. 5 - Energy diagram for level crossing.

than Δt, but is randomly changed during every time interval longer than Δt. From this fact, <u>the c.w. broad-band irradiation can be considered as equivalent to the succession of short incoherent pulses</u>, whose duration is of the order of Δt.

Suppose now that we work in a particular fixed value B of the magnetic field (different from B_c) for which the two levels of interest have the values E_2 and E_1 separated by the small interval $E_2 - E_1 = \hbar\omega_{12}$ (see Fig. 5). Each successive incoherent pulse produces a decaying and oscillating response $S_k(t)$ analogous to the one of Fig. 4. We represent on Fig. 6 these successive responses S_1, S_2, S_3, ..., which have the same frequency ω_{12} but random phases. At a particular time t, the whole fluorescence intensity $I(B,t)$ is obtained by summation of the intensities $S_1(t)$, $S_2(t)$, $S_3(t)$ emitted by different atoms, which have been

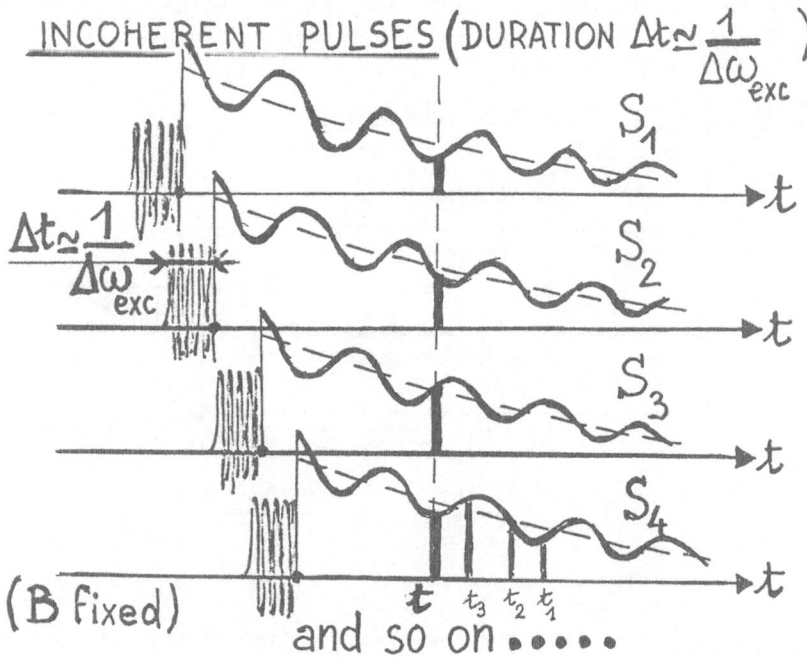

Fig. 6 - Excitation equivalent to a succession of incoherent pulses. Successive responses of the atoms versus the time (fluorescence intensity).

excited at different moments before t. That summation of different signals $S_k(t)$ at the same time t is equivalent to the summation of different values of the same signal $S(t)$ corresponding to various times t_k distributed randomly but with a mean distance Δt :

$$I(B,t) = \sum_k S_k(t) = \sum_k S(t_k) \simeq \frac{1}{\Delta t}\int_0^\infty S(t)dt = \Delta\omega_{exc}\int_0^\infty S(t)dt$$

Replacing $S(t)$ by its expression already written, we have now to calculate the integral :

$$\int_0^\infty e^{-(\Gamma+i\omega_{12})t}dt = \frac{1}{\Gamma+i\omega_{12}} = \frac{\Gamma-i\omega_{12}}{\Gamma^2+\omega_{12}^2}$$

One obtains a stationary solution where the resulting intensity does no more depend on time, but depends only on the frequency ω_{12}, i.e. on the magnetic field B. The calculus of the real part depends on the relative phases of the coefficients A_1 and A_2 ; if they have the same phase, we keep only the real part of the above integral, and we obtain :

$$I(B) = \frac{\Delta\omega_{exc}}{\Gamma} \{|A_1|^2 + |A_2|^2 + 2|A_1|.|A_2| \frac{\Gamma^2}{\Gamma^2+\omega_{12}^2}\}$$

The resulting fluorescence intensity varies versus the frequency ω_{12} with a Lorentzian shape, and passes through a maximum value when ω_{12} becomes zero, i.e. the magnetic field passes through the crossing value B_c (see Fig. 7). In the practice, this crossing value B_c is determined by recording the variation of the fluorescence intensity I(B) versus the values of the magnetic field B.

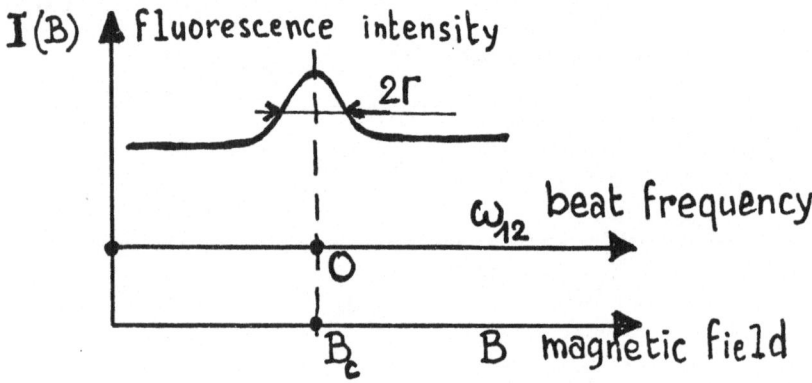

Fig. 7 - Level crossing = variation of the fluorescence intensity versus the atomic energy difference ω_{12}, corresponding to the magnetic field B.

It is possible to interpret in other words the result I(B) of the preceding calculation :
1- far from the crossing, the phase fluctuation of the oscillating responses produce an averaging to zero of the interference term between the two paths E_1 and E_2 ; the result is equivalent to the summation $|A_1|^2 + |A_2|^2$ of the intensities of the two paths.
2- At the exact crossing, the frequency ω_{12} of the oscillation becomes zero, we obtain stationary responses, whose interference term is no more destroyed by averaging : we must sum the amplitudes of the two paths, and then square the whole amplitude : $|A_1+A_2|^2$. So a level crossing is really a quantum beat at null frequency.

(Remark : a proper choice of the polarization permits to modify the real and imaginary parts of the matrix elements $\langle f|D_\epsilon|\Psi_1\rangle$ and $\langle f|D_\epsilon|\Psi_2\rangle$, that is to say of the coefficients A_1 and A_2 ; with different phases of the complex coefficients A_1 and A_2, it is possible to keep in the calculation the imaginary part of the above integral ; that gives a dispersion shape, instead of the absorption shape of Fig. 7).

1.2.3 <u>The modulated light irradiation</u> is another technique which does
not need Lasers but is closely related to quantum beats ; and we are in
the same energy configuration (see Fig. 2). Suppose that we irradiate
the atoms with periodic short pulses separated by the constant period
$T = 2\pi/\omega$ (see Fig. 8). We produce again succession of decaying and oscillating responses $S_k(t)$, but which are regularly spaced in time. Suppose, as in Fig. 8, that the period T of the pulses is equal to the period
$T_{12} = 2\pi/\omega_{12}$ of the oscillating responses ; it is evident that all the
oscillations are in phase, and their summation gives a sinusoïdal result
of strong amplitude. If the periodic excitation lasts a long time, we
obtain a stationary regime where the fluorescence intensity is modulated
with a steady state amplitude.

   If now the period T of the pulses changes and becomes smaller or
greater than the atomic period T_{12}, it produces dephasing between the
various atomic responses $S_k(t)$ and blurring of their summation : the
total fluorescence intensity, in stationary regime, is modulated with a
smaller amplitude. The amplitude of the fluorescence modulation passes
through a maximum when the excitation frequency ω becomes equal to the
atomic frequency ω_{12}. In practice, it is not necessary to use a pulse
excitation : it is necessary to use a broad band excitation ($\Delta\omega_{exc} > \omega_{12}$)
and to interrupt periodically the excitation light |13|.

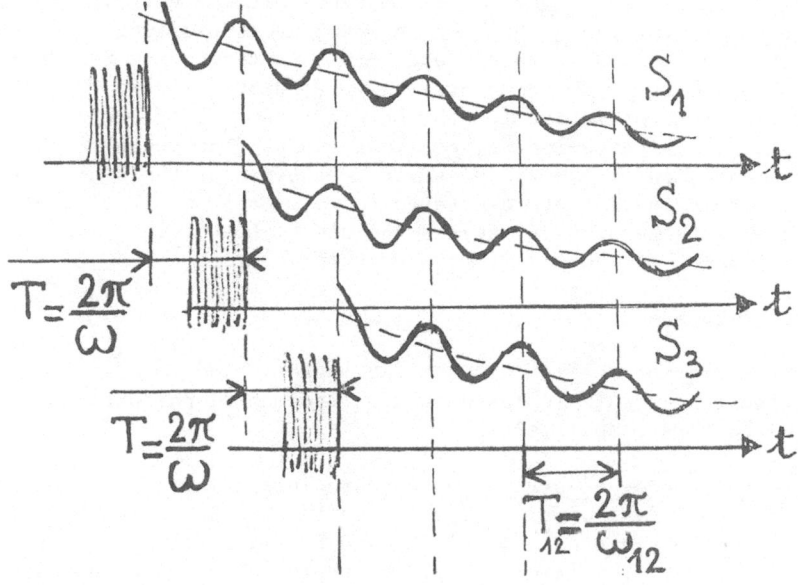

SUM OF COHERENT RESPONSES $S_k(t)$

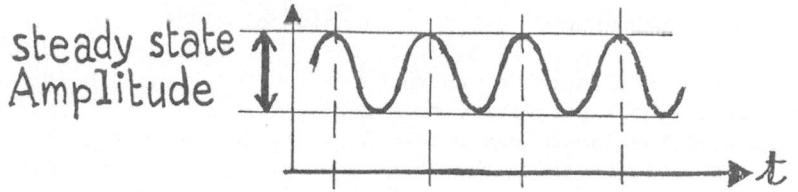

Fig. 8 - Excitation by periodic pulses.

2. IN OPTICAL DOMAIN

Lasers have really changed the situation in spectroscopy since 1972, when they became tunable. For the first time the physicists possessed light sources of tunable frequency, with a precision of the order of 1 MHz, that is to say comparable with the current natural width of many atomic levels. Using these precise light sources obliged the spectros-copists to study absorption spectroscopy rather than emission spectros-copy. But high precision of Laser frequencies was quite unuseful unless to find solutions permitting to overcome the Doppler broadening, curren-tly of the order of 1 000 MHz. Four solutions are developed ; the first of which received some applications before the Laser.

2.1 Atomic beam crossed at right angle

In this old and well known method, diaphragms inside a vacuum box select, among the atoms of an atomic or molecular vapour, the ones having velo-cities parallel to a particular direction. The Doppler frequency shift depends on the component of the velocity vector on the irradiation di-rection ; it is null if the light irradiation is perpendicular to the atomic beam. In practice, the difficulty is to concentrate enough light power in a light beam of very small aperture. It was very difficult be-fore the Lasers furnish high power in small apertures, which are curren-tly of the order of 10^{-3} radian, and permit to reduce the residual Doppler broadening to the same order as the natural linewidths. The same problem remains entire for the angular aperture of the atomic beam ; and in atomic beam experiments, one must always find a compromise bet-ween the reduction of the width and the reduction of the signal, as the number of atoms in the beam decreases to zero when the angular aperture is decreasing to zero. Nevertheless the use of Lasers has permitted a huge development of the atomic beam spectroscopy.

2.2 Saturation spectroscopy

Saturation spectroscopy is the most developed method in high pre-cision spectroscopy. It is based on two fundamental notions which will be explained first separately :

2.2.1 The phenomenon of saturation appears in the interaction of atoms

with a resonant light wave of high intensity. Let n_0 and n_1 be the po-pulations of the two atomic levels E_0 and E_1 interacting with that light wave of frequency $\omega_1 = (E_1 - E_0)/\hbar$. The evolution of these populations is given by the following differential equations :

$$\frac{dn_1}{dt} = -\frac{n_1}{\tau} + \sigma \frac{P}{S}(n_0 - n_1) \text{ and } \frac{dn_0}{dt} = +\frac{n_1}{\tau} - \sigma \frac{P}{S}(n_0 - n_1),$$

where τ is the mean lifetime of the excited level E_1, and P/S represents the intensity of the light wave (the power P divided by the cross-sec-tional area S of the laser beam). Note that $dn_1/dt = -dn_0/dt$ because $n_1 + n_0$ is constant, equal to the total number N of the atoms ($n_1 + n_0 = N$).

   In the right hand side of these equations, the first term n_1/τ
represents the loss or gain by spontaneous emission from the excited
level E_1 to the ground level E_0 ; the second term, proportional to the
light intensity P/S, represents the joint action of absorption and of
stimulated emission (the coefficient σ is proportional to the cross
section of the interaction between photons and atoms).

   In most experiments, one observes only the stationary regime where
the populations are constant. Setting $dn_1/dt = dn_0/dt = 0$, we obtain
easily the steady-state populations :

$$n_0 = N \frac{1/\tau + \sigma P/S}{1/\tau + 2\sigma P/S} \quad \text{and} \quad n_1 = N \frac{\sigma P/S}{1/\tau + 2\sigma P/S} \cdot$$

   They depend on the intensity P/S of the light wave ; that dependence is not linear as is shown in Figure 9. When the light intensity

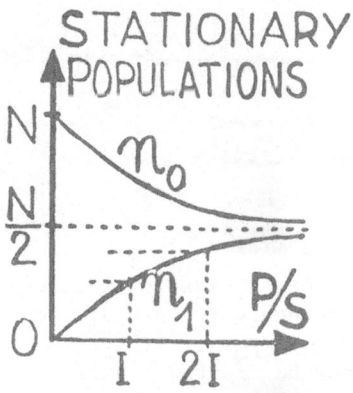

Fig. 9 - Variation of the atomic populations versus the light
        intensity P/S

P/S increases indefinitely, the two populations n_0 and n_1 tend toward
the same limit $N/2$: the saturation of the interaction. The important
point used in the Doppler-free technique is not the saturation itself
but the nonlinearity : the population n_1 of the excited state is less
than doubled when the light intensity P/S is doubled from I to 2I ; this
nonlinearity is stronger when one is approaching saturation.

2.2.2 The notion of velocity class permits to take into account the
Doppler shift, which depends on the component v_z of the atomic velocity
in the direction z of the wave propagation. Figure 10 represents the
Maxwell distribution of the atoms (number of atoms plotted against their
velocity component v_z). If the Laser emits a light wave of frequency ω_L
propagating in the z direction (forward wave), an atom of velocity component v_z sees in its own frame the frequency $(\omega_L - kv_z)$, where k is the
wave vector $(k = \omega/c)$. As a consequence, the Laser beam interacts only
with those atoms having a velocity v_z such that $\omega_L = \omega_1 + kv_z$. This
correspondence between ω_L and v_z is illustrated in Figure 10 by a frequency axis parallel to the v_z axis ; the frequency scale is chosen in
such a manner that the velocity zero corresponds to the exact frequency
ω_1 of the atomic transition, and the velocity v_z corresponds to the

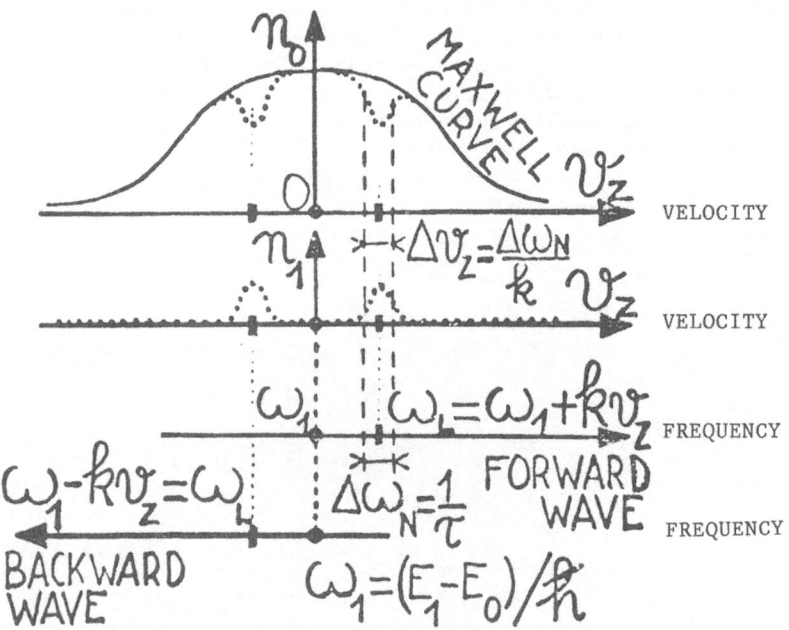

Fig. 10 - Velocity classes in interaction with the Laser beams at
frequency $_L$, propagating forward ($z > 0$) and backward ($z < 0$).
Laser frequency $\omega_L = \omega_1 + kv_z$. The Maxwell curve can be interpreted,
with this new horizontal ω axis, as the number of atoms interacting
with the Laser beam of frequency ω.

In fact the interaction frequency is not defined with an infinite
precision ; the interaction has some width on the frequency scale, which
is the natural width $\Delta\omega_n = 1/\tau$ (inverse of the lifetime). To this natural
width $\Delta\omega_n$ corresponds, on the velocity scale, some interval $\Delta v_z = \Delta\omega_n/k$.
All the atoms belonging to this interval Δv_z constitute the velocity
class interacting with the Laser beam propagating in the forward direction (z positive).

If the laser beam is reflected backwards with a mirror, the Doppler
shift is inverted for the backward wave, which interacts with the atoms
having a velocity v_z such that $\omega_L = \omega_1 - kv_z$. This new correspondence
between ω_L and v_z is illustrated in Figure 10 by another frequency axis
in the opposite direction : the same Laser frequency ω_L corresponds now
to the opposite value of v_z. That is to say : the same Doppler shift
$\omega_L - \omega_1$ is obtained by inverting simultaneously the sign of v_z and the
sign of k. As a consequence, the backward wave interacts with other atoms
constituting another velocity class. The dotted curves on Figure 10
represent the changes in the populations n_0 and n_1 resulting from the
interaction with the two waves.

2.2.3 <u>The principle of saturation spectroscopy</u> derives immediately
from the foregoing : when the Laser frequency ω_L is different from the
transition frequency ω_1, the two waves, forward and backward, interact
with different atoms, belonging to two different velocity classes. On

the other hand, if the Laser frequency becomes equal to the transition frequency, the two velocity classes merge into each other, i.e. the two waves now interact with the same atoms.

We obtain now : half the number of atoms (only one velocity class) interacting with twice the light intensity (the two waves together). Because of the nonlinearity of the interaction, the response of the system is less than for two independent velocity classes : fewer atoms leave the ground level and are excited to the upper level ; fewer photons are absorbed. We deduce that the transmitted light intensity through the absorbing cell is greater |14| ; or the fluorescence light re-emitted by the absorbing cell is less |15| (see Figure 11). This modified response of the absorbing medium is obtained over a frequency interval corresponding to the width of the velocity classes, i.e. of the order of the natural line width $\Delta\omega_n$: one observes a narrow curve superimposed on the broad Doppler curve (see Figure 11).

I have explained the principle of saturation spectroscopy for an absorbing medium (where $n_0 > n_1$). But in fact the phenomenon has been explained and observed for the first time for the amplifying medium that exists inside a Laser cavity (and where $n_0 < n_1$). In that case the nonlinearity of the interaction also gives a smaller response of the system when the two velocity classes merge each into the other, i.e. when the Laser frequency ω_L (determined by the cavity length) is exactly equal to the atomic frequency ω_1 of the amplifying transition. Then one observes less amplification, that is to say a smaller intensity of the Laser beam ; it is the so-called Lamb dip |16| illustrated in Figure 12. Many experiments have been done following the schemes of figures 11 and 12 for the purpose of frequency stabilization or spectroscopic measurements ; it is what is called Lamb-dip spectroscopy.

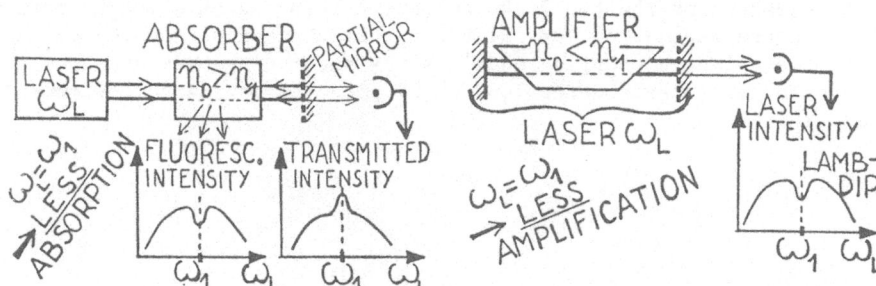

Fig. 11 - Saturated absorption and fluorescence inside an absorbing medium.

Fig. 12 - Lamb-dip inside an amplifying medium.

Both the sensitivity and the feasibility of the saturation technique have been strongly improved by the two modifications illustrated in Figure 13 (see reference |17|) :

(i) The intensities of the two counterpropagating light beams are no longer equal : one beam of strong intensity is called the saturating beam (or sometimes the pumping beam) ; the other beam of weak intensity is called the probe beam. As a consequence, the merging of the two

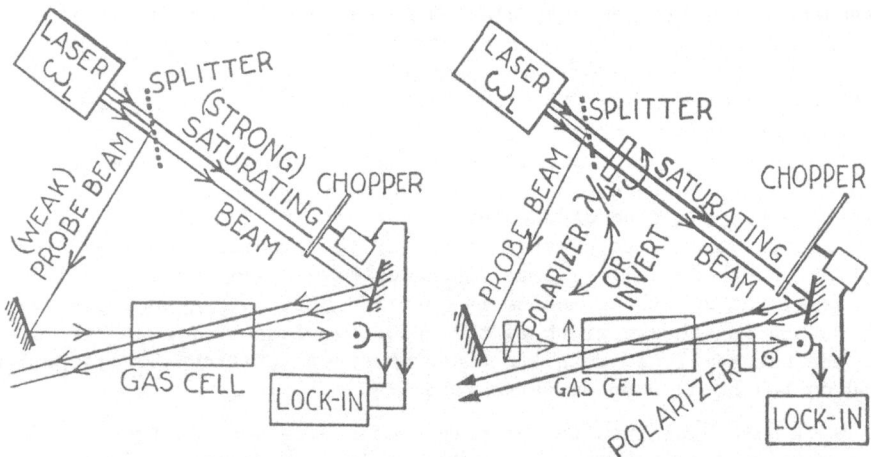

Fig. 13 - Practical set-up for saturated absorption spectroscopy.

Fig. 15 - Set-up for polarization spectroscopy.

velocity classes when $\omega_L = \omega_1$ is equivalent, for the probe beam, to a stronger change in intensity, and that magnifies the nonlinear effect.

(ii) The directions of the two beams can be slightly tilted in such a manner that one can measure separately the intensity of the weak probe beam after its passage through the absorbing cell. Thus one measures the absorption of the probe beam, which is greatly reduced by the interaction of the saturating beam, when $\omega_L = \omega_1$.

The small tilt angle changes slightly the definition of the velocity classes for the two light beams ; this produces a small broadening of the observed narrow curves. But the separation of the two beams permits the use of a chopping wheel on the saturating beam and the big improvement of lock-in detection, which produces narrow curves on a flat background.

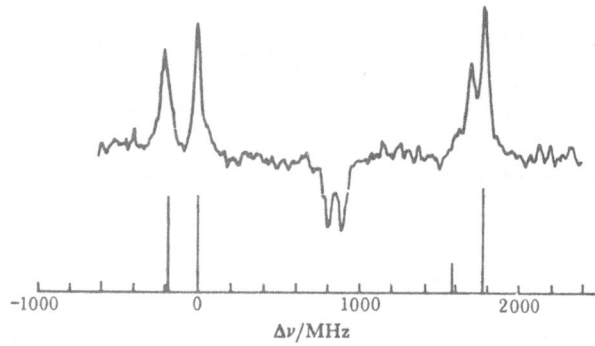

Fig. 14 - Saturated spectrum of the sodium D_1 line. The four theoretical peaks are shown on the bottom of the figure. The inverted peaks in the middle of the experimental curves are the parasitic crossover (reference |18|).

As an example, Figure 14 shows the hyperfine spectrum of the D_1 line of sodium obtained |18| using this technique : four components are visible, which are predicted between the two hyperfine sublevels (F = 0 and 1) of the ground level $3S_{1/2}$ and the two hyperfine sublevels (F = 0 and 1) of the excited level $3P_{1/2}$. The Doppler width is almost 2 000 MHz, i.e. the whole distance between the farthest peaks. The inverted peaks in the middle are parasitic signals, called crossovers. Such crossover peaks appear in all saturation spectra at mid-distance from the true peaks, when one velocity class corresponding to the first transition $(E_1 - E_0)$ coincides with one velocity class corresponding to the second transition $(E_2 - E_0)$ |19|.

2.2.4 <u>Saturated dispersion and polarization spectroscopy</u>. The general relation between absorption and dispersion is well known. The saturation of absorption also corresponds to a change in the index of refraction of the vapour, which is called saturated dispersion and was already predicted by Lamb in his theory of gas Lasers |16|. This saturated dispersion plays a role in the frequency stabilization of Lasers on the Lamb dip |20|. Clear evidence of saturated dispersion was given by using a ring interferometer |21|. But, in practice, the observation of saturated dispersion is much easier by interferences of polarized light. That introduces us to polarization spectroscopy.

The experimental scheme for polarization spectroscopy is very similar to the classical scheme for saturated absorption, as can be seen on Figure 15 ; but the two beams have different polarizations and one detects a change in the polarization of the probe beam induced by the saturating beam.

In the original scheme proposed by Wieman and Hänsch |22|, the saturating beam is circularly polarized by the use of a $\lambda/4$ plate ; it induces a circular anisotropy of the atomic vapour. The probe beam is linearly polarized, i.e. it is equivalent to the superposition of two circularly polarized counter-rotating waves. If these two counter-rotating polarizations have different absorptions or dispersions inside the saturated vapour, their superposition, at the exit of the vapour cell, no longer gives the initial linear polarization. The crossed polarizer introduced before the photoreceiver gives very sensitive detection of these changes in polarization.

In another scheme |23|, one interchanges the $\lambda/4$ plate and the polarizer ; the saturating beam is linearly polarized and induces in the vapour linear dichroism and birefringence ; on the other hand, the probe beam is circularly polarized. The probe beam is still analysed at the exit of the vapour cell by a polarizer ; it may be shown that the signal on the photoreceiver is a combination of dichroism and birefringence, depending on the orientation of that polarizer. A proper choice of its orientation and a subtracting technique permit one to obtain signals that represent either pure dichroism (Fig. 16b) or pure birefringence spectra (Figure 16c) ; one recognizes in this latter case the classical shape of the dispersion curve plotted against frequency. It is also possible to interpret these polarization signals as forward scattering of light by the atoms of the gas |24|.

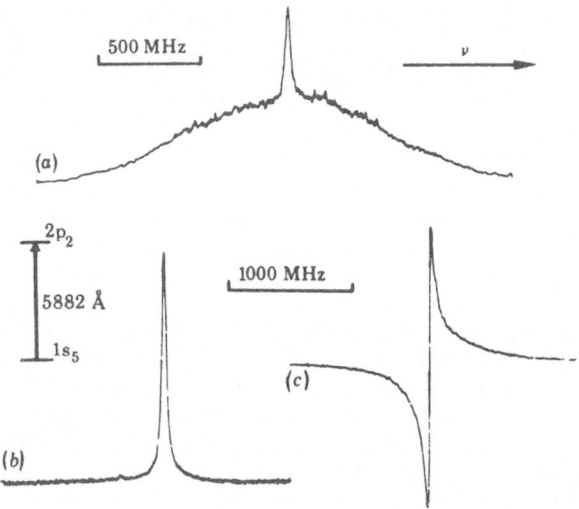

Fig. 16 - Examples of polarization spectrum (b) and (c) compared with
          saturated absorption spectrum (a). (Neon at λ = 5 882 Å,
          reference |23|).

Another advantage of polarization spectroscopy is the suppression
of the broad signal back-ground observed in saturated absorption (Fig.
16a), when collisions redistribute the velocities of the pumped atoms
over the Doppler profile, because these velocity-changing collisions
drastically reduce the laser-induced anisotropy (the curves in Figure
16 are obtained in the same experimental cell of neon). A good review
about Polarization Spectroscopy and related phenomena can be found in
ref. |25|.

2.3 Spectroscopy of accelerated ions

The importance of atomic or molecular ions in spectroscopy has grown
strongly with the development of plasma physics for fusion on the one
hand, with the discovery of many molecular ions in interstellar medium
on the other hand. Two different ideas are used for reducing the Doppler
broadening in that particular case.

2.3.1 <u>The reduction of velocity spreading through an acceleration</u> is
based on a very simple idea |26| : the well known parabolic law which
binds the kinetic energy E to the velocity v, and which is represented
on Figure 17. The problem of Doppler broadening is generally more severe
with ions than with neutrals because of the strong energy dispersion due
to discharge phenomena inside the ion sources. We call ΔE this energy
spreading ; it corresponds to a velocity spreading Δv_0 such that
$\Delta E = M(\Delta v_0)^2/2$ (see Fig. 17). If we now accelerate the ions of charge q
with a potential U, we increase their kinetic energy by the same amount
qU, whatever their initial velocity can be. At the output of the acce-

lerator, the kinetic energies of the ions are comprised between qU and
qU + ∆E, over the same energy interval ∆E as in the ion source. But the
slope of the parabola for the energy qU and the accelerated velocity
v_A is strongly increased ; and it is easy to remark on Fig. 17 that the
accelerated velocities are spread over an interval Δv_A strongly reduced,
compared to the initial spreading Δv_o.

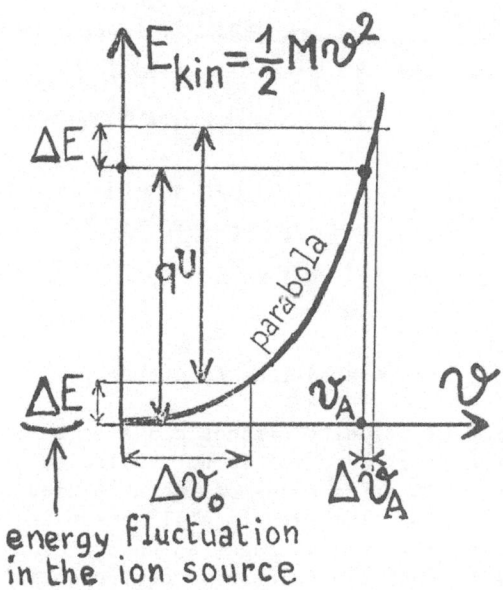

Fig. 17 - Principle of the reduction of velocity spreading.

That reduction can be estimated more precisely. The derivation of
the parabolic law, in the vicinity of the point of velocity v_A and
energy $qU = Mv_A^2/2$, gives : $\Delta E = Mv_A \Delta v_A$. Using the above relation between ∆E and Δv_o, one obtains :

$$\Delta v_A = \frac{\Delta E}{M \Delta v_A} = \frac{\Delta v_o^2}{2 v_A} \quad \text{or} \quad \frac{\Delta v_A}{\Delta v_o} = \frac{\Delta v_o}{2 v_A} = \frac{1}{2}\sqrt{\frac{\Delta E}{qU}}$$

The reduction factor can be currently comprised between hundred and
thousand. Fig. 18 shows the schematic experimental set-up using a colinear geometry for the ion beam and the Laser light beam. This colinear
geometry is easy to realize thanks to the curvature of the ion selector,
which removes the ion source aside from the accelerated beam axis. This
colinear geometry is absolutely necessary to insure a sufficient interaction time between the light and the fast ions ; it produces a
collective Doppler shift, $\delta v_D = v v_A/c$, almost identical for all the
ions, and which can be used as equivalent to a frequency tuning of the
Laser : the recording of a spectroscopic line can be done by sweeping
the potential U with fixed Laser frequency.

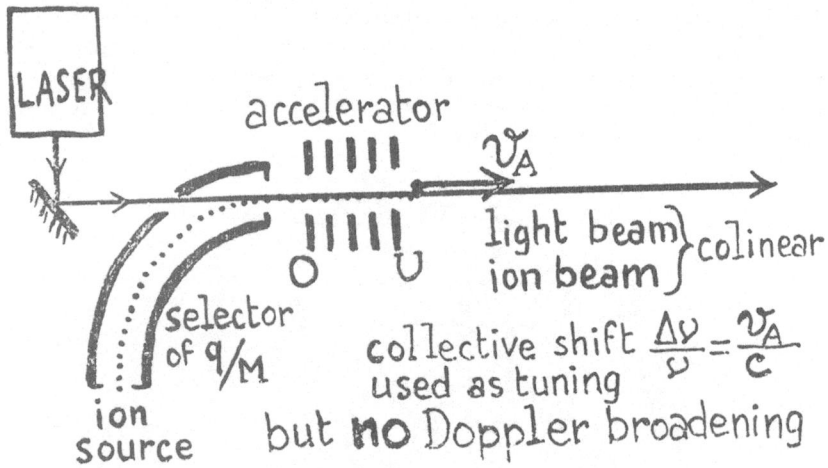

Fig. 18 - Set-up for accelerated ions spectroscopy.

2.3.2 <u>The saturation of velocity classes</u> can also be used either in cases where the preceding reduction is not sufficient (ion sources with big energy spreading ΔE) or in cases of instabilities of the high voltage U of the accelerator. The experimental set-up is analogous to the preceding one, except that the voltage can change along the accelerated beam, which passes through two successive equipotential chambers at slightly different voltages U and U' |27|. It is easy to stabilize the small energy difference U' - U (even if U is unstable) and to use it for the measurement of the small energy difference between two neighboring levels.

The ions see in the two chambers two different apparent light frequencies ν_A and ν_A' such that :

$$\nu_A' - \nu_A = \frac{\nu}{c} (\nu_A' - \nu_A) \simeq \frac{\nu}{c} \nu_A \frac{U' - U}{2U} = (\delta\nu_D) \frac{U' - U}{2U}$$

If there is some residual velocity spreading between the ions, there is the same spreading on the apparent frequency ν_A ; and the Laser in the first chamber interacts only with a part of the ions, having the good velocity. One generalizes the notion of velocity class introduced in § 2.2.2. But the frequency difference $(\nu_A' - \nu_A)$ depends only on the voltage difference (U' - U) and is precisely defined. When $(\nu_A' - \nu_A)$ is equal to the frequency difference between two neighboring spectral lines, the Laser in the second chamber interacts with the same ions (of the same velocity class) as in the first chamber. If the number of ions in the ground state has been reduced by the first interaction, the fluorescence signal in the second chamber is also reduced ; that permits to recognize when the apparent frequency difference $(\nu_A' - \nu_A)$ is exactly equal to the shift between the two spectral lines.

2.4 Two-photon spectroscopy

The two photon method uses the same remark as the saturated absorption method, namely that the sign of the Doppler shift changes with the sign of the wavevector k when the Laser beam is reflected back on itself by a mirror.

2.4.1 Principle of Doppler-free two-photon transitions.

We use now multi photon transitions where the atom absorbs simultaneously the energies of two photons $\hbar\omega_L$ to jump from the ground level E_0 to the excited level E_e. The energy conservation is written now : $E_e - E_0 = 2\hbar\omega_L$ (see Fig. 19a). The possibility of producing such transitions was calculated by Göppert-Mayer |28| at the beginning of quantum mechanics. It was one of the first applications of the perturbation theory to second order. This calcula-tion requires a summation on all the other levels. But we assume that only one term of this summation, corresponding to the intermediate level E_r, has a predominant role ; we can then interpret the calculation in the following manner : after absorption of the first photon, the atom is in the virtual state E_r with the energy defect $\hbar\Delta\omega_r = E_0 + \hbar\omega_L - E_r$ (see Fig. 19a). Because of the uncertainty principle, the atom remains in this virtual state during the short time $\Delta t_r \sim 1/\Delta\omega_r$; the atom has

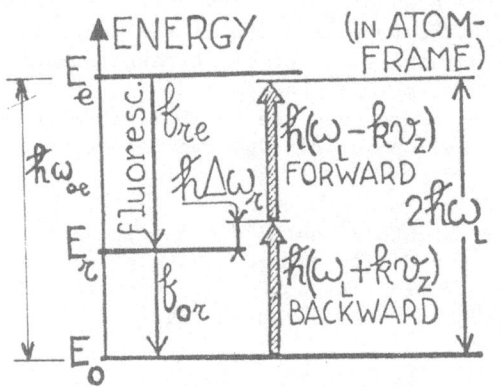

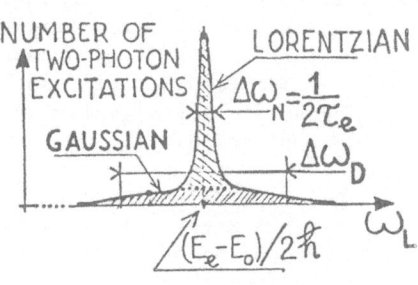

Fig. 19a - Principle of two-photon transition

Fig. 19b - Theoretical line-shape

a very small probability of absorbing a second photon during the very short time Δt_r unless the electromagnetic wave has a very high intensity. This indicates why the energy defect $\Delta\omega_r$ appears in the denominator of the two-photon probability, $\mathcal{P}^{(2)}$, which is also proportional to the square of the light intensity, P/S :

$$\mathcal{P}^{(2)} \sim C_t \ \tau_e (\omega_L/\Delta\omega_r)^2 \ f_{or} \ f_{re} \ (P/S)^2,$$

where τ_e is the lifetime of the E_e level and f_{or} and f_{re} are the os-cillator strengths of the one-photon transitions. Owing to the necessi-ty of high intensity, the multiphotonic transitions were observed first in the radiofrequency range, more than 20 years ago |29|. In the optical range it was necessary to wait for the high power of laser sources |30|.

Suppose now that the two-photon transition is produced with a La-ser beam reflected on itself with a mirror, as in Lamb-dip spectroscopy.

In the rest frame of one particular atom the two waves have slightly different frequencies, $(\omega_L - kv_z)$ and $(\omega_L + kv_z)$. If the atom absorbs one photon from each wave, the two opposite Doppler shifts cancel out and the total energy of the two photons is equal to $2\hbar\omega_L$, independent of the atomic velocity v_z (see Fig. 19a). The resonance condition of the two-photon transition is fulfilled at the same value of the laser frequency ω_L for all atoms, irrespective of their velocities |31| |32|.

In practice the atoms can also absorb two photons from the same travelling wave, but that absorption is still subject to Doppler broadening. As a consequence, when we sweep the Laser frequency ω_L, the number of two-photon excitation must vary, as indicated in Fig. 19b ; it is the superposition of two curves : (1) One Lorentzian curve of high intensity and narrow width, corresponding to the absorption of one photon of each counterpropagating light wave by all the atoms together (the width of this narrow curve is equal to the natural width $1/\tau_e$ for the energy gap $2\omega_L$, i.e. one half of this for the laser frequency ω_L) ; (2) One curve of weak intensity, a broad Gaussian curve (with Doppler width $\Delta\omega_D$), which corresponds to the absorption of two photons propagating in the same direction backward or forward by two symmetrical velocity classes ($+ v_z$ or $- v_z$).

If the two counterpropagating waves have the same polarization, a simple discussion shows that the area of the narrow curve is twice the area of the broad curve. Normally the Doppler width of the Gaussian curve is 100 or 1 000 times the natural width of the Lorentzian curve, and the Gaussian will appear as a very small background.

The detailed calculation in |32| showed the possibility of obtaining sufficient signal with the small power of c.w. dye Lasers in single mode operation.

2.4.2 <u>Experimental set-up</u>. The first experimental demonstrations, in Paris and Harvard |33|, were performed with pulsed dye Lasers ; however, the precision of the measurements is increased by the use of c.w. dye Lasers in monomode operation. The set-up is shown on Figure 20.

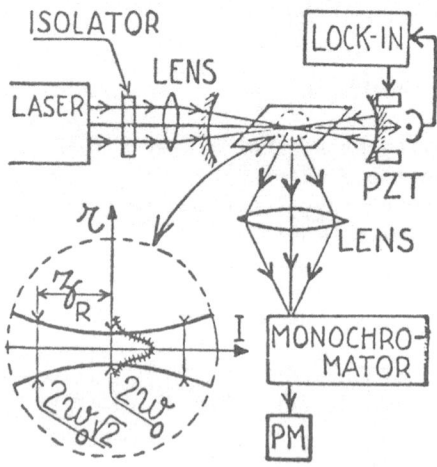

Fig. 20 - Experimental set-up for two-photon spectroscopy.

The light coming from the Laser is focused in the experimental cell with a lens, to increase the energy density P/S (the number of irradiated atoms decreases as S, but the two-photon probability increases as $(P/S)^2$ in such a manner that the signal increases as $1/S$). The transmitted light is refocused from the other side into the cell by a concave mirror, whose centre coincides with the focus of the lens.

The first concave mirror in Figure 20 is not essential, but in many experiments the energy density is increased by placing the experimental cell in a spherical concentric Fabry-Perot cavity. The windows of the experimental cell must be tilted at the Brewster angle to reduce the losses in the cavity. The length of this cavity is piezoelectrically locked on the Laser frequency, to maximize the signal transmitted through it.

In all cases, an optical isolator must be interposed between the Laser and the experiment, to prevent the return beam from perturbing the Laser (a $\lambda/4$ plate or a Faraday glass in an axial magnetic field achieves this).

The two-photon resonance is detected by collecting fluorescence photons emitted from the excited level at a frequency ω_{er}' which is different from the exciting frequency ω_L (Figure 19a) and can be selected with an interference filter or a monochromator. This allows the complete elimination of stray light from the Laser, despite its high intensity, and the observation of very small signals on a black background. For very highly excited states (near the ionization limit), one can use other, more sensitive techniques : thermoionic detection |34| or optogalvanic detection.

2.4.3 <u>Linewidth and transit time</u>. The detailed line shape is determined to a large extent by the transit time of the atoms through the Laser beam. We have seen the need to focus the light beam in order to increase the energy density. But this focusing reduces the time spent by the atom inside the Laser beam and broadens the two-photon line. It is possible to calculate exactly the effect of the varying electromagnetic field for a particular trajectory and afterwards the statistics on all trajectories. The calculation |35||36| leads to a very simple result ; the probability of the two-photon transition is proportional to :

$$\mathcal{P}^{(2)}_{(\omega)} \approx \int_{-\infty}^{+\infty} d\Omega \, \frac{\Gamma_e/2}{(\Omega-\omega_{oe})^2+\Gamma_e^2/4} \, e^{-\frac{|\Omega-2\omega|}{u/w_0}}$$

The line shape is the convolution of a Lorentzian curve with the natural width Γ_e of the excited level and a double exponential curve, whose width is determined by the inverse of the mean transit time w_0/u, where w_0 is the beam radius and u is the root mean square of the radial velocity, depending on the Maxwellian distribution and given by $u^2 = 2kT/M$ (ω_{oe} is the resonance center : $E_e - E_0 = 2 \hbar\omega_{oe}$).

As an example, I show in Figure 21 the recording of one component of the 3S-4D two-photon transition in sodium, where the photomultiplier current is plotted against the laser frequency |36|. That permits the evaluation of the broadening due to the short transit time of the atoms through the laser beam. The broader curve has been recorded with a waist radius of the Gaussian light beam $w_0 = 25$ µm, which gives a transit ti-

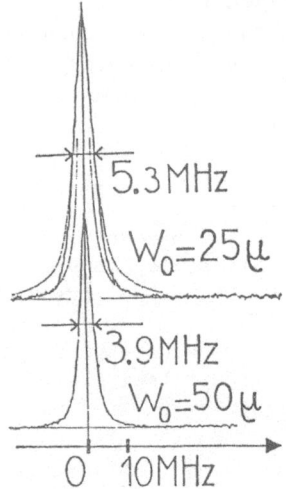

Fig. 21 - Experimental recordings of two-photon transitions depending
on the waist w_o of the Laser beam (fluorescence intensity
versus the Laser frequency).

me shorter than the lifetime. Its shape is quite different from a Lorentzian profile, as can be seen in Figure 21 (the dotted lines represent two Lorentzian curves, the first coinciding with the wings, the
second with the half-height of the experimental curve). The narrower
experimental curve has been recorded with a double waist radius (w_o =
50 µm). The transit time is double, and its shape is much closer to the
Lorentzian shape. That illustrates the compromise which must be found
between the signal intensity (necessity to focus) and the line narrow-
ness (avoid too small focus).

2.4.4 Experiments with two different light sources. If one uses two
different light sources of frequencies ω_1 and ω_2 instead of one, it is
possible, by keeping $\hbar\omega_1 + \hbar\omega_2$ constant and equal to the energy difference $E_e - E_o$, to increase the transition probability by a large factor
by reducing the energy defect $\hbar\Delta\omega_r$ (see Fig. 19a). The gain in transition probability is compensated for by the presence of a residual Doppler
broadening which is proportional to ($\omega_2 - \omega_1$). Such an experiment has
been performed on the 3S-4D two-photon transition of sodium by Bjorkholm
and Liao at the Bell Laboratories |37|. In this case, it has been possi-
ble to increase the transition probability by about seven orders of magnitude, while the residual Doppler broadening attains a value of the
order of 50 MHz :

$$\Delta\nu = \frac{v}{c}(\nu_1 - \nu_2) = \frac{v}{c}\nu_1 \times \frac{\nu_1 - \nu_2}{\nu_1} = \Delta\nu_D \times \frac{1}{40}$$

The two-photon spectrum is recorded by fixing the wave length λ_1
of the first absorbed photon and scanning the wave length λ_2 of the
second. The intensity of the two-photon line depends on the wave-length
λ_1. The variation of the two-photon transition rate plotted against λ_1

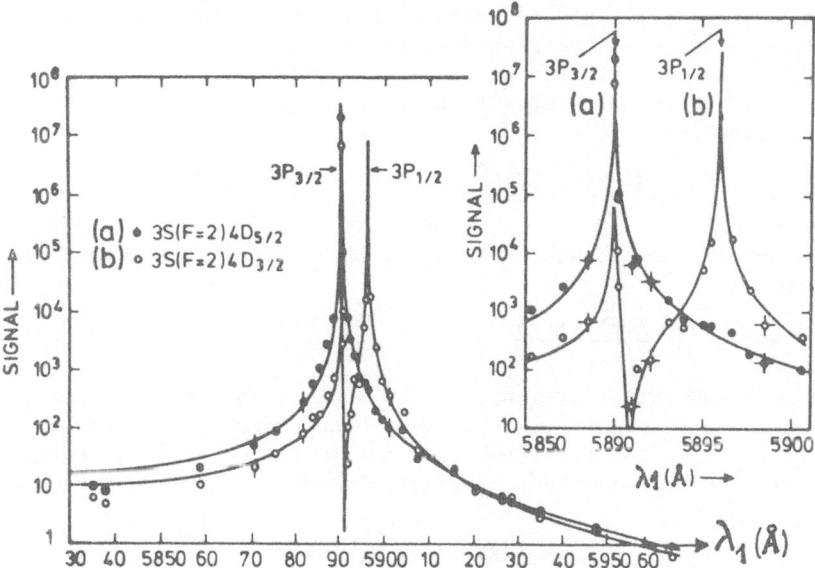

Fig. 22 - Two-photon transition rate in sodium using two unequal photons
as a function of the wavelength λ_1 of the fixed-frequency
Laser. The points are experimental and the curves theoretical.
The inset shows the behaviour of the central region with an
expanded scale (from ref |37|).

is represented in Fig. 22. When λ_1 is close to one of the resonance
lines of sodium, there is of course a sharp increase in the transition
probability. The variation of the transition rates with λ_1 are different in the cases of the excited states $4D_{3/2}$ and $4D_{5/2}$:

(1) For the transition to $4D_{5/2}$, only the $3P_{3/2}$ state can act as
an intermediate state since the $3P_{1/2} \rightarrow 4D_{5/2}$ single-photon transition
is forbidden. It follows that the corresponding transition rate presents
only one maximum.

(2) On the other hand, both $3P_{1/2}$ and $3P_{3/2}$ are possible intermediate states in the case of the two-photon transition to $4D_{3/2}$, and the
corresponding transition rate presents two maxima. Moreover, there is a
destructive interference between the two possible paths in this second
case for a value of λ_1 lying between 5890 Å and 5896 Å. Such a destructive interference between two terms of the calculation is evident
in other non-linear phenomena.

2.4.5 <u>Problem of the light-shifts</u>. Another important point to discuss
now is what is called the dynamical Stark effect of light shifts. These
light shifts are well understood since the pioneering work of Cohen-
Tannoudji in 1962 |38|. They are caused by every non-resonant irradiation
of the atoms, as we have in the intermediate steps of each multiphoton
transition.

These light shifts can be explained naively by virtual trans-
sitions, where the atom spends a very short time $1/\Delta\omega_r$ in the non-reso-

nant relay state r. The atom in the ground state shortly experiences the energy defect $\hbar\Delta\omega_r$ in this virtual state and this slightly changes the mean value of its ground-state energy E_0.

These modifications of the energy levels are given by the classical formula, where H represents the interaction Hamiltonian between atoms and light :

$$\frac{\delta E_0}{\hbar} = \sum_r \frac{\langle 0|H/\hbar|r\rangle\langle r|H/\hbar|0\rangle}{\Delta\omega_r}$$

One obtains the light shift δE_e of the excited state simply by changing 0 to e :

$$\frac{\delta E_e}{\hbar} = \sum_r \frac{\langle e|H/\hbar|r\rangle\langle r|H/\hbar|e\rangle}{\Delta\omega_r}$$

These light shifts are obtained by a perturbation calculation to the second order, exactly as the two-photon probability ; and that permits a comparison. One finds the same matrix elements in the formulae giving the two shifts and the probability (value at exact resonance when $\omega_L = \omega_{oe}$) :

$$\mathcal{P}^{(2)}_{oe}(res) = \frac{4}{\Gamma_e} \left| \sum_r \frac{\langle e|H/\hbar|r\rangle\langle r|H/\hbar|0\rangle}{\Delta\omega_r} \right|^2$$

In the simple case where it is possible to reduce the summation to one particular level r, the effect of which is predominating, we obtain this simple relation :

$$\frac{\delta E_g}{\hbar} \times \frac{\delta E_e}{\hbar} = \frac{\Gamma_e}{4} \times \mathcal{P}^{(2)}_{oe}(res)$$

The product of the two shifts is equal to the probability $\mathcal{P}^{(2)}_{oe}$ multiplied by the natural width Γ_e.

If the two photon transition is below the saturation, that is to say the probability is less than Γ_e, then the two shifts are also smaller than the natural width Γ_e and they can be neglected.

But this argument is not valid in the case of n-photon transitions (n greater than 2), the probabilities of which are calculated to higher order of the perturbation. This argument is also invalid in multiphoton ionization.

3. APPLICATIONS OF TWO-PHOTON SPECTROSCOPY

3.1 Short review of applications

In the limited pages of these lectures, we cannot explain in details all applications of the Doppler-free two-photon method, which can be found in the review papers |39| for the earliest experiments or |40| for more recent experiments.

In many experiments the Doppler-free two-photon transitions are used for their resolving power, which permits to distinguish close lying

levels, which are ordinarily confused inside the Doppler width : fine
or hyperfine structures, isotope shifts. The thermoionic detection method
|34| allows to easily detect highly excited Rydberg levels, near the
ionization limit. Large series of such structures have been studied
particularly in alkalis and alkaline-earths in Toronto |34||41|, Kiel
|42|, Berlin |43||44|, Mainz |45| and Göteborg |46|.

The multiphoton transitions have been used also to precisely measure the energies of some atomic levels. A large number of levels have
been determined with 10^{-8} precision in alkali metals : Rb in Boulder and
Toronto |47|, K in Kiel |48| and Cs in Washington |49|. Such a high pre-
cision is particularly interesting in simple system (two or three bodies)
which can be calculated ab initio with a comparable precision. It was
the case of experiments on Helium performed in Paris and which permitted
to attain seven highly excited levels from the metastable state |50|.
The agreement with theory |51| is very good for S levels, but not for D
levels, which seems to require a theoretical improvement.

A lot of experiments have been done also on collisions, measuring
broadenings and shifts due to collisions on buffer gases |39|. Concerning
the applications to molecules, we refer to other lectures in this
school.

3.2 Experiment on Hydrogen

3.2.1 Principle of the Hydrogen experiment. Obviously, Hydrogen is the
most interesting case for metrological purpose ; this is the reason why
in our laboratory F. Biraben and L. Julien undertook the Hydrogen experiment which we will describe now. Figure 23 shows a simplified energy
diagram of atomic hydrogen in relation with that description.

The interest of the 1S-2S two-photon transition was pointed out
from the beginning |32| since its theoretical linewidth is very small.
It is well known that the Stanford group succeeded in observing this
transition |52| and was able recently to obtain more precise measure-
ments |53|. Another measurement of this transition has been preformed
also in Southampton |54| ; these experiments provide new values of the
Rydberg constant. However the production of U.V. light (near 2430 Å)
presents large difficulties ; and the experimental linewidths remain up
to now relatively large (20 to 40 MHz), without any relation with the
theoretical limit. Moreover, as the Lamb shift of the 1S ground level
is not precisely known, the mere measurement of the 1S-2S transition
gives this Lamb-shift rather than the Rydberg constant.

The way chosen in Paris for the measurement of the Rydberg constant
is to study the two-photon transitions from the 2S metastable level toward highly excited Rydberg levels (Fig. 23). This choice presents three
advantages :
1- From the theoretical point of view : the 2S Lamb shift has been mea-
sured with high precision |55| ; and the calculation of the nuclear size
correction leaves on the 2S level (obviously on higher levels) a relative error much smaller than on the 1S ground level (3.10^{-12} instead of
5.10^{-11}).
2- The natural widths of the Rydberg levels are relatively small, since
they decrease as n^{-3} (not so small nevertheless as for the metastable

178

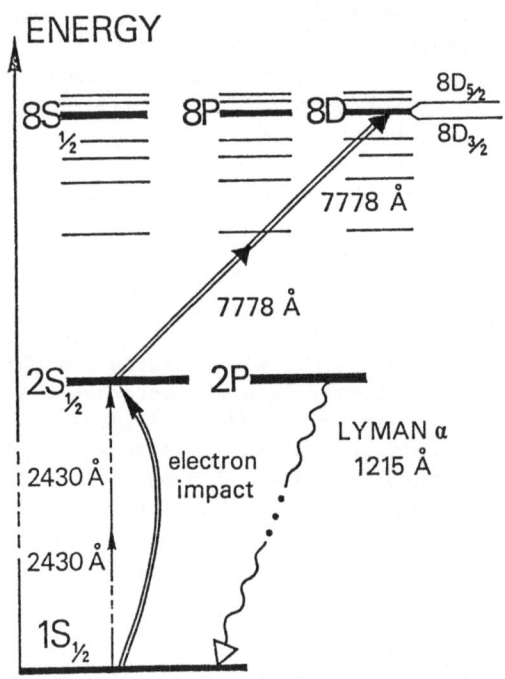

Fig. 23 - Two-photon transitions in the energy diagram of Hydrogen.
level). The width of the 8D level for example is 0.55 MHz, which corresponds to a relative linewidth smaller than 10^{-9}.
3- The range of wavelength concerned, at the limit of red and infra-red,
is covered by an efficient dye (LD700) which permits to obtain 1 Watt
in a c.w. dye Laser in single frequency operation.

3.2.2 <u>Set-up with an atomic beam</u>. In the current experiments, the two-
photon method permits to obtain the cancellation of the Doppler broadening in vapours and does not need the use of an atomic beam. But in that
particular experiment Biraben and Julien work on an atomic beam in order
to avoid broadening and shift due to collisions in the Rydberg levels.
That explains the experimental set-up |56| illustrated in Figure 24 :
    - Molecular Hydrogen is dissociated in a radiofrequency discharge,
and atomic Hydrogen is introduced through a narrow canal inside the vacuum box, where it makes an effusing beam.
    - Metastable 2S atoms are obtained by electronic bombardment of
the ground state 1S atoms beam inside the first vacuum chamber. A cylindrical grid at the ground potential delimites an equipotential volume
along the metastable beam axis, in order to protect metastable atoms
against electrical quenching. Nevertheless the efficiency of this production is very low ; only a very small percentage of the hydrogen atoms
are changed to metastable atoms.
    - Because of the inelastic collisions with electrons, the meta-
stable atomic beam is tilted with an angle of 20°, with respect to the
incident atomic beam. The tilting from the incident beam is used to make
the metastable beam colinear with the Laser beams. This colinear geometry

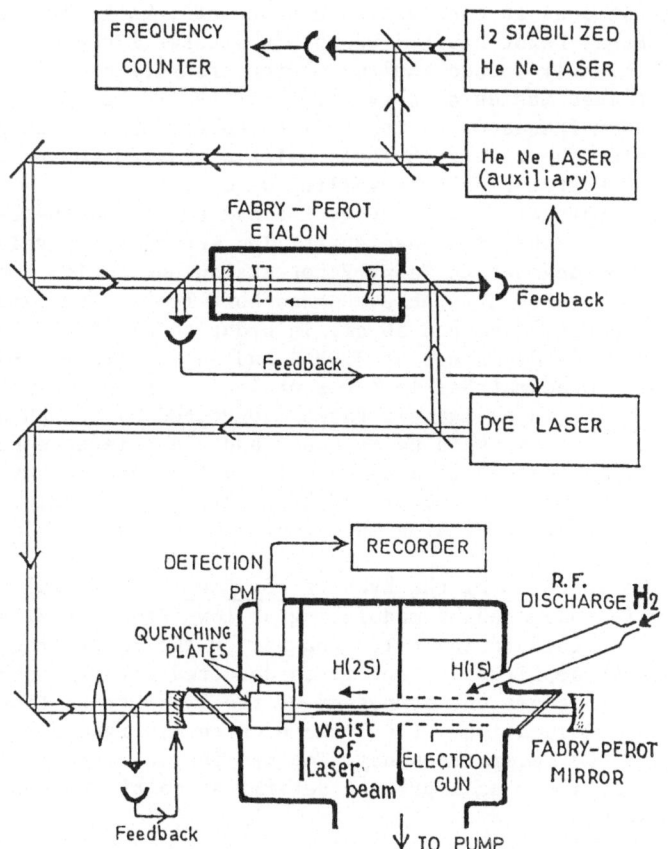

Fig. 24 - Experimental set-up for two-photon spectroscopy in a Hydrogen
          atomic beam.

has the advantage to suppress the line broadening due to the finite
transit time of atoms in the light beams |36| (the interaction between
light and atoms is longer than the lifetime of the excited Rydberg state). Optical excitation takes place in the second chamber of the vacuum
box where stray electric fields are reduced at best.

        - The third chamber is used to detect the metastable atoms : two
electrodes produce an electric field which quenches the 2S metastable
states : they are mixed to 2P state which radiates to ground state
emitting the Lyman α fluorescence. A photomultiplier measures the Lyman
α fluorescence which is proportional to the number of metastable atoms
in the beam. The metastable beam intensity can thus be estimated to be
about 10^7 atoms per second. The detection of the two-photon transition
is done by monitoring the population of the metastable level. It uses
the fact that when the Rydberg atoms fall down in cascade, about 90 %
of them come back to the ground state. So the metastable state is de-
populated by the two-photon transition.

        - The excitation beam is provided by a home-made c.w. ring dye
Laser |57|. At 7780 and 7600 Å (approximate wavelengths of the 2S-8D

and 2S-10D transitions) it can provide a power of 1 Watt. By locking it
on an external Fabry-Perot cavity, we obtain a Laser linewidth of about
150 kHz. To efficiently induce the two-photon transitions, the metastable
atomic beam is placed inside a Fabry-Perot cavity. The cavity length is
locked on the Laser frequency by monitoring the reflected beam polari-
zation |58|. Inside the cavity the beam waist w_0 is 570 μm and the light
power is about 40 Watt for each travelling wave.

    - The upper part of Fig. 24 shows the dye Laser and the experimen-
tal method used to measure its wavelength. The key of the wavelength
comparison is a non-degenerate Fabry-Perot etalon built with two silver
coated mirrors and enclosed in an evacuated box. Two etalon lengths have
been used, alternately 10 cm and 50 cm, in order to eliminate reflective
phase shifts |59|. The dye Laser at 7788 Å and an auxiliary He-Ne Laser
at 6328 Å are both mode matched into the etalon, in such a manner that
their wavelengths are in an integer ratio (corrected by the phase shifts).
The beat frequency between this He-Ne Laser and the reference He-Ne
Laser (stabilized on iodine) is measured with a frequency counter. During
a recording the dye Laser frequency is swept across the atomic resonance
by sweeping the Fabry-Perot etalon in the evacuated box.

3.2.2 Results. Measurement of the Rydberg constant.
The signal of the
photomultiplier is modulated by modulating at low frequency either the
quenching electric field in the third chamber or the laser frequency
(1 MHz peak to peak amplitude) ; and it is detected with a lock-in
amplifier. Figure 25 shows one component of the 2S-8D transition observed with frequency modulation : the lock-in output is recorded versus
the mean value of the Laser frequency. The two-photon line shape becomes
a derivative trace, allowing a precise determination of its center, but
which is somewhat broadened.

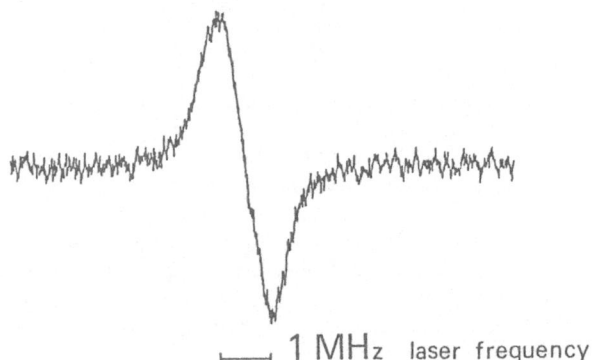

├─────┤ **1 MHz** laser frequency

Fig. 25 - Experimental recording of the two-photon transition $2S_{1/2}$
        $8D_{1/2}$ in Hydrogen (lock-in signal versus Laser frequency).

    The line center does not give exactly the 2S-8D energy because of
the small light shifts due to the non-resonant one-photon irradiation
|38|. For each transition, the center frequency has to be extrapolated
at null light power to eliminate these systematic shifts, whose the maximum value is 0.30 MHz.

    Precise measurements have been done for three transitions :

$2S_{1/2}$-$8D_{5/2}$ in Hydrogen and Deuterium, $2S_{1/2}$-$10D_{5/2}$ in Hydrogen |60|.
The detailed interpretation uses the theoretical work of Erickson |61|
and permits the determination of two fundamental constants : the ratio
M_p/m_e of the proton and electron masses (from the H and D measurements)
and the Rydberg constant : R_∞ = 109 737.315 69(6) cm^{-1}. This value is
compared in Figure 26 with other recent measurements obtained either in
the 1S-2S two-photon transition |53||54| or in the 2S-3P one-photon
transition in Stanford |62|, Teddington |63| and Yale |64|. The present
precision of the Paris experiment corresponds only to one half of the
linewidth, and it must be possible in the future to get improvement by
a factor 10.

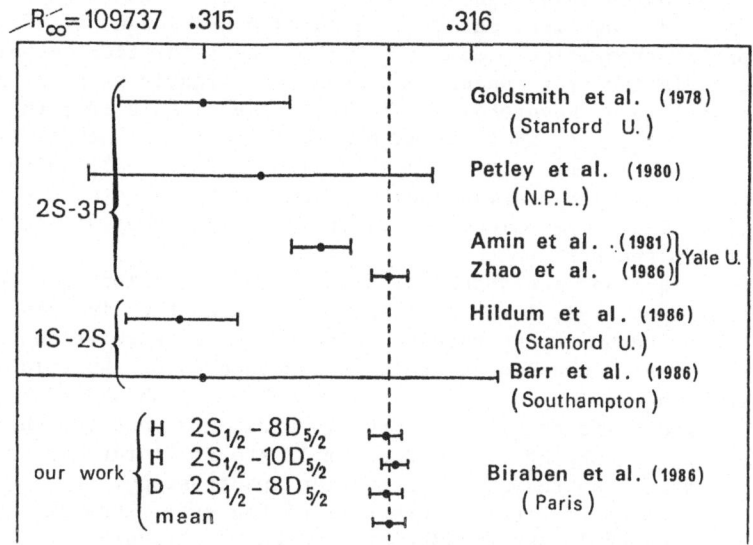

Fig. 26 -

Other metrologic experiments would be possible on Hydrogen in the
microwave range by studying direct transitions between adjacent Rydberg
levels. The experiments done in alkali atoms to produce "circular"
Rydberg states |65| could be generalized in Hydrogen. In these "circular"
states, the probability of the electron to be localized on the proton
is very small ; that permits to reduce the nuclear size correction and
to obtain a very precise calculation of the level energy. In the future,
may be, the Rydberg constant will be a theoretical tool allowing to
connect with high precision radiofrequencies and optical frequencies.

3.3 Conclusion. Comparison with other methods

In conclusion, we will try to compare succintly the two-photon method
with other Doppler-free methods. We first carry out the comparison with
the saturated absorption technique :
    1- In principle, the two techniques use the sign inversion of the
Doppler shift with the sign of the wave vector k, but in the two-photon
technique, the two opposite Doppler shifts cancel out by summation for

all the atoms, whereas in the saturation technique one selects the signal given by the small fraction of the atoms whose velocity component
$v_z = 0$.

2- The necessary light power is less for the saturation technique
than for the two-photon technique, but the power difference is not so
large because of the small fraction of the atoms which contribute to the
saturation signal and because of the high sensitivity of the two-photon
detection with a wave-length distinct from that of the laser. Some milli-
watts are enough for the current saturation experiments, but many two-
photon experiments have been done with about 100 mW. Moreover, it is
possible to increase the two-photon signal with powerful pulses, whereas
it is not possible in the case of the saturation technique.

3- The light shifts, caused by non-resonant irradiation of one-
photon transitions, do not exist in the saturation technique as the
irradiation is resonant, but they can slightly perturb the two-photon
transitions. Nevertheless, in most cases, it is easy to reduce these
light shifts so as to be much smaller than the natural line width.

4- The parasitic signal called "cross-over", on the contrary,
perturb the saturation spectra (and can be dangerous in the interpretation of complex molecular spectra), whereas they do not exist in two-
photon spectra.

5- The line shapes of two-photon transitions are very simple as
they are simply lorentzian curves, whereas the line shape in the saturation technique is quite complicated (its calculation involves the averaging of a non-linear effect which depends on the velocity component
v_z). In case of collisions, the two-photon line shape remains a lorentzian
one and it is easy to measure the broadening and the shifts, whereas the
velocity-changing collisions complicate still further the already complicated line shapes of saturation spectroscopy.

6- The finite transit time of the atoms through the Laser beam
produces analogous broadenings in the two techniques. The large light
beams (10 cm or more) are used in the saturation technique to reduce this
broadening. In the case of very narrow molecular transitions, it does
not seem to be possible to use two-photon transitions because the energy
density would be too small. But it is possible to use the Ramsey fringes
technique.

Finally, the two techniques have different advantages and disadvantages. In fact, they both have their own field of application - the
saturation technique for the lower excited levels, the two-photon technique for the higher excited levels. Moreover, the first technique must
be used between two levels of opposite parity, whereas the two-photon
technique must be used between two levels with the same parity.

The Doppler-free two-photon method must also be compared with other
classical Doppler-free methods, which can be used in vapours as the level
crossing technique or the double resonance technique.

1- These two classical methods do not necessitate the use of Laser
sources and can be performed using classical light sources. But, in practice, they have been greatly improved by the use of the c.w. tunable
Laser.

2- Concerning strictly the measurement of hyperfine intervals,
these two techniques maintain their useful role and many results have

recently been obtained in highly excited levels by the combination of these two techniques with the stepwise excitation using simultaneously two Lasers at different wave-lengths. It is difficult to compare the convenience of these two classical methods with that of the two-photon method and therefore it is a matter of personal preference only.

3- On the other hand, these two classical methods can not give any information about isotopic shifts or absolute measurements of wave-lengths as the two-photon and the saturation methods do.

REFERENCES

|1| R.H. DICKE, Phys. Rev. **89** (1953) 472.

|2| see for example A. ABRAGAM, Principles of Nuclear Magnetism (Clarendon Press, Oxford, 1960).

|3| J. BROSSEL and F. BITTER, Phys. Rev. **86** (1952) 308
G.W. SERIES, Reports on Progress in Physics **22** (1959) 280.

|4| A. KASTLER, Journal de Physique, **11** (1950) 255.
C. COHEN-TANNOUDJI and A. KASTLER "Optical Pumping" in Progress in Optics, **V** edited by E. Wolf (North Holland, 1966).

|5| E.B. ALEXANDROV, Opt. Spectrosc **17** (1964) 957.
J.N. DODD, R.D. KAUL, D.M. WARRINGTON, Proc. Phys. Soc. **84** (1964) 176
J.N. DODD, W.J. SANDLE, D. ZISSERMANN, Proc. Phys. Soc. **92** (1967) 497.

|6| T. HADEISHI, W.A. NIERENBERG, Phys. Rev. Lett. **14** (1965) 891.

|7| H.J. ANDRÄ, Phys. Rev. Lett. **25** (1970) 325.
H.J. ANDRÄ, Physica Scripta **9** (1974) 257.

|8| W. GORNIK, D. KAISER, W. LANGE, J. LUTHER, H.H. SCHULZ, Optics Comm. **6** (1972) 327.
P. SCHENCK, R.C. HILBORN, H. METCALF, Phys. Rev. Lett. **31** (1973) 189.
R. WALLENSTEIN, J.A. PAISNER, A.L. SCHAWLOW, Phys. Rev. Lett. **32** (1974) 1333.

|9| S. HAROCHE, J.A. PAISNER, A.L. SCHAWLOW, Phys. Rev. Lett. **30**(1973)948
J.S. DEECH,R. LUYPAERT, G.W. SERIES, J. of Phys. **B8** (1975) 1406

|10| S. HAROCHE, M. GROSS, M. SILVERMANN, Phys. Rev. Lett. **33** (1974) 1063
C. FABRE, M. GROSS, S. HAROCHE, Optics Comm. **13** (1975) 393.

|11| S. HAROCHE "Quantum Beats and Time Resolved Fluorescence Spectroscopy" in High Resolution Laser Spectroscopy edited by Shimoda, p. 253 (**13** of Topics in Applied Physics - Springer 1976).

|12| F.O. COLEGROVE, P.A. FRANKEN, R.R. LEWIS, R.A. SANDS, Phys. Rev. Lett. **3** (1959) 420.
G. ZU PUTLITZ in Atomic Physics **1** p. 227 (Plenum Press 1969 - Proceedings of the First International Conference on Atomic Physics 1968).

|13| A. CORNEY and G.W. SERIES, Proc. Phys. Soc. **83** (1964) 207.

|14| P.H. LEE and M.L.SKOLNIK, Appl. Phys. Lett. **10** (1967) 303.
V.S. LETOKHOV, Soviet Physics JETP Lett. **6** (1967) 567.

|15| C. FREED and A. JAVAN, Appl. Phys. Lett. **17** (1970) 53,541

|16| W.E. LAMB, Phys. Rev. **134** (1964) 1429
A. SZÖKE and A. JAVAN, Phys. Rev. **145** (1966) 137.

184

|17| C. BORDÉ, C.R. Ac. Sc. Paris **271** (1970) 371
M. OUHAYOUN and C. BORDÉ, C.R. Ac. Sc. Paris **274** (1972) 411
T.W. HÄNSCH, M.D. LEVENSON, A.L. SCHAWLOW and P. TOSCHEK, Bull.
Am. Phys. Soc. **16** (1971) 310 ; Phys. Rev. Lett. **26** (1971) 946.
|18| T.W. HÄNSCH, I.S. SHAHIN and A.L. SCHAWLOW, Phys. Rev. Lett. **27**
(1971) 707.
|19| H.R. SCHLOSSBERG and A. JAVAN, Phys. Rev. **150** (1966) 267.
|20| R.L. BARGER and J.L. HALL, Phys. Rev. Lett. **22** (1969) 4.
|21 C.BORDÉ, G. CAMY, B. DECOMPS and L. POTTIER, C.R. Ac. Sc. Paris
277 (1973) 381.
|22| C. WIEMAN and T.W. HÄNSCH, Phys. Rev. Lett. **36** (1976) 1170.
|23| J.C. KELLER and C. DELSART, Optics Comm. **20** (1977) 147.
|24| M. GAWLIK and G.W. SERIES in Laser Spectroscopy **IV** p. 210 (Springer
1979) - Proceedings of the Fourth International Conference on Laser
Spectroscopy).
|25| W. DEMTRÖDER "Polarization Spectroscopy" vol **5** of Springer Series
in Chemical Physics (1982).
|26| S.L. KAUFMAN, Optics Comm. **17** (1976) 309.
|27| H. WINTER and M. GAILLARD, J. of Phys. **B10** (1977) 2739
|28| M. GÖPPERT-MAYER, Annalen der Phys. (Leipzig) **9** (1931) 273.
|29| V. HUGHES, L. GRABNER, Phys. Rev. **79** (1950) 314 and 819
J. BROSSEL, B. CAGNAC, A. KASTLER, J. de Physique **15** (1954) 6
P. KUSCH, Phys. Rev. **93** (1954) 1022, and **101** (1956) 627
J.M. WINTER, Ann. de Physique **4** (1959) 745.
|30| I.D. ABELLA, Phys. Rev. Lett. **9** (1962) 453.
|31| L.S. VASILENKO, V.P. CHEBOTAYEV and A.V. SHISHAEV, JETP Lett. **12**
(1970) 161.
|32| B. CAGNAC, G. GRYNBERG and F. BIRABEN, J. de Physique **34** (1973) 845.
|33| F. BIRABEN, B. CAGNAC and G. GRYNBERG, Phys. Rev. Lett. **32** (1974)
643
M.D. LEVENSON and N. BLOEMBERGEN, Phys. Rev. Lett. **32** (1974) 645.
|34| K.C. HARVEY and B.P. STOICHEFF, Phys. Rev. Lett. **38** (1977) 537
D. POPESCU, M.L. PASCU, C.B. COLLINS, B.W. JOHNSON and I. POPESCU,
Phys. Rev. **A8** (1973) 1666.
|35| C. BORDÉ, C.R. Ac. Sc. Paris **282** (1976) B341.
|36| F. BIRABEN, M. BASSINI and B. CAGNAC, J. de Physique **40** (1979) 445.
|37| J.E. BJORKHOLM and P.F. LIAO, Phys. Rev. Lett. **33** (1974) 128.
|38| C. COHEN-TANNOUDJI, Ann. de Physique **7** (1962) 423
C. COHEN-TANNOUDJI and J. DUPONT-ROC, Phys. Rev. **A5** (1972) 968.
|39| G. GRYNBERG and B. CAGNAC, Rep. Progr. Phys. **40** (1977) 791
E. GIACOBINO and B. CAGNAC, in Progress in Optics, vol. **XVII**
(North Holland, Amsterdam, 1980) p. 87
G. GRYNBERG, B. CAGNAC and F. BIRABEN, in Advances in Coherent
Non-Linear Optics (Springer, 1980, vol. **21** of the series Topics in
Current Physics).
|40| B. CAGNAC, Hyper Int. **24-26** (1985) 19 and 43.
|41| B.P. STOICHEFF and E. WEINBERG, Can. J. Phys. **57** (1979) 2143.
|42| K. NIEMAX and K.H. WEBER, J. of Phys. **B11** (1978) L267
C.J. LORENZEN, K. NIEMAX and L.R. PENDRILL, Phys. Rev. **A28** (1983)
2051.

|43| R. BEIGANG, E. MATTHIAS and A. TIMMERMANN, Phys. Rev. Lett. 47 (1981) 326, and 48 (1982) 420
R. BEIGANG and A. TIMMERMANN, Phys. Rev. A25 (1982) 1496.

|44| H. RINNEBERG and J. NEUKAMMER, Phys. Rev. Lett. 49 (1982) 124
H. RINNEBERG, J. NEUKAMMER and E. MATTHIAS, Zeit. Phys. A306 (1982) 11.

|45| R. MENGES, G. HUBER, G. ULM and T. KÜHL, Zeit. Phys. A320 (1985) 575.

|46| S. HÖRBACH, A.M. and L. PENDRILL, and M. PETTERSON, Zeit. Phys. A318 (1984) 284.

|47| S.A. LEE, J. HELMCKE, J.L. HALL and B.P. STOICHEFF, Optics Lett. 3 (1978) 141
B.P. STOICHEFF and E. WEINBERGER, Can. J. Phys. 57 (1979) 2143.

|48| C.J. LORENZEN, K. NIEMAX and L.R. PENDRILL, Optics Comm. 39 (1981) 370.

|49| K.H. WEBER and C.J. SANSONETTI, J.O.S.A. B2 (1985) 1385.

|50| E. GIACOBINO and F. BIRABEN, J. of Phys. B14 (1982) L385.

|51| Y. ACCAD, C.L. PEKERIS and B. SCHIFF, Phys. Rev. A4 (1971) 516, and A11 (1975) 1479
P. BLANCHARD and G. DRAKE, J. of Phys. B6 (1973) 2495.

|52| S.A. LEE, R. WALLENSTEIN and T.W. HÄNSCH, Phys. Rev. Lett. 35 (1975) 1262
C. WIEMAN and T.W. HÄNSCH, Phys. Rev. Lett. 34 (1976) 1170.

|53| C.J. FOOT, B. COUILLAUD, R.G. BEAUSOLEIL and T.W. HÄNSCH, Phys. Rev. Lett. 54 (1985) 1913
E.A. HILDUM, U. BOESL, D.H. McINTYRE, R.G. BEAUSOLEIL and T.W. HÄNSCH, Phys. Rev. Lett. 56 (1986) 576.

|54| J.R.M. BARR, J.M. GIRKIN, J.M. TOLCHARD and A.I. FERGUSON, Phys. Rev. Lett. 56 (1986) 580.

|55| S.R. LUNDEEN and F.M. PIPKIN, Phys. Rev. Lett. 46 (1981) 232
YU L. SOKOLOV and V.P. YAKOVLEV, Sov. Phys. JETP 56 (1982) 7.

|56| F. BIRABEN and L. JULIEN, Opt. Comm. 53 (1985) 319.

|57| F. BIRABEN AND P. LABASTIE, Opt. Comm. 41 (1982) 49.

|58| T.W. HÄNSCH and B. COUILLAUD, Opt. Comm. 35 (1980) 441.

|59| H.P. LAYER, R.D. DESLATTES and W.G. SCHWEITZER Jr., Appl. Opt. 15 (1976) 734.

|60| F. BIRABEN, J.C. GARREAU and L. JULIEN, Europhys. Lett. 2 (1986) 925.

|61| G.W. ERICKSON, J. Phys. Chem. Ref. Data 6 (1977) 831.

|62| J.E.M. GOLDSMITH, E.W. WEBER and T.W. HÄNSCH, Phys. Rev. Lett. 41 (1978) 1525.

|63| B.W. PETLEY, K. MORRIS and R.E. SHAWYER, J. Phys. B13 (1980) 3099.

|64| S.R. AMIN, C.D. CALDWELL and W. LICHTEN, Phys. Rev. Lett. 47 (1981) 1234.

|65| R.G. HULET and D. KLEPPNER, Phys. Rev. Lett. 51 (1983) 1430.

ATOMIC LASER SPECTROSCOPY

P.E.G. Baird
Clarendon Laboratory
Parks Road
OXFORD
OX1 3PU

ABSTRACT. Tunable lasers make possible experiments on atoms with very
high precision and sensitivity. Examples of the application of lasers
to the study of fundamental atomic systems are given as well as to the
study of atoms only available in low concentration, e.g., radioactive
isotopes. New techniques such as those for cooling, trapping and
manipulating atoms are described together with recent experiments using
them. Finally, the field of sensitive laser polarimetry is reviewed
and throughout attention is drawn to phenomena occurring as a result of
the high spectral brightness of laser sources.

1. LASER SPECTROSCOPY OF FUNDAMENTAL SYSTEMS

1·1 Introduction

Experimental investigations of the simplest atomic systems have
always been at the forefront of theoretical developments in atomic
physics. The Bohr model of hydrogen gave way first to the development
of the quantum treatment of Schrödinger and Heisenberg and later to
relativistic quantum mechanics in the form of Dirac's treatment of the
electron. In 1947 Willis Lamb[1] discovered that the $2S_{1/2}$ and $2P_{1/2}$
levels were not in fact degenerate as predicted in Dirac's theory but
were separated by about a tenth of the fine structure interval between
$2P_{1/2} - 2P_{3/2}$ (~ 10 GHz). The explanation for this phenomenon was
subsequently given by Bethe and others in the development of the
quantum field theory of electromagnetic interactions, otherwise known
as quantum electrodynamics, or QED for short. To this day the
two-body, hydrogen-like system represents one of our most accurate
tests of bound state QED. I say 'bound state' because other
experiments, notably those which measure (g-2) for the muon and
electron, also provide stringent tests of QED of a single unbound
particle. Indeed, in the remarkable experiments conducted at the
University of Washington at Seattle, Dehmelt and co-workers[2] a few
years ago succeeded in trapping a single electron for a period of nine
months and have achieved an accuracy approaching 1 part in 10^{12} for the
value of (g-2) for the electron — a truly remarkable feat!

A. C. P. Alves et al. (eds.), Frontiers of Laser Spectroscopy of Gases, 187–239.
© 1988 by Kluwer Academic Publishers.

In the following I shall discuss a number of recent laser experiments on two-body systems namely, hydrogen, positronium and muonic atoms (μ He).   In describing these experiments I will be introducing several applications of Doppler-free laser spectroscopy and of frequency-doubled tunable radiation.   I will spend first a little time on the theory of the hydrogen atom, contrasting it with that of positronium and muonium.   I will then make one or two remarks about frequency calibration of Doppler-free spectra and then consider in some detail laser experiments performed in Oxford and Stanford on hydrogen.

1·2 Outline of the theory of the hydrogen atom

In the absence of hyperfine interaction due to the magnetic moment of the proton, the energy eigenvalues for a Dirac electron in a fixed Coulomb potential are given by

$$E_D(n,j) = \frac{2 R Z^2}{(\alpha Z)^2} \left\{ \left[1 + \left(\frac{\alpha Z}{n - (j+\frac{1}{2}) + \sqrt{(j+\frac{1}{2})^2 - \alpha^2 Z^2}} \right)^2 \right]^{-\frac{1}{2}} - 1 \right\}$$

where α is the fine structure constant, and

$$(1)$$

$$R = R_\infty \left(\frac{M}{m+M} \right) \tag{2}$$

is the reduced mass Rydberg constant (in cm^{-1}); R_∞ is the Rydberg constant for infinite nuclear mass and M and m are the proton and electron masses respectively.   Hydrogen itself, however, differs in a number of respects from this model.   Firstly, the proton is not infinitely massive and the relativistic Dirac equation cannot be written exactly in terms of a reduced mass — the hydrogen atom is really a two-body problem.   Inclusion of nuclear motion leads to a recoil term which contributes only to a level shift and not to the fine structure splitting (see, e.g., Bethe and Salpeter[3]) i.e.

$$E_R \sim - \left(\frac{Z^4 \alpha^2}{4n^4} \right) \cdot \left(\frac{m}{M} \right) \cdot R_\infty . \tag{3}$$

Next, the finite extent of the proton charge distribution has an effect for s-states and the correction to account for the nuclear size and structure takes the form:

$$E_N = \frac{\langle r^2 \rangle}{(\hbar/mc)^2} \frac{2(Z\alpha)^4 mc^2}{3n^3} \left[1 + C_{str} \right] \tag{4}$$

where $\langle r^2 \rangle$ is the mean square radius of the charge distribution and C_{str} is a nuclear structure factor.

In addition, quantum electrodynamic corrections to the Coulomb interaction modify the potential at short range; we refer here to the sum of all these corrections as the Lamb shift of a level, following Johnson and Soff[4] and Mohr[5] and note that this definition differs from that of Erickson[6].   In particular I shall be concerned with the

Lamb shift of the ground state (1s) of hydrogen which can be detected in the laser experiments I will describe, but which is of course not accessible by the sort of radiofrequency techniques applicable to $2s_{1/2} - 2p_{1/2}$ Lamb shift.

The form of these corrections then, to lowest order is

$$E_{QED} = \frac{8Z^4\alpha^3 R_\infty}{3\pi\ n^3}\ \{[\ell n\ \frac{1}{(Z\alpha)^2} + \frac{11}{24} - \frac{1}{5}]\delta_{\ell 0} + L_{n,\ell} + \frac{3}{8}(\frac{C_{\ell j}}{2\ell+1}) + H(Z\alpha)\} \tag{5}$$

made up respectively of the self-energy contribution, vacuum polarisation (the $-1/5$ in the first term), the mass- and $Z\alpha$-independent Bethe logarithm, the anomalous magnetic moment and higher-order binding corrections.

Finally, to take account of the hyperfine interaction in the states of interest we note that the contact interaction leads to

$$E_{hfs} = \frac{Z^3\alpha^2 g_I}{n^3}\ R_\infty\ (\frac{m}{M})\ \{\frac{f(f+1) - I(I+1) - j(j+1)}{j(j+1)(2\ell+1)}\} \tag{6}$$

Here $g_I = 5\cdot56$ and is the gyromagnetic ratio for the proton for which $I = \frac{1}{2}$. For the 1s state this gives a splitting of about 1420 MHz (actually the splitting is known to about 1/100 of a Hz from hydrogen maser measurements but that's another story!). Relativistic corrections to equation (6) come in at the level of about 30 parts per million and to first order the hyperfine interaction does not shift the centre of gravity of the pattern. Thus we may write for the energy level structure of the hydrogen atom,

$$E(n,j,\ell) = E_D(n,j) + E_R(n) + E_N(n,\ell) + E_{QED}(n,j,\ell)$$

$$+ E_{hfs}(n,j,I,f) \tag{7}$$

In table 1 is presented a summary of the contributions to the Lamb shift for the 1s, 2s and 4s levels (in accordance with Johnson and Soff's definition). Also given is the predicted value of

$$\delta = \Delta E\ (1s-2s) - 4\Delta E(2s-4s) \tag{8}$$

since in some of the laser work it is this quantity which can be measured. (Quite recently, however, the energy interval (1s-2s) has been determined against the calibrated $^{130}Te_2$ saturated absorption spectrum in the vicinity of 486 nm[7]. In this case the uncertainty in the Rydberg constant enters into the calibration, but with the present precision of about 3 parts in 10^{10} in R_∞ this gives rise only to an error of about 675 kHz, lower than the present uncertainty of 1 MHz (250 kHz at 486 nm) in the Te_2 reference itself[8].)

Contribution	1s	2s	4s
Dirac energy and two-body correction	-3288095031·33	-822026489·28	-205505425·06
Self-energy	8383·38	1071·29	134·96
Vacuum polarization	-214·82	-26·85	-3·36
High order (4th & 6th)	1·02	0·13	0·02
Finite nuclear size	1·07	0·13	0·02
Relativistic recoil	2·41	0·34	
Lamb shift	8173·07	1045·04	131·68

Calibration of the 1s-2s frequency:

$\frac{1}{4}$(1s-2s)[F=1 → F=1] — Te$_2$ line 'b$_2$': 1379·53 MHz

— 2s-4s[F=1 → F=1]: -4836·19 MHz

Table 1   Contributions in MHz to the total energy in 1s, 2s and 4s of hydrogen.

1·3 Muonic Atoms

The exchange of a muon for the electron in hydrogen leads to an atomic system having considerably enlarged QED effects. However, in the case of μp and μHe the major contribution to the Lamb shift comes now from the vacuum polarisation rather than the self-energy interaction as in hydrogen; this is because the muon penetrates the virtual e^+e^- cloud of the nucleus to a far greater extent.

1·4 Modification to the theory for the case of Ps and μ^+e^-

(a) Positronium (Ps)

The change in the reduced mass for this atomic system leads to an energy level structure similar to that of hydrogen but characterised by a Rydberg constant of half the magnitude. Of course, there are other differences as well, for example there is no real distinction any more between fine structure and hyperfine structure. Furthermore the positronium system is inherently unstable decaying into either 2 or 3 photons, depending on the angular momentum of the state concerned. The quantum electrodynamic corrections are therefore significantly different from those of atomic hydrogen as they include many virtual decay processes. Indeed, the Dirac equation itself is not an adequate starting point for the derivation of QED effects in Ps and the

Bethe-Salpeter formalism must be used. Unfortunately no known analytical solution to the fully covariant Bethe-Salpeter equation exists, although the exact Hamiltonian is believed known to high precision. To order α^4 the energy level structure is given by [9]

$$E(n,\ell,s,j) = -\frac{R_\infty}{2n^2} + \left\{\frac{11}{32n^4} + \left(E_{\ell sj} - \frac{1}{(2\ell+1)}\right)\frac{1}{n^3}\right\}\alpha^2 R_\infty \qquad (9)$$

where $E_{\ell sj}$ has the following values:

$$E_{\ell sj} = 0 \qquad \text{for } s=0 \text{ and } j=0$$

and

$$E_{\ell sj} = \frac{7}{6}\delta_{\ell o} + \frac{(1-\delta_{\ell o})}{2(2\ell+1)}
\begin{cases}
\dfrac{3\ell+4}{(\ell+1)(2\ell+3)} & ; \quad j = \ell+1 \\[2mm]
-\dfrac{1}{\ell(\ell+1)} & ; \quad j = \ell \\[2mm]
-\dfrac{3\ell-1}{\ell(2\ell-1)} & ; \quad j = \ell-1
\end{cases}$$

(b) Muonium

The muonium system resembles more closely the structure of hydrogen than does positronium; the reduced mass is not substantially changed $\left(M_R(\mu) = 0 \cdot 9952\ m_e;\ M_R(H) = 0 \cdot 99945\ m_e\right)$ and the 1s-2p transition remains in the vacuum ultraviolet. The energy level structure is summarised in figure 1 (see, e.g. C.J. Oram et al.[10])

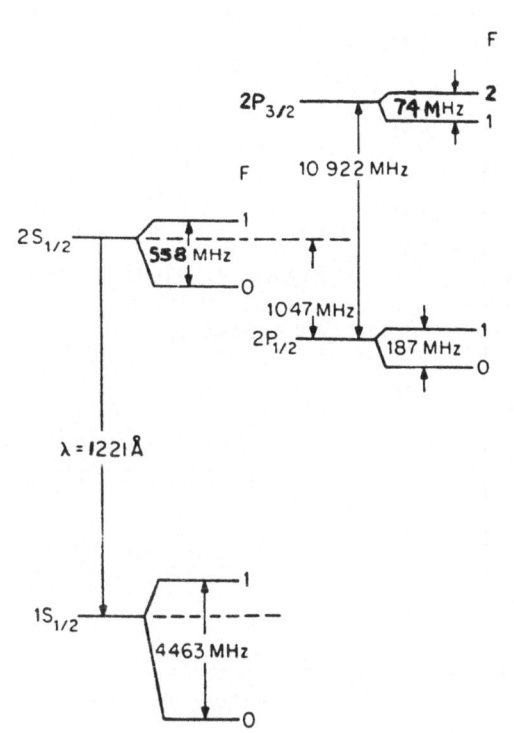

Figure 1. Structure of the n=1 and n=2 states of muonium.

Although measurements of the ground state hfs splitting and the n=2 Lamb shift have been made, the analogous two-photon laser experiment (1s-2s) to that in hydrogen has only recently been comtemplated because of developments in the production of slow muonium atoms[11]. It should be noted that both positronium and muonium are pure leptonic systems and therefore do not suffer from any uncertainty in nuclear size; in the case of hydrogen the present error in the proton size determination is ~ 4% (see equation (5)).

LASER EXPERIMENTS ON ATOMIC HYDROGEN

1•5 Linear absorption spectroscopy of Hα

Several investigations of the Doppler-free spectrum of the Hα-transition (λ656•3 nm) in atomic hydrogen have been carried out with the primary aim of determining the Rydberg constant to very high accuracy. The motivation for this stems from the fact that the Rydberg constant can be expressed solely in terms of fundamental constants and can in principle be determined to a precision limited only by the accuracy of the frequency standard itself. For such precise metrology three approaches have been used: saturated absorption spectroscopy, 2γ absorption and atomic beam, laser-resonance spectroscopy. Lichten and co-workers at Yale have for some years now adopted the last of these

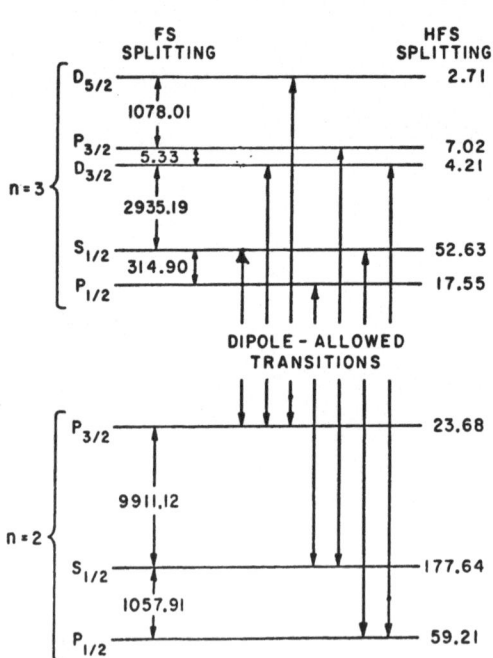

HYDROGEN BALMER-ALPHA

Figure 2. Term diagram for the Hα transition n = 2-3 in hydrogen.

techniques which is in general less prone to uncertainties introduced by collisions. The wavelengths of the four possible Hα and Dα lines from the 2S state were measured with respect to the calibrated frequencies of $^{127}I_2$, using an iodine stabilized, He—Ne laser and comparing the wavelengths using a very stable Fabry-Perot etalon. The value of the Rydberg deduced was $R_\infty = 109737 \cdot 31569(7)$ cm^{-1}, with a one standard deviation error of about 6·5 parts in 10^{10}[12]. Among the systematic effects that have to be considered at this level of precision are the photon recoil shift (~ 1 MHz) and the second order Doppler effect. Finally, we note that the atomic beam measurements have now been extended[13] to H_β which has a narrower natural width; these have yielded the improved value of $R_\infty = 109737 \cdot 31573(3)$ cm^{-1}.

1·6 Saturated absorption spectroscopy of Hα, Dα and Tα

The saturated absorption method has been applied by a number of authors to the study of the Hα transition. The technique is often preferred over its rival polarisation spectroscopy because the latter gives in general a mixture of symmetrical and asymmetrical lineshapes

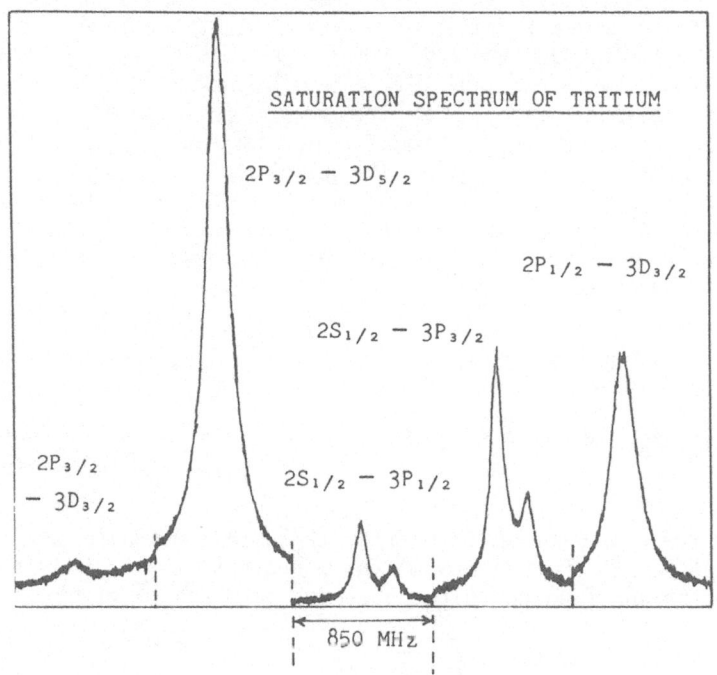

Figure 3. Doppler-free spectrum of Balmer-α in tritium. (The resolved splitting in the $2s_{1/2} - 3p_{1/2}/3p_{3/2}$ transitions is due mainly to the 177 MHz hfs of the 2s state.)

and can therefore easily lead to systematic shifts of the resonance frequency. In Oxford we have made use of the technique to obtain for the first time Doppler-free spectra of the radioactive isotope tritium[14]. The procedure used was as follows. A mixture of the three hydrogen isotopes (H_2, D_2 and T_2) was loaded into a cell which was subsequently sealed. Beta emission from the tritium component is weak and does not penetrate the glass walls of the cell. Metastable hydrogen atoms were produced using an r.f. exciter and an example of the saturation spectra obtained is shown in figure 3. Frequency calibration was provided by two confocal etalons of different lengths (20 cm and 1 m) each of which was held at a constant temperature in an evacuated chamber. The free spectral range of each was determined by scanning a large wavenumber interval, keeping track of the order number by means of a wavemeter. In this way the cavities were calibrated to an accuracy of ~ 5 parts in 10^7 and the observed drift rate of each was no more than about 15 MHz/hour. The observed H_α linewidth was attributable to a number of different sources: the natural width which varies from component to component, the finite uncrossing angle of pump and probe beams, pressure and Stark broadening and finally laser power broadening. To interpret the effects of laser power, scans were made at different pump intensities. The theoretical analysis of saturation spectroscopy for the case where the pump intensity is much greater than that of the probe is given by Baklanov and Chebotayev[15] and involves the solution of the density matrix equations for the interaction of the pump and probe beams with the atoms to obtain the polarisability χ of the medium and hence, from the imaginary part of χ, the absorption coefficient. The calculation for ρ_{21} is obtained to all orders in the pump field but only to terms linear in the probe. Within this framework the absorption coefficient comprises two parts, A and B, where the A-term gives the absorption coefficient K_S for the probe beam in the presence of the pump in the simple hole-burning limit. After velocity averaging, i.e.,

we find,
$$\int_{-\infty}^{\infty} dv_z \rho_{21}(\omega', v_z)$$

$$K_S(\omega') = K_0(\omega') \left\{ 1 - \left[1 - \frac{1}{(1 + \frac{I}{I_S})^{\frac{1}{2}}} \right] \frac{\tilde{\Gamma}^2}{[(\omega'-\omega) + 2(\omega-\omega_{21})]^2 + \tilde{\Gamma}^2} \right\}$$

(10)

where for the sake of generality the probe and pump are considered to have different frequencies ω' and ω respectively. I_S is the saturation intensity and $\tilde{\Gamma}$ which is the observed HWHM and is given by

$$2\tilde{\Gamma} = \Gamma \left[1 + (1 + \frac{I}{I_S})^{\frac{1}{2}} \right].$$

The B-term, which contributes when $I \gtrsim I_S$, describes interference between the probe and pump beams which sets up an oscillation in the populations $\rho_{11}(\omega'-\omega)$ and $\rho_{22}(\omega'-\omega)$; this causes coherent scattering of the pump beam at ω' which can interfere with the probe and affect its absorption. Thus, the lineshape contains two contributions, a Lorentzian curve of FWHM = 2Γ and depth ~ $\frac{1}{2}I/I_S$, and another Lorentzian of FWHM = $1/T_1$, where T_1 represents the decay of population. In our experiments although effects due to the B-term could be seen with

increased laser power (ca. 5 mW mm^{-2}), the spectra were recorded mainly
in the low saturation limit i.e. $I < I_s$.

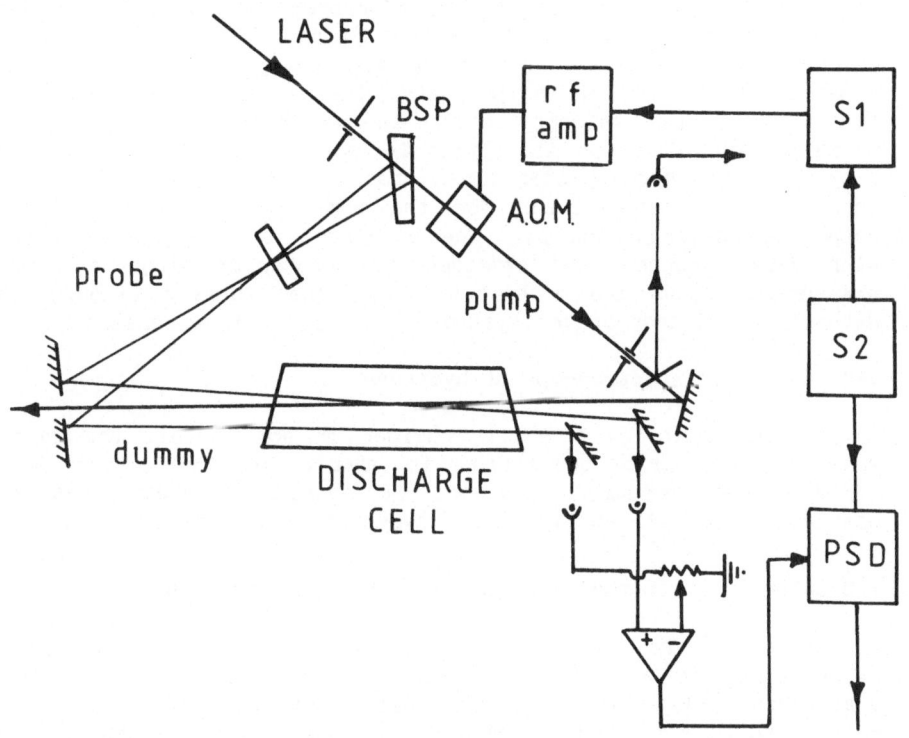

Figure 4. Apparatus for saturated absorption spectroscopy of tritium.

In the experiment (see figure 4) an acousto-optic modulator
provided the means for chopping the pump beam rapidly and controlling
its power. The 'dummy' probe allows for some cancellation of the noise
as follows. Let the probe and dummy signals in the absence of the pump
beam be given by I_P and I_D respectively, then

$$I_P = I_0 \exp\{- K_0(\omega)\ell\} \quad \text{where } K_0 \simeq K_\delta \text{ accounts for any}$$

$$I_D = \beta I_0 \exp\{- K_\delta(\omega)\ell\} \quad \text{slight difference in optical paths.}$$

ℓ is the length of the cell, I_0 is the fluctuating laser intensity and
β is a quantity which takes account of any difference in the
intensities of the two beams. The output of the difference amplifier
is thus

$$(I_P - I_D) \simeq I_0(1-\beta) - \ell \, I_0\big(K_0(\omega) - \beta K_\delta(\omega)\big).$$

By adjusting the relative gain the first term can be made equal to
zero. The output of the phase sensitive detector is then,

$$(I_P - I_D)_{I=0} - (I_P - I_D)_{I\neq 0} = \ell \, I_0 \big(K_S(\omega) - K_0(\omega)\big) \qquad (11)$$

Finally, in analysing the spectra due account had to be taken of the presence of cross-over resonances[16]. An interesting example of this occurs in the transition $2s_{1/2} - 3p_{1/2}$ where an improved fit was obtained by including a Stark induced cross-over which sits between the two partially resolved peaks. What happens is that the electric field in the discharge mixes the $3s_{1/2}$ and $3p_{1/2}$ levels which are only separated by 314 MHz and so provides an absorption route from 2s to 3s. For a particular velocity class in the cell, one laser beam can transfer population to the 3s state and the other beam, interacting with the same velocity class, can transfer population to the 3p state; there is thus competition for the same group of 2s atoms and a cross-over signal results. The existence of such effects, together with data recorded on hydrogen and deuterium alone at different pressures, allows the contributions to the observed linewidth to be determined and many of the systematic errors to be assessed.

1·7 Two-photon spectroscopy of hydrogen

In the following I shall consider in some detail the particular case of the 1s-2s transition in atomic hydrogen using cw laser radiation at 243 nm. However, it should be clear that similar considerations will apply to most two-photon experiments on atoms.

1·8 Calculation of the 2γ absorption rate for hydrogen

There are at least two theoretical papers dealing with this calculation: Gontier and Trahin[17] and Bassani, Forney and Quattropani[18]. These give the same result for the 1s-2s transition rate although some care is required in comparing the results. In the paper by Bassani, Forney and Quattropani the matrix element sum is calculated in both gauges (length and velocity) and the authors obtain the same answer, -11·7805, providing a valuable consistency check of the result. An outline of the calculation is given below. The transition rate, which is a second-order interaction[17], has the following form,

$$\Gamma_{2\gamma} \propto I_1 I_2 \sum_{\substack{np+ \\ \text{continuum}}} \left| \frac{\langle 2s|\epsilon_1.D|np\rangle\langle np|\epsilon_2.D|1s\rangle}{h\nu_L - (E_{np} - E_{1s})} \right|^2 \cdot g(\nu)$$

$$(12)$$

Or, in detail

$$\Gamma_{2\gamma} = 4.(2\pi)^2 \, r_0 c \, \frac{I_1 I_2}{I_0} \left(\frac{M}{E_p^2}\right) g(\nu) \qquad (13)$$

where r_0 is the classical electron radius,

$$I_0 = 1·4038 \times 10^{-17} \text{ W cm}^{-2} \quad \text{and} \quad E_p = h\nu_L.$$

The lineshape factor $g(\nu)$ for the 1s-2s case has to be chosen with some care since the natural width, which is about 1 Hz, is clearly not the dominant contribution. However, I will assume here for simplicity that the dominant lineshape is Lorentzian in form, caused by pressure broadening, although in reality contributions from the finite laser bandwidth and the transit time broadening[19] cannot so easily be dismissed in a continuous wave (cw) experiment. With this caveat,

$$g(\nu) = \frac{\Delta\nu_{1/2} / (2\pi)}{(\nu_{2s} - 2\nu_L)^2 + (\frac{\Delta\nu_{1/2}}{2})^2} \tag{14}$$

which has a maximum value at line centre of $2/\pi\Delta\nu_{1/2}$; $\Delta\nu_{1/2}$ is the FWHM. Evaluating this rate we have,

$$\Gamma_{2\gamma} = I_1 I_2 (\frac{2}{\pi\Delta\nu_{1/2}}) \times 2 \cdot 140 \times 10^{-7} \ m^4 \ J^{-2}$$

where the intensities I_1 and I_2 are in units of $W \ m^{-2}$.

For a width of 10 MHz the rate is

$$\Gamma_{2\gamma} = (1 \cdot 362 \times 10^{-14} \ sm^3 \ J^{-2}) \ I_1 I_2 = S \ I_1 I_2.$$

This is the rate per atom and so to obtain a value for the signal expected in an experiment we must consider the laser beam geometry and the number density n of hydrogen atoms. Here I will assume that the laser beam can be described by the lowest order TEM_{00} Gaussian mode. The power per unit area of each beam is then given by

$$I = \frac{2P}{\pi} \frac{1}{\omega^2} \exp(-\frac{2r^2}{\omega^2}) \tag{15}$$

with

$$\omega^2 = \omega_0^2 + (\frac{\lambda z}{\pi\omega_0})^2.$$

The integrated absorption is easily obtained as follows.

$$\text{Total rate} = \int n \ S \ I_1 \ I_2 \ d^3r$$

$$= \frac{\pi n \ S \ P_1 P_2}{\lambda} \tag{16}$$

At first sight this expression may appear peculiar since the spot size (ω_0) of the focussed laser beam is not included. The point is simply that the more tightly focussed the beam the fewer atoms are in the active volume. Of course, the length of the region over which say 50% of the two-photon absorption occurs decreases the more sharply focussed the beam, although at some point the increase in transit time width with decreasing spot size will affect the rate. The observable rate will clearly be less than the total rate and if we assume a waist size of ω_0 and an observation length ℓ we can determine the fraction of the total absorption taking place in this region. It is given by:

$$f = \frac{2}{\pi} \tan^{-1} \frac{\ell\lambda}{2\pi\omega_0^2}. \tag{17}$$

To take some typical values appropriate to a cw experiment we obtain for ω_0 = 100 μm and ℓ = 1 cm, f = 0·025. The total rate for an atomic hydrogen density of $10^{1.5}$ cm^{-3} and a UV power of 1 mW is of order 10^8 s^{-1} so in the observation region the absorption rate is about 10^6 s^{-1}. If we now further assume that every 2γ absorption is followed by a Lyman-α decay caused by collisions transferring population from $2s_{1/2}$ to $2p_{1/2}$, and make due allowance for the solid angle of collection (~ 10%), the efficiency of any Lα filter used (~10%) and assume a quantum efficiency for the detector ~ 10%, we obtain a signal count of order 10^3 s^{-1}. The crucial experimental questions are thus: what is the available UV power, what is the obtainable atomic hydrogen density and what is the background count rate? We will consider these questions in section 1·10.

1·9 Frequency doubling to 243 nm

I will briefly go through some of the important considerations in selecting a crystal for UV generation. Although I will illustrate this in the context of the hydrogen experiments the procedure for calculating the phase-matching angles is widely applicable.

(a) Urea

To generate efficiently UV radiation from non-linear crystals a number of conditions have to be met. Most obviously the crystal must be transparent at both fundamental and UV wavelengths otherwise heating and burning of the crystal will occur; the crystal also has to be cut in such a way as to enable the fundamental and UV beams to remain 'phase-matched' as they propagate through the crystal. Phase-matching is achieved in one of two ways: either by temperature tuning which makes use of the fact that the indices of refraction for the ordinary and extraordinary rays change differently with temperature, or alternatively by angle-tuning in which case the crystal is cut so that index surfaces for the two beams (visible and UV) intersect giving the same phase velocity. To illustrate these points let us consider frequency doubling in urea which is a positive ($n_e > n_o$) uniaxial crystal of symmetry $\overline{4}2$ m (i.e. tetragonal). For the case of interest, namely frequency doubling to 243 nm, the relevant non-linear coefficient is d_{14} which has the value $1·4 \times 10^{-12}$ m V^{-1}. Second harmonic polarisation can thus be generated by the following fundamental electric field components

$$P_x = 2d_{14}\varepsilon_0 E_y E_z$$
$$P_y = 2d_{14}\varepsilon_0 E_x E_z$$

and the resultant after resolving along the input light field direction is given by

$$(P_x + P_y) = 2\varepsilon_0 d_{14}\overline{E}_0^2 \tfrac{1}{2}\sin2\theta \left\{\sin\phi - \cos\phi\right\}. \tag{18}$$

Here θ is the angle to the z-axis (the optic axis) and φ the angle to the x-axis. Index-matching is possible in this configuration with the fundamental propagating as the extraordinary wave (plane-polarised in a

plane containing z) and the second harmonic travelling as the ordinary
wave (polarised orthogonal to a plane containing z). Furthermore, we
can evaluate the angle at which the refractive indices of the ordinary
and extraordinary waves at the required wavelengths (486 nm and 243 nm)
intersect. This is the phase-matching angle θ_m and is given by

$$\frac{\cos^2\theta}{n_{01}^2} + \frac{\sin^2\theta}{n_{e2}^2} = \frac{1}{n_{02}^2(\theta)} . \qquad (19)$$

In the present case $\theta_m = 74°$. Finally, the efficiency of 243 nm light
generated at this phase-matched angle varies with ϕ as can be seen from
the equation (18); it maximises for $\phi = 0$ or $\pi/2$ and is zero for
$\phi = \pi/4$. (It is interesting to note here that the converse is true for
negative uniaxial crystals such as ADA, ADP and its isomorphs.)

Since θ_m is not $\pi/2$, Poynting's vector for the extraordinary wave
is not collinear with $\underline{k}$ the wave-vector ($\underline{D}$ is not parallel to $\underline{E}$) and so
the fundamental beam gradually 'walks away' from the UV it has
generated. This leads to an astigmatic UV beam. The extent of the
walk-off is given by the angle ρ which can be evaluated from the
equation

$$\tan(\theta+\rho) = \left(\frac{n_0}{n_e}\right)^2 \tan\theta. \qquad (20)$$

For the present case $\rho = 2\cdot2°$. The optimum length for a crystal using
a focussed beam can be evaluated from the analysis given by Boyd and
Kleinman[20] i.e.

$$P^{SHG} = C k_1 \ell P_1^2 h (\xi,B) \qquad (21)$$

where P^{SHG} and P_1 are the powers of the second harmonic and fundamental
beams respectively. k_1 is the wavevector for the input light and
$h(\xi,B)$ stands for the integral over the active volume. Here the
focussing and double refraction parameters are

$$\xi = \frac{1}{k_1^2 \omega_0} \quad \text{and} \quad B = \rho(\ell k_1)^{1/2}/2.$$

Beyond the optimum length the integrated UV power continues to increase
but not nearly so rapidly, tending to a quasi-linear dependence on ℓ
(c.f. discussion of 2γ rate for hydrogen). The constant C contains the
non-linear coefficients in the form of d_{eff} which takes account of the
angles of propagation of the fundamental beam. Thus,

$$C = \frac{\omega_1^2 d_{eff}^2}{2\pi n_1^2 n_2 \varepsilon_0 c^3}$$

where n_1 and n_2 are the appropriate refractive indices and ω_1 the
angular frequency of the fundamental laser radiation.

To obtain the form of the intensity profile of the UV beam at the
output of the crystal we simply have to sum the contribution from each
element in the interaction volume within the crystal. An analysis of

this kind has been given by Fleck and Feit[21], but in our experiments using a 4 mm length crystal and a fundamental beam waist of ~ 30 μm the intensity distribution looks like an extended, flat-topped Gaussian. If, as in our case, it is required subsequently to couple this light into a cavity it is important to know what percentage of the beam can be represented as a TEM_{00} mode. To determine this theoretically we can make use of the completeness theorem for Gaussian beams, and calculate the overlap integral[22],

$$n_{TEM} \propto \int E^*(x,y) \, E_{TEM_{00}}(x,y) \, dxdy. \tag{22}$$

The results of such an analysis are indeed in good agreement with observations using a non-confocal cavity.

(b) β-Barium Borate

Very recently crystals of this compound, which has excellent properties for UV generation down to ~ 200 nm, have become commercially available. From reports in the literature and our own preliminary experiments this crystal is in many ways superior to urea: it is a harder material, has a higher damage threshold and a better d_{eff}. Its only drawback for SHG at 243 nm is the large walk-off angle, $\rho = 4.74°$. The results for this negative uniaxial crystal are given below for comparison. The symmetry is C_3 (planar ring structure) and

$$d_{eff} \text{ (type I)} = (d_{11}\sin\phi - d_{22}\cos\phi)\sin\theta + d_{31}\cos\theta \tag{23}$$

From the refractive index data of Kato[23], $\theta_m = 54°$ and assuming the same sign for d_{11}, d_{22} and d_{31} the optimum azimuthal angle is $\phi = 0$ which gives $d_{eff} = 2.18 \, d_{36}(KDP)$.

1.10 Observation of the two-photon transition (1s-2s) in hydrogen

Observation of the hydrogen two-photon signal in a cell using cw radiation has been reported by the group at Stanford[7] and now by the Oxford group[24]. In the Stanford experiment approximately 3 mW of 243 nm radiation generated by frequency summing in KDP was directed into a cell fed from a hydrogen dissociator. A loop of teflon pipe connecting the two serves both as an efficient material for the transfer of hydrogen atoms and as a 'light baffle' preventing radiation from the discharge reaching the Lyman-α detector. (Since reformation of molecular hydrogen is either a three-body process or involves the walls of the container, it is possible in such a flowing system for a substantial fraction of the hydrogen atoms to diffuse from the discharge region into the cell.) To enhance the UV power available the cell is placed inside a cavity which is servo-controlled so as to follow the wavelength changes of the incoming radiation. An intra-cavity power of order 10 mW was available and a peak Lyman-α count of 10,000 s^{-1} was observed. In the original experiment no calibration was available since neither of the input summing wavelengths (789 nm from an LD700 dye laser and 351 nm from an argon laser) could easily be measured to the precision required or compared straightforwardly with another hydrogen transition. Indeed ease of calibration is an important virtue of the frequency doubling approach adopted by the Oxford group since the fundamental frequency is very close to the H_β transition (ca. 4·8 GHz, see table 1). The Oxford experiment thus uses two coumarin 102 ring dye lasers, one for frequency doubling and one for calibration; the difference frequency at 486 nm is then measured directly on a fast photodiode. Furthermore recent developments in the frequency calibration of the Te_2 spectrum[8] in the vicinity of 486 nm make this also an attractive reference for calibrating the 1s-2s transition frequency. The present position then is that both the Stanford and Oxford groups have now measured the observed c.w. signal with respect to Te_2; experiments to assess the systematic error are currently underway in Oxford while Stanford have obtained[7] the value 2 466 061 413·8(1·5) MHz for the H(1s-2s) frequency from which the 1s Lamb shift is determined to be 8173·3(1·7) MHz, in good agreement with the theoretical value of 8172·94(9) MHz.

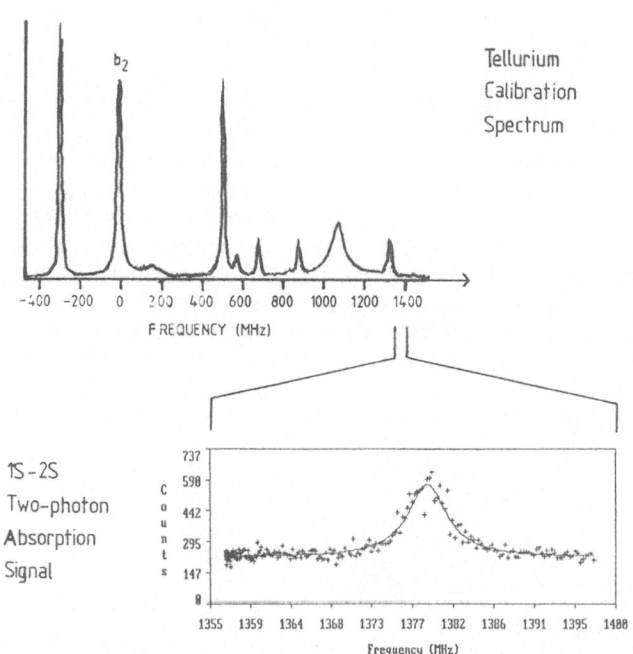

Figure 5. 1s-2s transition of hydrogen calibrated against the spectrum of $^{130}Te_2$.

A discussion of the feasibility of extending such experiments to the 1 Hz natural linewidth of this transition has been given by Beausoleil and Hänsch[25]. The collisional broadening is avoided by using a very cold hydrogen beam (ca. 1 mK) slowed in the earth's gravitational field. This 'fountain' of hydrogen atoms intersects a

243 nm radiation field twice, once on the way up and again on the way down. The 'double pulse' of excitation is analogous to the use of two coherent field regions in atomic-beam magnetic-resonance work (Ramsey[26]) and produces effectively a long interaction time which thereby reduces the transit time width. With a sufficiently narrow bandwidth laser, optical resolution approaching 1 in 10^{15} may one day be obtainable.

1·11 Two-photon spectroscopy of positronium

The development of high-intensity, pulsed, low-energy positronium emission from metal surfaces has in recent years made possible optical spectroscopy of this purely leptonic system. It is perhaps the most extreme example of the value of Doppler-free spectroscopy since even the slow positronium atoms (produced in the long-lived 1^3S state) having an energy of only 0·14 eV move with speeds of order 10^6 ms^{-1}. The transition $1^3S - 2^3S$ (energy interval equivalent to 243 nm) has a Doppler width in excess of 400 GHz. Indeed the second-order Doppler effect ($\simeq$ 100 MHz) and the profile asymmetry are by no means negligible in this case. In their first experiment using two-photon excitation at 486 nm Chu and Mills[27] obtained a linewidth of 1·5 GHz and a signal-to-noise ratio of 20:1. The transition was detected by subsequent photoionisation from the 2^3S_1 by a third photon at 486 nm and the e^+e^- pair detected in coincidence in using channel multiplier arrays. The light source was a pulsed dye laser system comprising a frequency-tripled, Q-switched Nd:YAG pump laser and a pressure-tuned, grazing incidence-grating dye laser with an intra-cavity etalon. Three subsequent stages of amplification produced tunable output of peak power 18 mJ in a 10 ns pulse with a bandwidth of 800 MHz. The Ps transition frequency was calibrated by using a stable Fabry-Perot, itself calibrated by recording the H_β-D_β isotope shift. The result obtained for $\frac{1}{2}E(1^3S - 2^3S)$ was (3/16) c R_∞ -41·4(5) GHz. This is in agreement with the $O(\alpha^4)$ calculation of Ferrell (41 GHz) — equation (9) — and the corrections given by Fulton and Martin (0·812 GHz)[28].

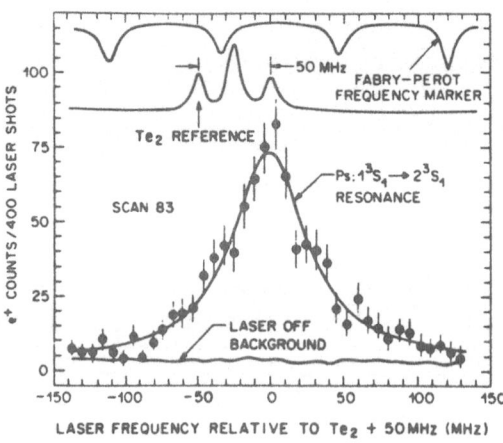

Figure 6. Two-photon excitation of positronium $1^3S - 2^3S$.

In a later experiment[29] the laser was replaced by a cw ring dye laser, the output of which was amplified in four stages using a XeCl excimer laser as pump. In this way the cw output (30 − 60 mW) at 486 nm was amplified into 25 mJ, 10 ns pulses with a repetition rate of 50 s^{-1}. In addition in this later experiment the frequency of the transition was determined with respect to the saturated spectrum of Te$_2$, a line of which lies within 50 MHz of half the $1^3S - 2^3S$ energy interval. Furthermore, an acousto-optic modulator was used to create a laser sideband at 50 MHz higher than the fundamental frequency so that each Te$_2$ absorption line consisted of 3 peaks, at ν_0, ν_0 + 50 MHz and ν_0 + 25 MHz, the cross-over resonance.

Using a calibrated interferometer the Te$_2$ line at ν_0 was measured with respect to D$_\beta$ and an improved value of the transition frequency which verifies the $\alpha^3 R_\infty$ corrections of Fulton[30] was determined, i.e.,

$$\frac{3}{8} R_\infty c - 83546\cdot3 \pm 10\cdot5 + 10\cdot6 = 1233607185 \pm 15 \text{ MHz.}$$

The major systematic effects in this measurement were: (i) the frequency offset between the cw dye laser and the pulsed beam resonant with the Fabry-Perot, (ii) the second-order Doppler effect and (iii) the a.c. Stark effect produced by the intense laser light.

Since the announcement of this result in 1984 the Te$_2$ absorption spectrum has been calibrated and the absolute frequency of the component near to the Ps transition measured[31].

1·12 Muonic hydrogen-like systems

The Lamb shift and fine structure ($2s - 2p_{1/2, 3/2}$) in the lightest muonic atoms corresponds to wavelengths in the vicinity of 6 μm for μH, 812 nm and 897 nm for μ^{4}He and 822 nm and 1061 nm for μ^{3}He. In experiments analogous to the rf measurements on n=2 of hydrogen the muonic measurements lead to precise tests of the vacuum polarisation and to the determination of the nuclear charge radii. Good experimental precision can be attained using laser resonance techniques, and the natural linewidth of the Lamb shift transition in μH for example is only 300 ppm while for μHe it is 1000 ppm.

To date the muonic Lamb shift has only been measured by Zavattini and co-workers in μ^{4}He using the infrared transition at 897 nm[32]. If the ^{4}He rms charge radius is taken from electron scattering experiments, the vacuum polarisation contribution can be verified to 1700 ppm which is at present one of the best tests of a specific QED effect. A similar experiment is currently underway at SIN.

Finally, for hydrogenic systems it has been noted[33] that dominant uncertainty in the nuclear size contribution arises from the normalisation of the electron scattering form factors F(q) which are used to determine the proton and deuteron charge radii. However, if the ratio of these factors F_d/F_p is used an improved value for the quantity $r_d^2 - r_p^2$ may be deduced leading to a reduction in the uncertainty in the theoretical value for the isotopic shift.

2. ATOMIC BEAMS, LASER COOLING, ATOMIC TRAPS AND RYDBERG ATOMS

2·1 Introduction

In this section I hope to show how the sensitivity of laser spectroscopy is exploited to obtain data on very low concentrations of atoms. In particular I will start off by considering a few laser atomic beam studies aimed at measuring optical isotope shifts and show how short-lived nuclei can be studied in this way. I shall also mention how it is possible to beat the natural linewidth and obtain 'supernatural spectra'. The discussion of laser studies at low atomic concentrations then leads me onto consider experiments on laser cooling and trapping of atoms and ions. In this context I will also mention some experiments using the 'shelved electron' idea to detect very weak transitions. Finally, I will say something about Rydberg atoms and the effects of atoms near metallic surfaces.

2·2 Optical isotope shifts in atoms

The change in mass and charge distribution of nuclei having the same charge, Z, but different mass number, A, leads to slight differences in the energy of corresponding levels of different isotopes of an element. This difference in energy for a transition can be expressed to reasonable accuracy in the following way

$$\Delta \nu_i = \delta(\Delta E)/h = F_i \Lambda_{AA'} + M_i g_{AA'} \tag{24}$$

where the two terms represent the field and mass contributions to the observed shift and

$$g_{AA'} = \left(\frac{M_A - M_{A'}}{M_A M_{A'}}\right) \quad ; \quad \Lambda_{AA'} = \left(\delta\langle r^2 \rangle + \beta\delta\langle r^4 \rangle \ldots\right)$$

F is a quantity related to the change in the total probability density of the electrons at the nucleus and the leading term in $\Lambda_{AA'}$, $\delta\langle r^2 \rangle$ is the change in the mean square radius of the nuclear charge distribution. (β increases with Z but is only ~ 8% for Z = 82 for typical nuclear models[34]). If a transition does not involve the change of one or more s-electrons the change in the probability density will be small and the field shift will be reduced accordingly. M represents the total mass shift and varies as $(\Delta M/M_1 M_2)$; it is therefore proportionately less important for heavier elements. It is made up of two components, the normal mass (NMS) and the specific mass shifts(SMS), as shown below

$$K.E. = \frac{P_N^2}{2M} = \frac{\sum_i P_i \cdot P_i}{2M} + \frac{\sum_{i \neq j} P_i \cdot P_j}{M} \tag{25}$$

where P_N is the nuclear momentum and P_i that of the electron. While the NMS can easily be calculated, the SMS is considerably more complicated, and is related to the type of transition being investigated. In most studies of isotope shifts it is the nuclear charge distribution which is of interest. The procedure used to extract the quantity $\delta \langle r^2 \rangle$ varies from case to case but in general it involves choosing a simple transition (i.e. where possible an alkali-like s-p transition) for which the SMS is likely to be small or alternatively may be estimated with reasonable precision. If such a transition is available but weak (i.e. it is not a resonance transition) then the shifts can be studied for the stable abundant isotopes in the weak transition and, using the linear form of the equation (24), can subsequently be related to other data on resonance transitions for low abundance (radioactive) isotopes. A plot of the expression relating the shifts in one transition to those in any other is known as a King plot[35].

The value of F for a transition can be obtained in favourable cases (e.g. alkali-like transitions again) from a knowledge of the contact hyperfine structure[36] or alternatively the field shift can be calibrated using muonic atom data[37].

2·3 Experimental studies of isotope shifts

In most laser studies of optical isotope shifts the crossed, laser-atomic beam method is used to avoid Doppler-broadening. In some cases short-lived radioactive isotopes are detected directly on-line, in others, by irradiating a target which is subsequently heated to form the source of an atomic beam. Figure 7 shows a spectrum obtained some years ago in Oxford on the resonance line of barium at 554 nm[38].

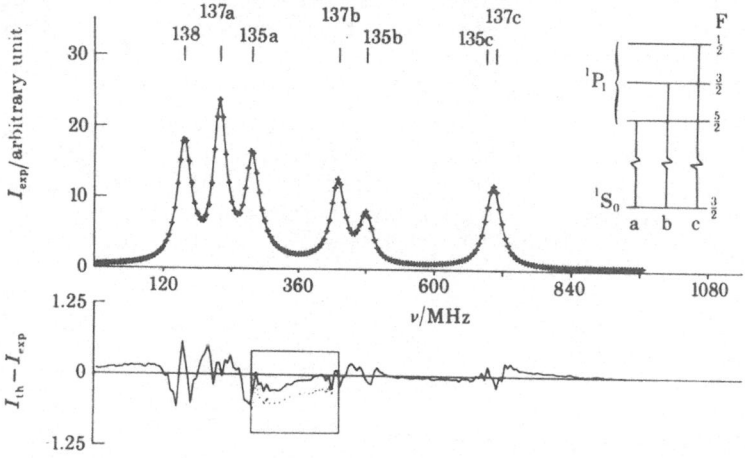

Figure 7. Laser-atomic beam technique applied to the λ554 nm transition of BaI. The input polarisation has been adjusted to reduce the contribution from the even isotopes (^{138}Ba is 71·6% abundant).

Several interesting features emerge from this study which involves a $J_g = 0 \rightarrow J_e = 1$ transition. The first is that the even isotopes can be made to disappear from the spectrum simply by using laser light linearly polarised, parallel to the direction of observation. The reason for this is that no radiation is emitted in the direction end on to a dipole. However, the additional nuclear angular momentum in the odd isotopes causes the J vector to precess around F allowing now some fluorescence to reach the detector. Conversely, if the plane of polarisation of the laser beam is aligned at right angles to the observation direction all isotopes contribute to the observed spectrum.

The second feature to note is that in the computer fit of the spectrum shown in figure 7, the region between the hyperfine components does not fit well unless due allowance is made for the interference between neighbouring hfs components; that is, the levels are not completely separated beyond the natural linewidth and a sum of transition amplitudes rather than a sum of squares of amplitude must be taken. Finally, there is the effect of optical pumping on the odd isotopes which alters the component ratios, and also the phenomenon of laser power broadening which can be seen on all spectral components when the laser intensity exceeds a few mW mm^{-2}.

Experiments conducted subsequently in Karlsruhe[39] have now extended the range of barium isotopes studied to include many short-lived, neutron-deficient isotopes.

In another experiment, also on the resonance line of barium[40], it was shown that the natural linewidth itself could be reduced, although at the cost of some signal. A low flux beam of barium atoms was used such that at any one time only one barium atom was in the laser interaction region. The detector recording the atomic fluorescence was gated so that only the signal from those atoms producing a burst of say 5 or more photons was recorded; this corresponded to atoms near the peak of the transition probability and the resulting spectrum contained narrowed components having a modified Lorentzian lineshape.

Studies of other elements that are worthy of mention here are calcium and tin. Both elements have magic numbers of nucleons and are of interest from the nuclear model point of view. Experiments in Oxford[40] (figure 8) on the two-photon transition $4s^2\ {}^1S_0 - 4s5s\ {}^1S_0$ of CaI realised a linewidth of just 4 MHz while experiments, again at

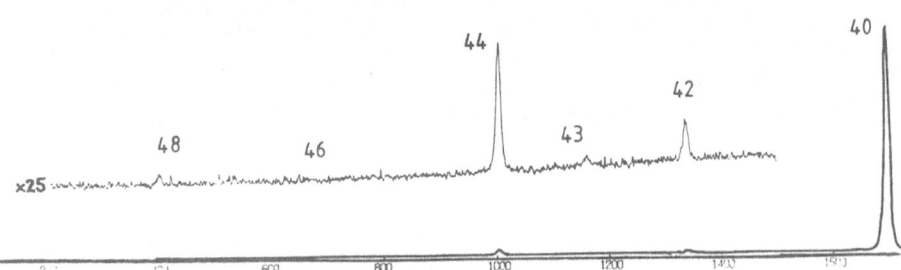

Figure 8. Two-photon spectrum of the transition $4s^2\ {}^1S_0 - 4s5s\ {}^1S_0$ in CaI, showing isotopic structure.

Karlsruhe, produced measurements on the resonance line at 422·7 nm for many of the radioactive isotopes. It is a similar story with the isotopes of tin (there are 10 stable ones representing the longest chain of stable isotopes in the periodic table) but here the resonance transition lies in the ultra-violet. At the Clarendon Laboratory[42] two dye lasers were used for this study, one locked to a line of I_2 to act as a reference for optical heterodyning and frequency calibration, and the second, whose output was frequency-doubled in ADA, was used to excite an atomic beam. Many other studies of this kind could be cited but the methods are very similar; optical heterodyning is in general replacing frequency calibration by interferometry and in the case of ions the collinear laser-atomic beam geometry is often used. In particular an accelerated ion beam produces a 'velocity bunching' effect which leads to narrowed resonances[43].

2·4 Laser cooling of atoms

I have already mentioned in passing in section 1 the trapping of a single, isolated electron. The extension of this to trapping ions and then laser cooling them was beautifully demonstrated by Toschek and his collaborators[44] who produced photographs of the fluorescence from a single, Ba[+] ion confined in an r.f. quadrupole trap. The trap geometry is shown in figure 9; the ring and two end cap electrodes, which are electrically connected, are hyperboloids of revolution. When a voltage is applied the potential produced is, from symmetry considerations, given by

$$\phi = \frac{U}{r^2}(x^2 + y^2 - 2z^2) \tag{26}$$

For a given sign of applied voltage U, the force on the charged particle on which the potential acts, is directed towards the centre either in the xy plane or in the z-direction, but is defocussing in the other. Stability in all three directions is achieved if the sign of the voltage is changed rapidly enough, and if the magnitude of this r.f. voltage is appropriately chosen. The values are given by solving the equation of motion for the ion and the range of amplitudes and frequencies depend on the size of the trap and mass of the particle.

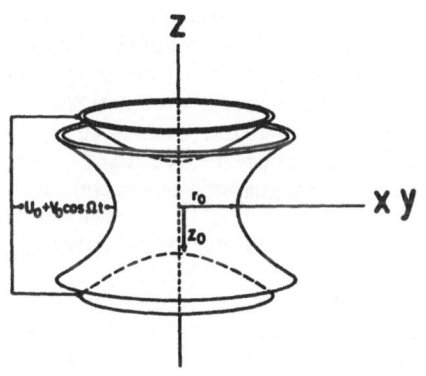

Figure 9. Geometry for an r.f. quadrupole trap.

An alternative to the rf trap for trapping ions is the Penning trap. Here instead of a high frequency voltage a static potential is used across the electrodes and the defocussing force of this voltage in the x-y direction compensated by a steady magnetic field along the symmetry axis of the trap. Thus, so long as the field strength exceeds the stability criterion

$$B^2 > 4\ mU/qr_0^2$$

a stable potential well can be created. r_0 is the trap radius and m and q the mass and charge of the particle respectively. The ion path in this configuration is a superposition of three harmonic motions: along z where the motion is independent of B and the frequency is $\omega_z = (eU_0/mr_0^2)^{1/2}$, in the x-y plane where the cyclotron motion is given by $\omega_c = eB/m$, slightly perturbed by the electric field, and the centre of the cyclotron orbit which circles around the trap centre with $\omega_m = E/B$. In general $\omega_c \gg \omega_z \gg \omega_m$.

Once an ion is electromagnetically trapped its temperature may be progressively reduced using laser radiation. The idea was proposed by Letokhov and others in Moscow[45], Schawlow and Hänsch in Stanford[46] and Wineland and Dehmelt in Seattle[47]. The idea is quite simple: a narrow bandwidth dye laser is tuned to the low energy side of a Doppler-broadened transition, at frequency $\omega_0 - \delta$ where ω_0 is the resonance frequency. Only those ions which travel in a direction opposite to the laser beam, with velocity $v = c\delta/\omega_0$, can therefore absorb the light; the re-emitted fluorescent light averaged over many absorption processes has frequency ω_0 and the energy difference is taken from the kinetic energy of the ions. The technique has been used by the group at Heidelberg on Ba+ ions held in a r.f. quadrupole trap (Neuhauser et al.[48]), by the National Bureau of Standards (USA) on Mg+ held in a Penning trap (Wineland et al.[49]) and again on Mg+ at the University of Washington (Nagourney et al.[50]).

The question of extending this trapping and laser cooling to neutral atoms has recently received a great deal of attention. The first point to note is that the well depth created in an ion trap is deep, of order 10000 K whereas a neutral atom without charge, possessing only a magnetic moment, can only be weakly trapped with a typical laboratory magnetic field, i.e. with B ~ 1 T the well depth is only ~ 1 K. Thus, to trap magnetically neutral atoms they must be exceedingly cold to start with and to achieve this the technique of laser cooling has been used. In a pioneering experiment Migdall, Prodan and Phillips[51] at the NBS and Bergeman and Metcalf at Stony Brook successfully cooled sodium atoms from a velocity 1000 ms^{-1} to only ~ 3 cm s^{-1}. The experimental arrangement is shown in figure 10. A beam of sodium atoms travels in the opposite direction to a laser beam through a magnetic field. The laser is right-circularly polarised (σ_+) and tuned so as to excite the transition $3S_{1/2}(F=2, M_F=2) \rightarrow 3P_{3/2}(F'=3, M_F'=3)$. The use of this Zeeman component avoids the problem of optical pumping which otherwise would after relatively few optical cycles terminate the cooling process by population transfer.

Furthermore, as the atom slows down it would, in the absence of the magnetic field, Doppler tune itself out of resonance with the laser; the specially designed solenoid prevents this happening by Zeeman tuning the atoms in synchronism with the velocity reduction i.e., the magnetic field in the direction of travel z of the atoms has the form:

$$B(z) = B_b + B_0 (1 - 2az/u^2)^{1/2} \qquad (27)$$

B_b is a steady bias field and B_0 is adjusted to match the Doppler shift for atoms undergoing a deceleration a from an initial velocity u. Both B_b and B_0 depend on the frequency and the g-factors for the transition and in the present case the typical values are $B_b = 0\cdot4$ T and $B_0 \simeq 0\cdot12$ T.

Once slowed down, Migdall et al. managed to trap the cold sodium atoms in a magnetic field the geometry of which was that of a 'spheroidal quadrupole'. This was created by two coils of diameter ~ 5·4 cm, separated by about 3·4 cm. The direction of the current in the coils was opposed, producing a potential well which creates a containing force twice as large along the axis of symmetry as along the radial direction of the coils. Storage times of order 1s were deduced with this arrangement, this being limited by collisions with the background gas, the pressure of which was $\leq 10^{-7}$ Torr.

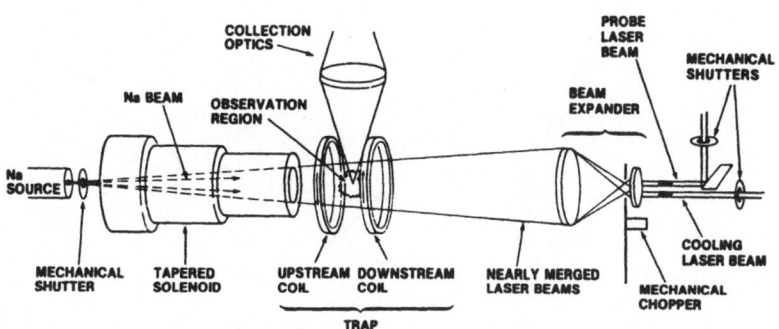

Figure 10. Laser cooling and trapping of neutral sodium atoms.

An alternative method for overcoming the Doppler tuning problem in laser cooling has been demonstrated by Ertmer et al.[52]. Instead of using a spatially varying magnetic field these authors controlled the frequency of the laser by means of an electro-optic modulator, driven by a radiofrequency source. Varying the frequency of the r.f. field applied produces sidebands to the laser frequency and these can be swept in step with the Doppler shift.

2·5 Light traps

The neutral atom traps that I have so far considered are only applicable to paramagnetic atoms i.e. those possessing a non-zero

210

ground state angular momentum. An alternative approach to trapping atoms, which was discussed theoretically some time ago (Ashkin[53]), is the use of the interaction between an atom and a laser light field. There are two forces which can be utilized in this context: the scattering force due to spontaneous emission and the 'ponderomotive' force which exists between an induced atomic dipole and an optical field gradient.

The scattering force has the form

$$F_S = \frac{1}{2} \left(\frac{h}{\lambda\tau}\right) [1 + 1/S(\nu)] \tag{28}$$

where $S(\nu) = (I/I_S) \cdot g(\nu)$ describes the effects of saturation and τ is the natural linewidth. For counter-propagating laser beams tuned slightly below resonance the effect of the atomic motion combined with the scattering force produces damping; the atoms see different frequencies depending on whether they are moving towards or away from the beams propagating to the left (L) or right (R) and so only absorb light from the opposing beam i.e.

$$(\nu_0 - \nu_R) = \left\{(\nu_0 - \nu) - \frac{v}{\lambda}\right\} \tag{29}$$

and

$$(\nu_0 - \nu_L) = \left\{(\nu_0 - \nu) + \frac{v}{\lambda}\right\}. \tag{30}$$

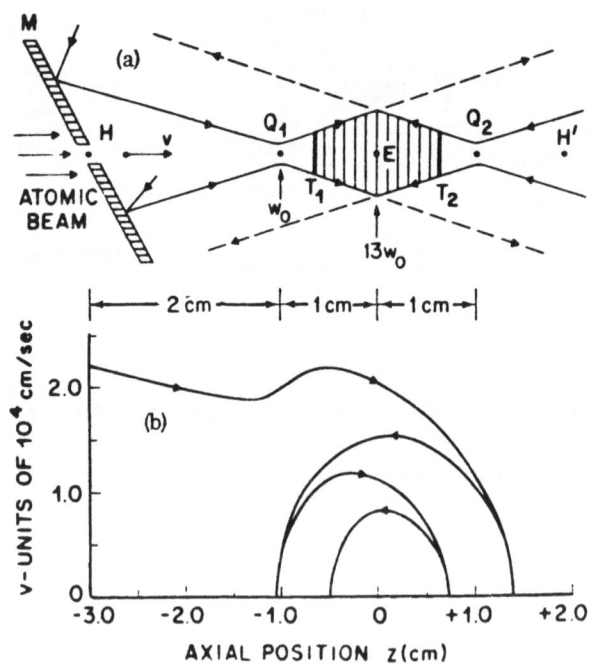

Figure 11. Ashkin's design for a 'light trap'.

On the other hand, the interaction of the optically induced atomic dipole with a Gaussian laser intensity distribution gives a pseudopotential of the type

$$U_D = h(\nu - \nu_0) \, \ell n[1 + S(\nu)]. \tag{31}$$

The ponderomotive force therefore looks like,

$$F_D = -\nabla U_D = \alpha(E) \, . \, S(\nu) \, [1 + S(\nu)]^{-1} \tag{32}$$

where $\alpha(E)$ depends on the lineshape function $g(\nu)$ and the saturation field strength E_S. The dipole gradient forces can be either repulsive or attractive allowing optical self-focussing or de-focussing. Ashkin's proposal uses the geometry shown in figure 11 where light from two opposing TEM_{00} mode beams is focussed at points Q_1 and Q_2. Each beam is tuned below the transition resonance and in this situation the point E is a stable point for an atom. Any displacement from E results in a restoring force; there is a restoring force in the axial direction from scattering due to the intensity imbalance, and a radial restoring force due to radial field gradients interacting with the atomic dipole. In order to trap an atom some damping of the motion is required; this is achieved by the Doppler shift, as we have already seen. Also, there is the effect of the standing-wave fringes in this trap which is as follows. Because of the dipolar force arising from the axial field gradient, atoms are attracted to the peaks of fringes with decreasing strength away from the point E. Beyond T_1 and T_2 the attractive force becomes less than the scattering force towards E. Therefore atoms executing damped oscillation about E can only be trapped on a loop if they come to rest inside the region between T_1 and T_2 (see figure 11). Once trapped however, atoms are progressively damped by the Doppler shift down to the level of about a single photon momentum. An encouraging step in realising a light trap has been announced recently by Chu et al.[54] who reported the viscous confinement and cooling of neutral sodium atoms in three dimensions.

2·6 The 'shelved electron' atom

The phrase 'shelved-electron' appears to have been coined by Dehmelt in a suggestion to make observable very weak transitions. The idea is as follows. Consider a 3-level atom or ion with states $|0\rangle$, $|1\rangle$ and $|2\rangle$ where $|0\rangle$ represents the ground state. The transition $|0\rangle - |1\rangle$ is strongly driven by a laser field and emits a stream of photons with an appropriate decay constant Γ_A. This represents the allowed transition; the transition $|0\rangle - |2\rangle$ is on the other hand forbidden and is therefore only weakly driven. If, however, the atom or ion makes the transition to state $|2\rangle$ the stream of photons from $|1\rangle$ to $|0\rangle$ will be switched off until the atom is returned to its ground state. A simple theoretical analysis of this process has been given by Cook and Kimble[55] who used a rate equation approach. They deduce that the probability density for the off-periods is

$$W_{off}(T) = R_- \exp(-R_-T) \tag{33}$$

and for the on-periods

$$W_{on}(T) = R_+ \exp(-R_+T) \tag{34}$$

where the time constants are given by $R_+ = \frac{1}{2}B_2U_2$ and $R_- = A_2 + B_2U_2$ with U representing the spectral energy density. A and B are the Einstein coefficients.

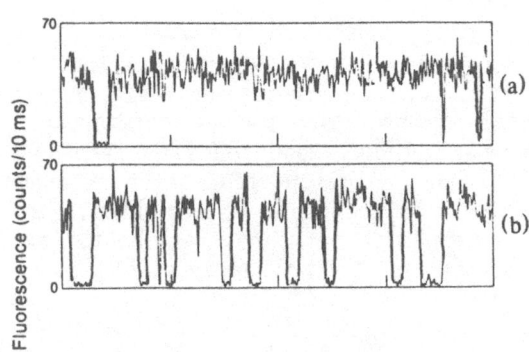

Figure 12. (a) Fluorescence from the allowed transition 6s-6p of HgII. (b) When laser light tuned to the forbidden (E2) transition $5d^{10}6s - 5d^96s^2$ simultaneously acts on the ion the fluorescence on the allowed transition switches on and off.

The analysis shows the Einstein A coefficient for a very weak transition can be measured by observing the on-off signal rate and that the distribution of 'on' times and 'off' times gives information on the photon correlations which is dependent on the characteristics of the exciting radiation. (The above results are generally valid only in the case of incoherent excitation. A more detailed analysis involving solution of the density matrix equations and an interpretation in terms of the second-order correlations of the light field has been given by Pegg, Loudon and Knight[56].)

2·7 Atoms near conducting surfaces

In recent years considerable interest has been focussed on the influence of cavities on the energy level structure and the decay processes of atoms. I shall consider these in turn. The interaction of an atom with the conduction electrons of a metallic surface can be described in terms of an image-atom charge behind the surface. This simple picture is a very good approximation for a Rydberg atom having relatively low characteristic Bohr frequencies ~ 100 GHz (for n ~ 30) which are much lower than the plasma frequency ~ $10^{15} - 10^{17}$ Hz. The induced dipole coupling between the atom and its image can then be written as a perturbation:

$$V_{VW} = - \frac{q^2}{16d^3} (x^2 + y^2 + 2z^2) = - \frac{q^2}{16d^3} (r^2 + z^2) \qquad (35)$$

where d is the distance from the atom to the surface (assumed infinite) and x and y are displacements parallel, and z perpendicular, to the surface.

For atomic ground states the expression above is only an approximation since a metal cannot be a perfect reflector for the virtual photons exchanged between the atom and surface in the optical region. Indeed even for an ideal, perfect conductor the expression ceases to be valid when the atom-surface distance is greater than a few wavelengths. At large distance the van der Waals force is modified by retardation effects and becomes asymptotically proportional to d^{-5}. Unfortunately since the interaction is then itself very small, retardation effects are very difficult to demonstrate by atomic beam deflection.

Compared to ground states then, the energy shifts produced by the van der Waals interaction for high n-states have two distinguishing characteristics: the shifts scale as n^4 so the deflection of atoms can be observed at much larger distances and secondly the shifts can lead to observable splittings.

2·8 Modification of the radiation rate

The van der Waals interaction as described above can be interpreted as an effect caused by the modification of the boundary conditions imposed on the electromagnetic field around the atom. This affects the level width as well as its position; the radiation rate is proportional to the density of vacuum modes which in turn depend on the boundary. Thus, an excited atom connected to lower states by an E1 transition moment parallel to the metallic surface will have its radiation rate reduced. If the distance d to the surface is less than λ the rate is given approximately by

$$\frac{16\pi^2}{5} (\frac{d}{\lambda})^2 \, \Gamma_0 \qquad (36)$$

where Γ_0 is the free decay rate. The effect can be exaggerated by the presence of two plane parallel conductors with an atom placed centrally between the two. Suppression of radiation being very strong when the gap is less than $\lambda/2$.

Conversely, for radiation polarised normal to the mirrors the image fields should add constructively and increase the decay rate by a factor of $3\lambda/4g$ where g is the gap distance. Similar effects can of course occur in traps and have already been detected in the electron (g-2) experiments[57].

Observation of cavity-induced enhancement and inhibition of spontaneous decay have been reported by a number of groups for microwave transitions between Rydberg levels of alkali atoms. In a very recent observation a group at Yale University[58] observed suppression of spontaneous emission on the 3·49 μm transition in caesium ($5^2D_{5/2} - 6^2P_{3/2}$) when the atoms were passed between two metallic mirrors spaced 1·1 μm apart. The substates with maximum angular momentum were found to survive without substantial decay for times of order 13 natural lifetimes.

Enhanced spontaneous emission of radiation from Rydberg atoms was reported in 1983 by Goy et al.[59]. In this experiment excited sodium atoms (in the 23S state) were formed within a superconducting cavity resonant at 340 GHz. This frequency closely corresponds to the transitions $23S_{1/2} - 23P_{1/2}$ (ν = 340·967 GHz) and $23S_{1/2} - 23P_{3/2}$ (ν = 340·396 GHz). By tuning the cavity spacing the natural decay rate from the 23S state could be increased from Γ_0 = 150 s^{-1} up to Γ_{cav} = 8×10^4 s^{-1}.

2·9 The Rydberg atoms

In the experiments described so far I have not really emphasized the peculiar properties of Rydberg atoms. For example, radiative transitions between highly-excited states of these atoms are easily induced by the background thermal radiation. The reason for this is easy to see once the scaling laws for Rydberg atoms have been established[60]. To start with then let us consider how the wavelength and the decay constant vary for low ℓ, highly eccentric orbits and high ℓ, circular orbits. In both cases in the classical limit the period of the Kepler orbit varies as

$$T \propto (n^2 a_0)^{3/2} \sim n^3 \tag{37}$$

so that the characteristic radiation frequencies and wavelengths are proportional to, n^{-3} and n^3 respectively. The radiative power is proportional to the square of the acceleration and so for low ℓ orbits we have

$$\Gamma_{n\ell} \propto n^{-3} \tag{38}$$

since the electron radiates mostly when close to the nucleus For high ℓ states the electron continuously radiates throughout the orbit. In this case as the acceleration is proportional to the square of the orbit radius (i.e. n^{-4}), the acceleration scales as n^{-8}. The energy radiated in each 'quantum step' is h/T so that the characteristic time to radiate down each step varies as

$$\Gamma_{n\ell} \propto n^{-5} \tag{39}$$

This is considerably slower than that for low ℓ states.

Now let us consider the power required to saturate a Rydberg transition. This can be obtained by saying that the monochromatic field irradiating the atom must produce a nutation frequency for the atomic dipole, qa_0, equal to the radiative decay rate, i.e.,

$$E_s = \hbar\Gamma/qa_0 \, n^2 \tag{40}$$

For low ℓ states this is

$$E_s ~\sim~ \alpha^3 \, n^{-5} \, E_a \tag{41}$$

which gives a value for E_s of order 10^{-3} V m^{-1} for n = 30 (E_a is the atomic field $= q/4\pi\varepsilon_0 a_0^2$). For high ℓ states on the other hand

$$E_s ~\sim~ \alpha^3 \, n^{-7} \, E_a \tag{42}$$

which gives $E_s \sim 10^{-6}$ V m^{-1} for n = 30.

The corresponding power fluxes are respectively proportional to n^{-10} and n^{-14}.

2·10 Rydberg masers and superradiance

The threshold condition for collective emission (superradiance) or maser action (gain in the medium exceeding the losses) is much lower for Rydberg atoms than it is for the same number density of atoms in low n states. To obtain an order of magnitude estimate for the point at which collective effects occur we estimate the amplitude of the electric dipole field $\bar{E}$ radiated by an atom at a distance corresponding to a neighbouring atom. That is, if we let L represent a linear dimension in the sample which contains N atoms,

$$\bar{E} ~=~ \frac{D}{4\pi\varepsilon_0} \; \frac{1}{\lambda^2_{if} \cdot L}. \tag{43}$$

Here $D = n^2 q a_0$ and is the typical dipole moment of a Rydberg atom. The collective field radiated is thus of order $N\bar{E}$ and for this to dominate the individual dipoles must be flipped by the self-radiated field in a time shorter than the spontaneous decay rate or other damping processes. The threshold condition is therefore,

$$N ~\geq~ 4\pi\varepsilon_0 \, \hbar \, \Gamma \, \lambda^2_{if} \, L \, \frac{1}{q^2 a_0^2 n^4} \tag{44}$$

i.e. $N \propto n^{-4} . \lambda^2_{if}$. For ordinary atomic systems in low lying states, N is of order $10^{11} - 10^{12}$ cm^{-3} for collective emission in the wavelength range $0·1 - 1$ mm, whereas for a Rydberg atom in n $\sim$ 30 N$\sim 10^5 - 10^6$ cm^{-3} is sufficient.

With the use of a resonant cavity the threshold for collective emission is lower by the magnitude of the cavity finesse. Using a cavity finesse of a 100 or so Rydberg maser action on sodium has been observed[61]. In this case the field ionisation signal was used to monitor the population as a function of time for the states 27S, 26P and 25P corresponding to transitions Δn = 1 at $\lambda = 1·470$ mm and Δn = 2 at $\lambda = 0·466$ mm. The threshold conditions were of order 5000 atoms cm^{-3} for the Δn = 1 transition and $\sim 10^5$ cm^{-3} for Δn = 2, in good agreement with theory. However, in a recent report, Meschede et al.[62] announced operation of a single atom maser. A superconducting cavity with a Q of 8×10^8 at a temperature of 2K was employed and signals could still be detected with an average number of atoms of only 0·06. Although I have already mentioned that enhancement of spontaneous decay has been observed, in this experiment it is the oscillatory exchange of energy between the atom and field that was detected.

2·11 Rydberg wave packets

An interesting proposal to observe Rydberg wave packets has been suggested by Parker and Stroud[63] and by Alber, Ritsch and Zoller[64]. Such a spatially localised wave packet is achieved by pulsed excitation of a manifold of Rydberg states resulting in a coherent superposition of states. As the high lying states are long-lived the existence and temporal development of the wave packet can be monitored by probing with a laser whose pulse is delayed by varying amounts with respect to the initial excitation. The second pulse could be arranged to lead either to subsequent field ionisation of the atom or to cause stimulated emission to a lower state from which the decay could be detected. Thus, as the packet moves around a Kepler orbit its extent and position may be deduced. This topic is clearly closely related to that of quantum beats; the rapid time variation of the spontaneous decay can be interpreted in terms of the motion and spreading of the packet. The classical motion of the packet as well as the quantum-mechanical spreading, destruction and revival may also be studied therefore by measurement of the quantum beat signal. To discuss this in a little bit more detail consider laser excitation of some Rydberg states from the ground state in terms of the interaction picture with amplitude coefficients $a_n(t)$ and $a_g(t)$ respectively.

$$\dot{a}_g = -\tfrac{1}{2} i \sum \Omega_n a_n(t) f(t) \exp(-i\Delta_n t) \tag{45}$$

$$\dot{a}_n = -\tfrac{1}{2} i \Omega_n a_g(t) f(t) \exp(-i\Delta_n t). \tag{46}$$

The pulse envelope of the laser is given by $f(t)$, the Rabi frequencies for the various transitions are Ω_n and the laser detuning is denoted by Δ_n. Solution of these equations by numerical integration has been given by Parker and Stroud for the case of a laser pulse of duration $6 - 10$ psec with a central frequency tuned to $n = 85$. The sum over states involves $60 \leq n \leq 100$. After the excitation the amplitudes will be constant except for a very slow spontaneous decay. The time evolution to a good approximation will therefore be the oscillatory free evolution i.e.

$$\Psi_R = \sum a_n(t) \exp(-i\omega_n t) a_n(r) \tag{47}$$

where ω_n stands for the various frequencies back to the ground state and $a_n(r)$ are the hydrogenic radial functions. The form of the wave packet is shown in figure 14. The well-defined wave packet $r^2 |\psi_R(r,t)|^2$ moves out to the classical turning point where it narrows; the uncertainty in $\Delta r \Delta p$ is of order $0·53 \, \hbar$ — very close to the minimum allowed. Beyond the turning point the packet reverses direction and accelerates back towards the nucleus where upon it is dispersed by the strong Coulomb potential. This cycle of events continues with a frequency given by the period of the Kepler orbit, resulting in first progressive dispersion followed by reformation.

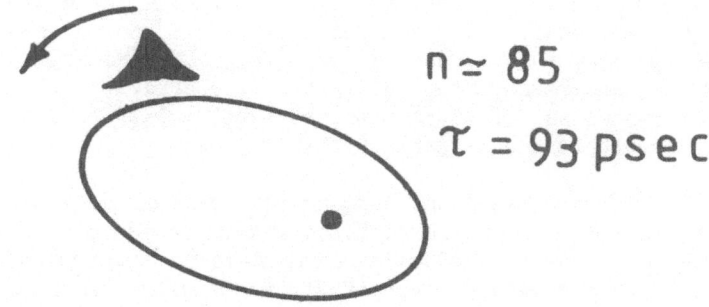

Figure 13. Kepler orbit for a Rydberg atom.

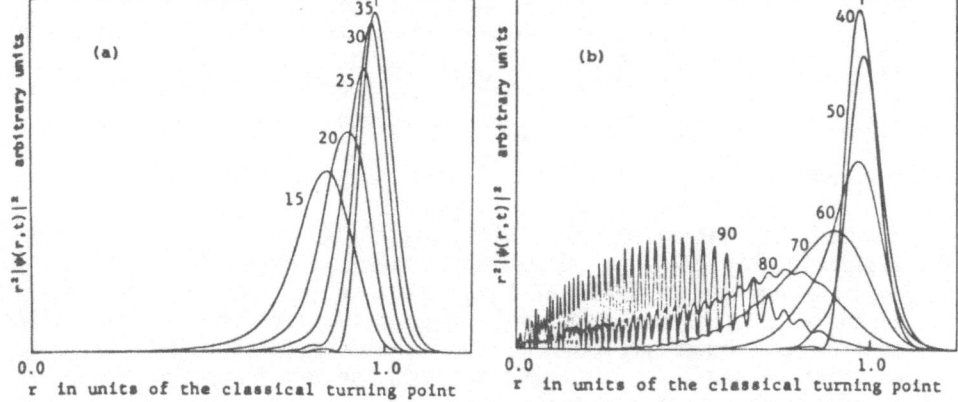

Figure 14. Temporal development of a Rydberg wave packet in the first 100 ps after formation.

3. OPTICAL ROTATION PRODUCED BY ATOMS

3.1 Introduction

This field of study although long established received an enormous stimulus in the early 1970's when it was shown that it ought to be feasible to detect the presence of weak interactions in atoms. Glashow, Weinberg and Salam[65] postulated in the late sixties a model for the unification of weak and electromagnetic forces in which the existence of a 'neutral weak current' was predicted. Evidence for this new aspect of the weak interaction was soon provided by neutrino scattering experiments conducted at CERN[66] and Fermilab[67] and the consequences of this new interaction in atomic physics were considered. In particular, in 1974 Marie Anne and Claude Bouchiat[68] showed that a sizeable enhancement of the parity non-conserving (PNC) effect (the signature of the weak interaction) could be expected in heavy atoms where the fractional left-right asymmetry could, for some transitions, be as high as 10^{-3}. One manifestation of the induced left-right asymmetry is a minute optical rotation (~ 1 μrad in typical cases) and this has prompted an era of precise, highly-sensitive, laser polarimetry. I should perhaps mention that although the aim in many of the new optical rotation experiments was to detect and measure the PNC effect, the study itself has led to a critical re-appraisal of other sources of optical effects such as those induced by electric and magnetic fields[69].

In the following I want to attempt a sort of unification of different sources of optical rotation and dichroism and show that far from being a narrow specialist's area of laser spectroscopy it is an enormously rich and varied field of study. I will therefore take as my starting point the famous and well-known dispersion relations and develop from these the form of the Faraday, Stark and PNC optical rotation. I shall also consider very briefly the extension of these ideas to the case of Doppler-free polarimetry and later I shall discuss how the use of lasers themselves brings in a variety of problems, in particular that of saturation. Finally, I will say something about the form of the weak interaction in so far as it enters the atomic Hamiltonian as a weak (no pun really intended!) perturbation.

3·2 The Dispersion Relations

In the limit of low light intensities and low to moderate absorption the familiar classical dispersion relations suffice to give an adequate description of the variation of the real and imaginary parts of the refractive index with optical frequency. The real part (giving a dispersion lineshape) corresponds to the variation of the propagation velocity of light through the medium while the imaginary part represents the absorption i.e.

$$(n - 1) \quad \propto \quad \frac{(\omega - \omega_0)}{(\omega_0 - \omega)^2 + \Gamma^2/4} \tag{48}$$

$$k \quad \propto \quad \frac{\Gamma/2}{(\omega_0 - \omega)^2 + \Gamma^2/4} . \tag{49}$$

An important point to note is that away from an atomic resonance the imaginary part varies approximately as $1/(\omega_0-\omega)^2$ while the real part has long wings varying as $1/(\omega_0-\omega)$. In the absence of weak interactions within the atom, or externally applied fields, there is no optical rotation, i.e. a plane-polarised light beam may suffer attenuation by absorption in the medium but there is no rotation of the plane of polarisation. However, in the presence of an axial magnetic field (i.e. collinear with the laser propagation direction) a simple $J_g = 0$ to $J_e = 1$ transition produces two dispersion curves centred at $\omega_0 \pm g_J \mu_B B$; one corresponds to right-handed circularly polarised light, the other to left-handed. Optical rotation can now be observed: the

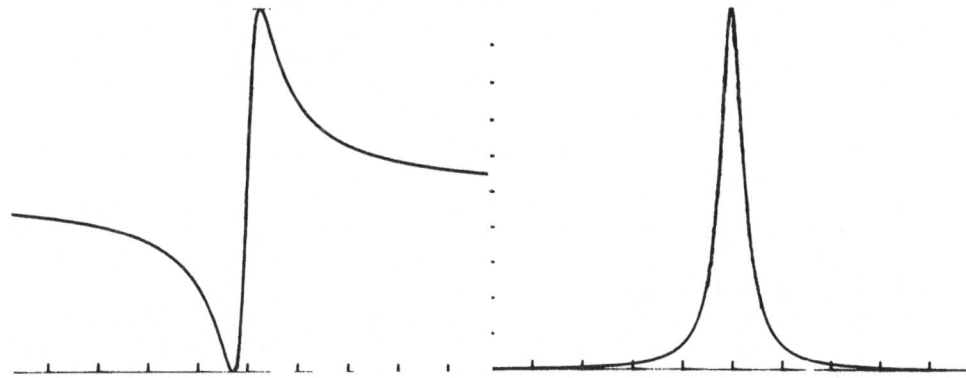

Figure 15. Classical dispersion (a) and absorption (b) curves.

incident plane-polarised light can be represented as equal left- and right-handed circularly polarised components which propagate differently through the atomic medium. Far from any optical resonance ($\geq$ 100 GHz) the absorptive part of the refractive index plays little part and so the right- and left-handed components have very nearly equal amplitudes after passage through the medium. However there will have accrued a relative phase difference due to the different propagation velocities. This can easily be shown to lead to plane-polarised light emerging which is rotated through an angle ϕ given by

$$\phi = \frac{\omega}{2c} \int_0^L \left(n_L(\omega) - n_R(\omega) \right) dL \tag{50}$$

where ω is the angular frequency of the light and L the length of the medium. The difference in refractive indices depends on the optical frequency and the strength of the magnetic field, i.e. neglecting for a moment Doppler broadening,

$$(n_L - n_R) \propto \left[\frac{(\omega_0 - \omega + g\mu_B B)}{(\omega_0 - \omega + g\mu_B B)^2 + \Gamma^2/4} - \frac{(\omega_0 - \omega - g\mu_B B)}{(\omega_0 - \omega - g\mu_B B)^2 + \Gamma^2/4} \right]$$

which in the limit of small fields (i.e. $\Delta\omega_{nat} > g\mu_B B$) is the differential of a dispersion lineshape. The sign used in equation (50) is given by the convention that it is positive (clockwise) when viewed at a fixed point on the axis of propagation looking towards the source (hence perhaps the reason for calling this the Blindman's Convention!).

So far then what we have described is the normal (off-resonance) Faraday effect and this describes the principle of operation of the 'Faraday modulator' used in the laser polarimeter I will describe in a later section.

As the frequency of the light approaches that of a resonance the magnitude of the rotation clearly grows, although now we are no longer justified in ignoring the absorption contribution. The emerging light in this case is no longer plane-polarised: differential absorption of the RH and LH components of the incident beam, in addition to the phase accrued from the differential propagation velocities results in elliptically polarised light, the major axis of which is rotated in the manner described above.

We shall consider the details of the atomic calculation of the Faraday rotation later; here we simply make some further remarks about electric fields and PNC neutral weak currents. Firstly, from a knowledge of the Stark effect we can expect a perpendicular electric field to induce dichroism (differential absorption) between light plane-polarised along and perpendicular to the field. The maximum dichroism effect occurs if the incident plane of polarisation is at 45° to the field, giving equal amplitude components travelling along and perpendicular to the field. The output light is elliptically polarised and there is a 'pseudorotation' of the plane of polarisation corresponding to the orientation of the major axis of the ellipse. To illustrate this consider figure 21 which shows how a differential absorption in the x- and y-axes produces an apparent rotation of the plane of polarisation. A similar effect would occur if the perpendicular electric field were replaced with a magnetic field although the two effects are distinguishable by their different frequency dependences in the vicinity of an absorption (see table 2).

In contrast to both the effects produced by electric and magnetic fields the neutral weak current interaction gives rise to different dispersion curves for LH and RH polarised light but without any energy shift. Therefore, unlike the on-resonance Faraday rotation which gives a symmetric line profile, the PNC optical rotation is antisymmetric having the form of an ordinary dispersion profile. We can therefore see in a rather general way how the different optical phenomena can in principle be distinguished. Before considering the theory of these effects in more detail I will give a brief description of the sort of laser polarimeter which can be used to study these phenomena.

3·3 Principles of the polarimeter

A schematic diagram of the apparatus used in Oxford for measurement of PNC effects[70] is shown in figure 16. It comprises three parts: a tunable laser light source L, the polarimeter itself and an on-line computer for control and data acquisition. The optical arrangement is as follows. Between two crossed calcite polarisers (P1 and P2) sit a Faraday modulator F (a silica glass rod in a solenoid) and the oven. This comprises two tubes, one containing the atomic vapour under investigation and the other an empty tube used for comparison. Light emerging from the polarimeter is directed onto a low-noise, p-i-n photodiode after passing through an interference filter to remove much of the thermal radiation from the hot oven, and the signal is digitised by the voltage to frequency converter V. Two points of technique adopted as a result of experience are probably worth mentioning here. Firstly, in scanning a tunable laser the output beam characteristics (direction and intensity cross-section) are microscopically altered.

The problem is significantly alleviated by decoupling the laser from the polarimeter by passing the beam through a single-mode, optical fibre: changes in beam characteristics then appear as changes in intensity. This technique was first successfully applied by the group working in Moscow on their PNC experiment on bismuth[71] and has subsequently been used in our own experiments at Oxford. Secondly, changing the laser's wavelength can give rise to interference effects within the polarisers and other optical components, altering the intensity profile of the laser beam within the polarimeter. For this and other reasons high quality polarisers in the form of calcite prisms with apex angles of order 30° are used.

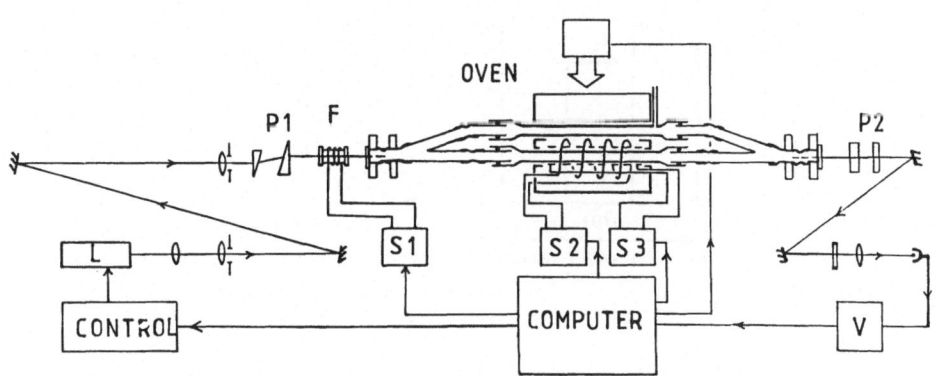

Figure 16. Laser polarimeter used for detecting PNC optical rotation in atomic bismuth.

To extract the optical rotation signal with optimum signal to noise ratio, a square-wave modulation pattern is applied to the solenoid of the Faraday modulator as shown in figure 17(a). To cope with rapid changes in the current, the solenoid is wound to have low inductance and the drive amplifier has a current sensing feedback arrangement to improve further the rise time of the current waveform. The intensity has the form shown in figure 17(b). If the polarisers are exactly crossed the intensities I_1 and I_2 are equal; an optical rotation ϕ_R appears then as a difference between I_1 and I_2 and to obtain a 'normalised' angle measurement the on-line computer evaluates in real time the quotient

$$Q = \frac{I_1 - I_2}{I_1 + I_2 - 2I_3} = \frac{\tan 2\phi_R}{\tan A} \tag{51}$$

where the modulation amplitude is A, and

$$\begin{aligned}
I_1 &= I_L(\omega) \sin^2\left[\phi_R(\omega) + A\right] \\
I_2 &= I_L(\omega) \sin^2\left[\phi_R(\omega) - A\right] \\
I_3 &= I_L(\omega) \sin^2\left[\phi_R(\omega)\right].
\end{aligned} \tag{52}$$

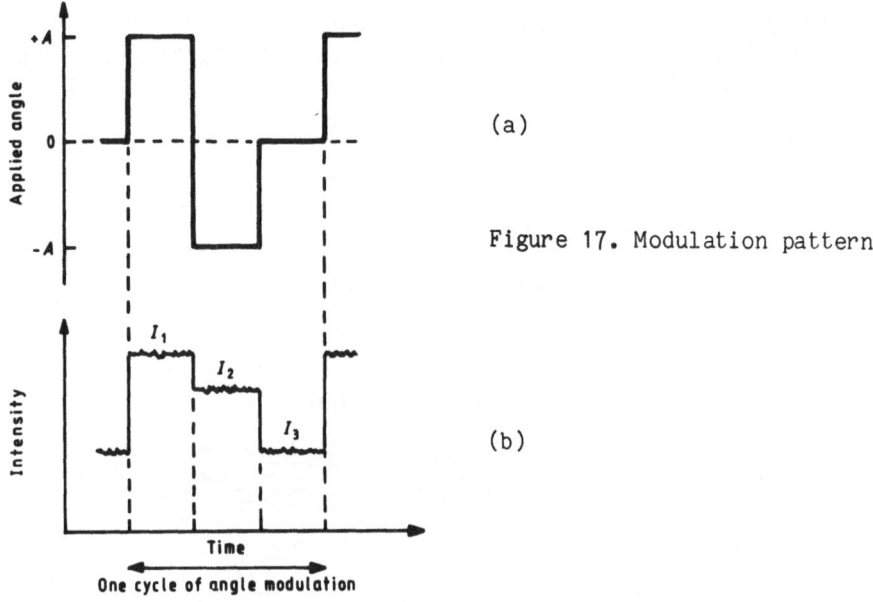

(a)

Figure 17. Modulation pattern.

(b)

        I say normalised because a variation in the laser intensity or the absorption in the vapour will alter the sensitivity of measurement if only the difference $I_1 - I_2$ is computed. The factors which determine the magnitude of A depend on the particular study in hand. However, we can see some of the important criteria by writing down an expression for the ratio of the signal (I_1-I_2) to background for the case of small angles ϕ_R.

$$\frac{S}{B} \sim \frac{\phi_R(\omega).A}{[A^2 + B] + C/I_t} \cdot \qquad (53)$$

In this model, the symbols have the following meaning. $\phi_R(\omega)=\alpha_0\phi_R D(\omega)$ is the optical rotation produced by the vapour and depends on the number of absorption lengths α_0 and on the lineshape function $D(\omega)$ which takes the form of a Doppler-broadened dispersion curve; for magnetic and electric field induced rotation ϕ_R will depend on the strength of the field and $D(\omega)$ on the direction and type of field (see table 2).   The transmitted intensity $I_t = I_L \exp[-\alpha_0 G(\omega)]$ where the lineshape function $G(\omega)$ for a single spectral component can usually be accurately described by a Doppler-broadened Lorentzian curve.   Finally the terms B and C in equation (53) represent respectively the finite extinction ratio of the polarisers and a laser independent contribution, arising from stray oven light, detector electronics etc. We can now make the following points regarding the choice of the magnitude of A. First the signal to noise ratio clearly varies as the laser is scanned through the absorption spectrum of the vapour.   In regions of low absorption the background is dominated by the laser intensity and the fluctuations on this represent the noise limit; in regions of very high absorption little laser light reaches the detector and the term C becomes important. Second, it is important to realise that the output intensity variations of the dye laser are usually very much greater than the photon shot noise limit; were this not so a modulator would not offer any signal to noise advantage and the intensity transmitted through polarisers set close to 45° could be used instead. As it is, for small angles ϕ_R the optimum value for A is close to $\sqrt{B}$, as can be seen from equation (53).

3·4 Faraday modulation polarisation spectroscopy

The Faraday modulator used in the polarimeter has another important function other than providing an easy way of optimising the signal to noise ratio; it allows the real and imaginary parts of the refractive indices to be separated. This stems from the fact that in the modulator there is very little absorption i.e. the effect is only to modulate the angle of rotation of the linearly polarised laser beam with negligible modulation of the phase or ellipticity of the light.

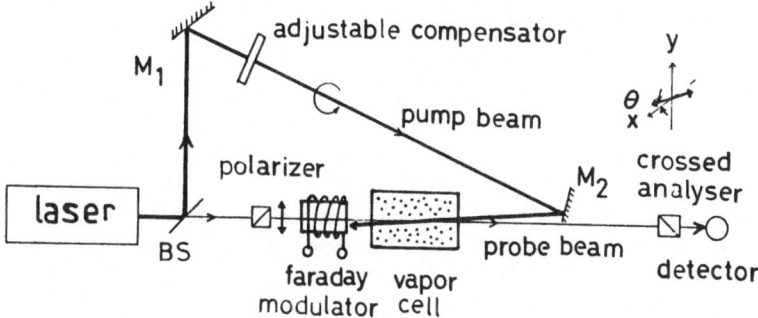

Figure 18. Experimental arrangement for Faraday modulated polarisation spectroscopy.

Thus, a potentially very useful application of the modulator is in polarisation spectroscopy (figure 18) where for uncrossed polarizers the signal is in general a mixture of dispersion and Lorentzian lineshapes. With a modulator the dispersion contribution alone can be extracted thus avoiding the systematic error in the determination of the linecentre[72].

3·5 Theory of the 'Normal' Faraday effect

By the 'normal' Faraday effect I refer to the simplest form of the theory in which effects of optical pumping and saturation are specifically excluded[73]. The treatment is likely to be valid either when using classical light sources of low temporal coherence or when studying forbidden transitions (e.g. magnetic dipole and electric quadrupole transitions in the visible region). I shall choose the axis of quantization to be that of the magnetic field which is coincident with the propagation direction of the laser beam. In this case the

Faraday rotation arises from two distinct mechanisms caused by the field: energy shifts of the magnetic substates and mixing of hyperfine states having the same M_F but different F. For low field strengths (i.e. where the energy level splitting is much less than the Lorentzian width) the former gives rise to a symmetric lineshape (i.e. the differential of a dispersion shape, as I have mentioned previously) and the latter to an asymmetric dispersion-shaped curve. Effects arising from the thermal distribution of population among the lower magnetic substates are in general of little consequence since normally $g_J\mu_B B \ll KT$ (see e.g. Roberts et al.[73]). Furthermore, higher order effects of the field (the next order is, on symmetry grounds, $O(B^3)$) can also be shown under normal circumstances to be negligible i.e. for field strengths $\lesssim 10$ G.

The form of the rotation for the symmetric ϕ_S and antisymmetric ϕ_a components of the rotation can be written in terms of reduced matrix elements and the sum over M_F and F performed using standard techniques. When this is done we find,

$$\phi_S = -\frac{\mu_0 \pi B_z}{2\hbar^2}\left(\frac{\omega L}{c}\right)\frac{N(F_i)}{(2F_i+1)}(U_{11} + 2U_{12}X + U_{22}X^2)Q^2\, D_0'(\omega_0-\omega) \tag{54}$$

$$\phi_A = \frac{\mu_0 \pi B_z}{\hbar^2}\left(\frac{\omega L}{c}\right)\frac{N(F_i)}{(2F_i+1)}(V_{11} + V_{12}X + V_{22}X^2)Q^2\, D_0(\omega_0-\omega) \tag{55}$$

where $X = \langle j||T^{(1)}||i\rangle\, /\, \langle j||T^{(2)}||i\rangle$ and $Q = \langle j||T^{(1)}||i\rangle$.

The U and V coefficients are simple products of 6j symbols with U_{12} and V_{12} representing interference terms between the transition amplitudes $T^{(1)}$ and $T^{(2)}$ (e.g. between M1 and E2); U_{11} and V_{11}, and U_{22} and V_{22} result solely from the individual $T^{(1)}$ and $T^{(2)}$ components of the mixed transition. We will examine the form of these angular coefficients in the next section. Here, however, we note the existence of certain sum rules which can be anticipated by 'switching off' the hfs interaction causing the spectrum to degenerate into a single component, viz,

$$\sum_{FF'}(V_{11}) + (V_{12}X) + (V_{22}X^2) = 0 \qquad \begin{aligned}\sum V_{11} &= 0\\ \sum V_{12}X &= 0\\ \sum V_{22}X^2 &= 0\end{aligned} \tag{56}$$

$$\sum_{FF'}\phi_S(F_jF_i) + \sum\phi_a(F_jF_i) = \phi_S(J_jJ_i) \tag{57}$$

3·6 Effects of Optical Pumping

If the distribution of population in the magnetic substates is significantly altered by optical pumping a straightforward summation over states as described above will not represent the true experimental situation, rather the individual components in the summation have to be weighted appropriately by the population residing in the particular M substate. We examine the consequences of this now. For an allowed E1 transition (for which effects of optical pumping can be expected to be appreciable) the two components ϕ_S and ϕ_R of the rotation are of the form:

$$\phi_S \; \alpha \; U_{11} \; (M_i, M_j) \;=\; (-1)^n \; (2F_i+1) \; (2F_j+1)^2 \begin{pmatrix} F_j & k & F_i \\ -M_j & q & M_i \end{pmatrix}^2$$

$$\times \begin{pmatrix} F_j & 1 & F_j \\ -M_j & 0 & M_j \end{pmatrix} \begin{Bmatrix} J_j & F_j & I \\ F_i & J_i & K \end{Bmatrix}^2 \begin{Bmatrix} J_j & F_j & I \\ F_j & J_j & 1 \end{Bmatrix} \tag{58}$$

with $k = 1$ and the incident plane-polarised, laser beam described by $q = \pm 1$. The phase factor is given by $n = J_j + I + k - M_j + 2F_j$.

When summed over the M_i, M_j substates a sum rule on the product of 3 j-symbols yields

$$\begin{pmatrix} k & 1 & k \\ q & 0 & -q \end{pmatrix} \begin{Bmatrix} k & 1 & k \\ F_j & F_i & F_j \end{Bmatrix} \;=\; (-1)^{k-q} \; \frac{q}{\left[(2k+1)(k+1)k \right]^{1/2}} \; \begin{Bmatrix} k & 1 & k \\ F_j & F_i & F_j \end{Bmatrix} \tag{59}$$

which produces a result identical to that quoted by Roberts et al.[73].

Similarly for the V-coefficients, we have

$$\phi_A \; \alpha \; V_{11}(M_i, M_j, M_k) \;=\; (-1)^m \; (2F_k+1) \; (2F_j+1) \; (2F_i+1)$$

$$\times \begin{pmatrix} F_k & 1 & F_j \\ -M_k & 0 & M_j \end{pmatrix} \begin{pmatrix} F_j & k & F_i \\ -M_j & q & M_i \end{pmatrix} \begin{pmatrix} F_i & k & F_k \\ -M_i & -q & M_k \end{pmatrix} \begin{Bmatrix} J_j & F_k & I \\ F_j & J_j & 1 \end{Bmatrix} \begin{Bmatrix} J_j & F_j & I \\ F_i & J_i & k \end{Bmatrix}$$

$$\times \begin{Bmatrix} J_j & F_i & I \\ F_k & J_k & k \end{Bmatrix} \tag{60}$$

where $m = 3I + J + J + J_k - M - M_j - M_k + 2(F_i+F_j+F_k)$. Since μ_z is diagonal in J we may write $J_k = J_j$ and then by summing over the M_i, M_j and M_k substates we again arrive at the 'normal' Faraday effect result.

However, given in the form of equation (58) and (60) the Faraday rotation can now be computed for the case where there is optical pumping once the equilibrium populations in the magnetic substates have been determined using rate equations. I should perhaps emphasize here that I am not referring to the development of atomic coherences but only to population changes. Finally, we should also note that if optical pumping is present the rotation lineshape will become distorted since pumping near linecentre will be more efficient than in the wings.

In the context of the effects produced by optical pumping it is perhaps worth going over some of the considerations that are important in deciding whether or not optical pumping is likely to play a significant role. Clearly if the ground state has $J = 0$ and there is no nuclear spin, no net alignment ($|M_i|$) or orientation (M_i) in the ground state is possible. To sustain a population imbalance in the excited state requires in general, for optical transitions, a substantial excitation rate to compete with collisions and spontaneous decay. In these circumstances with strong light fields the development of atomic coherences may become important and a model for the atomic alignment based solely on rate equations is likely to be inadequate. However at the opposite extreme, if the ground state of an atom possesses only nuclear angular momentum (i.e. a transition J_g=0 — J_e=1 with I ≠ 0) then appreciable alignment can be generated since destruction of nuclear alignment in atomic collisions can be a slow process with time constants in the region of 100—1000s[74]. For alkali metals, the ground states of which possess electronic spin in addition to nuclear spin, the rate of destruction of alignment can be more rapid (~ 10^{-3} s), although still not sufficient to prevent effective pumping.

To conclude, two further complications may alter dramatically the efficiency of any optical pumping cycle: one is the question of whether or not an atom in the excited state can be re-aligned by a collision before spontaneous decay — this is sometimes referred to as Dehmelt pumping[75] — and the second is the branching ratio from the excited state to other lower levels just above the ground state. In this situation thermalising collisions can be important, as well as the magnitude of any hyperfine structure in these states.

3·7 Faraday rotation obtained with strong light fields

Dramatic changes to the Faraday rotation recorded on an E1 transition of samarium have been observed[76] using a polarimeter similar to that described in section 3·3. Under conditions of low buffer gas pressures and moderate light levels (ca 10mW/mm^2 in a bandwidth of 1 MHz) the Faraday rotation produced by the even Sm isotopes was markedly altered from its 'normal' form; the magnitude of the rotation at a given magnetic field strength increased by over a factor of 50 and the lineshape was dramatically transformed from a differential of a Doppler-broadened dispersion shape to a Voigt form. This effect can be ascribed to the development of optical and Zeeman coherences and an exact treatment based on the solution of the steady state density matrix equations is possible in certain cases. Here I shall give only an outline of the theory of the effect for $J_g = 0 \rightarrow J_e = 1$ transition appropriate to the case of the even samarium isotopes (see figure 19).

We start by writing the bulk polarisation of the medium as follows,

$$\begin{pmatrix} P_+ \\ P_- \end{pmatrix} = \varepsilon_0 \begin{pmatrix} \chi_{++} & \chi_{+-} \\ \chi_{--} & \chi_{--} \end{pmatrix} \begin{pmatrix} E_+ \\ E_- \end{pmatrix} \tag{61}$$

where E_+ and E_- are the electric field amplitudes for left- and right-circularly polarised light. The corresponding complex refractive indices $\tilde{n}_L$ and $\tilde{n}_R$ are related to χ by the relations,

$$\tilde{n}_L = \left[1 + \chi_{++} + \chi_{+-} \left(\frac{E_-}{E_+} \right) \right]^{\frac{1}{2}} \tag{62}$$

$$\tilde{n}_R = \left[1 + \chi_{--} + \chi_{-+} \left(\frac{E_+}{E_-}\right) \right]^{\frac{1}{2}}. \tag{63}$$

Atomic coherences appear here as non-zero off-diagonal elements χ_{+-} and χ_{-+}. To solve for χ we use the ensemble density matrix and compute the elements from the equations of motion (see e.g. Cohen-Tannoudji[77])

$$\dot{\sigma}_{++} = 0 = -\Gamma_1 \sigma_{++} - iv(\rho_{01} - \rho_{10}) \tag{64a}$$

$$\dot{\sigma}_{--} = 0 = -\Gamma_1 \sigma_{--} + iv(\rho_{0-1} - \rho_{-10}) \tag{64b}$$

$$\dot{\sigma}_{-+} = 0 = (2i\omega_e - \Gamma_1)\sigma_{-+} + iv(\rho_{01} + \rho_{-10}) \tag{64c}$$

$$\dot{\rho}_{01} = 0 = -i(\omega - \omega_0 - \omega_e)\rho_{01} - (\Gamma_2/2)\rho_{01}$$
$$- iv(\sigma_{++} - \sigma_{00} - \sigma_{-+}) \tag{64d}$$

$$\dot{\rho}_{0-1} = 0 = -i(\omega - \omega_0 - \omega_e)\rho_{0-1} - (\Gamma_2/2)\rho_{01}$$
$$+ iv(\sigma_{--} - \sigma_{00} + \sigma_{+-}). \tag{64e}$$

The total number of atoms is $N = \sigma_{00} + \sigma_{++} + \sigma_{--}$ and the Rabi frequency is given by $v = ed.[\Sigma |E_\mu|^2]^{1/2}/\hbar$; ed is the transition dipole matrix element and E_μ is a polarisation component of the laser's electric field. Γ_1 and Γ_2 represent the decay of population and coherence respectively. In general $\Gamma_1 = \Gamma_e + \gamma_{coll}$ and $\Gamma_2 \geq (\Gamma_1 + \Gamma_g)$; the

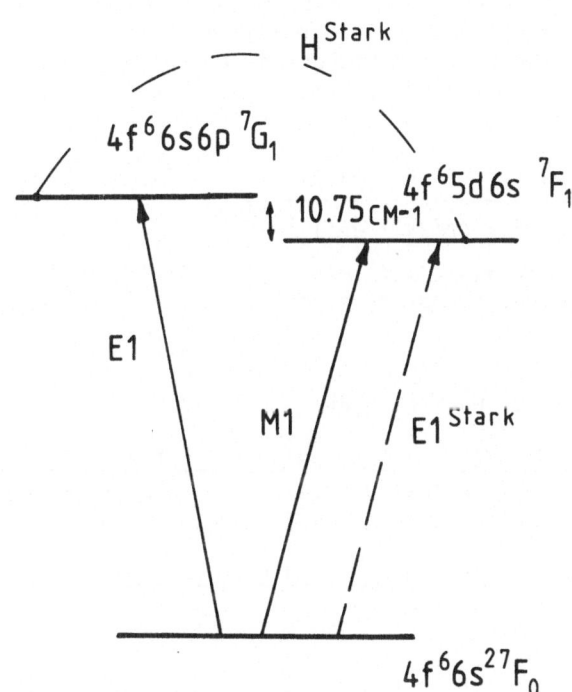

Figure 19. Partial term diagram of SmI.

inequality applying when there are dephasing collisions. Here we assume no distinction between relaxation of probability (i.e. population) and relaxation of alignment or polarisation. Γ_g and Γ_1 are the relaxation constants of the ground and excited states respectively and γ_{coll} the collisional rate. Γ_e represents spontaneous emission.

To determine the Faraday rotation obtained with a strong monochromatic light field we have to solve the above equations for ρ_{01}^R and ρ_{0-1}^R since the angle of rotation is now given by:

$$\phi = \frac{\omega L}{4c} \frac{e}{V\epsilon_0 E_0} \left[|d_{10}|\rho_{01}^R + |d_{-10}|\rho_{0-1}^R \right]. \tag{65}$$

where V represents the volume of the atomic vapour. The dipole matrix elements are defined by

$$d_{10} = \langle J=1|r_+|J=0\rangle \qquad d_{-10} = \langle J=1|r_-|J=0\rangle. \tag{66}$$

For the case of $\Gamma_1=\Gamma_2$ and $\omega_e \ll \Gamma$ we find near the onset of saturation, $v/\Gamma \leq 1$ that

$$\rho_{01}^R = \frac{vN\ (\omega - \omega_0 - \omega_e)}{(\omega - \omega_0 - \omega_e)^2 + (\Gamma^2/4 + 4v^2)}. \tag{67}$$

and,

$$\rho_{0-1}^R = \frac{-vN\ (\omega - \omega_0 + \omega_e)}{(\omega - \omega_0 + \omega_e)^2 + (\Gamma^2/4 + 4v^2)}. \tag{68}$$

These represent slightly power-broadened dispersion lineshapes and the predicted form (for low magnetic fields) is obtained as in the 'normal' case by the derivative of the lineshape function, i.e.,

$$\delta D = \frac{\partial D}{\partial \omega} \delta \omega$$

so that

$$\phi_F(\omega) \propto \frac{(\omega - \omega_0)^2 - (\Gamma^2/4 + 4v^2)}{\left[(\omega - \omega_0)^2 + (\Gamma^2/4 + 4v^2) \right]^2} g_J \mu_B B. \tag{69}$$

Conversely, in the high intensity limit where $4v^2 \gg \Gamma^2/4$ and where the excited and ground state populations are given by,

$$\sigma_{++} \sim \sigma_{--} = N/\alpha \qquad \text{and} \qquad \sigma_{00} = N(1-2\alpha)$$

where $\alpha(\omega) \sim 4$ for $(\omega-\omega_0) \sim 0$, we find that

$$\rho_{01}^R \sim \frac{v\omega_e N}{2\Gamma} \frac{\Gamma/2}{[\Gamma^2/4 + (\omega-\omega_0-\omega_e)^2]} - \left(\frac{4}{\alpha} - 1\right) \frac{vN(\omega-\omega_0-\omega_e)}{[\Gamma^2/4 + (\omega-\omega_0-\omega_e)^2]} \tag{70}$$

and

$$\rho_{0-1}^R \sim \frac{v\omega_e N}{2\Gamma} \frac{\Gamma/2}{[\Gamma^2/4 + (\omega-\omega_0+\omega_e)^2]} + \left(\frac{4}{\alpha} - 1\right) \frac{vN(\omega-\omega_0+\omega_e)}{[\Gamma^2/4 + (\omega-\omega_0+\omega_e)^2]} \tag{71}$$

The Lorentzian lineshape function is brought in and the Faraday profile changes form i.e.

The rotation is then to first order,

$$\phi_S(\omega) \sim \frac{\omega L}{4c} \frac{e^2\pi}{V\varepsilon_0 E_0} \frac{2d_{01}^2}{\hbar} \frac{E_0^2(g_J\mu_B B)N}{2\Gamma_{eff}} \frac{\Gamma_{eff}/2\pi}{\left[\Gamma_{eff}^2/4 + (\omega - \omega_0)^2\right]}$$

$$(72)$$

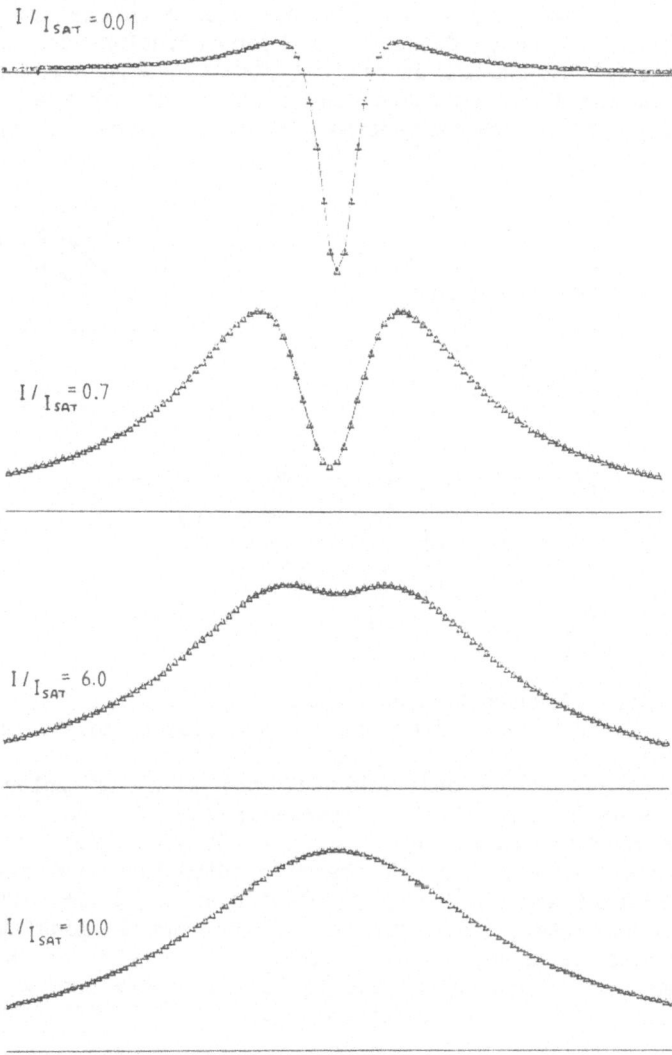

I / I_{SAT} = 0.01

I / I_{SAT} = 0.7

I / I_{SAT} = 6.0

I / I_{SAT} = 10.0

Figure 20. The effect of optical and Zeeman coherence on the Faraday rotation (theory).

where $\Gamma_{eff} = \Gamma(v)$ i.e. a power-broadened width.   Now, although the magnitude of the Lorentzian function for a single atom is comparable to that of the dispersion curve, Doppler averaging increases the symmetric contribution enormously: the rotation for the asymmetric lineshape is as often positive as negative from atom to atom, while the symmetric lineshape represents a rotation in the same sense for all atoms.

3·8 Optical rotation produced by electric fields

If the atomic vapour within the polarimeter is now subjected to a transverse electric field (i.e. perpendicular to the propagation direction of the laser) an optical rotation signal proportional to the square of the field strength can be obtained.   This arises through the imaginary part of the refractive index as shown in figure 21.   Of

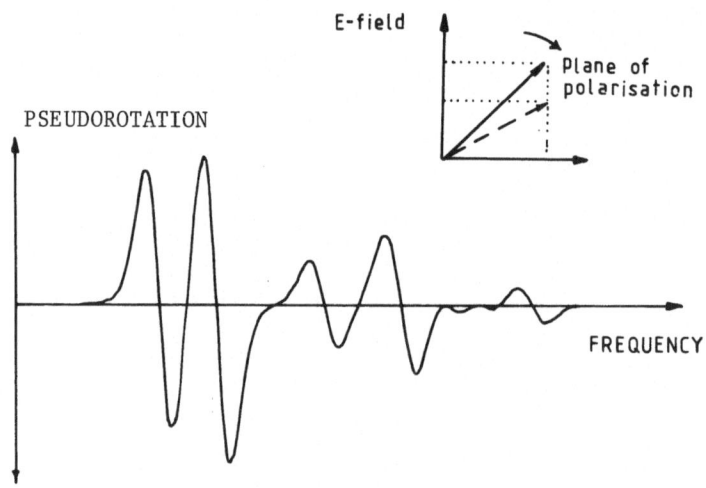

Figure 21.   Pseudorotation in Sm produced by a transverse electric field.   The structure is due to the presence of several isotopes.

course no effect is observable if the plane of polarisation of the laser is either parallel or perpendicular to the field, but it is interesting to note that with a Faraday modulator within the polarimeter the real part of the refractive index is picked out with an axial B-field whereas for a perpendicular B- or E-field the imaginary part is extracted. Furthermore, if the Faraday modulator is replaced by a phase-modulator (i.e. either a Pockels or Kerr cell) the complementary component of the refractive index can be obtained.   The situation is summarized in table 2.

To evaluate the magnitude of the pseudorotation expected in any particular case we must, as in the case of the Faraday effect, evaluate the relevant matrix elements.   To illustrate the procedure I shall again consider specifically the case of the samarium E1 transition for in this case there is a near degeneracy with a state of opposite parity which both enhances the effect and facilitates calculation of it. I shall consider only the 'normal' form although effects of saturation and optical pumping are easily visible here too. As before then there are two mechanisms: energy shifts and state mixing. The energy shift ΔE_j for a state $|j\rangle$ is given by standard second-order perturbation theory as

$$\Delta E_j = \sum_k \frac{\langle j|x|k\rangle\langle k|x|j\rangle}{\hbar\omega_{jk}} e^2 E_x^2 = -\tfrac{1}{2}\gamma E^2 \qquad (73)$$

where γ is the polarisability and the field is applied along the x-direction which is here taken to be the quantization axis. The differences in the absorptive part of the refractive index k_x and k_y are then obtained using

$$\delta k_q(1) = \frac{\pi}{2\hbar\epsilon_0} \sum_{\substack{F_k \\ M_k}} \sum_{\substack{F_j F_i \\ M_j M_i}} \frac{N(F_i)}{(2F_i+1)} |\langle j|T_q^1|i\rangle|^2 E_x^2 e^2$$

$$\times \{\frac{|\langle k|x|j\rangle|^2}{\hbar\omega_{kj}}\} \frac{\partial}{\partial\omega_{ji}} L_G(\omega_{ji}-\omega) \qquad (74)$$

in which

$$\delta k_x(1) = \delta k_0(1) \quad \text{and} \quad \delta k_y(1) = \frac{\delta k_{+1}(1) + \delta k_{-1}(1)}{2}.$$

The normalised pseudorotation can therefore be written

$$\phi = \frac{\delta k_x(1) - \delta k_y(1)}{4k(0)} \sin 2\theta \qquad (75)$$

where θ is the angle between the plane of polarisation and the field, and $k(0)$ is the linecentre absorption coefficient. As in the Voigt effect, described in the next section, the summations over 3j-symbols do not reduce to zero and to a good approximation we may take the energy differences ω_{kj} for the odd Sm isotopes to be the same for all hyperfine components between the two levels.

The Stark mixing term is obtained as follows

$$|j'\rangle = |j\rangle + \sum_k \frac{\langle k|x|i\rangle}{\hbar\omega_{kj}} e E_x |k\rangle. \qquad (76)$$

The mixing affects the transition probability to the different M-substates of the excited state only (remember $J_g = 0$ for the case being considered) and so once again differential absorption occurs producing a pseudorotation. The relative importance of the mixing term, $\delta k(2)$, to the energy shift term, $\delta k(1)$, depends on the relative strengths of the transitions which are mixed. For example in the present case of an allowed E1 transition, $\delta k(2)$ is a factor of $\sim \Gamma\omega_{kj}$ smaller than $\delta k(1)$. Even with such a near degeneracy ($\sim$ 300 GHz), $\delta k(2)$ only contributes about 1%. Finally, it should be emphasized that optical polarimetry is a very sensitive way of

Type of field	Direction	Name of effect	Type of effect	Lineshape with Faraday modulator and crossed polarizers
Magnetic field	Longitudinal	Faraday	Optical rotation & circular dichroism	
	Transverse	Voigt	Pseudorotation	Second derivative of a Lorentzian convolved with a Gaussian
Electric field	Longitudinal	No effect		
	Transverse	Stark induced rotation	Pseudorotation	Lorentzian derivative convolved with a Gaussian

Table 2 Optical rotation effects in atoms

determining atomic polarisabilities; the alternative of measuring ΔE_j spectroscopically usually requires very high optical resolution (i.e. Doppler-free signals) and very high electric field strengths.

3·9 Voigt Effect

A closely related, although distinguishable, effect is that due to a perpendicular magnetic field. Here we take as the axis of quantization the direction of the B-field which coincides with the x-axis. The energy shift term gives for light polarised along x ($q = 0$) the absorption coefficient

$$k_x(1) = \frac{\pi}{2\hbar\epsilon_0} \sum \frac{N(F_i)}{(2F_i+1)} |\langle j|T_0^1|i\rangle|^2 \left[L_G(\omega_{ji}-\omega) + \frac{\partial L_G(\omega_{ji}-\omega)}{\partial\omega_{ji}} \right.$$

$$\left. \times g_{ji} B_x + \frac{\partial^2 L_G(\omega_{ji}-\omega)}{2\partial\omega_{ji}^2} g_{ji}^2 B_x^2 \right] \tag{77}$$

where $g_{ji} = [\langle j|\mu_x|j\rangle - \langle i|\mu_x|i\rangle]$.

The y-polarisation interacts with both $q = 1$ and $q = -1$ components of the absorption equally giving

$$k_y(1) = \frac{k^{(1)}_{q=1} + k^{(1)}_{q=-1}}{2} \tag{78}$$

The linear B terms in the expressions for $k_x(1)$ and $k_y(1)$ are proportional to the 3j-symbol

$$\begin{pmatrix} 1 & 1 & 1 \\ -q & q & 0 \end{pmatrix}$$

which is equal to $(-1)^{1-q}.q/\sqrt{6}$ so that all first order contributions in B_x are zero. Thus, $[\delta k_y(1) - \delta k_x(1)]$ is proportional to B_x^2, this term remaining finite because it is proportional to

$$\begin{pmatrix} Fi & 1 & F_j \\ -M_i & q & M_j \end{pmatrix}^2 \begin{pmatrix} F & 1 & F \\ -M & 0 & M \end{pmatrix}^2$$

State-mixing terms on the other hand are evaluated by making the substitution

$$|i\rangle \rightarrow |i\rangle - \sum \frac{\langle k|\mu_x|j\rangle}{\hbar\omega_{ji}} \cdot |k\rangle B_x + \text{higher order terms.} \tag{79}$$

Terms linear in B_x are once again proportional to

$$\begin{pmatrix} 1 & 1 & 1 \\ -q & q & 0 \end{pmatrix}$$

and so disappear in the summations, leaving the second order term to be evaluated. I shall not attempt this here because of the complexity but it is worth noting that the Voigt effect is smaller than the Faraday effect by the ratio $B_x g_{ji}/\Gamma$ which is of order 10^{-3} for $B_x = 1$ G and $\Gamma = 100$ MHz. Furthermore, the pseudorotation produced by the Voigt effect in the low field limit has a lineshape which is a second differential of a Lorentzian curve convolved with a Gaussian (Doppler) function.

3·10 Parity Non-conserving Neutral Weak Interaction

The presence of a neutral weak current interaction between the electrons and nucleons in an atom gives rise to a parity non-conserving part of the atomic Hamiltonian which can be written as

$$H_{weak} = H^{(1)}_{PNC} + H^{(2)}_{PNC} = \frac{G_F}{\sqrt{2}} \left[C_{1N}\gamma^5_e \delta(r_{eN}) + C_{2N}\sigma_N \cdot \alpha_e \delta(r_{eN}) \right]. \tag{80}$$

The first term which is independent of nuclear spin dominates by several orders of magnitude in heavy elements over the second, nuclear spin dependent, term. In the above expression α and γ^5 are the standard Dirac matrices and σ is the Pauli spin matrix (see e.g. Sakurai[78]). N stands for any nucleon either proton or neutron and the coupling constants C take on the following form in lowest order in the 'standard', Glashow-Weinberg-Salam model for electroweak unification:

$$C_{1p} = \tfrac{1}{2}(1 - 4\sin^2\theta_W) \qquad C_{2p} = g_A \tfrac{1}{2}(1 - 4\sin^2\theta_W)$$

$$C_{1n} = -\tfrac{1}{2} \qquad\qquad C_{2n} = -g_A \tfrac{1}{2}(1 - 4\sin^2\theta_W)$$

(81)

g_A is the ratio of vector to axial coupling and θ_W is the Weinberg angle, a free parameter in the model.

A few points at this stage are worthy of note.

(i)   It is customary to write the C_1 contribution in terms of a so-called 'weak charge' $Q_W = -N + Z(1 - 4\sin^2\theta_W)$ since the individual nucleon contributions add coherently.

(ii)   Since $\sin^2\theta_W$ has been found to have the value $0\cdot23(1)$[79] from high energy experiments, $C_{1p} \sim 0\cdot08 \times \tfrac{1}{2}$. This means that PNC effects in atomic transitions involving electrons which overlap with the nucleus are sensitive mainly to C_{1n}.

(iii)   The non-relativistic form of H_{weak} can be obtained straightforwardly giving,

$$\gamma_e^5 \rightarrow \frac{\sigma_e \cdot p_e}{2mc} + h.c.$$

$$\sigma_N \cdot \alpha_e \rightarrow \left[\frac{\sigma_N \cdot p_e}{2mc} + i\,\frac{\sigma_N \cdot \sigma_e \times p_e}{2mc} \right] + h.c.$$

3·11 Effects in Atoms of H_{weak}

Consider a transition between two states $|I\rangle$ and $|J\rangle$ which are parity admixed. The transition matrix element is given by

$$\langle I|T|J\rangle = \langle i|T|i\rangle + \sum_k \frac{\langle i|T|k\rangle\langle k|V_{PNC}|j\rangle}{E_k - E_j} + \sum_k \frac{\langle i|V_{PNC}|k\rangle\langle k|T|j\rangle}{E_i - E_k}$$

(82)

Writing $\langle I|T|J\rangle$ as $A_{PC} + A_{PNC}$ we see that the transition probability is proportional to

$$|\langle I|T|J\rangle|^2 = |A_{PC}|^2 + |A_{PNC}|^2 + \left(A_{PC}A_{PNC}^* + A_{PC}^*A_{PNC}\right)$$ (83)

The fractional parity non-conserving part is thus,

$$F = \frac{\left(A_{PC}A_{PNC}^* + A_{PC}^*A_{PNC}\right)}{|A_{PC}|^2 + |A_{PNC}|^2} \sim \frac{A_{PNC}}{A_{PC}}$$

(84)

3·12 Enhancement of the PNC asymmetry

The parity mixing of states can be enhanced in two ways

(a)   The size of the H^{PNC} matrix element.   The spin independent part of the electron-nucleus, weak interaction scales approximately as Z^3, giving a considerable enhancement for heavy atoms.   To see how this comes about we first write down the matrix element which looks like

$$\langle H_{PNC}^{(1)} \rangle \ = \ \langle s_{1/2} \ | \ \frac{G_F}{2\sqrt{2}} \ Q_W \ \rho(r) \ \gamma_e^5 \ | \ p_{1/2} \rangle \qquad (85)$$

This gives three factors of Z as follows

i)   The weak charge Q_W is roughly proportional to Z.

ii)   The electron momentum appears in the operator (i.e. in the non-relativistic approximation $\gamma_e^5 \rightarrow \sigma_e \cdot p_e / 2mc$) and the momentum is proportional to Z.   This operator, $-i\hbar\nabla_r$, also makes the matrix element pure imaginary as required for time-reversal invariance.

iii) The short range nature of the weak interaction means that the PNC matrix element depends on the value of the valence electron wavefunctions at the origin.

   On the other hand the spin-dependent weak interaction $H_{PNC}^{(2)}$ contains no enhancement factor Q_W because the nuclear spins cancel in pairs in the sum over nucleons.

(b)   The energy denominator.   In certain circumstances considerable enhancement can be obtained because of a near degeneracy between opposite parity states.   In practice, the conditions for this enhancement can only be found in the hydrogen atom, hydrogen-like ions and in the complex structure of rare-earth elements. Experiments to date all use one or other enhancement technqiue, but not both simultaneously.   Thus, in the case of atomic hydrogen (Z = 1) PNC experiments look for radiofrequency transitions within the $2^2S_{1/2}$ hyperfine multiplet.   The $2S_{1/2}$ and $2P_{1/2}$ levels are separated by the Lamb shift (~1058 MHz) which is about 10^5 times less than the typical energy denominator of ~ 10^4 cm^{-1} in heavy atom experiments.   The denominator may be further reduced by using a static magnetic field to drive the levels to a crossing point; this gives a further enhancement of 20 but it is questionable whether or not this is necessarily advantageous in view of the attendant systematic effects due to $v \times B$ motional fields.   By contrast heavy atoms benefit by a factor of ~ 10^6 from the Z^3 scaling.   The mixing is therefore not so different but for hydrogen the n = 2 metastable state is required and the obtainable number density will be lower and the sensitivity to systematic effects produced by fields greater because of the near degeneracy.   We note also that the spin-dependent term in heavy elements may not be reduced by the full factor of $Z^{-1}(1 - 4\sin^2\theta_W)$ compared to the spin-independent term because the nucleus may possess a so-called 'anapole moment'. This P-odd magnetic moment μ_W can be pictured as being produced by a solenoid bent round in the form of a torus; the field is totally enclosed giving no net external field[80] i.e.

$$\mu_W = -\pi \int r^2 j(r) d^3 r. \qquad (86)$$

where $j(r)$ is the electromagnetic current density. Calculations[81] suggest that the anapole moment can induce a spin-dependent PNC effect of the same form as $H_{PNC}^{(2)}$ but larger by a factor of between 6—8.

Finally, the spin-dependent contribution is in principle separable from the spin-independent term because it gives an optical rotation which varies in a distinctive way from one hfs component to another.

3·13 The Experiments

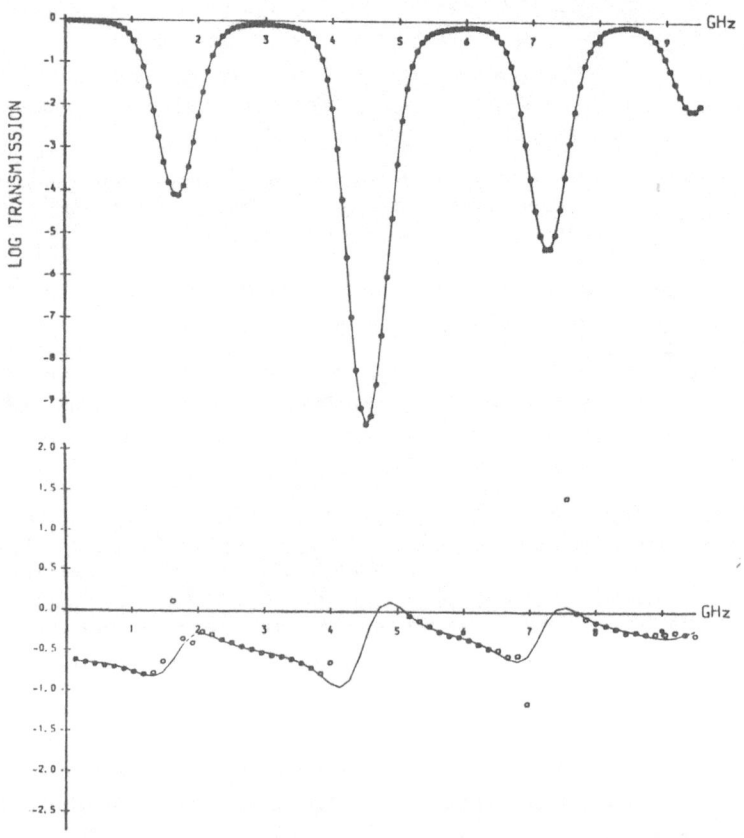

Figure 22. Part of the absorption spectrum of the M1 transition $\lambda876$nm of BiI (upper curve); measured PNC optical rotation (lower curve).

Experiments on heavy elements are of one of two types: optical rotation on allowed M1 transitions (amplitude $\sim \mu_B$) using a polarimeter such as that described in section 3·3, or fluorescence experiments performed on Stark-assisted, forbidden M1 transitions (amplitude $\sim \alpha^2 \mu_B$). In the latter case the dependence of the fluorescent intensity on the helicity (circular polarisation) of the incident exciting radiation is measured; in the former the variation of optical rotation with wavelength. As an example some typical data obtained from the optical rotation experiment performed in Oxford[70] is given in figure 22; the PNC rotation is clearly visible although it is only of order 10^{-6} radians.

Acknowledgements

Many of the experiments and calculations that have been described in these lectures have been carried out by visitors and members of the Atomic Physics Group at the Clarendon Laboratory. In particular I wish to acknowledge the work of M. Boshier, I. Davies, C. Foot, E. Hinds, M. Macpherson, J. Nicol, M. Plimmer, C. Palmer, D. Stacey, D. Tate and G. Woodgate. I am also indebted to J. Deeny, P. Ewart, S. Nakayama and P. Sandars for valuable and stimulating discussions.

References
1. W. Lamb and R. Retherford, Phys. Rev. 72, 241 (1947).
2. R. van Dyck, P. Schwinberg and H. Dehmelt, Atomic Physics 9, (Proc. 9th Int. Conf. on Atomic Physics), p.53, ed. R. van Dyck and E. Fortson (World Scientific 1985).
3. H. Bethe and E. Salpeter, 'Quantum Mechanics of One and Two Electron Atoms' (Springer 1957).
4. W. Johnson and G. Soff, Atomic and Nuclear Data Tables 33, 407 (1985).
5. G. Erickson, J. Phys. Chem. Reference Data 6, 831 (1977).
6. P. Mohr, Ann. Phys. 88, 52 (1974).
7. R. Beausoleil, D. McIntyre, C. Foot, B. Couilland, E. Hildum and T. Hänsch, submitted to Phys. Rev. A for publication.
8. J.R.M. Barr, J.M. Girkin, A.I. Ferguson, G.P. Barwood, P. Gill, W.R.C. Rowley and R.C. Thompson, Opt. Comm. 54, 217 (1985).
9. R.A. Ferrell, Phys. Rev. 84, 858 (1951).
10. C.J. Oram et al., Phys. Rev. Lett. 52, 910 (1984).
11. A.P. Mills, J. Imazato, S. Saitoh, A. Uedono, Y. Kawashima and K. Nagamine, Phys. Rev. Lett. 56, 1463 (1986).
12. P. Zhao, W. Lichten, H. Layer and J. Bergquist, Phys. Rev. (Rapid Comm.) 34A, 5138 (1986).
13. P. Zhao, W. Lichten, H. Layer and J. Bergquist, Phys. Rev. Lett. 58, 1293 (1987).
14. D. Tate, P. Baird, M. Boshier, E. Hinds, D. Stacey and G. Woodgate, in preparation for J. Phys. B.
15. E. Baklanov and V. Chebotayev, Sov. Phys. JETP 33, 300 (1971).
16. V. Letokhov and V. Chebotayev, 'Non-Linear Laser Spectroscopy' (Springer 1977).

238

16. V. Letokhov and V. Chebotayev, 'Non-Linear Laser Spectroscopy' (Springer 1977).
17. Y. Gontier and M. Trahin, Phys. Lett. $\underline{36A}$, 463 (1971).
18. F. Bassini, J. Forney and A. Quattropani, Phys. Rev. Lett. $\underline{39}$, 1070 (1977).
19. F. Biraben, M. Bassini and B. Cagnac, J. de Phys. 40, 445 (1979).
20. G. Boyd and D. Kleinman, J. Appl. Phys. $\underline{39}$, 3597 (1968).
21. J. Fleck and M. Feit, J. Opt. Soc. Am. $\underline{73}$, 920 (1983).
22. W. Leeb, Appl. Phys. $\underline{6}$, 267 (1975).
23. K. Kato, IEEE J. of Quant. elect. $\underline{QE-22}$, 1013 (1986).
24. [This Summer School] and M. Boshier, P. Baird, C. Foot, E. Hinds, M. Plimmer, J. Swan, D. Stacey, D. Tate, D. Warrington and G. Woodgate, Invited contribution to the Eighth International Conf. on Laser Spectroscopy held in Åre, Sweden (June 1987).
25. R. Beausoleil and T. Hänsch, Phys. Rev. $A\underline{33}$, 1661 (1986).
26. N.F. Ramsey, 'Molecular Beams' (Oxford University Press 1955).
27. S. Chu and A. Mills, Phys. Rev. Lett. $\underline{48}$, 1333 (1982).
28. T. Fulton and P. Martin, Phys. Rev. $\underline{95}$, 811 (1954).
29. S. Chu, A. Mills and J. Hall, Phys. Rev. Lett. $\underline{52}$, 1689 (1984).
30. T. Fulton, Phys. Rev. $A\underline{26}$, 1794 (1982).
31. D. McIntyre and T. Hänsch, Phys. Rev. (Rapid Comm.) (1986).
32. G. Carbonni et al., Phys. Lett. $\underline{73B}$, 229 (1978).
33. I. Sick, private communication.
34. S. Blundell, P. Baird, C. Palmer, D. Stacey and G. Woodgate, to appear in J. Phys. B.
35. W. King, J. Opt. Soc. Am. $\underline{53}$, 638 (1963).
36. S. Blundell, P. Baird, C. Botham, C. Palmer, D. Stacey and G. Woodgate, J. Phys. $B\underline{17}$, 53 (1984).
37. E. Shera et al., Phys. Rev. $C\underline{14}$, 731 (1976).
38. P. Baird, R. Bramley, K. Burnett, D. Stacey, D. Warrington and G. Woodgate, Proc. R. Soc. $A\underline{365}$, 567 (1979).
39. K. Bekle, A. Andle, G. Göring, A. Hanser, G. Nowicki, H. Rebel and G. Schatz, Z. Phys. $A\underline{291}$ (1979).
40. G. Greenlees, D. Clark, S. Kaufman, D. Lewis, J. Tonn and J. Broadhurst, Opt. Comm. $\underline{23}$, 236 (1977).
41. C. Palmer, P. Baird, J. Nicol, D. Stacey and G. Woodgate, J. Phys. $B\underline{15}$, 993 (1982); also, C. Palmer et al., J. Phys. $B\underline{17}$, 2197 (1984).
42. P. Baird, S. Blundell, G. Burrows, C. Foot, G. Meisel, D. Stacey and G. Woodgate, J. Phys. $B\underline{16}$, 2485 (1983).
43. K. Anton et al., Phys. Rev. Let. $\underline{40}$, 642 (1978).
44. W. Neuhauser, M. Hohenstall, P. Toschek and H. Dehmelt, Phys. Rev. $A\underline{22}$, 1137 (1980).
45. W. Letokhov and V. Minogun, Appl. Phys. $\underline{17}$, 99 (1978); also, V. Letokhov, V. Minogun and B. Pavlik, Opt. Comm. $\underline{19}$, 72 (1976).
46. T. Hänsch and A. Schawlow, Opt. Comm. $\underline{13}$, 68 (1975).
47. D. Wineland and H. Dehmelt, Bull. Am. Phys. Soc. $\underline{20}$, 637 (1975).
48. W. Neuhauser, M. Hohenstatt, P. Toschek and H. Dehmelt, Phys. Rev. Lett. $\underline{41}$, 232 (1976).
49. D. Wineland, R. Drullinger and F. Walls, Phys. Rev. Lett. $\underline{40}$, 1639 (1978).

50. W. Nagourney, G. Janik and H. Dehmelt, Proc. Nat. Acad. Sci. (USA) 80, 643 (1983).

51. A. Migdall, J. Prodan, W. Phillips, T. Bergeman and H. Metcalf, Phys. Rev. Lett. 54, 2596 (1985).

52. W. Ertmer, R. Blatt, J. Hall and M. Zhu, Phys. Rev. Lett. 54, 996 (1985).

53. A. Ashkin, Phys. Rev. Lett. 40, 729 (1978).

54. S. Chu., L. Hollberg, J. Bjorkholm, A. Cable and A. Ashkin, Phys. Rev. Lett. 55, 48 (1985).

55. R. Cook and H. Kimble, Phys. Rev. Lett. 54, 1023 (1985).

56. D. Pegg, R. Loudon and P. Knight, Phys. Rev. 33A, 4085 (1986).

57. G. Gabrielse and H. Dehmelt, Phys. Rev. Lett. 55, 67 (1985).

58. W. Jue, A. Anderson, E. Hinds, D. Meschede, L. Moi and S. Haroche, Phys. Rev. Lett. 58, 666 (1987).

59. P. Goy, J. Raimond, M. Gross and S. Haroche, Phys. Rev. Lett. 50 1903 (1983).

60. See e.g., S. Haroche, Atomic Physics 7 (1981) [Proc. Int. Conf. on Atomic Physics (Plenum).]

61. M. Gross, P. Goy, C. Fabre, S. Haroche and J. Raimond, Phys. Rev. Lett. 43, 343 (1979).

62. D. Meschede, H. Walther and G. Muller, Phys. Rev. Lett. 54, 551 (1985).

63. J. Parker and C. Stroud, Phys. Rev. Lett. 56, 716 (1986).

64. G. Alber, H. Ritsch and P. Zoller, Phys. Rev. 34A, 1058 (1986).

65. S. Weinberg, Phys. Rev. Lett. 19 1264 (1967); A. Salam in 'Elementary Particle Theory', Proc. 8th Nobel Symp. (1968).

66. F. Hasert et al., Phys. Lett. 46B, 138 (1973).

67. A. Benvenuti et al., Phys. Rev. Lett. 32, 800 (1974).

68. M. Bouchiat and C. Bouchiat, Phys. Lett. 48B, 111 (1974).

69. See e.g. E. Fortson and L. Lewis (and references therein) Phys. Rep. 113, 289 (1985).

70. M. Macpherson, D. Stacey, P. Baird, J. Hoare, P. Sandars, K. Tregidgo and W. Guowen, submitted to Europhysics Letters.

71. G. Birich et al., Sov. Phys. JETP 60, 442 (1984).

72. S. Nakayama and P. Baird, Jap. J. Appl. Phys. 26, 1765 (1987).

73. G. Roberts, P. Baird, M. Brimicombe, P. Sandars, D. Selby, and D. Stacey, J. Phys. B13, 1389 (1980).

74. B. Cagnac, Ann. de Phys. 6, 467 (1960).

75. W. Franzen and A. Emslie, Phys. Rev. 108, 1453 (1957).

76. I. Davies, P. Baird and J. Nicol, J. Phys. B20, 5371 (1987).

77. C. Cohen-Tannoudji, Atomic Physics 4 (Proc. 4th Int. Conf. on Atomic Physics) p.589, Ed. G. zu Putlitz, E. Weber and A. Winnacker (Plenum)

78. J. Sakurai, 'Advanced Quantum Mechanics' (published by Addison-Wesley 1967).

79. W. Marciano and A. Sirlin, Phys. Rev. D27, 552 (1983).

80. Ya. Zeldovich, Sov. Phys. JETP 6, 1184 (1957).

81. V. Flambaum, I. Khriplovich and O. Sushkov, Phys. Lett. 146B, 367 (1984).

LASER RAMAN SPECTROSCOPY

L. D. Barron
Chemistry Department
The University
Glasgow G12 8QQ
U.K.

ABSTRACT. These lectures present a survey of modern Raman spectroscopy with emphasis on applications to gases. The semi-classical theory of conventional linear Raman scattering is surveyed, including the use of irreducible tensor methods to calculate rotational Raman intensities and to deduce selection rules. Novel aspects of linear Raman spectroscopy discussed include antisymmetric scattering in resonance Raman processes, Raman EPR associated with spin-flip transitions in molecules with Kramers degeneracy, and pure rotational Raman optical activity from chiral symmetry tops. Semi-classical theories of nonlinear Raman processes are then developed, concentrating on coherent Raman phenomena since these can provide very high resolution gas phase spectra. The current range of coherent Raman techniques is reviewed, including CARS, RIKES, stimulated Raman gain and loss spectroscopy, and PARS.

1. INTRODUCTION

1.1. General

Next year, 1988, marks both the centenary of the birth of the Indian scientist C.V. Raman and the diamond jubilee of the discovery of the Raman effect in 1928. Raman spectroscopy has turned out to be one of the most valuable and versatile of all spectroscopic techniques, and the Indian scientific community, and others, plan to celebrate.

Most of the vibrational and rotational spectra obtained before the second world war were measured using Raman methods. Interest in Raman then declined as infrared and microwave absorption instrumentation developed, but the introduction of visible lasers in the early 1960s has led to a dramatic renaissance in Raman spectroscopy. As well as decreasing the acquisition time and increasing the sensitivity of conventional Raman spectra by orders of magnitude, the high power and coherence properties of laser radiation has spawned a host of new nonlinear Raman spectroscopies, some of which can be performed without a

A. C. P. Alves et al. (eds.), Frontiers of Laser Spectroscopy of Gases, 241–280.
© 1988 by Kluwer Academic Publishers.

spectrometer.

    In these lectures I have presented a survey of modern Raman
spectroscopy, emphasising the applications to gases. Since it is
necessary to have some idea of the theory in order to fully appreciate
the various applications, I have outlined the semi-classical theory of
both conventional and nonlinear Raman processes. Quite a lot of space
is devoted to the basis of conventional Raman scattering, not only
because it provides an essential foundation for the semi-classical
theory of nonlinear processes, but also to emphasise that, using
modern electro-optical technology, conventional Raman is still very
much the preferred technique for many applications, and that the new
nonlinear techniques should only be used when conventional Raman fails
or when very high resolution is essential.

1.2. Linear Raman Spectroscopy

Unlike most other forms of spectroscopy, which usually rely on
absorption or emission of radiation, Raman is based on the inelastic
scattering of a visible light beam. A laser beam is focused into a
sample and the spectrum of the scattered light analysed. Most of the
scattered light has the same frequency as the incident laser beam, and
this is the Rayleigh component. But a tiny fraction has frequencies
shifted from the Rayleigh component by amounts corresponding to
rotational, vibrational and sometimes electronic transition frequencies
of the molecules in the scattering medium, and these of course are the
Raman components.

    The corresponding photon-matter interactions are illustrated in
figure 1.1. In Rayleigh and Raman scattering at <u>transparent</u> frequencies

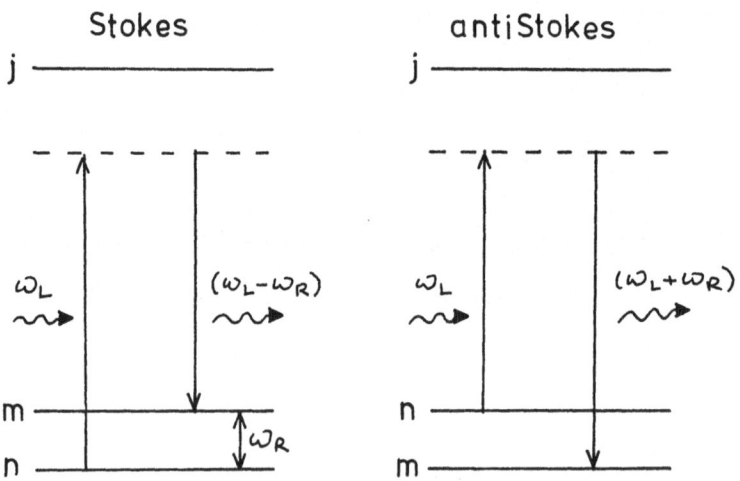

Figure 1.1.   Energy level diagrams for spontaneous Raman scattering.

using visible excitation the energy $\hbar\omega_L$ of the incident laser photon is usually less than the energy separation of the ground and first excited electronic level, so the scattering proceeds via a virtual intermediate level. The Stokes Raman scattering process results in excitation of the molecule from an initial state $|n\rangle$, usually within the ground level, to a final state $|m\rangle$, of higher energy; whereas the antiStokes process takes the molecule from an initially excited state $|n\rangle$ to a final state $|m\rangle$ of lower energy, usually within the ground electronic level. The two steps in the scattering process are each electric dipole in character, so the initial and final Raman states $|n\rangle$ and $|m\rangle$ have the same parity.

If the incident photon energy $\hbar\omega_L$ happens to coincide with an electronic absorption frequency of the molecule, a tremendous enhancement (up to 10^6) of the Rayleigh and Raman scattering intensity can occur. This is called the resonance Raman effect.

The Raman effect therefore allows observation of low-energy molecular transitions using excitation with visible light. The Raman spectrum is analysed with a visible spectrometer and the frequency shifts $\Delta\omega$ of bands relative to the exciting line measured (figure 1.2). Since the Stokes Raman process originates in ground-state molecules, the Stokes bands maintain intensity throughout the

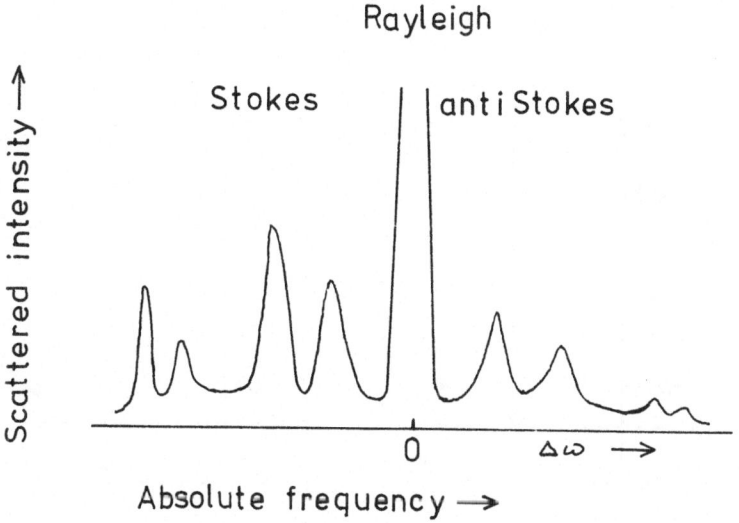

Figure 1.2. The general appearance of a Raman spectrum.

spectrum; whereas the antiStokes band intensities drop off rapidly with increasing $\Delta\omega$ in accordance with the Boltzmann distribution of molecules in excited states.

So the great advantage of conventional spontaneous Raman spectro-scopy is its simplicity, with complete rotational and vibrational spectra obtainable in a single run on a simple instrument using visible optics. The disadvantages include the inherent weakness of the scattering process, sample fluorescence obscuring the Raman spectrum, and the limitations on frequency resolution imposed by the visible spectrometer.

References [1 - 4] contain useful reviews of conventional Raman spectroscopy.

1.3. Nonlinear Raman Spectroscopy

The high intensity and coherence of laser radiation can lead to more elaborate photon scattering processes than those involved in the conventional Raman effect. The simplest example is second harmonic generation (hyper-Rayleigh scattering) and the associated hyper-Raman effect in which two laser photons of frequency ω_L interact simultaneously with the molecule to produce a scattered photon at frequency $2\omega_L$ (hyper-Rayleigh), or at $2\omega_L - \omega_R$ (Stokes hyper-Raman) or at $2\omega_L + \omega_R$ (antiStokes hyper-Raman). As illustrated in figure 1.3, these processes involve two virtual intermediate excited states.

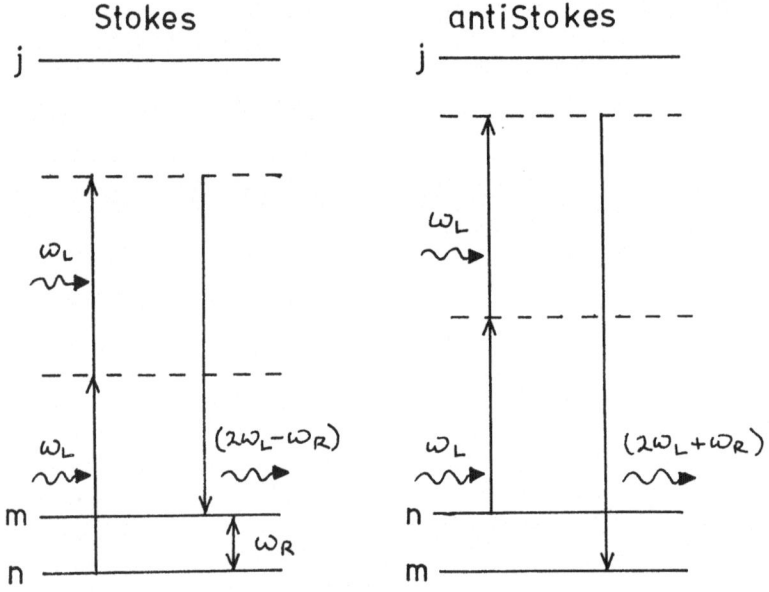

Figure 1.3. Hyper-Raman scattering processes.

These second harmonic processes are still spontaneous, just like the conventional Raman effect, which means that there is no definite phase relationship between the Raman waves scattered from separate molecules: the scattering is incoherent. Furthermore, the hyper-Raman intensity is several orders of magnitude weaker than conventional Raman, so measurements are very difficult.

However, there is in addition a range of nonlinear processes which

generate coherent Raman waves with intensities comparable with those of
the exciting laser beams themselves, and the associated coherent Raman
techniques are much more important in gas phase studies. All of these
coherent Raman phenomena involve a four-photon process of the general
type illustrated in figure 1.4. Up to three independent fields with

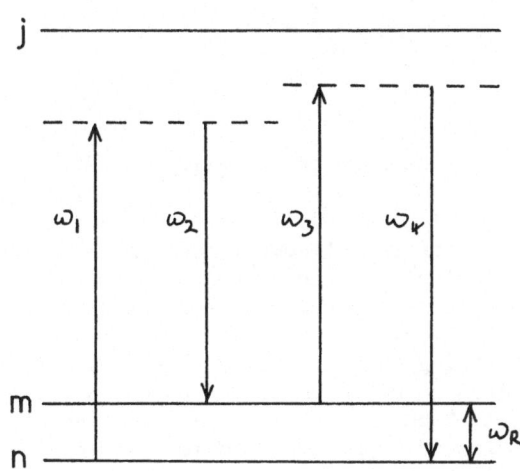

Figure 1.4. The general coherent Raman scattering process.

angular frequencies ω_1, ω_2 and ω_3 may be incident on the molecule,
and generate a fourth field at $\omega_4 = \omega_1 - \omega_2 + \omega_3$ which is phase-
coherent relative to the input fields. In coherent antiStokes Raman
spectroscopy (CARS), for example, the fields at ω_1 and ω_3 originate
in the same laser of fixed frequency ω_L, and the field at ω_2
originates in a second, tunable, laser at ω_3 so that $\omega_4 = 2\omega_L - \omega_3$.
If ω_3 is tuned until $\omega_L - \omega_3$ equals the Raman transition frequency
ω_R (so that ω_3 becomes the associated Stokes frequency), a tremendous
enhancement of ω_4 occurs.

This is an appropriate point to mention the important matter of
phase-matching in coherent Raman processes. As well as the frequency
condition $\omega_4 = \omega_1 - \omega_2 + \omega_3$, the coherent Raman wave must also
propagate in a specific direction with wave vector

$$\underline{k}_4 = \underline{k}_1 - \underline{k}_2 + \underline{k}_3 . \qquad (1.1)$$

The wave vector is a function of the refractive index n at the
particular frequency,

$$|\underline{k}| = n\omega/c \qquad (1.2)$$

which means that, when the frequency dispersion is significant, as in
condensed media, a **collinear** arrangement of all four beams is not
possible. The input fields must overlap at the specific angles which

satisfy (1.1), as illustrated in figure 1.5.

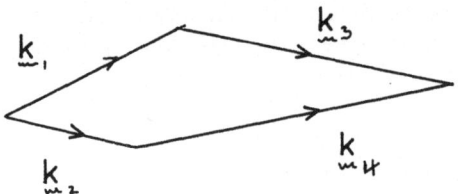

Figure 1.5. Phase-matching (momentum conservation) condition for a coherent Raman process.

Although this condition can be troublesome to set up experimentally, it does have the advantage that all of the signal can be easily collected. But for gas-phase studies where the refractive index dispersion is small, **collinear** beams can usually be used.

Thus as well as returning the molecule to its original state, the four-wave process transfers no momentum to the molecule if the phase matching condition is met. Such a mechanism is known generally in nonlinear optics as a parametric process, and is associated with optimum conversion efficiency.

These coherent nonlinear Raman techniques come into their own for high resolution gas phase studies: no spectrometer is involved, the resolution being determined solely by the line widths of the input laser beams. Figures 1.6 and 1.7 reproduce an example, given by Kiefer [4], of the improvement in resolution that is possible. Figure 1.6 shows a conventional Raman spectrum of NO at a pressure of 100 Torr obtained with a commercial 0.85m double monochromator and excited with an argon-ion laser beam of 1.5W power. The resolution is about 2 cm^{-1}. Figure 1.7 shows a CARS spectrum with a resolution of about 0.05 cm^{-1} excited with a nitrogen laser pumped dye laser system: although the time-averaged total power was reduced by a factor of about 2000, a forty-fold improvement in resolution was obtained. However, even higher resolution is possible.

Comprehensive reviews of coherent nonlinear Raman spectroscopy can be found in references [4 - 6].

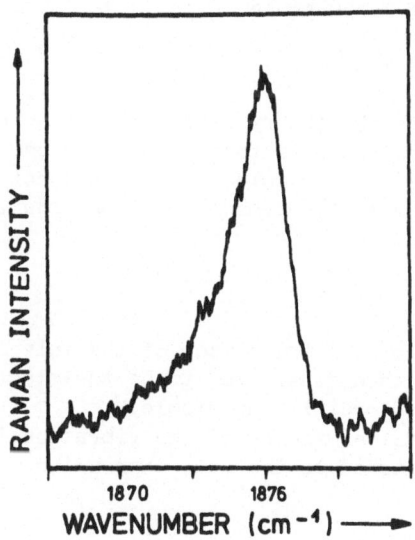

Figure 1.6. Conventional Raman spectrum of the Q-branch of NO at 100 Torr pressure [4].

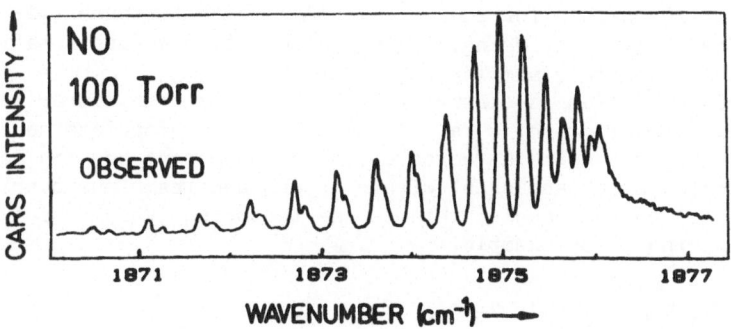

Figure 1.7. CARS spectrum of the Q-branch of NO at 100 Torr pressure [4].

2. THEORY OF LINEAR RAMAN PROCESSES

2.1. The Induced Dipole Moment

The simplest approach to the theory of light scattering in general and Raman scattering in particular is to introduce an oscillating electric dipole moment $\underline{\mu}$ induced in the molecule by the electric field vector $\underline{E}$ of the light wave. In cartesian tensor notation, the α-component is written

$$\mu_\alpha = \alpha_{\alpha\beta} E_\beta , \qquad (2.1)$$

where $\alpha_{\alpha\beta}$ is the polarizability tensor of the molecule. The origin of the scattered light is then considered to be the electromagnetic field radiated by this induced oscillating dipole.

In the simple classical theory of the vibrational Raman effect [1] the electric vector is taken as the real expression

$$E_\alpha = E_\alpha^{(0)} \cos\omega t , \qquad (2.2)$$

where ω is the angular frequency, and the polarizability is expanded as a Taylor series in the normal vibrational coordinates $Q_p = Q_p^{(0)} \cos\omega_p t$. This yields terms of the form

$$\frac{1}{2} \left(\frac{\partial \alpha_{\alpha\beta}}{\partial Q_p} \right)_0 Q_p^{(0)} \cos (\omega \pm \omega_p) t \qquad (2.3)$$

which are responsible for Stokes ($\omega - \omega_p$) and antiStokes ($\omega + \omega_p$) vibrational Raman scattering. It is possible to describe rotational Raman scattering in a similar fashion [7,8].

Tensor notation is central to the theory of both linear and non-linear light scattering, so a brief summary is appropriate here. A Greek subscript denotes a vector or tensor component and can be x,y or z. A repeated Greek subscript within a term denotes a sum over all three cartesian components: this is the tensor equivalent of a scalar product so that, for example,

$$a_\alpha b_\alpha = a_x b_x + a_y b_y + a_z b_z \equiv \underline{a}.\underline{b} .$$

The unit symmetric tensor $\delta_{\alpha\beta}$ takes the value zero when $\alpha \neq \beta$ and is unity when $\alpha = \beta$. For example, the scalar product above could be written $a_\alpha b_\alpha = a_\alpha b_\beta \delta_{\alpha\beta}$. The unit alternating tensor $\varepsilon_{\alpha\beta\gamma}$ is defined such that $\varepsilon_{\alpha\beta\gamma} r_\beta p_\gamma$ is the α-component of the vector product $\underline{r} \times \underline{p}$; that is, $\varepsilon_{\alpha\beta\gamma}$ is equal to +1 or −1 if α, β, γ is an even or odd permutation of x,y,z and is zero if any two subscripts are the same. The scalar product of a vector with a vector product would thus be written

$$a_\alpha \varepsilon_{\alpha\beta\gamma} b_\beta c_\gamma = a_x (b_y c_z - b_z c_y) + a_y (b_z c_x - b_x c_z)$$
$$+ a_z (b_x c_y - b_y c_x) \equiv \underline{a}.(\underline{b} \times \underline{c}) .$$

The electric vector of a plane-wave light beam travelling in the direction of the propagation vector $\underset{\sim}{n}$, is, in complex notation,

$$\tilde{E}_\alpha = \tilde{E}_\alpha^{(o)} \, e^{-i(\omega t - \underset{\sim}{k} \cdot \underset{\sim}{r})} \tag{2.4}$$

where $\underset{\sim}{k} = \underset{\sim}{n}\,\omega/c$ is the wave vector and a tilde denotes a complex quantity. The magnitude of $\underset{\sim}{n}$ is equal to the refractive index, so that $\underset{\sim}{n}$ becomes a unit propagation vector is free space. In terms of the charges e_i of particles i with position vectors $\underset{\sim}{r}_i$ relative to an arbitrary origin within the molecule, the electric dipole moment is defined as

$$\mu_\alpha = \sum_i e_i r_{i\alpha} \; . \tag{2.5}$$

Quantum-mechanical expressions for the polarizability and other higher-order molecular response tensors are obtained by taking expectation values of the operator equivalent of the electric dipole moment (2.5) using molecular wavefunctions perturbed by the light wave (2.4). This particular semi-classical approach avoids the complications of formal time-dependent perturbation theory: it has a respectable pedigree, being found in Placzek's famous treatise on the Raman effect [9], and also in the books by Born and Huang [10] and Davydov [11]. Further details of the particular version outlined here can be found in my own book [12].

The periodically-perturbed wavefunctions ψ_n' are obtained by solving the time-dependent Schrödinger equation

$$\left(i\hbar \frac{\partial}{\partial t} - H \right) \psi_n' = V \psi_n' \tag{2.6}$$

where H is the unperturbed molecular Hamiltonian and V is that describing the interaction of the molecule with the light wave. It is convenient to write V as a series displaying the coupling of the molecular multipole moment operators with appropriate real fields of the light wave:

$$V = -\mu_\alpha E_\alpha + \cdots \; . \tag{2.7}$$

In what follows, the real field is written as a sum of the complex field and its complex conjugate:

$$E_\alpha = \tfrac{1}{2} \left(\tilde{E}_\alpha + \tilde{E}_\alpha^* \right) . \tag{2.8}$$

We assume that, when the stationary non-degenerate eigenfunction

$$\psi_n = \psi_n^{(o)} \, e^{-i\omega_n t} \tag{2.9}$$

of the unperturbed molecule is subjected to a small harmonic perturbation of angular frequency ω from a plane wave radiation field, the corresponding perturbed eigenfunction can be written in the form

$$\Psi'_n = \left\{ \Psi_n^{(0)} + \sum_{j \neq n} \left(\tilde{a}_{jn_\beta} \tilde{E}_\beta + \tilde{b}_{jn_\beta} \tilde{E}_\beta^* + \cdots \right) \right\} e^{-i\omega_n t} . \tag{2.10}$$

Using (2.10) and (2.7) in the Schrödinger equation (2.6), we find

$$\tilde{a}_{jn_\beta} = \langle j | \mu_\beta | n \rangle / 2\hbar (\omega_{jn} - \omega) , \tag{2.11a}$$

$$\tilde{b}_{jn_\beta} = \langle j | \mu_\beta | n \rangle / 2\hbar (\omega_{jn} + \omega) . \tag{2.11b}$$

The required expectation value of the induced oscillating electric dipole moment is thus

$$\mu_\alpha = \langle n' | \mu_\alpha | n' \rangle$$

$$= \langle n | \mu_\alpha | n \rangle + \frac{2}{\hbar} \sum_{j \neq n} \frac{\omega_{jn}}{\omega_{jn}^2 - \omega^2} \, \text{Re} \left(\langle n | \mu_\alpha | j \rangle \langle j | \mu_\beta | n \rangle \right) E_\beta$$

$$- \frac{2}{\hbar} \sum_{j \neq n} \frac{\omega}{\omega_{jn}^2 - \omega^2} \, \text{Im} \left(\langle n | \mu_\alpha | j \rangle \langle j | \mu_\beta | n \rangle \right) \frac{1}{\omega} \dot{E}_\beta + \cdots \tag{2.12}$$

where $\omega_{jn} = \omega_j - \omega_n$. The first term is the permanent electric dipole moment; and from the second term we can identify the conventional polarizability with

$$\alpha_{\alpha\beta} = \frac{2}{\hbar} \sum_{j \neq n} \frac{\omega_{jn}}{\omega_{jn}^2 - \omega^2} \, \text{Re} \left(\langle n | \mu_\alpha | j \rangle \langle j | \mu_\beta | n \rangle \right) = \alpha_{\beta\alpha} . \tag{2.13a}$$

From the third term we can identify an additional imaginary polarizability with

$$\alpha'_{\alpha\beta} = -\frac{2}{\hbar} \sum_{j \neq n} \frac{\omega}{\omega_{jn}^2 - \omega^2} \, \text{Im} \left(\langle n | \mu_\alpha | j \rangle \langle j | \mu_\beta | n \rangle \right) = -\alpha_{\beta\alpha} . \tag{2.13b}$$

This imaginary polarizability is usually neglected in standard treatments, but in recent years it has been shown to give important new contributions to light scattering from atoms and molecules in degenerate states, particularly Kramers degeneracy associated with an odd number of electrons.

Notice that $\alpha_{\alpha\beta}$ is symmetric with respect to interchange of the tensor subscripts, whereas $\alpha'_{\alpha\beta}$ is antisymmetric. This follows from the fact that μ_α is a Hermitian operator, so that

$$\langle n | \mu_\alpha | j \rangle \langle j | \mu_\beta | n \rangle = \langle n | \mu_\beta | j \rangle^* \langle j | \mu_\alpha | n \rangle^* , \tag{2.14a}$$

$$\text{Re} \left(\langle n | \mu_\alpha | j \rangle \langle j | \mu_\beta | n \rangle \right) = \text{Re} \left(\langle n | \mu_\beta | j \rangle \langle j | \mu_\alpha | n \rangle \right), \tag{2.14b}$$

$$\text{Im} \left(\langle n | \mu_\alpha | j \rangle \langle j | \mu_\beta | n \rangle \right) = -\text{Im} \left(\langle n | \mu_\beta | j \rangle \langle j | \mu_\alpha | n \rangle \right). \tag{2.14c}$$

It is convenient to present these results in a complex

representation. Introducing the complex polarizability

$$\tilde{\alpha}_{\alpha\beta} = \alpha_{\alpha\beta} - i\alpha'_{\alpha\beta} = \tilde{\alpha}^*_{\beta\alpha} , \qquad (2.15)$$

where the minus sign arises from our choice of a negative exponent in the complex light wave (2.4), a corresponding complex induced dipole moment can be written as

$$\qquad\qquad\qquad\qquad (2.16)$$
$$\tilde{\mu}_\alpha = \tilde{\alpha}_{\alpha\beta}\tilde{E}_\beta + \ - \ - \ - \ .$$

More details of this development, including the extension to the higher-order tensors containing magnetic dipole and electric quadrupole transition moments that are resonsible for optical activity phenomena, can be found elsewhere [12].

2.2. The Raman Transition Polarizability

The Raman components of the scattered waves have frequencies different from, and unrelated in phase to, the incident wave, so the polarizability tensors given above must be replaced by corresponding transition tensors which take account of the different initial and final molecular states. It is possible to generalize the previous treatment by invoking a transition electric dipole moment between initial and final molecular states $|n\rangle$ and $|m\rangle$ [9,10,12], and this leads to the following generalization of (2.16):

$$(\tilde{\mu}_\alpha)_{mn} = (\tilde{\alpha}_{\alpha\beta})_{mn}\tilde{E}_\beta + \ - \ - \ - \ . \qquad (2.17)$$

Although the complex transition polarizability $(\tilde{\alpha}_{\alpha\beta})_{mn}$ can be expressed as a sum of real and imaginary parts analogous to (2.15), it is usually best to express it in the form

$$(\tilde{\alpha}_{\alpha\beta})_{mn} = \frac{1}{\hbar}\sum_{j\ne n,m}\left[\frac{\langle m|\mu_\alpha|j\rangle\langle j|\mu_\beta|n\rangle}{\omega_{jn}-\omega} + \frac{\langle m|\mu_\beta|j\rangle\langle j|\mu_\alpha|n\rangle}{\omega_{jm}+\omega}\right]. \qquad (2.18)$$

The subsequent development depends on the particular application; for example transparent or resonant scattering, and rotational, vibrational or electronic Raman transitions.

This general complex transition polarizability has no definite symmetry with respect to permutation of the tensor subscripts, unlike the ordinary polarizability which has a pure symmetric real part (2.13a) and a pure antisymmetric imaginary part (2.13b).

2.3 Intensity, Depolarization Ratio and Cross Section

We now require an expression for the intensity radiated by the induced oscillating molecular dipole derived above. In SI, the intensity (mean rate of energy flow) of a plane wave is [12]

$$I = \frac{1}{2}\left(\frac{\epsilon\epsilon_0}{\mu\mu_0}\right)^{\frac{1}{2}} E_{(0)}^2 . \qquad (2.19)$$

252

The radiated electric vector in the wave zone (which refers to distances $R \gg \lambda$) is given by [12]

$$\tilde{E}_\alpha = \frac{\omega^2 \mu_0}{4\pi R} e^{i\omega(R/c - t)} (\tilde{\mu}_\alpha^{(0)} - n_\alpha^d n_\beta^d \tilde{\mu}_\beta^{(0)}) , \qquad (2.20)$$

where n^d is the propagation vector of the detected wave. The calculation of polarized intensity components for arbitrary scattering angles becomes rather complicated, and the general expressions will not be given here. They can be found in [3,9 and 12], for example. We shall consider explicitly just the case of 90° scattering; this is particularly simple, and is the most common experimental configuration.

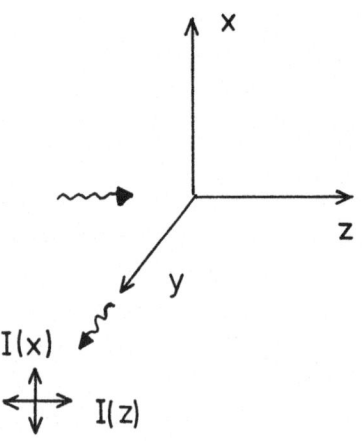

Figure 2.1. Axis system for light scattering at 90°.

Using the axis system shown in figure 2.1 and substituting (2.16) for the induced oscillating dipole moment into (2.20), we find the following expressions for the intensity components scattered at 90° and linearly polarized along x and z using incident light linearly polarized along x:

$$I(x) = \frac{1}{2} \left(\frac{\epsilon \epsilon_0}{\mu \mu_0}\right)^{\frac{1}{2}} \left(\frac{\omega^2 \mu_0}{4\pi R}\right)^2 \tilde{\alpha}_{xx} \tilde{\alpha}_{xx}^* \tilde{E}_x^{(0)} \tilde{E}_x^{(0)*}, \qquad (2.21a)$$

$$I(z) = \frac{1}{2} \left(\frac{\epsilon \epsilon_0}{\mu \mu_0}\right)^{\frac{1}{2}} \left(\frac{\omega^2 \mu_0}{4\pi R}\right)^2 \tilde{\alpha}_{zx} \tilde{\alpha}_{zx}^* \tilde{E}_x^{(0)} \tilde{E}_x^{(0)*}. \qquad (2.21b)$$

The depolarisation ratio is then

$$\rho = \frac{I(z)}{I(x)} = \frac{\tilde{\alpha}_{zx} \tilde{\alpha}_{zx}^*}{\tilde{\alpha}_{xx} \tilde{\alpha}_{xx}^*} . \qquad (2.22)$$

These results apply to scattering from a molecule in a fixed

orientation with the polarizability tensor components referred to the
space-fixed axes of figure 2.1. If the molecules are undergoing free,
unquantized, rotations, a classical orientational average is taken with
the results expressed in molecule-fixed axes. The space-fixed axes α,
β, γ,.. are related to molecule-fixed axes α', β', γ',.. using direction
cosines such as $l_{\alpha\alpha'}$ between the α and α' axes, so that

$$\alpha_{\alpha\beta} = l_{\alpha\alpha'} \, l_{\beta\beta'} \, \alpha_{\alpha'\beta'} \,. \tag{2.23}$$

We then make use of isotropic averages such as $\langle \cos^4\theta \rangle = 1/5$ to
obtain the following averages of products of four direction cosines [12]:

$$\langle i_\alpha i_\beta i_\gamma i_\delta \rangle = \langle j_\alpha j_\beta j_\gamma j_\delta \rangle = \langle k_\alpha k_\beta k_\gamma k_\delta \rangle$$
$$= \tfrac{1}{15}\left(\delta_{\alpha\beta}\delta_{\gamma\delta} + \delta_{\alpha\gamma}\delta_{\beta\delta} + \delta_{\alpha\delta}\delta_{\beta\gamma} \right), \tag{2.24a}$$

$$\langle i_\alpha i_\beta j_\gamma j_\delta \rangle = \langle j_\alpha j_\beta k_\gamma k_\delta \rangle = \langle i_\alpha i_\beta k_\gamma k_\delta \rangle$$
$$= \tfrac{1}{30}\left(4\delta_{\alpha\beta}\delta_{\gamma\delta} - \delta_{\alpha\gamma}\delta_{\beta\delta} - \delta_{\alpha\delta}\delta_{\beta\gamma} \right), \tag{2.24b}$$

where $i_\alpha = l_{x\alpha}$, $j_\alpha = l_{y\alpha}$, $k_\alpha = l_{z\alpha}$ are the direction cosines
between the space-fixed axes x,y,z and a molecule-fixed axis α (as
usual, $\underset{\sim}{i},\underset{\sim}{j},\underset{\sim}{k}$ are unit vectors along x,y,z). It is convenient to write
the tensor $\alpha_{\alpha\beta}$ as a sum of three irreducible parts,

$$\alpha_{\alpha\beta} = \alpha^0 \delta_{\alpha\beta} + \overset{a}{\alpha}_{\alpha\beta} + \overset{s}{\alpha}_{\alpha\beta} , \tag{2.25}$$

where α^0 is a scalar, $\overset{a}{\alpha}_{\alpha\beta}$ is a second-rank antisymmetric tensor (i.e.
an axial vector) and $\overset{s}{\alpha}_{\alpha\beta}$ is a second-rank traceless symmetric tensor:

$$\alpha^0 = \tfrac{1}{3}\alpha_{\alpha\alpha} , \tag{2.26a}$$

$$\overset{a}{\alpha}_{\alpha\beta} = \tfrac{1}{2}\left(\alpha_{\alpha\beta} - \alpha_{\beta\alpha} \right), \tag{2.26b}$$

$$\overset{s}{\alpha}_{\alpha\beta} = \tfrac{1}{2}\left(\alpha_{\alpha\beta} + \alpha_{\beta\alpha} \right) - \alpha^0 \delta_{\alpha\beta} . \tag{2.26c}$$

Using these results in (2.22), the depolarization ratio for an isotropic
sample becomes

$$\rho = \frac{3B + 5C}{45A + 4B} , \tag{2.27}$$

where A,B and C are the fourth-rank tensor invariants

$$A = \alpha^{0\,2}, \tag{2.28a}$$

$$B = \tfrac{3}{2} \overset{s}{\alpha}_{\alpha\beta} \overset{s}{\alpha}_{\alpha\beta} , \tag{2.28b}$$

$$C = \tfrac{3}{2} \overset{a}{\alpha}_{\alpha\beta} \overset{a}{\alpha}_{\alpha\beta} . \tag{2.28c}$$

If the molecule is a symmetric top, the polarizability tensor can
be written as follows in terms of components referred to principal
axes:

$$\alpha_{\alpha\beta} = \alpha_\perp \delta_{\alpha\beta} + (\alpha_{||} - \alpha_\perp)K_\alpha K_\beta, \tag{2.29}$$

where $\alpha_\perp = \alpha_{XX} = \alpha_{YY}$ and $\alpha_{||} = \alpha_{ZZ}$ denote polarizability components perpendicular and parallel to the top axis Z, and K is the unit vector along Z (we denote internal molecule-fixed axes with capital letters). The invariant B then takes the familiar form of the anisotropy invariant,

$$B = (\alpha_{||} - \alpha_\perp)^2. \tag{2.30}$$

An important quantity for relating the conventional spontaneous Raman scattering we are now considering to the nonlinear coherent Raman processes considered later is the differential scattering cross section $d\sigma/d\Omega$. The quantity $d\sigma$ is called the effective scattering cross section [3,13] and is defined as the ratio of the energy dW scattered per second into a solid angle $d\Omega$ to the incident energy flux density (i.e. the incident intensity). dW is obtained by multiplying the scattered intensity at R by $R^2 d\Omega$ (since this gives the quantity of radiant energy passing in unit time across the surface element $R^2 d\Omega$). Again, the general expressions for arbitrary scattering angles and polarizations can be found elsewhere [3,13]. Here we give the result for scattering into the forward direction with the same linear polarization direction for the incident and scattered beam since this is required later on in the CARS theory:

$$d\sigma = dw/I = \left(\frac{\omega^4 \mu_0^2}{16 \pi^2}\right) \alpha_{xx} \alpha_{xx}^* d\Omega. \tag{2.31}$$

For an isotropic sample, the average (2.24a) gives

$$\left(\frac{d\sigma}{d\Omega}\right)_x = \left(\frac{\omega^4 \mu_0^2}{16 \pi^2}\right)\left(A + \frac{4}{45}B\right). \tag{2.32}$$

This result also applies to the x-polarized component of the light scattered at 90°. The total differential cross-section for unpolarized scattered light in x-polarized incident light is obtained by adding the following result for the y-polarized scattered component:

$$\left(\frac{d\sigma}{d\Omega}\right)_y = \left(\frac{\omega^4 \mu_0^2}{16 \pi^2}\right)\frac{3}{45}B. \tag{2.33}$$

2.4. Rotational Raman Scattering

Gas phase Raman measurements can, of course, resolve individual rotational transitions, so it is necessary to consider rotational-vibrational electronic states in the general transition polariz-ability (2.18). The classical isotropic averages derived in the previous section provide a useful background because, in accordance with van Vleck's 'principle of spectroscopic stability' [14], a quantum-statistical average over all allowed transitions should yield the classical isotropic result, which corresponds to the situation where the individual transitions are not resolved.

In the Born–Oppenheimer approximation, each state is written as a product of electronic, vibrational and rotational parts:

$$|i\rangle = |j_e j_v j_r\rangle = |j_{int} j_r\rangle , \qquad (2.34)$$

where $|j_{int}\rangle$ is the internal molecular vibronic state. In the usual theory of rotational Raman scattering [7,9,15], the rotational contributions to the transition frequencies in the transition polarizability are neglected and the closure theorem invoked with respect to the complete set of rotational states associated with every electronic-vibrational state, which enables the transition polarizability (2,18) to be written

$$(\tilde{\alpha}_{\alpha\beta})_{mn} = \langle m_r|(\tilde{\alpha}_{\alpha\beta})_{m_{int} n_{int}}|n_r\rangle , \qquad (2.35)$$

where $(\tilde{\alpha}_{\alpha\beta})_{m_{int} n_{int}}$ is an internal transition tensor that acts as an operator on the rotational states. The space-fixed axes $\alpha , \beta , \dots$ can then be related to molecule-fixed axes $\alpha', \beta', \dots$ using direction cosines so that

$$\langle m_r|(\tilde{\alpha}_{\alpha\beta})_{m_{int} n_{int}}|n_r\rangle = (\tilde{\alpha}_{\alpha'\beta'})_{m_{int} n_{int}} \langle m_r|l_{\alpha\alpha'} l_{\beta\beta'}|n_r\rangle \qquad (2.36)$$

with the rotational transition being effected by the direction cosine operators.

In recent years a more sophisticated approach using irreducible spherical tensor operators has been found advantageous [13,16–18]. Symmetric top matrix elements can be obtained by using the following extension of the Wigner–Eckart theorem to axially symmetric systems (within the Condon and Shortley phase convention)[16]

$$\langle n', J'K'M'|\hat{T}^k_q|n,JKM\rangle = i^{J'+J-K'-K}(-1)^{J-M}[(2J'+1)(2J+1)]^{\frac{1}{2}}$$

$$\times \begin{pmatrix} J' & k & J \\ -K' & K'-K & K \end{pmatrix} \begin{pmatrix} J' & k & J \\ -M' & q & M \end{pmatrix} \langle n'|\overline{\hat{T}^k_{K'-K}}|n\rangle , \qquad (2.37)$$

where J,K,M are the usual set of symmetric top rotational quantum numbers, n denotes the internal (vibrational-electronic) states, and $\hat{T}^k_q$ is the q th component of the rank-k set of tensor operators expressed in irreducible spherical form with respect to space-fixed axes. A bar over the operator indicates that it is defined with respect to molecule-fixed axes, so that $\langle n'|\overline{\hat{T}^k_{K'-K}}|n\rangle$ is an internal matrix element and would be an analogue, in a spherical basis, of $(\tilde{\alpha}_{\alpha'\beta'})_{m_{int} n_{int}}$ in (2.36) if the same approximations are invoked.

Taking $\hat{T}^k_q$ to be the general polarizability operator expressed in irreducible spherical tensor form, and making use of the well-known properties of the 3j symbols, the Raman intensity is found to be [16–18]

$$\left(I_\nu^k\right)_{J'K',JK} = \frac{1}{2k+1} \left|\langle n'|\widehat{\alpha}_{K'-K}^{k}|n\rangle\right|^2 D_{J'K',JK}^{k} \,, \qquad (2.38)$$

where k takes values 0,1,2 to denote the rank-k irreducible spherical components and $D_{J'K',JK}^{k}$ is the associated intensity factor. When k = 2, this intensity factor equals the factor $b_{J'K'}^{JK}$ of Placzek and Teller [19]. For k = 0 only $D_{JK,JK}^{0} = 1$ is allowed, and for k = 1 and 2 we find the factors listed in Tables 1 and 2, respectively [18].

Table 1. Rotational Raman factors, $D_{J'K',JK}^{1}$

	J−1	J	J+1
K'			
K−1	$\dfrac{(J+K)(J+K-1)}{2J(2J+1)}$	$\dfrac{(J+K)(J-K+1)}{2J(J+1)}$	$\dfrac{(J-K+1)(J-K+2)}{2(J+1)(2J+1)}$
K	$\dfrac{J^2-K^2}{J(2J+1)}$	$\dfrac{4K^2}{J(J+1)}$	$\dfrac{[(J+1)^2-K^2]}{(J+1)(2J+1)}$
K+1	$\dfrac{(J-K)(J-K-1)}{2J(2J+1)}$	$\dfrac{(J-K)(J+K+1)}{2J(J+1)}$	$\dfrac{(J+K+1)(J+K+2)}{2(J+1)(2J+1)}$

Only the following internal irreducible spherical polarizability operator components survive [18]:

$$\overline{\alpha}_0^0 = -\frac{1}{\sqrt{3}}\left(\widehat{\alpha}_\| + 2\widehat{\alpha}_\perp\right), \qquad (2.39a)$$

$$\overline{\alpha}_0^1 = \frac{i}{\sqrt{2}}\left(\widehat{\alpha}_{xy} - \widehat{\alpha}_{yx}\right), \qquad (2.39b)$$

$$\overline{\alpha}_0^2 = \frac{2}{\sqrt{6}}\left(\widehat{\alpha}_\| - \widehat{\alpha}_\perp\right). \qquad (2.39c)$$

A consequence of (2.39) is that $K'-K=0$ in the intensity factors $D_{J'K',JK}^{k}$ in (2.38). These results lead to the well-known selection rules for conventional pure rotational Raman scattering originating in the symmetric polarizability operators (2.39a and c), namely $\Delta J = 0, \pm 1, \pm 2$; $\Delta K = 0$; with $\Delta J = \pm 1$ forbidden if $K = 0$ [16,18]. Transitions with $\Delta J = +2, +1, 0, -1, -2$ generate the S,R,Q,P,O branches of the rotational Raman spectrum. The treatment outlined here is a little more general than usual since we have retained the antisymmetric operator contribution (2.39b), which generates the less familiar selection rules for antisymmetric scattering, namely $\Delta J = 0, \pm 1$; $\Delta K = 0$; with $\Delta J = 0$ forbidden if $K = 0$ [18].

The selection rules are less restrictive for rotation–vibration transitions. Thus while $\Delta K = 0$ still holds for totally symmetric vibrational fundamentals, $\Delta K = \pm 1, \pm 2$ transitions are possible for certain non–totally symmetric fundamentals [16].

Murphy [17] has given an extension of the irreducible spherical

Table 2. Rotational Raman factors $D^2_{J'K',JK}$

J'	$J-2$	$J-1$	J
$K-2$	$\dfrac{(J+K)(J+K-1)(J+K-2)(J+K-3)}{4J(J-1)(2J+1)(2J-1)}$	$\dfrac{(J+K)(J+K-2)[J^2+(K-1)^2]}{2J(J+1)(J-1)(2J+1)}$	$\dfrac{3[J^2-(K-1)^2][(J+1)^2-(K-1)^2]}{2J(J+1)(2J-1)(2J+3)}$
$K-1$	$\dfrac{(J^2-K^2)(J+K-1)(J+K-2)}{J(J-1)(2J+1)(2J-1)}$	$\dfrac{(J+K)(J+K-1)(J-2K+1)^2}{2J(J+1)(J-1)(2J+1)}$	$\dfrac{3(2K-1)^2(J+K)(J-K+1)}{2J(J+1)(2J-1)(2J+3)}$
K	$\dfrac{3[J^2-K^2][(J-1)^2-K^2]}{2J(J-1)(2J+1)(2J-1)}$	$\dfrac{3K^2(J^2-K^2)}{J(J+1)(J-1)(2J+1)}$	$\dfrac{[J(J+1)-3K^2]^2}{J(J+1)(2J-1)(2J+3)}$
$K+1$	$\dfrac{(J^2-K^2)(J-K-1)(J-K-2)}{J(J-1)(2J+1)(2J-1)}$	$\dfrac{(J-K)(J-K-1)(J+2K+1)^2}{2J(J+1)(J-1)(2J+1)}$	$\dfrac{3(2K+1)^2(J-K)(J+K+1)}{2J(J+1)(2J-1)(2J+3)}$
$K+2$	$\dfrac{(J-K)(J-K-1)(J-K-2)(J-K-3)}{4J(J-1)(2J+1)(2J-1)}$	$\dfrac{(J-K)(J-K-2)[J^2-(K+1)^2]}{2J(J+1)(J-1)(2J+1)}$	$\dfrac{3[J^2-(K+1)^2][(J+1)^2-(K+1)^2]}{2J(J+1)(2J-1)(2J+3)}$

J'	$J+1$	$J+2$
$K-2$	$\dfrac{[(J+1)^2-(K-1)^2](J-K+1)(J-K+3)}{2J(J+1)(J+2)(2J+1)}$	$\dfrac{(J-K+1)(J-K+2)(J-K+3)(J-K+4)}{4(J+1)(J+2)(2J+1)(2J+3)}$
$K-1$	$\dfrac{(J+2K)^2(J-K+1)(J-K+2)}{2J(J+1)(J+2)(2J+1)}$	$\dfrac{[(J+1)^2-K^2](J-K+2)(J-K+3)}{(J+1)(J+2)(2J+1)(2J+3)}$
K	$\dfrac{3K^2[(J+1)^2-K^2]}{J(J+1)(J+2)(2J+1)}$	$\dfrac{3[(J+1)^2-K^2][(J+2)^2-K^2]}{2(J+1)(J+2)(2J+1)(2J+3)}$
$K+1$	$\dfrac{(J-2K)^2(J+K+1)(J+K+2)}{2J(J+1)(J+2)(2J+1)}$	$\dfrac{[(J+1)^2-K^2](J+K+2)(J+K+3)}{(J+1)(J+2)(2J+1)(2J+3)}$
$K+2$	$\dfrac{[(J+1)^2-(K+1)^2](J+K+1)(J+K+3)}{2J(J+1)(J+2)(2J+1)}$	$\dfrac{(J+K+1)(J+K+2)(J+K+3)(J+K+4)}{4(J+1)(J+2)(2J+1)(2J+3)}$

tensor approach to general <u>asymmetric</u> tops.

3. SOME NEW ASPECTS OF LINEAR GAS PHASE RAMAN SPECTROSCOPY

Conventional linear Raman spectroscopy is still widely used to obtain pure rotation, and rotation-vibration, gas phase spectra. There are many references describing such applications in detail [e.g. 1,2], so nothing more will be said here. Instead we shall consider briefly some new aspects.

3.1. Resonance Raman and Antisymmetric Scattering

It can be seen from the general transition polarizability (2.18) that if the laser frequency ω_L coincides with an electronic absorption frequency $\omega_{j_i n}$ of the molecule, considerable enhancement of the Raman scattering is possible because the first term "blows up" (a more complete treatment includes a damping factor in order to describe the resonance situation [12,13,15]).

An additional characteristic of resonance Raman scattering is that, in certain circumstances, the antisymmetric parts of the transition polarizability (2.18) can make significant contributions. As Placzek [9] showed, the antisymmetric contributions vanish for scattering at transparent wavelengths. Furthermore, for Rayleigh and pure rotational Raman scattering, and also for most cases of vibrational Raman scattering, degeneracy is required in either the initial or final state (degeneracy is not required for Raman scattering in a fundamental transition of a mode of vibration that transforms according to an anti-symmetric irreducible representation)[12,20]. In particular, it has been shown that the M_J-degeneracy associated with rotational quantum states can generate real antisymmetric transition polarizability components [12,20,21].

Depolarization ratio anomalies are the hallmark of antisymmetric scattering. It can be seen from the general expression (2.27) for the depolarization ratio that, for pure antisymmetric scattering arising from the invariant C alone, the depolarization ratio is infinite. This compares with zero for pure isotropic scattering and ¾ for pure aniso-tropic scattering. Thus a depolarization ratio larger than expected indicates antisymmetric contributions.

Zeigler [22] has recently obtained some beautiful experimental examples of large antisymmetric contributions to pure rotation, and rotation-vibration, resonance Raman scattering in NH_3. The spectrum shown in figure 3.1b, for example, was obtained using ultraviolet excitation at 208.8 nm produced by anti-Stokes Raman shifting in H_2 the second harmonic of a Nd:YAG laser. This wavelength is in resonance with the $\nu_2' = 2$ vibronic level, associated with the ν_2 "umbrella" coordinate, of the $\tilde{A} \leftarrow \tilde{X}$ Rydberg transition of ammonia (this ν_2 activity is due to the change in nuclear equilibrium geometry from pyramidal in the ground state to planar in the excited electronic state). It is estimated that antisymmetric scattering contributes about 25% of the total resonance Raman intensity at 208.8 nm. The main purpose of

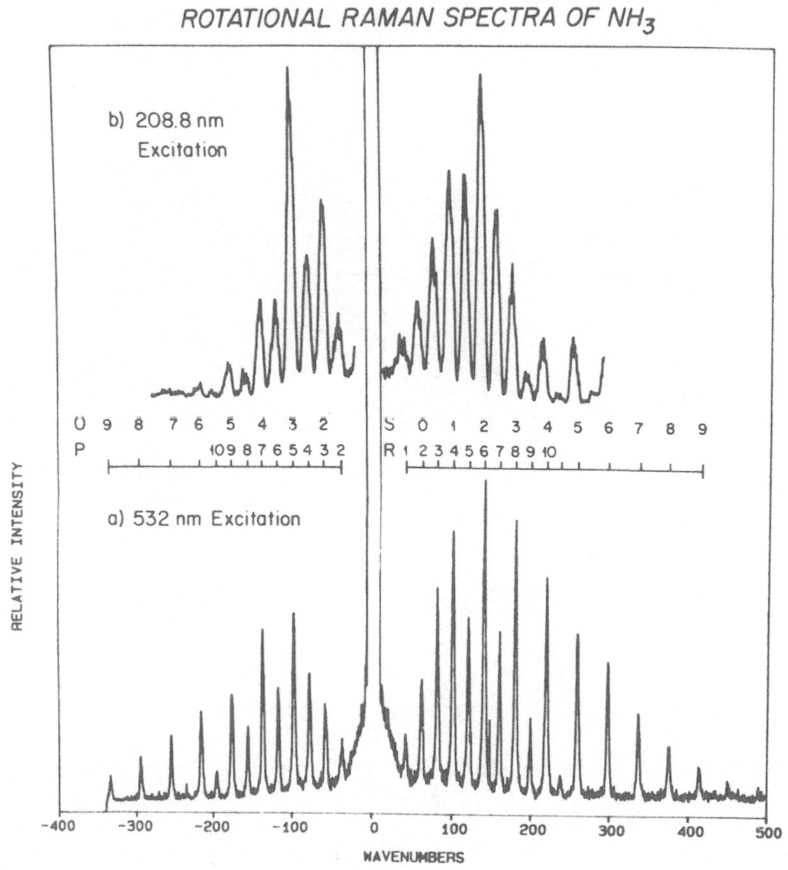

Figure 3.1. Rotational Raman spectra of NH_3 excited (a) off resonance
at 532 nm and (b) on resonance at 208.8 nm [22].

the study was to demonstrate the use of resonance rotational and
vibrational Raman scattering as a probe of excited state potential
energy surfaces: detailed analysis of the bands determines excited
state vibronic lifetimes on the sub-picosecond timescale.

3.2. Raman Electron Paramagnetic Resonance (EPR)

Resonance Raman and antisymmetric scattering are involved in a novel
technique involving spin-flip Raman transitions in paramagnetic
molecules that can function as Raman electron paramagnetic resonance.
Figure 3.2a shows a conventional vibrational Stokes resonance Raman
process, while 3.2b and 3.2c show the polarization characteristics
of the two distinct spin-flip Raman processes for scattering at 90°

that are generated if a twofold Kramers degeneracy in the initial and final levels is lifted by a magnetic field parallel to a laser beam incident along the z-direction. Each scattering pathway can be

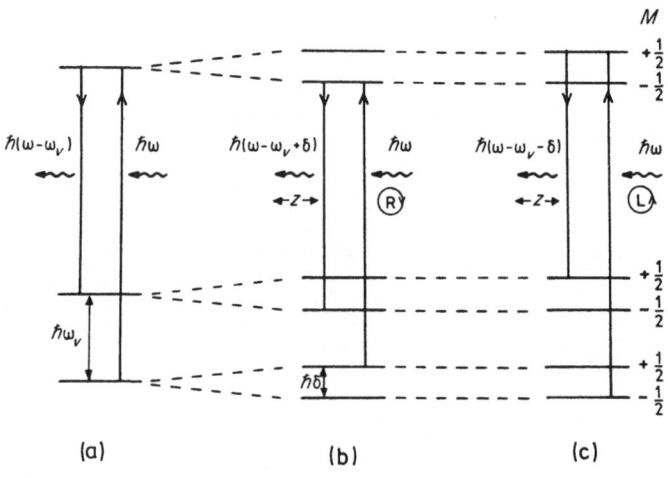

Figure 3.2. Spin-flip Raman scattering pathways.

envisaged as the **longitudinal** and transverse Zeeman effects back-to-back: the incident circularly polarized photon generates a $\Delta M = \pm 1$ change in the molecule, and the 90°-scattered z-polarized photon a $\Delta M = 0$ change. Two depolarized Raman lines, shifted in frequency by the Zeeman splitting δ on either side of the Stokes vibrational Raman line at $(\omega - \omega_v)$, are thereby generated: right circularly polarized incident light generates the line shifted to lower Stokes frequency $(\omega - \omega_v + \delta)$, left circularly polarized light generates the line shifted to higher Stokes frequency $(\omega - \omega_v - \delta)$. Because the molecule suffers a $\Delta M = \pm 1$ change overall, the corresponding transition tensor is pure antisymmetric since it must transform like an axial vector. A related feature is that a transition between spin states has been effected by a scattering operator that contains no spin operator, but simply two spatial electric dipole moment operators.

Even if the resolution is insufficient to detect these magnetic field-induced shifts directly, they can be detected by measuring a Raman difference spectrum in right and left circularly polarized incident light. A residual positive-negative Raman circular intensity difference (CID) couplet will be generated from incomplete cancellation of the two partially overlapping bands. This Raman EPR effect was first observed by Barron and Meehan in 1979 [23] in resonance Raman scattering from dilute solutions of odd-electron transition metal complexes. It should be readily observable in paramagnetic gases, but no such observations

have been reported to date. Raman EPR can give new information about
the magnetic structure of ground and low-lying excited states,
including the sign of the g-factor, and how the magnetic structure
changes when the molecule is in an excited non-totally symmetric
vibrational state (these changes can be dramatic for degenerate
vibrations).

To give an idea of the potential of this technique, figure 3.3
shows the depolarized magnetic resonance-Raman circular intensity
difference spectrum of a very dilute aqueous solution of the low-spin
d^5 complex $IrCl_6^{2-}$ recorded recently on the Glasgow multichannel Raman
instrument [24]. The three Raman-active fundamentals of the octahedral
MX_6 structure are assigned to $\nu_1(A_{1g}, 341 cm^{-1})$, $\nu_2(E_g, 290 cm^{-1})$
and $\nu_5 (T_{2g}, 161 cm^{-1})$. The ground electronic state belongs to the
E_g'' $(^2T_{2g})$ Kramers doublet of O_h^*, and the excited charge-transfer
resonant level to $U_u' (^2T_{1u}^{(1)})$ when 488.0 nm laser excitation

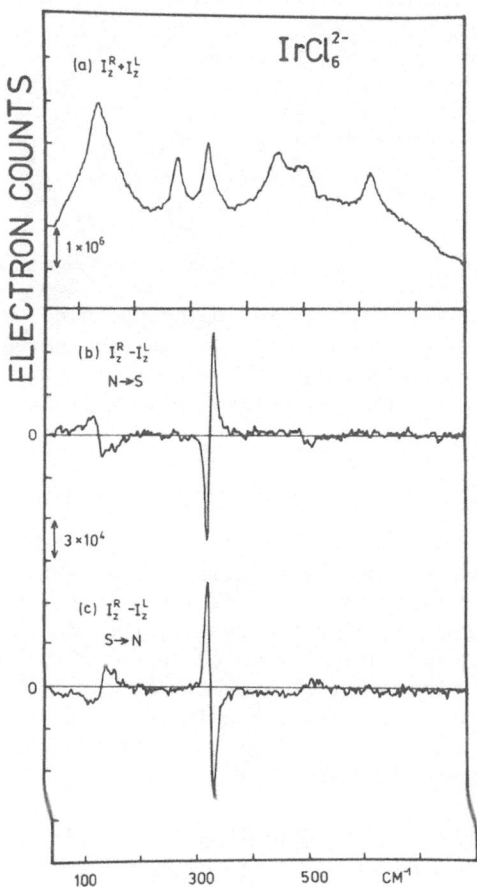

Figure 3.3. Depolarized multichannel magnetic Raman circular intensity
sum and difference spectra of $IrCl_6^{2-}$ in dilute aqueous $HClO_4$ solution
[24] with a field strength of 1 T.

is used. The sharp couplet in the A_{1g} band at 341 cm^{-1} originates in the spin-flip mechanism of figure 3.2 and, since the lower frequency component is negative in the positive magnetic field (N→S), it follows directly that the ground state g-factor is <u>negative</u>. This unusual result is expected theoretically in iridium IV hexahalides [25]. The opposite sign in the T_{2g} couplet probably originates in a large positive g-factor in the first excited degenerate vibrational state.

Raman EPR is in fact one particular manifestation of the more general phenomenon of magnetic Raman optical activity (see reference [26] for a detailed review).

3.3. Raman Optical Activity (ROA)

In addition to the magnetic Raman circular intensity difference, outlined above, that all molecules can show in a magnetic field, <u>chiral</u> molecules can show a circular intensity difference without a magnetic field. This <u>natural</u> Raman optical activity was first observed by Barron, Bogaard and Buckingham in 1973 [27], and provides detailed stereochemical information by measuring vibrational optical activity. Reference [28] is the most recent review of natural ROA: also, detailed theories of both natural and magnetic ROA can be found in my book [12].

Natural ROA offers the interesting prospect of measuring optical activity in pure rotational transitions of gas phase chiral molecules. Although such observations have not yet been reported, the detailed theory of rotational ROA in chiral symmetric tops has been published [18], and the experiment should be feasible using existing technology such as optical multichannel detection. It is also possible that one of the coherent Raman techniques discussed below could be advantageous.

Natural ROA originates in interference between waves scattered via the polarizability and optical activity tensors of the molecule. The relevant experimental quantity is a dimensionless circular intensity difference

$$\Delta_\alpha = \frac{I_\alpha^R - I_\alpha^L}{I_\alpha^R + I_\alpha^L}, \tag{3.1}$$

where I_α^R and I_α^L are the scattered intensities with α-polarization in right and left circularly polarized incident light. The numerator is the pseudoscalar optical activity observable, and the denominator is the conventional scattered intensity which serves to normalize the data. For scattering at 90°, the explicit expressions for the polarized and depolarized CIDs Δ_x and Δ_z (i.e. transmitted through an analyzer perpendicular and parallel, respectively, to the scattering plane yz in the axis system of figure 2.1) are

$$\Delta_x = \frac{2(7\alpha_{\alpha\beta}G'_{\alpha\beta} + \alpha_{\alpha\alpha}G'_{\beta\beta} + \tfrac{1}{3}\omega\alpha_{\alpha\beta}\varepsilon_{\alpha\gamma\delta}A_{\gamma\delta\beta})}{c(7\alpha_{\gamma\mu}\alpha_{\gamma\mu} + \alpha_{\lambda\lambda}\alpha_{\mu\mu})}, \tag{3.2a}$$

$$\Delta_z = \frac{4(3\alpha_{\alpha\beta}G'_{\alpha\beta} - \alpha_{\alpha\alpha}G'_{\beta\beta} - \tfrac{1}{3}\omega\alpha_{\alpha\beta}\varepsilon_{\alpha\gamma\delta}A_{\gamma\delta\beta})}{2c(3\alpha_{\gamma\mu}\alpha_{\gamma\mu} - \alpha_{\lambda\lambda}\alpha_{\mu\mu})}. \tag{3.2b}$$

$\alpha_{\alpha\beta}$ is the familiar real polarizability tensor (2.13a), and $G'_{\alpha\beta}$ and $A_{\alpha\beta\gamma}$ are tensors responsible for natural optical activity. They are given quantum-mechanically by

$$G'_{\alpha\beta} = -\frac{2}{\hbar} \sum_{j\neq n} \frac{\omega}{\omega_{jn}^2 - \omega^2} \, \text{Im} \, (\langle n|\mu_{\alpha}|j\rangle \times \langle j|m_{\beta}|n\rangle) \,, \qquad (3.3a)$$

$$A_{\alpha\beta\gamma} = \frac{2}{\hbar} \sum_{j\neq n} \frac{\omega_{jn}}{\omega_{jn}^2 - \omega^2} \, \text{Re} \, (\langle n|\mu_{\alpha}|j\rangle \times \langle j|\textcircled{H}_{\beta\gamma}|n\rangle) \,, \qquad (3.3b)$$

where m_{α} and $\textcircled{H}_{\alpha\beta}$ are the magnetic dipole and electric quadrupole moment operators.

Using the version (2.37) of the Wigner-Eckart theorem for axially symmetric systems to develop products of rotational Raman transition tensors of the form

$$\langle JKM| \widehat{\alpha}_{\alpha\beta} | J'K'M' \rangle \langle J'K'M'| \widehat{G}'_{\alpha\beta} | JKM \rangle,$$

the following expressions are found for the CIDs associated with $\Delta J = \pm 1, \pm 2 \; ; \; \Delta K = 0$ transitions of a chiral symmetric top [18]:

$$\Delta_x = \frac{2[7(G'_{\parallel} - G'_{\perp}) + \frac{\omega}{2}(A_{xyz} - A_{yzx})]}{7c\,(\alpha_{\parallel} - \alpha_{\perp})} \,, \qquad (3.4a)$$

$$\Delta_z = \frac{2[(G'_{\parallel} - G'_{\perp}) - \frac{\omega}{6}(A_{xyz} - A_{yzx})]}{c\,(\alpha_{\parallel} - \alpha_{\perp})} \,.$$

Measurements of pure rotational ROA could therefore provide values for the anisotropies $(G'_{\parallel} - G'_{\perp})$ and $(A_{xyz} - A_{yzx})$ of the optical activity tensors, about which very little is known at present.

4. THEORY OF NONLINEAR RAMAN PROCESSES

4.1. Hyper-Rayleigh and Hyper-Raman Scattering

Hyper-Rayleigh and hyper-raman scattering can be described by extending the linear relation (2.1) between the induced oscillating electric dipole moment and the electric field of the light wave to include higher-order terms:

$$\mu_{\alpha} = \alpha_{\alpha\beta} E_{\beta} + \beta_{\alpha\beta\gamma} E_{\beta} E_{\gamma} + \gamma_{\alpha\beta\gamma\delta} E_{\beta} E_{\gamma} E_{\delta} + --- \,. \qquad (4.1)$$

$\beta_{\alpha\beta\gamma}$ and $\gamma_{\alpha\beta\gamma\delta}$ are called the first and second hyperpolarizability tensors. By writing the electric vector as

$$E_{\perp} = E_{\perp}^{(0)} \cos \omega t \qquad (4.2)$$

it can be seen that the second term contains contributions of the form

$$\beta_{\alpha\beta\gamma} E_{\beta}^{(0)} E_{\gamma}^{(0)} \cos 2\omega t \qquad (4.3)$$

which gives rise to hyper–Rayleigh (i.e. second harmonic) scattering. By expanding the hyperpolarizabilities in the normal vibrational coordinates $Q_p = Q_p^{(o)} \cos \omega_p t$ we obtain the terms of the form

$$\frac{1}{2} \left(\frac{\partial \beta_{\alpha\ell\gamma}}{\partial Q_p} \right)_o Q_p^{(o)} \cos(2\omega \pm \omega_p) t \tag{4.4}$$

which are responsible for vibrational hyper–Raman scattering.

It is possible to develop expressions for hyper–Rayleigh and Raman intensity components in terms of sixth–rank tensor invariants, analogous to the familiar fourth–rank invariants given above, together with quantum–mechanical expressions for transition hyperpolarizability tensors. However, these expressions are too complicated to give here: the articles by D.A. Long in reference [4] should be consulted for further details.

The main interest in hyper–Raman spectroscopy is that certain vibrational modes that are inactive in both the infrared and in conventional Raman can be active in the hyper–Raman. A good example is the $\nu_4 (A_u)$ torsional mode of ethylene. Unfortunately, hyper–Raman scattering is many orders of magnitude less intense than conventional Raman, so measurements take a long time and sensitivity can be low. Gas–phase studies are therefore unfavourable, and only a few gases have been studied so far, giving just vibrational spectra at modest resolution.

4.2. The General Third–Order Susceptibility

Much more important for gas phase spectroscopy than the hyper–Raman effect are the various coherent Raman effects, so we shall develop the theory of coherent Raman scattering in rather more detail. The usual starting point is the bulk polarization $\underline{P}$ of the medium expressed as a function of the electric field vectors of the various light waves present simultaneously in the medium (SI)

$$P_\alpha = \epsilon_o [\chi_{\alpha\beta} E_\beta + \chi_{\alpha\beta\gamma} E_\beta E_\gamma + \chi_{\alpha\beta\gamma\delta} E_\beta E_\gamma E_\delta + \cdots]. \tag{4.5}$$

The bulk polarization can be related to the induced electric dipole moments of the constituent molecules through

$$\underline{P} = N \overline{\underline{\mu}}, \tag{4.6}$$

where N is the number density of molecules and the bar denotes a statistical average appropriate to the particular medium.

Coherent Raman effects originate in the third–order susceptibility $\chi_{\alpha\beta\gamma\delta}$ (a fourth–rank tensor) which is the bulk version of the second molecular hyperpolarizability $\gamma_{\alpha\beta\gamma\delta}$ in (4.1). For the general non–linear process depicted in figure 1.4, the induced time–dependent bulk polarization which radiates at angular frequency ω_4 is given by the real part of

$$\tilde{P}_\alpha(\omega_4,t) = \epsilon_0 \tilde{\chi}_{\alpha\beta\gamma\delta}(-\omega_4, \omega_1, -\omega_2, \omega_3)$$
$$\times \tilde{E}_\beta(\omega_1,t)\tilde{E}_\gamma^*(\omega_2,t)\tilde{E}_\delta(\omega_3,t). \tag{4.7}$$

Many theories have been published which relate the third-order susceptibility to molecular properties. These theories range from full quantum electrodynamics treatments to simple classical models. The simple semi-classical treatment given below is sufficient for our purposes since it will expose the basic physics of the coherent scattering process and will also give us an expression in terms of the conventional spontaneous Raman transition polarizabilities.

Detailed theories of various kinds can be found in references [4-6]. The book by Eesley [6] is especially useful. Also, general works on nonlinear optics, such as Bloembergen's classic book [29] or Shen's new book [30], contain much valuable material on the theory of coherent Raman processes.

4.3. The Harmonic Oscillator Model of Coherent Raman Scattering

In the harmonic oscillator model, attention is focused on the effect of incident laser waves on the normal vibrations of the collection of molecules in the medium. The equation of motion for an oscillator with normal coordinate Q, angular frequency ω_σ and reduced mass μ is [31]

$$\frac{\partial^2 Q}{\partial t^2} + \Gamma \frac{\partial Q}{\partial t} + \omega_\sigma^2 Q = \frac{F}{\mu}, \tag{4.8}$$

where Γ is the damping constant responsible for the linewidth and F is the driving force acting on the oscillator. We shall henceforth take Q to be mass-weighted so that μ is replaced by unity in (4.8).

The force can be deduced from the potential energy of the induced electric dipole moment in the inducing fields themselves. Thus the induced dipole moment $\mu_\alpha = \alpha_{\alpha\beta}E_\beta$ is associated with a potential energy [29,30]

$$U = -\tfrac{1}{2} \alpha_{\alpha\beta} E_\alpha E_\beta, \tag{4.9}$$

so expanding $\alpha_{\alpha\beta}$ in the normal vibrational coordinate Q,

$$\alpha_{\alpha\beta}(Q) = (\alpha_{\alpha\beta})_0 + \left(\frac{\partial\alpha_{\alpha\beta}}{\partial Q}\right)_0 Q + \text{---}, \tag{4.10}$$

the force $F = -\nabla U$ is

$$F = -\frac{\partial}{\partial Q}U = \frac{1}{2}\left(\frac{\partial\alpha_{\alpha\beta}}{\partial Q}\right)_0 E_\alpha E_\beta. \tag{4.11}$$

Using (4.11), the damped driven oscillator equation (4.8) becomes

$$\frac{\partial^2 Q}{\partial t^2} + \Gamma \frac{\partial Q}{\partial t} + \omega_\nu^2 Q = \frac{1}{2} \left(\frac{\partial \alpha_{\alpha\beta}}{\partial Q} \right)_0 E_\alpha E_\beta . \qquad (4.12)$$

The electric vector is written as the sum of the driving waves present in the medium, which depends on the particular coherent Raman process of interest. We shall develop the case for CARS, since this has been the most widely used technique. In the CARS process, illustrated in figure 4.1, the molecule interacts with two waves at ω_1 which

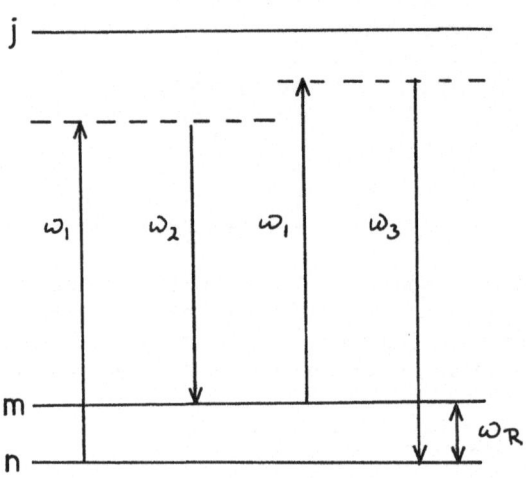

Figure 4.1 The CARS four-wave mixing process.

originate in the same laser of fixed frequency ω_L, and a third wave at ω_2 which originates in a second tunable laser tuned to the Stokes frequency ω_s. The fourth wave at $\omega_3 = 2\omega_1 - \omega_2$ is the antiStokes wave ω_{AS} generated by the polarization induced via the third-order susceptibility. The driving waves are therefore at ω_1 and ω_2, so the real electric vector to be used in (4.12) is

$$E_\alpha = \frac{1}{2} \left[\tilde{E}_{1\alpha}^{(0)} e^{-i(\omega_1 t - \underline{k}_1 \cdot \underline{r})} + \tilde{E}_{2\alpha}^{(0)} e^{-i(\omega_2 t - \underline{k}_2 \cdot \underline{r})} + c.c. \right]. \qquad (4.13)$$

The oscillators are driven at various frequencies, but we are particularly interested in the contribution at $\omega_R = \omega_1 - \omega_2$ since this corresponds to the antiStokes wave at $\omega_3 = 2\omega_1 - \omega_2$. Assuming a solution of (4.12) of the form

$$Q = \frac{1}{2} \left\{ \tilde{Q}^{(0)} e^{-i[(\omega_1 - \omega_2)t - (\underline{k}_1 - \underline{k}_2) \cdot \underline{r}]} + c.c. \right\}, \qquad (4.14a)$$

we find

$$\tilde{Q}^{(0)} = \frac{(\partial \alpha_{\alpha\beta}/\partial Q)_0 \, \tilde{E}_{1\alpha}^{(0)} \, \tilde{E}_{2\beta}^{(0)*}}{4 [\omega_\nu^2 - (\omega_1 - \omega_2)^2 - i\Gamma(\omega_1 - \omega_2)]} . \qquad (4.14b)$$

Now introduce (4.6) for the bulk polarization, with the induced electric dipole moment written in terms of the polarizability:

$$P_\alpha = N\left[(\alpha_{\alpha\beta})_o + \left(\frac{\partial \alpha_{\alpha\beta}}{\partial Q}\right)_o Q + \cdots\right] E_\beta . \tag{4.15}$$

Using (4.14) for the normal coordinate driven at the difference of the two laser fields, $\omega_1 - \omega_2$, the second term of (4.15) gives the required polarization at $\omega_3 = 2\omega_1 - \omega_2$ if we take just $\underset{\sim}{E}_1$ from the total electric vector (4.13) for use in (4.15):

$$P_\alpha(\omega_3, t) = \frac{N(\partial \alpha_{\alpha\beta}/\partial Q)_o (\partial \alpha_{\gamma\delta}/\partial Q)_o}{16\left[\omega_\sigma^2 - (\omega_1 - \omega_2)^2 - i\,T(\omega_1 - \omega_2)\right]}$$

$$\times \tilde{E}_{1\beta}^{(0)}\, \tilde{E}_{1\delta}^{(0)}\, \tilde{E}_{2\delta}^{(0)*}\, e^{-i[(2\omega_1 - \omega_2)t - (2k_1 - k_2)\cdot \underset{\sim}{r}]} + c.c. \tag{4.16}$$

Comparing this with (4.7) we find the complex third-order susceptibility to be

$$\epsilon_o \tilde{\chi}_{\alpha\beta\gamma\delta}(-\omega_3, \omega_1, -\omega_2, \omega_1) =$$

$$\frac{N}{8}\left(\frac{\partial \alpha_{\alpha\beta}}{\partial Q}\right)_o \left(\frac{\partial \alpha_{\gamma\delta}}{\partial Q}\right)_o \frac{\Delta_\sigma}{\left[\omega_\sigma^2 - (\omega_1 - \omega_2)^2 - i\,T(\omega_1 - \omega_2)\right]} + \epsilon_o \tilde{\chi}_{\alpha\beta\gamma\delta}^{NR} . \tag{4.17}$$

The factor Δ_σ has been included to account for the difference in population between the lower state (usually the ground state) and the excited vibrational state: thus Δ_σ approaches unity at low temperature and zero at high temperature. The term $\tilde{\chi}^{NR}$ has been added to describe the non-resonant background from other Raman transitions.

The susceptibility expression (4.17) can be developed further by writing the spontaneous Raman transition polarizability for the vibrational transition as

$$\langle m_\sigma | \alpha_{\alpha\beta}(Q) | n_\sigma \rangle = (\alpha_{\alpha\beta})_o \delta_{m_\sigma n_\sigma} + \left(\frac{\partial \alpha_{\alpha\beta}}{\partial Q}\right)_o \langle m_\sigma | Q | n_\sigma \rangle + \cdots . \tag{4.18}$$

Thus for the fundamental transition $1_\sigma \leftarrow 0$,

$$\langle 1_\sigma | \alpha_{\alpha\beta}(Q) | 0 \rangle = \left(\frac{\hbar}{2\omega_\sigma}\right)^{\frac{1}{2}} \left(\frac{\partial \alpha_{\alpha\beta}}{\partial Q}\right)_o . \tag{4.19}$$

This is known as Placzek's approximation and is only valid at transparent laser frequencies, i.e. well away from electronic absorption resonances [9,10,12]. Thus the susceptibility (4.17) becomes

$$\epsilon_o \tilde{\chi}_{\alpha\beta\gamma\delta}(-\omega_3, \omega_1, -\omega_2, \omega_1) =$$

$$\left(\frac{\omega_\sigma N\Delta_\sigma}{4\hbar}\right) \frac{\langle 1_\sigma | \alpha_{\alpha\beta}(Q) | 0 \rangle \langle 1_\sigma | \alpha_{\gamma\delta}(Q) | 0 \rangle}{\left[\omega_\sigma^2 - (\omega_1 - \omega_2)^2 - i\,T(\omega_1 - \omega_2)\right]} . \tag{4.20}$$

In deriving (4.20), we ignored rotational structure for simplicity, but this can easily be incorporated if necessary. For isotropic media, still ignoring rotational structure, the susceptibility must be averaged over all orientations. Different components give different averages. Typically, the incident laser beams and the CARS beam will be propagating in virtually the same direction, say along z (provided the medium is diffuse so that phase-matching problems are minimized), and the laser beams at ω_1 and ω_2 will be linearly polarized in the same direction, say x. This means that the CARS beam will also be polarized along x (since $\langle \chi_{yxxx} \rangle = 0$), so the relevant isotropic average is, using (2.24a),

$$\langle \epsilon_0 \tilde{\chi}_{xxxx} \rangle = \left(\frac{\omega_\sigma N}{4\hbar} \right) \left(A + \frac{4}{45} B \right) \left[\frac{\Delta_\sigma}{\omega_\sigma^2 - (\omega_1 - \omega_2)^2 - i\Gamma(\omega_1 - \omega_2)} \right], \quad (4.21)$$

where A and B are the isotropic and anisotropic polarizability invariants (2.28). In terms of the differential scattering cross-section (2.32),

$$\langle \epsilon_0 \tilde{\chi}_{xxxx} \rangle = \left(\frac{4\pi^2 N}{\omega^4 \mu_0^2 \hbar} \right) \left(\frac{d\sigma}{d\Omega} \right)_x \left[\frac{\omega_\sigma \Delta_\sigma}{\omega_\sigma^2 - (\omega_1 - \omega_2)^2 - i\Gamma(\omega_1 - \omega_2)} \right]. \quad (4.22)$$

This result is the SI version of the c.g.s. equation (44) in the article by W.M. Tolles and A.B. Harvey in reference [5].

4.4. Coherent Scattering from Bulk Samples

So far, the theory has concentrated on the molecular scattering aspect of coherent Raman scattering. There is still a lot more physics to consider in order to complete the description of the phenomenon in bulk samples; but we shall make do with a brief sketch. From Maxwell's equations, the following complex expression can be obtained describing the propagation of plane light waves in the medium [6]:

$$\nabla^2 \tilde{E}_\alpha - \frac{1}{c^2} \frac{\partial^2 \tilde{E}_\alpha}{\partial t^2} = \mu_0 \frac{\partial^2 \tilde{P}_\alpha}{\partial t^2}, \quad (4.23)$$

where the medium is assumed to be free of current, charge and magnetization. From this it is possible to deduce the following expression for the complex electric vector of an unfocused CARS beam at $\omega_3 = 2\omega_1 - \omega_2$ having traversed an interaction length L of the sample in the z-direction [6]:

$$\tilde{E}_{3x} \approx i\omega_3 \left(\frac{\mu_0 \epsilon_0}{\epsilon_3} \right)^{\frac{1}{2}} \langle \tilde{\chi}_{xxxx} (-\omega_3, \omega_1, -\omega_2, \omega_1) \rangle$$

$$\times \tilde{E}_{1x}^{(\omega)2} \tilde{E}_{2x}^{(\omega)*} L \frac{\sin(\Delta k L/2)}{(\Delta k L/2)}, \quad (4.24)$$

where ϵ_3 is the dielectric constant at the antiStokes frequency ω_3 and

$$\Delta k = |(2\underline{k}_1 - \underline{k}_2 - \underline{k}_3) \cdot \hat{\underline{z}}|, \quad (4.25)$$

$\hat{\underline{z}}$ being the unit vector along z.

From (2.19), the intensity of the detected CARS wave is

$$I_3(z) = \frac{4\omega_3^2 \mu_0^2}{\Lambda_1^2 \Lambda_2 \Lambda_3} |\langle \widetilde{\chi}_{xxxx}(-\omega_3, \omega_1, -\omega_2, \omega_1)\rangle|^2$$

$$\times I_1(z)^2 I_2(z) L^2 \left[\frac{\sin(\Delta k L/2)}{(\Delta k L/2)}\right], \qquad (4.26)$$

where $\Lambda_i = \epsilon_i^{1/2}$ is the refractive index of the medium at ω_i (assuming the medium is magnetization-free so that $\mu_i = 1$). This is equivalent to equation (3.11) of reference [6]. If $\Delta k \neq 0$, this result shows that the CARS signal will vary sinusoidally with path length (remember that the theory was set up for all input and output beams collinear along z). The <u>coherence length</u> is defined as the path length to reach maximum conversion efficiency, which occurs when $\Delta k L/2 = \pi/2$ or $\Delta k = \pi L$, where

$$\Delta k = \frac{1}{c}(2\Lambda_1 \omega_1 - \Lambda_2 \omega_2 - \Lambda_3 \omega_3). \qquad (4.27)$$

For gases at low pressure, Δk is very small and the coherence length can be many centimetres; but in condensed media, where the refractive indices for the different frequencies differ significantly, the coherence length is very short. However, in the latter case phase-matching can be achieved by crossing the two input laser beams at the appropriate angle.

An important feature is that χ contains real and imaginary parts χ' and χ'', as well as the nonresonant contribution χ^{NR}. With no χ^{NR} contribution, a plot of $|\chi|^2$ versus $\omega_1 - \omega_2$ has a Lorentzian line-shape; but if χ^{NR} is significant, the cross-term $\chi'\chi^{NR}$ can lead to a considerable distortion of the lineshape [4,6].

5. COHERENT RAMAN SPECTROSCOPY

We shall now review the current range of coherent Raman techniques. Although the basic theory of coherent Raman effects was developed above with specific reference to CARS, similar general features apply to other techniques: they are all basically the same phenomenon generated via a third-order nonlinear susceptibility, but use different measurements to detect the vibrational resonance in the four-wave mixing process. The common feature of the coherent Raman techniques is the excitation of Raman-active molecular vibrations or rotations in the field of one laser beam at fixed frequency ω_L and another with tunable frequency ω_S with an appropriate signal being generated when their difference $\omega_L - \omega_S = \omega_R$, the frequency of the molecular transition. More details, together with some additional techniques, can be found in references [4-6].

5.1. Coherent AntiStokes Raman Scattering (CARS)

As we have seen, in CARS the observable is the intensity of the

coherent antiStokes beam of frequency $\omega_{AS} = 2\omega_L - \omega_S$, or
$\omega_3 = 2\omega_1 - \omega_2$ in the notation of figure 4.1, which is scattered
into one direction lying in the plane of the two input laser beams. The
actual direction is determined by the phase-matching condition
$\underline{k}_3 = 2\underline{k}_1 - \underline{k}_2$. A typical CARS apparatus is drawn schematically in
figure 5.1. The input laser beams at ω_1 and ω_2 are overlapped at the

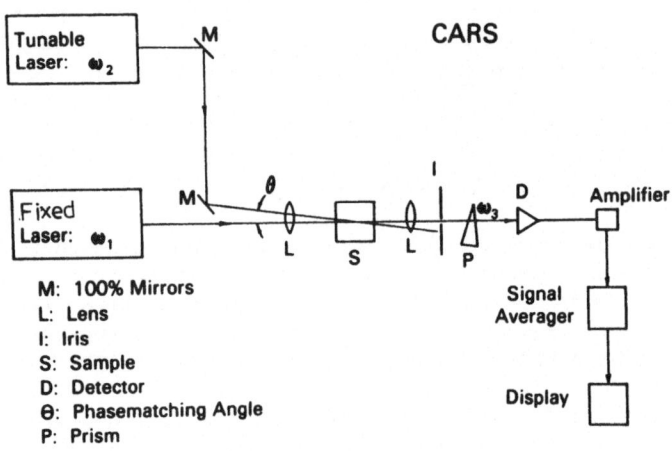

Figure 5.1. Schematic diagram of CARS apparatus [6].

phase-matching angle θ . The laser at ω_1 is fixed in frequency, while
the other at ω_2 is tuned so that the output frequency parametrically
generated at ω_3 is changing. The iris aperture I provides the initial
spatial discrimination of the antiStokes wave at ω_3 from ω_1 and ω_2 .
The detector D is often a broadband diode or photomultiplier tube,
followed by waveform-shaping electronics and an amplifier, a signal-
averager such as a boxcar integrator, and a chart recorder. Scattering
from windows, lenses, etc. can lead to contamination of the spatially-
filtered coherent antiStokes beam at ω_3 by components of the ω_1 and
ω_2 beams, which can be suppressed using a dispersive element such as a
prism P or a spectrometer in front of the detector.

   We saw in section 4 above that the CARS intensity depends quad-
ratically on the number density of molecules N. Hence whilst
vibrational CARS spectra of liquids can be obtained with cw lasers,
giant pulsed lasers producing megawatt pulses are often required to
obtain rotation-vibration spectra of gaseous samples. A frequency-
doubled Nd: YAG laser at 532 nm is a common choice, and by splitting

the beam it can also be used to pump the tunable dye laser. The technique is attractive for gas phase studies because of its high resolution capability and its sensitivity at pressures low enough to reduce collision broadening. With pulsed lasers the resolution is limited to the pulsed laser linewidth of ~0.1 cm^{-1}; but using cw lasers resolutions of ~0.01 cm^{-1} are obtainable, although the reduced signals using cw excitation make it difficult to go to sufficiently low pressures to take full advantage of the 0.001 cm^{-1} laser linewidths that are available (but see the SRS techniques below).

We have already seen in figures 1.6 and 1.7 an example of the dramatic resolution improvement in CARS over spontaneous Raman [4]. The doublet appearance of the lower-frequency Q-branch components of NO in figure 1.7 arises from the overlap of two series originating in $\Delta J = 0$ transitions starting at given J levels in the $^2\Pi_{1/2}$ and $^2\Pi_{3/2}$ electronic states. The stronger bands come from molecules in the $^2\Pi_{1/2}$ ground state, and the weaker from the $^2\Pi_{3/2}$ state which lies 124 cm^{-1} above the ground state. At higher wavenumbers the Q-branches from each of the two series become increasingly coincident.

One important practical application of CARS is in combustion and other high-temperature reaction diagnostics, for which spontaneous Raman methods are unsuitable because of the strong fluorescence background (see the article by A.C. Eckbreth and P. Schreiber in reference [5]).

Although the basic CARS experiment can give very good results, there are several inherent problems that can sometimes spoil the spectra. As mentioned previously, there is always a nonresonant background signal which can obscure weak bands; and interference between the resonant and nonresonant contributions can cause a considerable distortion of the Raman band lineshapes, which can even take on a "derivative" appearance. Several variations of CARS have therefore been developed to overcome these difficulties. For example, in coherent antiStokes Raman ellipsometry (CARE) introduced by Akhmanov and Koroteev [32], advantage is taken of different polarization characteristics of the resonant and nonresonant contributions: thus the pump and probe laser polarizations are set at pre-determined angles, and the antiStokes signal is transmitted through a polarizer set at an appropriate angle to block the non-resonant signal. Figure 5.2 demonstrates the significant improvement that CARE can give over CARS. The example is the spectrum of a methane-air flame in the region of the CO Q-branch, obtained by Kahn, Zych and Mattern [33]. The nonresonant CARS background in 5.2b is completely eliminated in the CARE spectrum above. Most of the features in this spectrum can be interpreted in terms of known Raman transitions in CO and N_2.

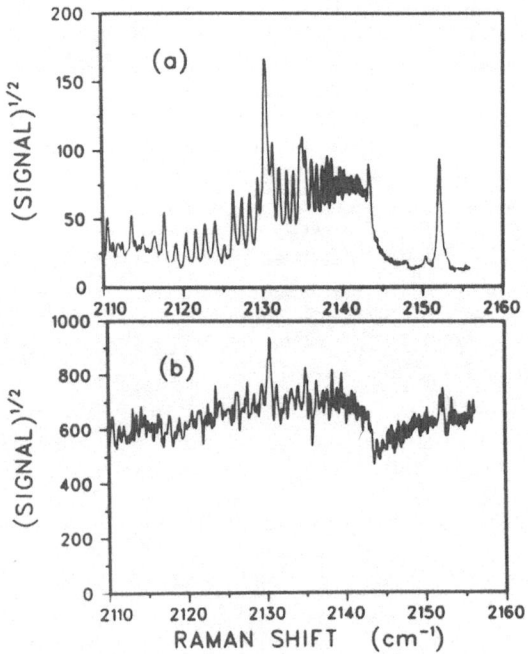

Figure 5.2. (a) CARE and (b) CARS spectra of a methane-air flame in the region of the CO Q-branch [33].

5.2. Raman - Induced Kerr Effect Spectroscopy (RIKES)

The Raman-induced Kerr effect, introduced by Heiman et al. [34], provides another method of suppressing the non-resonant background, but has the added advantage of dispensing with the phase-matching condition. It is based on a Raman version of the optical Kerr effect in which the electric vector of an intense laser beam can induce significant bi-refringence in an isotropic sample. For example, if a linearly polarized probe laser beam is transmitted through a sample with its plane of polarization inclined at 45° to the plane of polarization of a second, intense, laser beam the probe beam will emerge elliptically polarized. Thus the effect of the birefringence is to induce a component in the probe beam that is linearly polarized perpendicular to the polarization direction of the initial beam, and this results in partial transmission through a polarizer that would normally block the beam.

The RIKES process is illustrated in figure 5.3

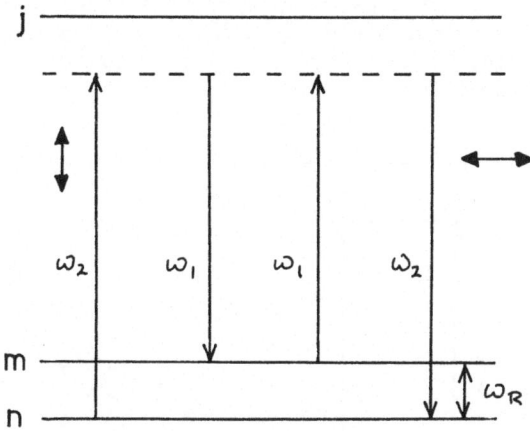

Figure 5.3. Energy level diagram for the RIKES process.

and is described by a third-order susceptibility
$\tilde{\chi}_{\alpha\beta\gamma\delta}(-\omega_2, \omega_2, -\omega_1, \omega_1)$. The "virtual" birefringence induced by
the strong pump wave at ω_1 is sampled by the weak probe wave at ω_2,
and exhibits resonances when $|\omega_1 - \omega_2|$ approaches a Raman-active
frequency. Usually ω_2 is fixed and ω_1 is tuned through the Raman
resonance. Since the detected polarization component has the same
frequency ω_2 as the probe laser, the phase-matching factor
$\sin(\Delta k L/2)$ in (4.26) vanishes:

$$\Delta \underline{k} = \underline{k}_2 - \underline{k}_1 + \underline{k}_1 - \underline{k}_2 = \underline{0}.$$

The nonresonant background is suppressed by using a <u>circularly</u> polarized pump wave: the detailed explanation for this can be found in
references [6,30,34], but basically it arises because the nonresonant
third-order susceptibility $\chi^{NR}_{\alpha\beta\gamma\delta}$ has a higher permutation symmetry than the resonant part $\chi_{\alpha\beta\gamma\delta}$, leading to complete cancellation in isotropic media of the associated nonresonant circularly
polarized component of the probe wave.

The absence of a phase-matching condition means that it is also
possible to use a fixed frequency pump laser with a broadband probe
laser. This gives a complete Raman spectrum over the bandwidth of the
probe laser ($\sim$1000 cm^{-1}) which can be analysed using a spectrometer
placed after the blocking polarizer, followed by an optical multichannel analyzer.

5.3. Stimulated Raman Spectroscopy (SRS)

The most venerable of the nonlinear Raman effects is stimulated Raman
scattering, first observed by Woodbury and Ng in 1962. The basic

stimulated Raman effect is not a parametric process since the molecule
does not return to its original state, but it is still coherent. As
illustrated in figure 5.4, Stokes photons generated by spontaneous
Raman scattering and travelling in the direction of the pump beam

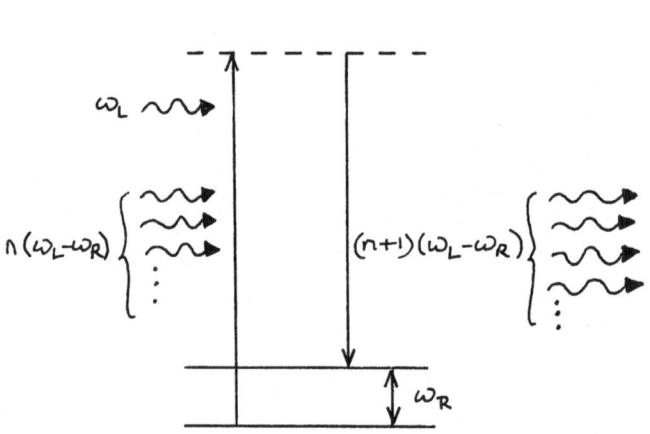

Figure 5.4. Schematic diagram of the stimulated Raman process.

undergo avalanche multiplication due to stimulated emission of like
Stokes photons, thus creating a Stokes beam in the same direction as
the pump beam and with almost the same power. Whilst interesting and
important in itself, and also as a means of generating frequency-
shifted laser beams, the basic stimulated Raman effect is not generally
very useful in spectroscopy because the strongest Raman-active mode
usually steals all the intensity (e.g. the ring breathing mode at
992 cm^{-1} in liquid benzene). There are, however, versions of this
effect employing two separate laser beams that avoid this restriction,
and the area of stimulated Raman spectroscopy has now become very
important in high resolution gas phase studies.

Stimulated Raman spectroscopy relies on the use of a strong pump
laser to induce a gain or loss in the intensity of a weak probe laser
when the frequency difference of the two lasers is equal to a Raman
transition frequency. In the case of stimulated Raman gain spectroscopy,
illustrated in figure 5.5a, the pump frequency ω_1 is greater than the
probe frequency ω_2 so that the probe laser beam enjoys an intensity
gain due to stimulated Stokes emission. In the case of inverse Raman
spectroscopy, illustrated in figure 5.5b, the pump frequency is less
than the probe frequency so that the strong Stokes emission stimulated
by the pump laser beam leads to a loss of intensity in the probe laser
beam. In fact the inverse Raman effect was observed by Jones and
Stoicheff as long ago as 1964 [35], but it was more than a decade
before the spectroscopic potential of the phenomenon started to be
exploited, mainly because of the unavailability of reliable narrowband

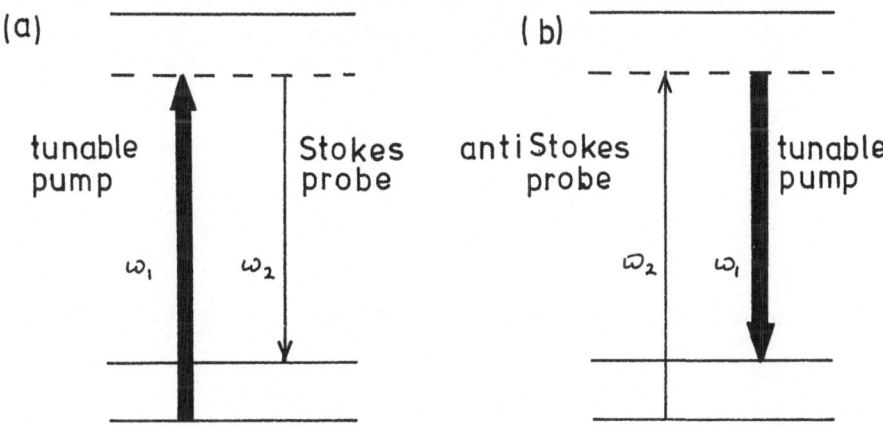

Figure 5.5. Energy level diagrams for (a) stimulated Raman gain and (b) inverse Raman absorption processes.

tunable lasers (which also held up the development of CARS spectroscopy).

Although not immediately apparent from the simple description just given, the basic SRS process also involves four-wave mixing described by a third-order susceptibility $\tilde{\chi}_{\alpha\beta\gamma\delta}$ $(-\omega_2, \omega_2, -\omega_1, \omega_1)$. In fact the process is very similar to that in RIKES, except that it is the loss or gain in the probe beam at ω_2 that is measured, rather than the orthogonal polarization component of the transmitted beam at ω_2. For this reason the SRS observables depend only on the <u>imaginary</u> part of $\tilde{\chi}_{\alpha\beta\gamma\delta}$, which means that the non-resonant contribution is automatically excluded, and also that there is no phase-matching requirement. Furthermore, the absence of a contribution from the real part of $\tilde{\chi}_{\alpha\beta\gamma\delta}$ eliminates complications due to nonlinear dispersion, resulting in a simple Lorentzian lineshape.

The basic features of an SRS apparatus are illustrated in figure 5.6. This is essentially equivalent to the RIKES apparatus except for the absence of a polarizer prior to detection. Since no phase-matching is required, the beams can be overlapped **collinearly** using a dichroic mirror M_2. A frequency dispersing element P spatially separates the pump and probe beams which are then spatially filtered using the iris I. Since the full probe laser power falls on the detector D, the use of a photomultiplier is precluded, and a solid state photodiode is usually used. Since the requirement for high degrees of polarization as in RIKES is greatly relaxed in SRS, a multipass sample cell can be used to amplify the signal. As well as acquiring the Raman spectrum by scanning ω_1 with respect to ω_2 , it is also possible, as in RIKES, to use a broadband probe laser and so acquire a large portion of the Raman spectrum using a single pump laser shot.

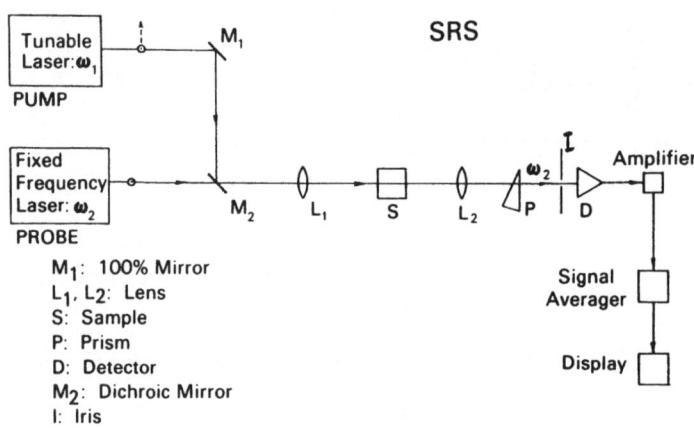

Figure 5.6. Schematic diagram of SRS apparatus [6].

Although the original SRS experiments were performed with high-power pulsed lasers, it was subsequently found to be possible to obtain high-quality SRS spectra of gases using low-power cw lasers for both pump and probe, leading to the acquisition of some very impressive high resolution Raman spectra of gases. For example, figure 5.7 shows a "quasi-cw" Raman gain spectrum obtained by Esherick and Owyoung [36]: the 0.002 cm^{-1} instrumental bandwidth fully resolves the rotational structure in the Q-branch of the fundamental breathing vibration of SF_6 at 3.8 Torr.

It can sometimes be highly advantageous to detect the SRS signal indirectly. Thus in photoacoustic Raman spectroscopy (PARS), introduced by Barrett and Berry [37], a microphone or piezoelectric transducer is used to measure the acoustic wave amplitude generated in the sample when the vibrational or rotational excitation, created by the SRS process, relaxes non-radiatively into translational (heat) energy. Figure 5.8 shows the pure rotational PARS spectrum of CO_2, taken from the article by J.J. Barrett, D.R. Siebert and G.A. West in Reference [4]. Although the resolution of ~ 0.3 cm^{-1} is not remarkable, the

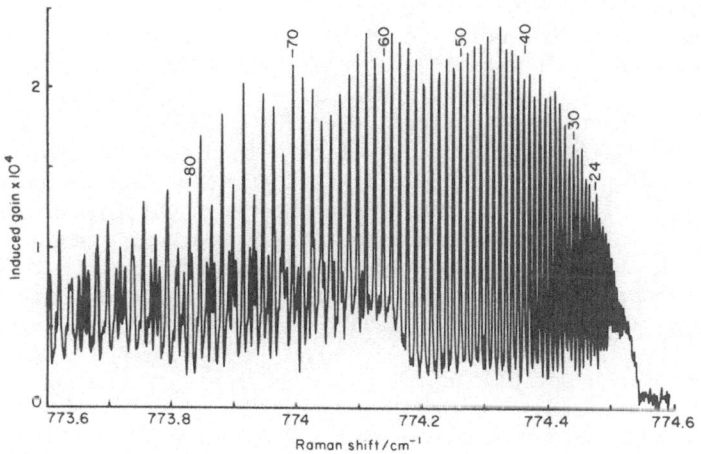

Figure 5.7. Raman gain spectrum of SF_6 at 3.8 Torr, showing ν_1 fundamental Q(J) transitions, and underlying $\nu_1 + \nu_6 \leftarrow \nu_6$ hot band transitions [36].

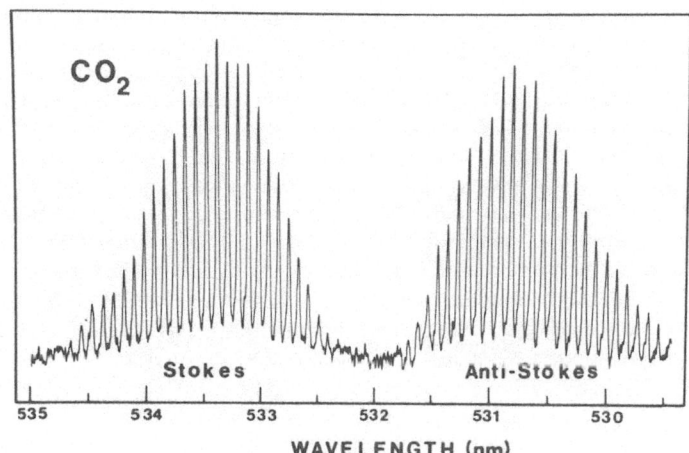

Figure 5.8. Pure rotational PARS spectrum of CO_2 at 500 Torr [4].

striking feature of this spectrum is the complete absence of a
Rayleigh component at the pump laser wavelength. This is a consequence
of the acoustic detection scheme which is only sensitive to optical
energy deposited in the molecules irreversibly. Rayleigh lines are
usually many orders of magnitude stronger than Raman lines, and
interference with low-frequency Raman lines is often an intractable
problem with conventional optical detection methods.

Finally, another indirect detection method for SRS involves ultraviolet ionization. The vibrationally-excited molecules are selectively
ionized by a UV laser beam, and a Raman spectrum obtained by detecting
the variation in the ionization signal while scanning one of the Raman
lasers. Using this method, Esherick and Owyoung [38] claim to have
achieved an increase in sensitivity over conventional SRS detection of
three orders of magnitude without any loss of resolution!

5.4 Coherent Raman Optical Activity

In view of the inherent weakness of the Raman optical activity
phenomena outlined above (typical Δ-values are $10^{-3} - 10^{-5}$), coherent
nonlinear techniques might at first sight appear to offer the best
means of making gas-phase measurements of both natural ROA from chiral
molecules, and ROA from achiral molecules in a magnetic field
(including Raman EPR). One note of caution, however, is that, as Eesley
[6] has discussed in detail, large signal intensity does not
necessarily translate into high signal-to-noise, so it might still not
be possible to extract the very small ROA signals. Although the
signals are much weaker, conventional spontaneous Raman techniques can
often provide better signal-to-noise than coherent techniques.

The simple semi-classical theory of CARS and related phenomena
outlined above is eminently suitable for describing coherent ROA
phenomena since the products of two transition polarizabilities in
expressions such as (4.20) for the third-order susceptibility are
replaced by appropriate transition polarizability-transition optical
activity products in expressions for corresponding higher-order
contributions to the third-order susceptibility. Such expressions
have been given by Oudar, Minot and Garetz [39], together with a
discussion of the various polarization configurations that could be
used in coherent ROA experiments. No measurements of coherent ROA in
gases have yet been reported, but coherent (stimulated) ROA has been
observed in α-quartz [40].

REFERENCES

1. Long, D.A., Raman Spectroscopy, McGraw-Hill, New York, 1977.

2. Weber, A., ed., Raman Spectroscopy of Gases and Liquids, Topics in
    Current Physics, Vol. 11, Springer, Berlin, 1979.

3. Sushchinskii, M.M., Raman Spectra of Molecules and Crystals,
    Israel Program for Scientific Translations, Jerusalem, 1972.

4. Kiefer, W., and Long, D.A., eds., Non-Linear Raman Spectroscopy and its Chemical Applications, Reidel, Dordrecht, 1982.

5. Harvey, A.B., ed., Chemical Applications of Nonlinear Raman Spectroscopy, Academic Press, New York, 1981.

6. Eesley, G.L., Coherent Raman Spectroscopy, Pergamon Press, Oxford, 1981.

7. Bhagavantam, S., Scattering of Light and the Raman Effect, Chemical Publishing Company, New York, 1942.

8. Fabelinskii, I.L., Molecular Scattering of Light, Plenum Press, New York, 1968.

9. Placzek, G., in Handbuch der Radiologie, ed. E. Marx, vol. 6, part 2, p. 205. Akademische Verlagsgesellschaft, Leipzig, 1934.

10. Born, M., and Huang, K., Dynamical Theory of Crystal Lattices, Clarendon Press, Oxford, 1954.

11. Davydov, A.S., Quantum Mechanics, Pergamon Press, Oxford, 1976.

12. Barron, L.D., Molecular Light Scattering and Optical Activity, Cambridge University Press, Cambridge, 1982.

13. Berestetskii, V.B., Lifshitz, E.M., and Pitaevskii, L.P., Quantum Electrodynamics, Pergamon Press, Oxford, 1982.

14. Van Vleck, J.H., The Theory of Electric and Magnetic Susceptibilities, Oxford University Press, London, 1932.

15. Koningstein, J.A., Introduction to the Theory of the Raman Effect, Reidel, Dordrecht, 1972.

16. Papoušek, D., and Aliev, M.R., Molecular Vibrational-Rotational Spectra, Elsevier, Amsterdam, 1982.

17. Murphy, W.F., J. Raman Spectrosc., 11, 339 (1981).

18. Barron, L.D., and Johnston, C.J., J. Raman Spectrosc., 16, 208 (1985).

19. Placzek, G., and Teller, E., Z. Phys., 81, 209 (1933).

20. Barron, L.D., and Nørby Svendsen, E., Advances in Infrared and Raman Spectroscopy, 8, 322 (1981).

21. Baranova, N.B., and Zel'dovich, B. Ya., J. Raman Spectrosc., 7, 118 (1978).

280

22. Zeigler, L.D., J. Chem. Phys., 84, 6013 (1986).

23. Barron, L.D., and Meehan, C., Chem. Phys. Lett., 66, 444 (1979).

24. Barron, L.D., Cutler, D.J., and Torrance, J.F., J. Raman Spectrosc., in the press.

25. Abragam, A., and Bleaney, B., Electron Paramagnetic Resonance of Transition Ions, Clarendon Press, Oxford, 1970.

26. Barron, L.D., and Vrbancich, J., Advances in Infrared and Raman Spectroscopy, 12, 215 (1985).

27. Barron, L.D., Bogaard, M.P., and Buckingham, A.D., J. Am. Chem. Soc., 95, 603 (1973).

28. Barron, L.D., and Vrbancich, J., Topics in Current Chemistry, 123, 151 (1984).

29. Bloembergen, N., Nonlinear Optics, Benjamin, New York, 1965.

30. Shen, Y.R., The Principles of Nonlinear Optics, Wiley, New York, 1984.

31. Landau, L.D., and Lifshitz, E.M., Classical Mechanics, Pergamon Press, Oxford, 1969.

32. Akhmanov, S.A., and Koroteev, N.I., Sov. Phys. Usp., 20, 899 (1977).

33. Rahn, L.A., Zych, L.J., and Mattern, P.L., Opt. Comm., 30, 249 (1979).

34. Heiman, D., Hellwarth, R.W., Levenson, M.D., and Martin, G., Phys. Rev. Lett., 36, 189 (1976).

35. Jones, W.J., and Stoicheff, B.P., Phys. Rev. Lett., 13, 657 (1964).

36. Esherick, P., and Owyoung, A., Advances in Infrared and Raman Spectroscopy, 9, 130 (1982).

37. Barrett, J.J., and Berry, M.J., App. Phys. Lett., 34, 144 (1979).

38. Esherick, P., and Owyoung, A., Chem. Phys. Lett., 103, 235 (1983).

39. Oudar, J.-L., Minot, C., and Garetz, B.A., J. Chem. Phys., 76, 2227 (1982).

40. Klein, M., Maier, M., and Prettl, W., Phys. Rev. B, 28, 6008 (1983).

TECHNIQUES IN MOLECULAR LASER SPECTROSCOPY

R.N. Dixon
School of Chemistry,
The University,
Bristol BS8 1TS

1. INTRODUCTION

Experiments in molecular spectroscopy involve the three main steps of sample preparation, scanning or dispersing the photon frequencies over the range to be studied and signal detection. There then follows the assignments of spectral lines or the measurement of the dynamics of the quantum states. Unfortunately, in classical linear spectroscopy with spectrometers the optimisation of one part of the experiment often degrades the performance of another part. For example, in emission spectroscopy high spectral resolution conflicts with high sensitivity and fast time resolution. Furthermore, many molecular spectra have such a congested line density and complex structure that they may defy all attempts at unambiguous assignment.

The introduction of laser-based techniques into molecular spectroscopy has lessened many of the conflicting requirements of classical spectroscopy. Furthermore, as has already been discussed in the context of atomic spectroscopy, the laser has led to the development of many new experiments that assist the spectroscopist in his theoretical analysis.

Lasers have largely replaced conventional spectral lamps as light sources for spectroscopic experiments because the great enhancements of spectral power density, monochromaticity, and low divergence all bestow very high sensitivity on laser-based experiments, so that far fewer molecules constitute an acceptable sample. The high power reduces noise problems from detectors, and also makes possible new non-linear processes such as multiphoton excitation and saturation spectroscopy. The narrow bandwidth can ensure that the resolution limit is defined by the molecule rather than by the light source. The low divergence can facilitate the use of long path lengths to enhance weak absorption, or can be used to concentrate the power into a small volume, providing an ideal linear fluorescent volume of gas which can be efficiently imaged on to a spectrometer slit.

These advantages have particularly benefited the spectroscopic study of free radicals, molecular ions, and molecules in weakly bound complexes; and also molecules in 'collision-free' environments of low

A. C. P. Alves et al. (eds.), Frontiers of Laser Spectroscopy of Gases, 281–307.
© 1988 by Kluwer Academic Publishers.

total pressure which permits the separate study of intra- and inter-molecular relaxation processes. Thus most spectroscopic knowledge prior to about 1970 concerning polyatomic free radicals was obtained using the technique of flash photolysis. The typical sample pressures (1-100 Torr) and flash durations (1-10µs) resulted in extensive relaxation before detection of radicals formed in excited states, precluding detailed studies of the nascent distribution of photodissociation products. In contrast, with laser-based experiments using sample pressures of 1-100 mTorr or molecular beams, and a time resolution of 1-10ns, nascent distributions may be probed before any relaxation can occur. This capa-bility has revolutionised the study of the primary processes of photo-chemistry, which is discussed in other lectures. It is sufficient to note here that free radicals and ions for spectroscopic study may readily be generated in reactions initiated by H,F,N or O atoms in fast flowing gases, by electrical discharges, or by photochemical dissociation or ionisation.

The sensitivity of detection of excitation by lasers is frequently enhanced, over that obtained from measurement of the absorption loss from the beam, by monitoring a process which is initiated by the absorbed energy. Examples are laser induced fluorescence (LIF), the production of ions, usually involving the multiple absorption of photons (MPI), the degradation of the absorbed energy to heat and thus sound in the optoacoustic effect, and the polarisation of a gas on resonance leading to the transmission of a beam through crossed polarisers either side of the sample. With continuous-wave lasers the use of a.c. modul-ation techniques may further improve signal to noise ratios, as in many other types of spectroscopy which employ electronic methods of detection. The inter-modulated fluorescence technique of saturation spectroscopy is one such example.

This chapter is concerned with the following techniques in molecular laser spectroscopy: (i) laser-Stark spectroscopy and electric field spectroscopy; (ii) laser-Zeeman, or laser-magnetic-resonance spectros-copy (LMR); (iii) dispersed laser-induced fluorescence; and (iv) double resonance spectroscopy.

2. LASER STARK SPECTROSCOPY

The earliest lasers operating in the infrared and far infrared regions of the spectrum, where almost all molecules have vibration-rotation or rotation spectra, were lasers operating on transitions in gas-phase molecules, and in particular the CO_2 laser. These lasers are not continuously tunable, although they can be step-tuned from line to line. In consequence two main methods were introduced in order that these lasers could be used for molecular spectroscopy. In these the resonance condition is achieved by tuning the transitions rather than the laser frequency, using either an electric field (the Stark effect) or a magnetic field (the Zeeman effect). Although these methods were originally regarded as stopgaps until truly tunable infrared lasers became available, they have survived because they have high sensitivity for detecting low concentrations of reactive species, because of the ease of calibration, and because the high power and narrow bandwidth of these lasers can be harnessed through saturation spectroscopy to give extremely high spectral resolution.

2.1 The Stark effect

For a molecule with a permanent electric dipole moment μ the Hamiltonian for interaction with an electric field is:

$$H' = -\mu.E$$

The leading term in the expression for the Stark splitting of the levels in a weak to moderate field will accord with one of the three following cases:

(a) A symmetric top level with rotational quantum numbers J and K, with a space-fixed projection M, has a first order perturbation:

$$\Delta E^{(1)} = - \frac{KM\mu E}{J(J+1)} \tag{2}$$

(b) For K=0, or for a linear molecule, the splitting is second order in the field with:

$$\Delta E^{(2)} = \frac{\mu^2 E^2}{2hB(2J-1)(2J+3)} \left[1 - \frac{3M^2}{J(J+1)} \right]$$

or

$$\Delta E^{(2)} = - \frac{\mu^2 E^2}{6hB} \quad , \quad \text{for } J=0, M=0 \tag{3}$$

where B is the rotational constant.

(c) For levels of a nearly symmetric top, where $|K_q|$ is a good quantum number (q = a or c) but there is a zero-field splitting 2δ:

$$\Delta E = \pm \left[\delta^2 + \frac{K_q^2 M^2 \mu_q^2 E^2}{J^2(J+1)^2} \right]^{\frac{1}{2}} \tag{4}$$

which is quadratic in E at low E but linear at high E. These formulae are illustrated in figure 1.

Given 1 Debye in a field of 1 kV cm^{-1}, $\mu E/h$ = 503.4 MHz $\equiv$ 0.01679 cm^{-1}. It is possible to generate uniform electrostatic fields of up to 50 kV cm^{-1} in cells with parallel electrodes having spacings of the order of mm, which is sufficient to cause splittings of several GHz in the lines of gases with a near first order Stark effect. This can lead to a rich structure of resonances, particularly with the step-tuned CO_2 or CO lasers. However, at these high fields it is important to take account of higher order interactions in quantitative analyses (Duxbury, 1985).

284

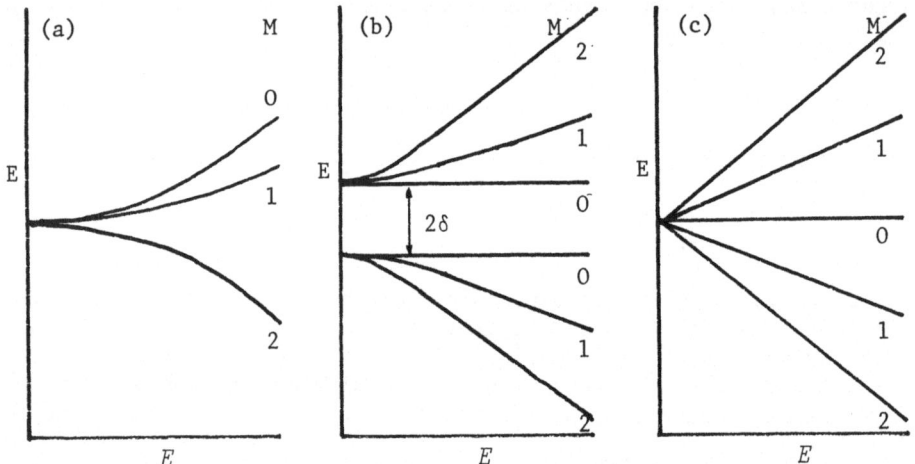

Fig.1   Stark splittings for levels with J=2. (a) K=0 (eqn 3);
         (b) K=1 with slight asymmetry doubling (eqn 4);   and (c) K=2
         with negligible doubling (eq.2); but not to a common scale.

2.2 Experimental methods

In laser Stark spectroscopy fixed-frequency laser radiation is
passed through a parallel plate Stark cell across which the field is
varied. At certain field-strengths specific M components of molecular
vibration-rotation or rotational transitions come into resonance,
giving rise to an electric resonance spectrum. This is usually
detected as a change in the laser power transmitted through the cell,
so that long path lengths enhance sensitivity. It is also advantageous
to modulate the Stark field at an audio frequency and use phase-sensitive detection to enhance the signal to noise ratio.

Figure 2 presents a spectrum of formaldehyde obtained in this way
using a CO laser line at 1746.307 cm^{-1}, which is close to the centre
of the ν_2 band (Johns and McKellar, 1973). The electric vector of the
laser beam has been arranged perpendicular to the E-field, giving the
selection rule $\Delta M = \pm 1$. Many components of the ^qQ branches come into
resonance in this spectrum. Those with K$\geqslant$2 have the simple pattern
predicted from eqn. (2), but for those with K=0 or 1 the Stark shifts
are no longer first order in E (see eqns 3 and 4). Figure 3 shows, for
the case J=2, how the gross tuning is generated by the overall Stark
effect, whereas the splitting into a close group of 2J transitions
relies on the (small) difference between the Stark splittings in the
upper and lower vibrational levels.

The upper limit to the fields used in these experiments is set by
the breakdown voltage of the gas, and is thus a function both of pressure and of the interelectrode spacing.

2.3 The use of saturation spectroscopy

Recent studies have generally made use of multiple pass cells to

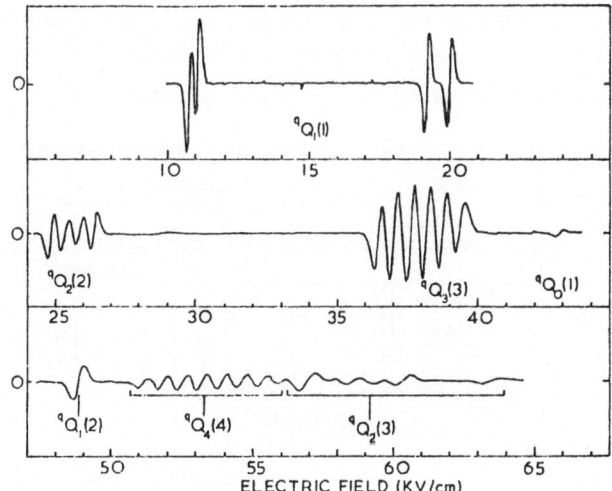

Fig.2   Laser Stark spectrum of the ^qQ branches of the ν_2 band of
H$_2$CO obtained with the 14-13P(16) line of the CO laser at
1746.307 cm^{-1}.   (Reproduced by permission from Journal of
Molecular Spectroscopy, 1973, 48, 364.)

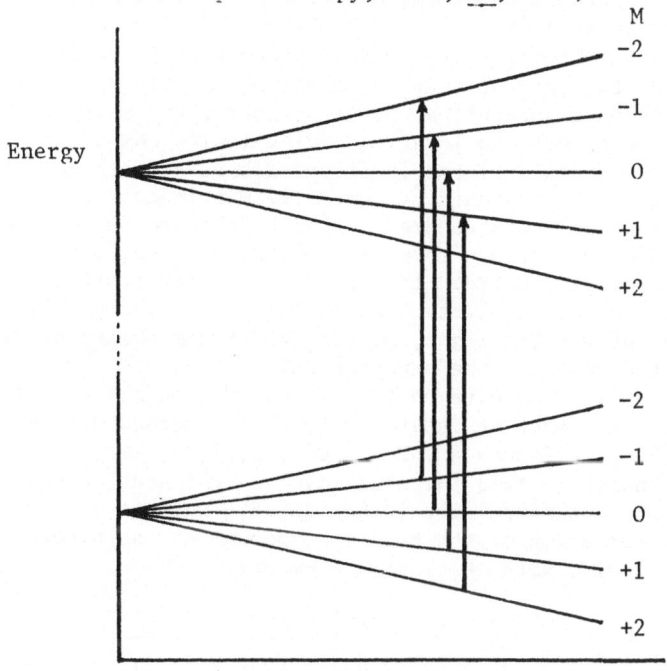

Fig.3   Stark energy level diagram for a $^qQ(2)$ transition of a
molecule with a small difference between the dipole moments
for the upper and lower vibrational levels, with $\Delta M = \pm 1$.

increase the path length, or have included the Stark cell within the laser cavity to enhance the absorption losses at resonance by gain spoiling (figure 4). This not only leads to greater sensitivity, but

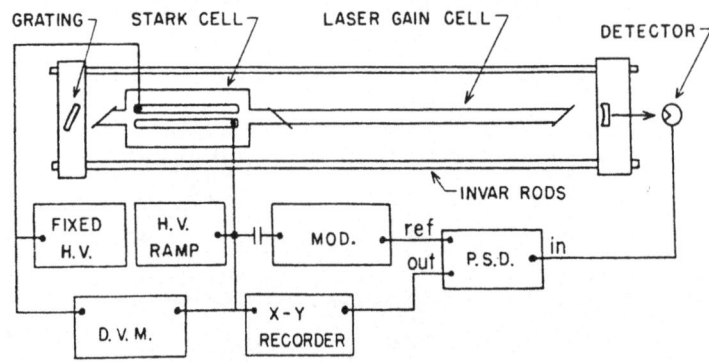

Fig.4　　Schematic diagram of an intracavity laser Stark spectrometer. PSD stands for phase sensitive detector, DVM for digital voltmeter, HV for high voltage, and MOD for modulation source. (Reproduced by permission from Journal of Chemical Physics, 1977 $\underline{66}$, 127)

the presence of intense counterpropagating beams in the sample can saturate many vibration-rotation transitions, leading to the production of sub-Doppler 'Lamb-dips', giving an enhancement of resolution. These dips occur when the laser frequency coincides exactly with the trans-ition frequency of a molecule at rest, so that both beams are competing to excite molecules with no longtitudinal velocity component. Thus, whereas the resonances depicted in figure 2 have a width determined by the heterogeneous Doppler shift (fwhm = 117 MHz in this instance), the Doppler-free resonance widths are limited by the greater of the laser bandwidth or the homogeneous molecular width, governed mainly by pressure broadening. In order to reach this latter limit of ~1MHz it is essential that the Stark electrodes should be flat and parallel to one or two fringes of visible light, so that the Stark shifts are adequately uniform over the whole of the experimental volume.

Figure 5 shows Lamb dips on the various M components of a line in the ν_3 band of the reactive intermediate HNO, recorded with a CO laser line at 1500.69 cm^{-1} (Johns and McKellear, 1977). A major contribution to the Stark tuning in this instance involves the near coincidence of the ground state rotational levels 6_{06} and 5_{15} which are coupled through the interaction of the b axis component of the dipole moment (μ_b) with the field, with the matrix element:

$$\langle J,K,M|H'|J-1,K+1,M\rangle = \frac{\mu_b E}{J}\left[\frac{(J+K+1)(J+K+2)(J^2-M^2)}{(2J-1)(2J+1)}\right]^{\frac{1}{2}} \tag{5}$$

The use of electric field modulation in recording this spectrum greatly enhances the peak heights of the sharp Doppler-free Lamb dips relative

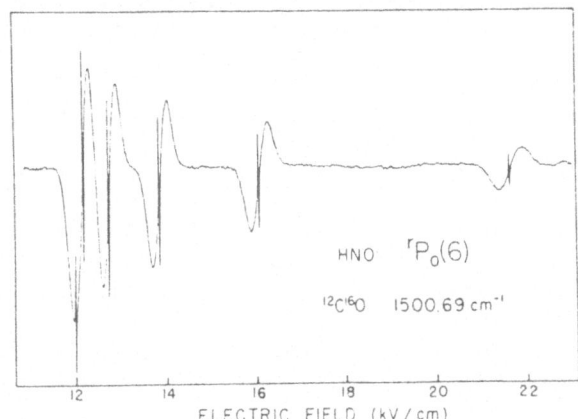

Fig. 5   Intractivity laser Stark spectrum of the ν_3 band of HNO
recorded with the 24-23 (P(16) line of a CO laser and
parallel polarization. (Reproduced by permission from
Journal of Chemical Physics, 1977, 66, 1220.)

to the broader Doppler profiles. The first two M-components are
clearly resolved as Lamb dips, but completely overlapped in the Doppler
broadened line profiles, thus emphasizing the resolution advantage of
saturation spectroscopy.

The detectors used in these experiments are quantum-limited liquid
nitrogen cooled devices, PbSnTe or CdHgTe in the 10μm region, and InSb
or Au doped Ge in the 5μm region. Such laser spectrometers have very
high sensitivity, and it has been estimated that in an intracavity
spectrometer with a pressure of 4m Torr in a 20 cm long absorption cell
only about 10^6 molecules are responsible for the Lamb dips seen on a
particular vibration-rotation transition. In consequence laser Stark
spectroscopy has proved invaluable in the study of the vibration-
rotation bands of a number of short-lived molecules generated in
chemical reactions, such as HCO, CH_2NH, H_2CS and HNO_2 (Duxbury 1985).

In the experiments described above the Stark splitting is a necessary adjunct to bringing a molecular transition into resonance with a
fixed frequency laser, and the observation at high resolution of many
different M components makes for reliable extrapolation back to the
zero field transition frequencies. Given many laser lines, as with CO,
CO_2 or N_2O lasers, it is possible to build up a detailed picture of a
vibration rotation band from the glimpses obtained through the many
"windows" that this experiment provides.

Duxbury (1985) has recently published an extensive review of laser
Stark spectroscopy. Table 1 lists most of the molecules that have been
studied in this way. The majority of these have closed-shell singlet
ground states, and their Stark splittings follow eqns 2-5, at least for
weak fields. However, one of the more reactive species, HCO, has a
doublet ground state. The presence of the unpaired electron spin complicates the energy level pattern, but the same principles hold as for
a singlet state.

Despite the successes of these studies with fixed frequency lasers, there are nevertheless a number of advantages to the use of truly tunable lasers.

Table 1   Molecules studied by laser Stark spectroscopy[*]

Molecule	Laser	Molecule	Laser	Molecule	Laser
DBr	CO	AsH_3	CO_2	SiH_3F	CO_2
NO	CO	H_2CO	CO	CH_2F_2	CO_2
$DCN^\dagger$	CO	HDCO	CO	POF_3	CO_2
CO_2	F-centre	D_2CO	CO, CO_2	HCOOH	CO
OCS	CO_2	HFCO	CO	$CH_2F_2CF_2$	CO_2, N_2O
FCN	CO_2	CH_3D	CO_2	CH_3OH	CO_2, HCN
D_2O	HCN	SiH_4	CO_2	$CH_3CN^\dagger$	CO_2
$H_2O^\dagger$	CO	GeH_4	CO_2	CH_3CCH	CO_2
FNO	CO	$CH_3F^\dagger$	CO_2, HeNe	HNO	CO
ClO_2	CO_2, N_2O	CD_3F	CO_2	DNO	CO
HCCF	CO_2	CH_3Cl	CO_2	HCO	CO
DCCF	CO_2	CD_3Cl	CO_2	CH_2NH	CO, CO_2
$NH_3^\dagger$	D_2O, CO, CO_2, N_2O	CH_3Br	CO_2, N_2O	H_2CS	CO_2
NH_2D	CO_2	CD_3Br	CO_2	D_2CS	CO_2
ND_3	DCN, CO_2	CD_3I	CO_2, N_2O	HNO_2	CO
PH_3	CO_2, N_2O	CDF_3	CO_2	CH_2NOH	CO_2

[*] See Duxbury (1985) for details and references.
[†] and other isotopers

2.4   Electric field spectroscopy

Infrared diode lasers are now available which operate in the spectral region of vibration-rotation bands, and permit continuous tuning. The power is usually insufficient to saturate transitions, so that the resolution is then limited by the Doppler effect. In some molecules the line density in a vibration-rotation band is sufficiently high that the average spacing is less than the Doppler width of each line so that only a band envelope is obtained. Even where this is not the case, it may prove impossible to give an unambiguous analysis from a simple absorption spectrum.

However, it is well known from classical studies with optical or microwave spectrometers that the J,K and branch dependence of Stark splittings is a valuable aid to band assignment (Bridge, Haner and Dows 1968). Equations 2-4 indicate that for a symmetric or near symmetric top the level splittings are greatest for levels with J=K. If the electric vector of the laser beam is parallel to the Stark field ($\Delta M=0$) the line-strength factors for the various M components are highest for high $|M|$ in Q branches, but for low $|M|$ in P or R branches. The Stark splittings are therefore most apparent for Q lines with J ≈ K. Conversely, in perpendicular linear polarisation ($\Delta M = \pm 1$) the line strength factors are highest for low $|M|$ for Q branches, but for high $|M|$ in P or R branches, and the Stark splittings are most apparent for P and R lines. This selective behaviour is the basis of electric field spectroscopy. In this technique the electric field between a pair of electrodes is generated using an a.c. voltage supply, and the modulated component of the beam transmitted through the sample is detected using a lock-in amplifier. When the tunable laser is scanned across a vibration-rotation band, the strongest signals will be centred on those lines that conform to the rules given above. The ability to use both parallel and perpendicular polarisations has a clear diagnostic value.

Figure 6 presents an application of this technique to the analysis of a vibration-rotation band of nitric acid vapour (Webster, May and Gunson, 1985). The congested nature of the centre of this band is

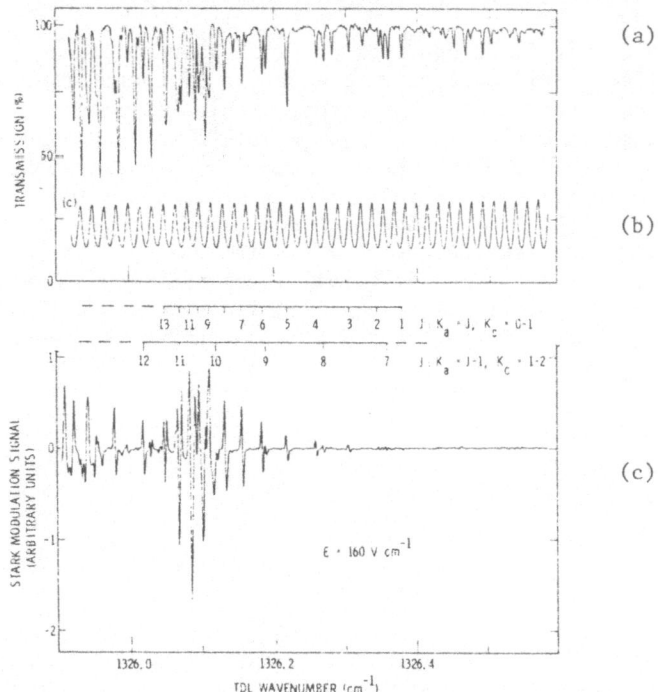

Fig.6   (a) TDL direct absorption, (b) etalon transmission, and (c) Stark modulation spectra of HNO_3 near 1326 cm^{-1}. The fringe FSR is 0.0160 cm^{-1}, and the peak-to-peak applied field for the lower trace is 160 V cm^{-1}. (Reproduced by permission from Chemical Physics Letters, 1985, 121, 429)

evident from the large number of lines in the 0.7 cm^{-1} section near the strong central Q-branch, recorded here with a tunable diode laser and a 20 cm absorption cell. To record the Stark modulation spectrum a zero based square wave voltage at 2 kHz was applied to 10 cm electrodes, giving a peak field of up to 800 V cm^{-1}. This shows an obvious simplification compared with the absorption spectrum. First derivative line-shapes are associated with lines showing Stark shifts, whereas second derivative line-shapes arise from more symmetrical Stark splittings. The electric dipole moment of HNO_3 in its ground state is very large

(2.17 D) with components μ_a = 1.99 D and μ_b = 0.88 D. Consideration of the magnitude of the asymmetry doublings δ for various J and K_a show that the strong electric field activity near 1326.1 cm^{-1} arises from Q-branches with K_a=J and J-1. This provides the key to the band analysis. Variation in the magnitude of the electric field brings up further Stark modulated lines, adding to the list of definitive assignments. However, in this experiment the laser output was unpolarised, so that the ability to discriminate between Q and P or R branches on polarisation characteristics was not available in this instance.

2.5   Excited state dipole moments

Electric field spectroscopy with a tunable laser uses the Stark effect to provide selective modulation of specific branches, but does not permit the direct measurement of dipole moments. There are a number of molecules in which a tunable laser has been used to scan Stark-split line profiles of lines of known assignment. One such example is illustrated in figure 7: this shows the five M components of the $^P P_3$ (3) line in the electronic 0-0 band of the HCF radical near 17238.4 cm^{-1}, recorded using a single-mode dye laser and an electric field of 19.25 kV cm^{-1} in parallel polarisation. The analysis of many Stark-split spectra using eqns 2-4 leads to the following values for the a-axis component of the dipole moment in the two states:

$$\mu_a{}' = 0.488 \pm 0.005 \text{ D} ; \quad \mu_a{}'' = 0.061 \pm 0.005 \text{ D}. \qquad (6)$$

(Dixon and Wright 1983). The large change on excitation gives valuable information concerning the electronic structure of HCF.

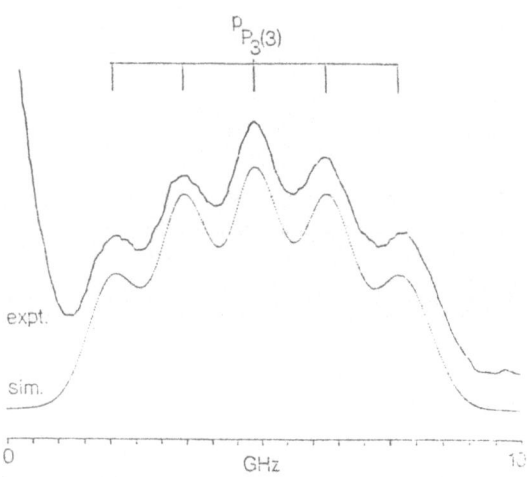

Fig.7   The $^P P_3$(3) transition at 17238.4 cm^{-1} in parallel polarisation at 19.25 kV/cm with a simulation using the derived dipole moments. This shows the five M_J components ($|M_J|$=2,1,0). (Wright, Ph.D. thesis, Bristol University, 1985)

Similar studies have been made for the species HNO (Dixon and Noble 1980), NHD (Brown, Chalkley and Wayne, 1979), HSO (Webster, Brucat and Zare, 1982), and H_2CS (Dixon and Gunson, 1983), each using a tunable single-mode c.w. dye laser.

3. LASER MAGNETIC RESONANCE

The second method of tuning molecular transitions into coincidence with a fixed frequency laser is to make use of the Zeeman effect. The magnetic moment of a molecule consists of two types of contribution. Those from unpaired electron spin angular momentum $\underset{\sim}{S}$ and the electron orbital angular momentum $\underset{\sim}{L}$ are of the order of magnitude of the Bohr magneton $\beta = 1.3996$ MHz/G or 0.4669 cm^{-1}/T. In an applied magnetic field $\underset{\sim}{B}$ the leading term in the Zeeman hamiltonian is given by:

$$H_Z^e = g_S \beta \underset{\sim}{S}.\underset{\sim}{B} + g_L \beta \underset{\sim}{L}.\underset{\sim}{B} \qquad (7)$$

where $g_S = 2.0023$ and $g_L = 1.0$. In addition there are contributions from molecular rotation and nuclear spin angular momenta, but these are of order of magnitude of the nuclear magneton β_N, which is only $5.15 \times 10^{-4}\beta$. There is no first order energy contribution from H_Z^e for molecules in closed-shell singlet states.

For practical purposes the available magnetic fields (~2T) therefore restrict the application of Zeeman tuning to open-shell molecules with non-zero spin or orbital quantum numbers, which molecules are generally very reactive. Laser magnetic resonance spectrometers are therefore designed to have very high sensitivity, usually by employing intracavity absorption cells. The early spectrometers used far infrared (FIR) HCN and H_2O discharge lasers as sources. More recent developments have been the use of CO_2, N_2O and CO lasers as mid-IR sources for the detection of vibration-rotation transitions, and optically pumped FIR lasers (of which nearly 2000 laser lines are now known) for pure rotational transitions.

A CO_2 laser pumped FIR LMR spectrometer is sketched in figure 8. In this design (Radford and Litvak 1975, Evenson 1981) the CO_2 laser beam is introduced into the FIR gain cell through a side window and is multiply reflected within a copper or gold-plated-glass tube to increase the gain. The absorption cell is placed between the pole pieces of a magnet, with a thin polypropylene sheet at Brewster's angle to separate it from the gain cell. This window determines the polarisation of the FIR laser with respect to the magnetic field. The electric vector parallel to $\underset{\sim}{B}$ gives π electric dipole transitions ($\Delta M_J = 0$); perpendicular gives σ transitions ($\Delta M_J = \pm 1$). The transient magnetically active species is normally generated by the reaction of a stable molecule with the discharge products of a second species, the reaction mixture being pumped rapidly through the cell. The wavelengths of FIR laser transitions are sufficiently long, with narrow gain profiles, that it is necessary to control the cavity length with a micrometer, the setting of which will determine the active laser transition. The use of a modulation coil giving a few G peak-to-peak at 1-100 kHz permits the

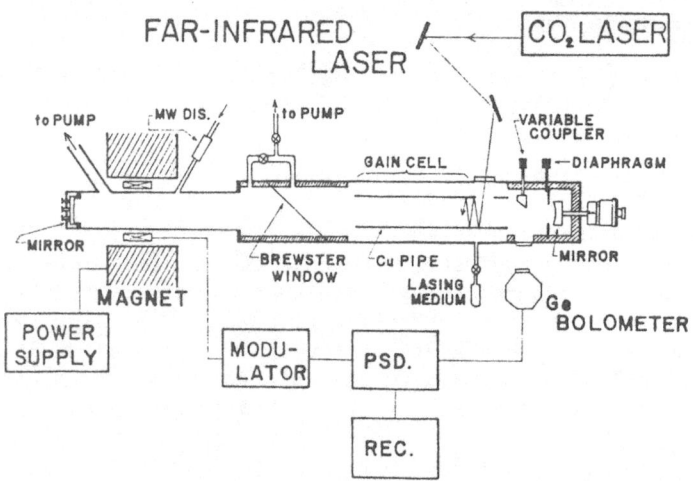

Fig. 8   Block diagram of a FIR LMR spectrometer

use of phase-sensitive detection of the FIR output.   The sensitivity
of such spectrometers has been estimated as 10^6 OH radicals/cm^3.

Mid-IR discharge operated LMR spectrometers are of very similar
design, but with a CO_2, N_2O or CO laser gain tube in place of the FIR
gain cell, and a grating instead of the micrometer-controlled mirror.

3.1   The Zeeman effect

A molecular energy level with total angular momentum J splits into
(2J+1) Zeeman components in a magnetic field $\underset{\sim}{B}$.   If this level is well
removed from other levels of the same parity the Zeeman splitting will
be linear in the field, with:

$$\Delta E = g\beta BM.$$ (8)

g is the Landé factor which depends on the coupling scheme of the
different angular momenta.   In such a case the frequency of a transition $(v',J',M') - (v'',J'',M'')$ is tuned by the magnetic field from its
unperturbed value ν_o according to:

$$\nu = \nu_o + \beta(g'M' - g''M'')B/h$$ (9)

This is closely analogous to the electric field tuning illustrated
in figure 3, but with one important difference.   The Stark hamiltonian
(eq.1) couples levels of opposite parity, so that a first order Stark
splitting is only obtained for degenerate pairs of levels of opposite
parity, such as K-doublets.   In contrast the Zeeman hamiltonian can
have diagonal matrix elements with defined-parity wavefunctions, and
such a degeneracy is not necessary.   It should be stressed that most
LMR spectra involve <u>electric</u> dipole transitions, even though the tuning
is magnetic.

One class of molecules that conforms to this simple first-order pattern comprises linear molecules in $^1\Pi$ or $^1\Delta$ states. For example, mid-IR LMR has been used to record the vibration rotation spectrum of SO, which species was generated by the reaction of OCS with a large excess of discharged O_2 (Yamada et. al. 1978).

The magnetic behaviour of many molecules with unpaired electron spins is qualitatively different from that described above. If the molecule is linear and in a Σ state, or non-linear, the rotational levels can be described in terms of a quantum number N for the total angular momentum apart from electron spin (this is Hund's case b). Each state with N>0 is then split by spin-rotation interaction, and spin-spin interaction if $S>\frac{1}{2}$, into (2S+1) close-lying J-states. Each J-state will have a linear Zeeman splitting into its various M-components for very weak fields. However, at the typical fields of LMR experiments the spin fine-structure splittings and the Zeeman splittings are often comparable in magnitude, leading to strong non-linearity in the Zeeman tuning. At high fields the electron spin becomes decoupled from N - the Paschen-Back effect - so that M_S becomes a good quantum number, and the magnetic sub-levels separate into (2S+1) groups.

The LMR spectrum of the $6_{43}-5_{32}$ rotational transition of the $\tilde{X}\,^2B_1$ ground state of the PH_2 radical (figure 9a) illustrates this behaviour. (Davies, Russell and Thrush, 1976). The appropriate energy level diagram is given in figure 9b. At 14 kG the two groups of levels corresponding to $M_S=\pm\frac{1}{2}$ are well separated, and the small splittings derive from the fine-structure interactions. Figure 9 also illustrates a further characteristic of LMR spectra, namely hyperfine splitting. For this pair of para rotational levels the two proton spins are coupled to zero, so that the hyperfine splitting arises solely from coupling of the ^{31}P nuclear spin to the unpaired electron spin, giving doublets. The observation of hyperfine splittings in molecular spectra not only provides valuable data concerning electronic structure, but is also an important diagnostic in identifying the molecular species responsible for previously unknown spectra.

3.2 The analysis of LMR spectra

The variety of coupling energies in open-shell molecules is such that the analysis of LMR spectra is accomplished by numerical fitting of the observed transitions to the eigenvalues of a model hamiltonian, with many more parameters than for equivalent closed shell singlet states. Nevertheless, the availability of a large number of laser lines both in the mid infrared and in the far infrared has permitted the full parameterisation of the spin/rotational hamiltonian in many cases. For example, for NH_2 $\tilde{X}^2B_1$ there are 3 rotational constants, 5 centrifugal distortion constants, 3 spin-rotation interaction constants, 4 Zeeman parameters, and 3 hyperfine constants for each of ^{14}N and ^1H, plus some higher order terms (Davies et. al., 1977). Hirota (1985) has given a detailed account of the setting up of the hamiltonian for such molecules, and summarises the analysis of a number of LMR spectra.

294

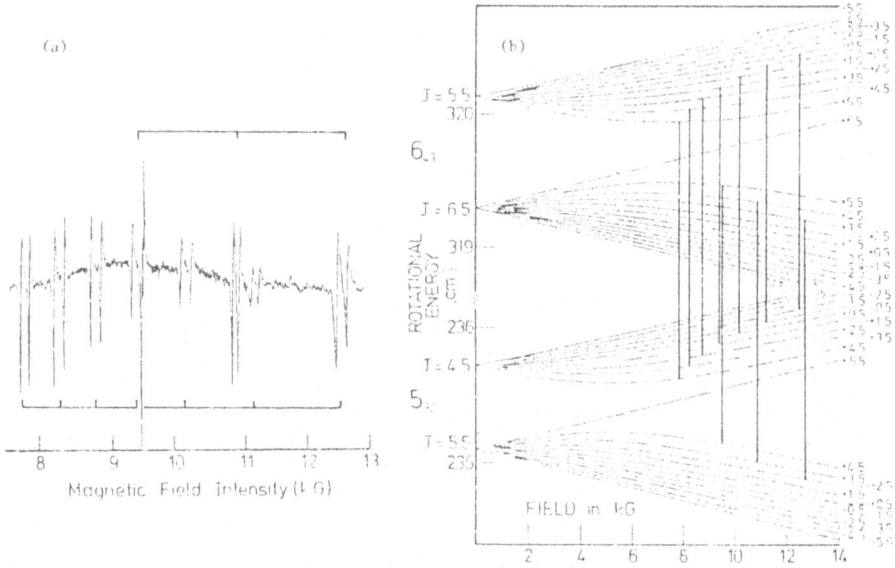

Fig.9   An LMR spectrum of PH$_2$ obtained with the 118.6 μm water
vapour laser. (a) Spectrum recorded with the electric vector
of the laser radiation perpendicular to the magnetic field
($\Delta M=\pm1$). The doublets arise from ^{31}P hyperfine splitting.
(b) Assignment as transitions between the Zeeman components
of the 5_{32} and 6_{43} rotational levels of PH$_2$. (After Davies
et. al. 1976)

3.3 Zeeman modulation of infrared spectra

We have already discussed the use of electric field modulation as
a means of providing the selective detection of those molecular absorption lines having the strongest Stark effects, using tunable lasers as
sources (§2.4). As in microwave spectroscopy, where field modulation
is routine, magnetic fields may also be used for this purpose, as was
demonstrated by Urban and Herrmann (1978). The spectrum of NO was
recorded with extremely high sensitivity using Zeeman modulation with
a spin-flip Raman tunable infrared laser.

Zeeman modulation has a great advantage over other techniques in
the study of reactive intermediates generated in chemical reactions or
discharges. Whereas all but centro-symmetric molecules have a dipole
moment, and therefore have levels with significant Stark effects, only
molecules with open shells have magnetic moments of the order of Bohr
magnetons, with the accompanying strong Zeeman effect. Zeeman modulation therefore discriminates strongly in favour of the detection of
free radicals, even though these may be present in very low mole fract-
ions compared to the precursors and products of their reactions.

An example is provided in the detection of the ν_1 (OH stretching)
band of HO$_2$ near 3400 cm^{-1} by Yamada et. al. (1983). The Zeeman effect

is fairly large for this spectrum, because the spin-rotation splitting is large and the selection rules are b type ($\Delta K_a = \pm 1$), making the recorded line shape almost a first derivative. However care must be taken in the rotational analysis to take account of non-linearity in the Zeeman splittings which may slightly displace the observed peaks from their zero-field frequencies.

The spectrum near 600 cm^{-1} of the ν_2 band of the CH$_3$ radical generated by a glow discharge in di-tert-butylperoxide (Yamada et. al. 1981) provides a second example of this selectivity. With Zeeman modulation all the observed lines arise from CH$_3$, but with source frequency modulation many additional lines from diamagnetic species are detected in the same spectral region.

4. LASER INDUCED FLUORESCENCE

The detection of laser induced fluorescence, first introduced to molecular spectroscopy near twenty years ago, has now become one of the standard techniques for recording the electronic spectra of gases, or of molecules trapped in inert matrices. Line sources, such as the argon ion laser, can only excite levels which give accidental coincidence with their essentially fixed frequencies, but dispersion of the emission can give extensive information on the higher levels of molecular ground states. Although limited tuning via the Stark or Zeeman effects have been used with these line sources, almost all current studies employ tunable dye lasers. The highest resolution (~1MHz) is only obtained with continuous wave (cw) dye lasers, the various dyes giving a coverage from about 400 to 900 nm. On the other hand pulsed dye lasers give peak powers sufficient for a wide range of non-linear processes, including multiphoton excitation and harmonic generation of ultraviolet wavelengths. In addition they permit the direct measurement of the temporal evolution of the molecular system following an excitation pulse. A recent compromise between these two types of dye laser has been the use of a dye amplifier pumped by a pulsed Nd:YAG or excimer laser with injection seeding from a narrow-band c.w. dye laser. Such a system can produce laser light with a bandwidth close to the Fourier transform of the duration of the pulsed pump laser, that is ~100 MHz (Riedle et. al. 1982), and with a peak power sufficient to permit harmonic generation of u.v. light followed by multiphoton excitation at sub-Doppler resolution (Ashfold et. al. 1986).

A block diagram of a typical LIF experiment using a c.w. dye laser is given in figure 10. An unstable radical (in this case HSiF, Dixon and Wright 1985) is generated in a glass cell by the reaction of discharged fluorine with silane. Brewster angle windows and trains of baffle discs serve to minimise the scattering of dye laser light within the cell, and the fluorescent strip of excited gas is imaged onto the photomultiplier detector through a mask. Most molecular gases fluoresce over an extended wavelength range, and it is common to use a long pass filter to further block dye laser scatter. Even though this partially attenuates the fluorescence reaching the detector, it usually enhances the signal to noise ratio, as does chopping the beam and using phase sensitive detection. A wavemeter which measures the ratio of the dye

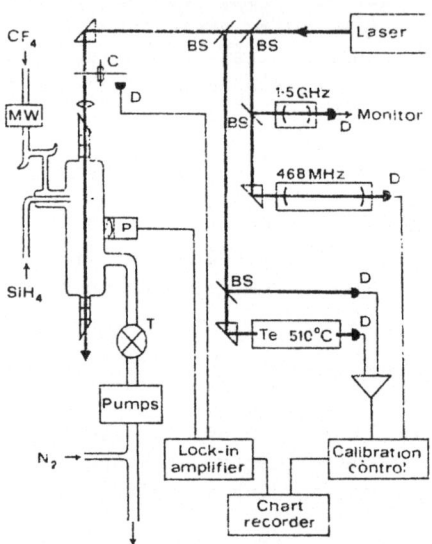

Fig.10   Schematic diagram of the experimental arrangement for laser
          induced fluorescence. BS, beam splitter; C, chopper; D, photo-
          diode; MW, microwave discharge; P, photomultiplier; T, throttle
          valve.

laser wavelength to that of a reference HeNe laser provides a convenient
method of setting the dye laser, but is less useful while scanning.
The absorption spectra of I_2 (Gerstenkorn and Luc 1978,79) and of
$^{130}Te_2$ (Cariou and Luc 1980) have been measured in the visible region
with a precision of 10^{-3} to 10^{-4} cm^{-1} by Fourier transform spectroscopy
and serve collectively to give excellent secondary wavenumber standards
between 14000 and 24000 cm^{-1}.   Interferometer fringes provide convenient
interpolation between these calibration lines.
     Pulsed dye laser experiments are not greatly different in design.
The laser pulses are typically ~10 ns duration, as is the time response
of the detection system.   Thus for excited state lifetimes greater than
about 50 ns, it is possible to avoid the detection of scattered light
by the use of a Boxcar amplifier operating in a delayed linear gate
mode.   Wavenumber calibration is more of a problem outside the ranges
for I_2 or Te_2.   However, most pulsed dye lasers employ diffraction
gratings with sine-bar drives, facilitating linear interpolation
between atomic calibration lines detected optogalvanically.
     In the arrangement described above the fluorescence serves merely
to indicate that excitation has taken place, the only requirement being
that a photon is emitted within the pass band of the detection system.
A very large number of molecular spectra have been recorded in this
manner and analysed in detail, but in some cases it has proved invaluable to obtain additional information by dispersing the fluorescence.

4.1 Dispersed laser induced fluorescence

     The dispersion of laser induced fluorescence has three main

applications; (i) the detection of emission to high lying levels of the lower state, (ii) the study of intramolecular or collisional processes that transfer population from the initially excited level to others, leading to a time evolution of the fluorescence spectrum, and (iii) the use of narrow band detection to deconvolute overlapping excitation spectra. We will describe two examples of this third application.

The BO_2 radical has an extended $\tilde{A}\ ^2\Pi_u$-$\tilde{X}\ ^2\Pi_g$ band system in the blue and green regions of the visible spectrum which was first characterisation using flash photolysis absorption spectroscopy (Johns 1964), and which has been reinvestigated in more detail by a number of authors using dye laser excitation. Some of these studies have used narrow band dye lasers to give a resolution limited only by the Doppler effect, but this is still not sufficient to avoid overlapping lines and bands. Dixon et. al. (1977) dispersed the emission from BO_2 excited in band heads to give unambiguous vibronic assignments, and were thereby able to analyse the Fermi resonance between the (100) and (020) vibronic levels of the $\tilde{X}\ ^2\Pi_g$ state. Beaudet et. al. (1979) have used a higher resolution monochromator to deconvolute unresolved rotational structure in a similar manner. Each band has a dense region of overlapping R branch lines, but a more open P branch. Thus by monitoring successive P branch lines through the monochromator it is possible to resolve the R branch excitation profiles one line at a time. Figure 11 shows the

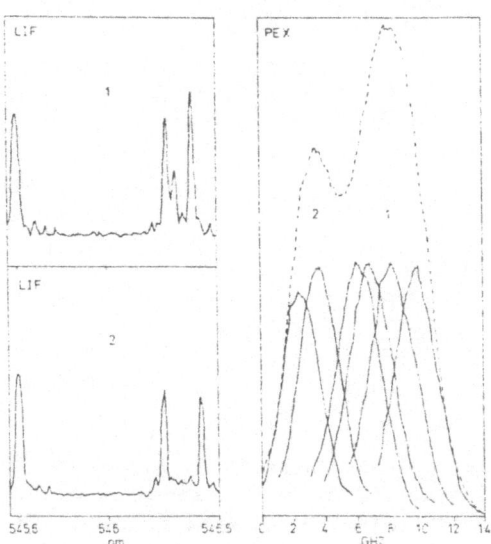

Fig. 11   Left panel: Dispersed LIF spectra of $^{11}BO_2$ excited at two different points within the head at 579 nm of the $\tilde{A}$ - $\tilde{X}$ (000) - (100) band, with emission of the (000)-(000) band. Right panel: Simple excitation spectrum of the band head (dotted curve), and its deconvolution into separate R branch transitions by monitoring individual P lines (solid curves).(Reproduced with permission from Chem. Phys. Lett. 1979, 60, 489)

experimental deconvolution of a band head of $^{11}BO_2$ at 579 nm into six separate rotational lines.

A very different example is provided by the spectrum of the Rydberg states of ammonia. The ground $\tilde{X}$ $^1A'$ state of ammonia is pyramidal, whereas all its lower excited states are planar, so that each electronic excitation is accompanied by a long progression in the out of plane bending vibration v_2. This causes considerable congestion, with low v_2-0 transitions of one band system overlapped by higher v_2-0 bands of an electronic transition with a lower system origin. Some of these states have been found to fluoresce, despite lying well above the lowest dissociation limit, the emission being predominantly inter-Rydberg in character. Two examples are the $\tilde{C}'$ $^1A_1'$ - $\tilde{A}$ $^1A_2''$ and the $\tilde{B}$ $^1E''$ - $\tilde{A}$ $^1A_2''$ band systems, both of which involve a $3p\rightarrow3s$ one-electron transition. In contrast to the absorption spectrum, these transitions are between pairs of planar states, and are dominated by vibrational transitions with $\Delta v=0$, giving narrow regions of emission near 570 nm for $\tilde{C}' \rightarrow \tilde{A}$ and near 740 nm for $\tilde{B} \rightarrow \tilde{A}$ irrespective of the value of v_2. Stickland (1987) has used a small monochromator to disperse the fluorescence of NH_3 or ND_3 following two-photon excitation with pulsed dye laser light at a wavelength of 330 nm. By first monitoring emission near 570 nm, and then near 740 nm, he has completely deconvoluted the overlapping C'-X and B-X excitation bands.

4.2 The polarisation of laser induced fluorescence

The molecules of a gas sample in thermal equilibrium will be isotropically oriented in space, as will their angular momentum vectors. If, however, such a sample is excited with a directional laser beam, and particularly a polarised laser beam, the vectorial nature of the interaction between the molecule and the light will usually give rise to an anisotropic excited state distribution. Subsequent fluorescence will then result in polarised emission. Greene and Zare (1982,1983) have shown how this polarisation can be related to the moments of the angular momentum distribution, with the dominant contribution given by the second moment $A_0^{(2)}$ (the alignment) when using linearly polarised one-photon excitation, where for each J_i

$$A_0^{(2)}(J_i) = <(J_i|3J_z^2 - J^2|J_i)>/J_i(J_i+1) \qquad (10)$$

For a given experimental arrangement the expression for the polarisation involves terms which are products of moments of the distribution, transition dependent factors, and geometrical factors. In general Q branch transitions have an opposite behaviour from P and R transition, as has also been noted in section 2.4. The moments of the distribution can be determined either by comparing intensities of lines in different branches for one geometrical arrangement, or by comparing the polarisation and intensity of light emitted in different directions.

For prompt fluorescence in a collision-free environment the moments are directly related to the anisotropy of the molecule-light interactions, so that the polarisation can be predicted a priori. However, when an additional dynamical process intervenes between excitation and

fluorescence the measurement of the polarisation of the emission can
lead to detailed vectorial information about this process. Two important applications are to collisional energy transfer and to photodissociation. A good example of the latter process concerns the alignment of
the OH radicals generated by photodissocation of HONO in the near
ultraviolet (Vasudev et. al. 1984). In this case the OH radicals are
produced in the ground state, and their rotational anisotropy is probed
via LIF. Two sets of measurements were made with the same LIF arrangement, but in one the electric vector of the photolysis light was directed towards the detector, whereas it was perpendicular to this direction
for the second set. The OH showed a modest degree of rotational excitation, with negligible alignment at low J becoming more and more positive as J increased. Thus the OH is produced with its rotational vector
preferentially parallel to the direction of the transition moment of the
parent HONO. This latter is known to be perpendicular to the molecular
plane ($^1A''$ - $^1A'$ transition). Thus it can be concluded that the fragmentation process produces in plane rotation of the OH radical.

4.3 Competitive decay routes for excited states

An LIF excitation spectrum is only equivalent to an absorption
spectrum if the fluorescence quantum yield is the same for all excited
levels. This will not necessarily be the case if other dynamic processes can compete with the decay of the excited state population through
fluorescence emission. Some of the possibilities are:

Molecular excited state → Lower state + photon (fluorescence)
→ Lower state (Collisional quenching)
→ Fragments (Photodissociation)
→ Ionic state (Multiphoton excitation)

(11)

Figure 12 gives an example where dissociation provides the competitive
pathway provided that the excited state energy exceeds a threshold.
(Dixon et. al. 1981). The LIF excitation spectrum of HNO in figure 12b
exhibits a "breaking off" of the rotation structure above K'=4, J'=11 in
the (100) excited state, whereas the structure continues to higher K'
and J' in the absorption spectrum of figure 12a. The analysis of many
such breaking off limits has resulted in the evaluation of the ground
state dissociation energy D_0^0(HNO→H+NO) to within ±10 cm^{-1}.

The competition between dissociation and fluorescence produces
such a dramatic effect in this case because the radiative lifetime is
very long (~23 μsec). The $\tilde{C}$ $^1B_1$ state of H_2O furnishes a very different
example. The photodissociation dynamics of this state have been probed
using multiphoton excitation, with detection either of ions or of
fluorescence. The 3+1 resonance enhanced multiphoton ionisation
spectra of the $\tilde{C}$ states of H_2O and D_2O are dominated by levels with low
K_a, and particularly K_a=0 (Ashfold, Bayley and Dixon 1984). The two-
photon excitation spectrum for parent molecule fluorescence at 420 nm
($\tilde{C} \rightarrow \tilde{A}$) shows the same characteristics (Docker et. al. 1986). In
contrast, parent molecule transitions to K_a=0 are completely absent in
the excitation spectrum for emission near 310 nm by OH photofragments.
The explanation of these observations is that, although the $\tilde{C}$ state of

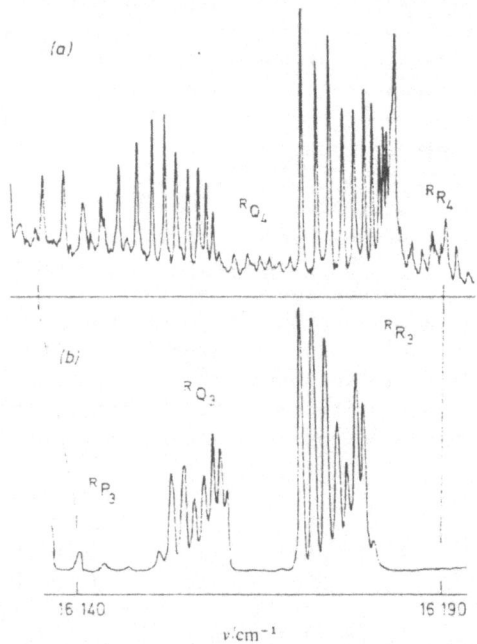

Fig. 12   The (100)-(000) K=4-3 sub-band of the Ã $^1A''$ - X̃ $^1A'$ band
system of HNO. (a) Microdensitometer trace of a photograph
of the absorption spectrum. (b) LIF excitation spectrum. The
weak $^rQ_4$ and $^rR_4$ branches in (a) are of the (020)-(000) K =
5-4 sub-band. (After Dixon et. al. 1981)

$H_2O(D_2O)$ lies well above the threshold for dissociation to H+OH, A $^2\Sigma^+$,
this dissociation is symmetry forbidden in a non-rotating molecule,
but increasingly competes with the other decay routes when the molecule
rotates about its a-axis.

The rate of dissociation from an excited state of a polyatomic
molecule may also be modified to different extents by the excitation of
its vibrational modes. This has the result that the relative intensities
of the various vibrational peaks in an excitation spectrum depends on
the channel that is being monitored.

4.4.   Resonance Raman scattering

It is common practice to refer to "resonance fluorescence" when a
molecule is excited to emission within the width of a single rovibronic
transition, and Raman scattering when the excitation is off-resonance.
The signature of the first processes is that it only contains rovibrational transitions which emanate from the specifically populated level,
whereas the second gives a complete rotational distribution with intensities related to the ground state Boltzmann factors. Ziegler (1985,
1987) has recently obtained gas-phase resonance Raman scattering spectra
of ammonia   intermediate in character between these two limits, and

has described a detailed theory for their interpretation.

The essential aspect of these experiments is that the rovibrational levels of the Ã ^{1}A" state of NH$_3$ (ND$_3$) are sufficiently short-lived that the bands are rotationally diffuse, but not sufficiently so as to wash out all structure from the band envelopes. In consequence, a fixed laser excitation frequency within the band profile can encompass a range of broadened lines, with the strongest scattering through those levels closest to resonance. Furthermore, the interference between the scattering through different excitation branches (e.g. P↑Q↓ and Q↑P↓) requires the introduction of an antisymmetric scattering tensor as well as the more usual symmetric tensors. The beauty of these experiments is that the breadth of the Ã state levels becomes manifest through the relative intensities of the fully resolved rotational structure in the Raman spectrum. From these spectra Ziegler has made a detailed study of the dissociation rates in NH$_3$ and ND$_3$.

5. DOUBLE RESONANCE SPECTROSCOPY

In sections 4.2 and 4.4 emission spectroscopy provided more information than could be obtained from an excitation spectrum. Even so, the weakness of such emission makes it difficult to obtain high resolution in its dispersed spectrum, and data are only obtained about levels lying below that excited. Two-colour double resonance can remove both these limitations in favourable cases. We may envisage three basic schemes for optical-optical double resonance using two lasers.

In figure 13a two sequential excitations lead to detection of a

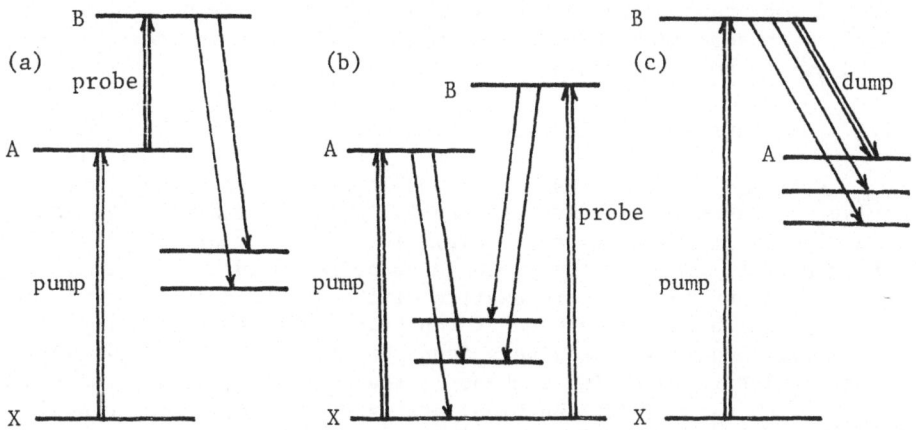

Fig. 13   Level schemes for two-colour optical-optical double resonance: Either the pump laser or the probe laser may be scanned with the frequency of the other laser held fixed. (a) Sequential OODR. (b) Competitive excitation. (c) Stimulated emission pumping.

higher excited state. By holding the pump laser frequency fixed on a known transition and scanning the probe laser a simplified spectrum of the third state will be obtained, consisting of only those transitions

which connect with the known intermediate level. Detection of fluorescence from the uppermost level, or of resonance enhanced multiphoton
ionisation through this level will usually give a signal against a zero
background, and hence a good signal to noise ratio. If, however, the
uppermost state is too short-lived to be detected in this way, then
resonance of the probe transition may be detected via the depletion of
fluorescence from the intermediate state, albeit with poorer signal to
noise ratio.

Sequential double resonance excitation requires that both transitions fall within the range of suitable laser sources. However, it is
often desirable to use population labelling to assign a single band
system where this may not be possible. Figure 13b uses competitive
excitation to seek pairs of transitions with a level in common. In this
case it is necessary to partially saturate the pump transition in order
that it may affect the fluorescence intensity of excitation via the
probe laser. With an extended band system and c.w. lasers the pump
beam may be chopped, with modulation detected on the fluorescence
excited by the unchopped probe, a filter having been used to block the
fluorescence from the pump.

Finally in figure 13c we depict the scheme for stimulated emission
pumping, which is the stimulated equivalent of dispersed LIF. When the
probe is on resonance this stimulates emission in the direction of the
probe laser beam, thereby depleting spontaneous fluorescence. The resonance condition can therefore be sought through detection of fluorescence
side-light without the losses inherent in the use of a monochromator.
This technique has mainly been used with pulsed lasers, and it has
proved important to ratio the fluorescence intensity on a shot-to-shot
basis with some measure of the pump beam intensity in order to obtain
good signal to noise ratio.

5.1. Sequential optical-optical double resonance

King and coworkers used two tunable dye lasers pumped by a single
pulsed N_2 laser to implement sequential OODR in studies of I_2 (Danyluk
and King 1977; King et. al. 1981a), ICl (King et. al. 1979) and Na_2
(King et. al. 1981b). The probe laser pulse could be optically delayed
by up to 24 ns to permit relaxation of the intermediate state if so
desired, the beams being counterpropagated and focussed in the centre of
the cell. Many new states were found in this way. In particular, in
centrosymmetric molecules such as I_2 the overall selection rules are
u-g-u or g-u-g, so that the final states are not directly accessible via
one-photon transitions from the initial states. Many of the new states
found in this way are ion-pair states with very long bond lengths. The
Franck-Condon factors to the lower levels of these states are much more
favourable via an intermediate state than in equivalent transitions from
the ground state where these are symmetry allowed.

Each of the bands in these double resonance spectra consists either
of a single P and a single R line, or has an additional Q branch line,
provided that the pump laser has a sufficiently narrow bandwidth
(0.1 cm^{-1}) to populate a single intermediate level. These lines are
therefore well separated, and even with a modest resolution of the probe

laser (~1 cm⁻¹) the upper state B-values can be determined with good
accuracy because the J-values are known. By recording several such
spectra – to check for perturbations or other abnormalities and to
locate band origins – a fairly detailed analysis of the complete manifold of levels of a new electronic state can be achieved, at least for
diatomic molecules. This ability to correlate lines in an unknown
spectrum with assigned lines in a known spectrum of the same molecule
is one of the great advantages of double resonance spectroscopy.

Similar experiments have been carried out using c.w. dye lasers
to probe excited states of BaO generated in a flame of Ba with CO_2
diluted in Argon (Field 1981). In this case the lasers were both
single mode, so that the pump laser excited a velocity-selected population and the second excitation by the probe laser was therefore sub-
Doppler. The schemes $C^1\Sigma^+$ and $D^1\Sigma^+ \leftarrow A^1\Sigma^+ \leftarrow X^1\Sigma^+$ were used to probe
velocity changing collisions and rotational energy transfer in the A
state (Gottscho et. al. 1980a,b) as well as detailed analyses of the
upper states. Figure 14 shows the dispersed $C^1\Sigma^+ \rightarrow X^1\Sigma^+$ uv fluorescence

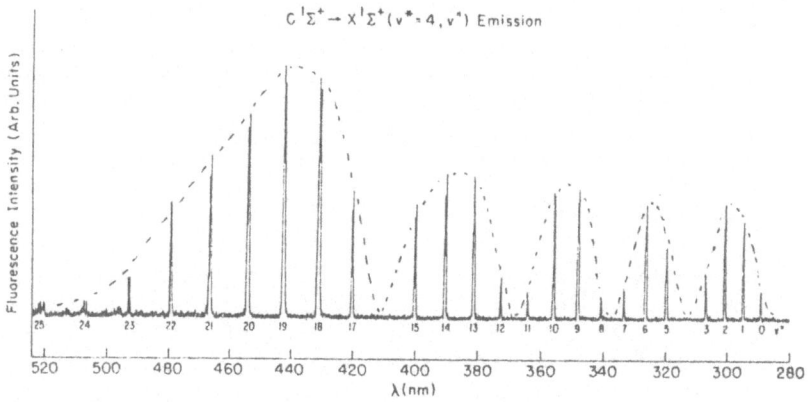

Fig.14   A dispersed emission spectrum of BaO, excited to the C $^1\Sigma^+$
state by double resonance via the A $^1\Sigma^+$ state (Reproduced
with permission from J. Mol. Spectrosc. 1980, 82, 283)

following one such excitation. Each band is a doublet consisting of a
P and an R line. The four minima in the vibrational intensity distribution reflect the pattern of Franck-Condon factors associated with the
oscillations of the upper state vibrational wavefunction, and indicate
that v'=4.

The sub-Doppler nature of double resonance with single mode c.w.
lasers has been used to resolve hyperfine structure in an infrared-
optical study of NH_2 (Amano et. al. 1982). A fluorescence cell was
placed inside the cavity of a CO_2/N_2O laser and between the poles of a
15" electromagnet capable of fields up to 22 kG. A dye laser beam was
introduced into the cell through a ZnSe mirror, and excited NH_2 fluorescence through transitions to levels of the $v_2' = 9$ or 10 states. At
magnetic fields which Zeeman tuned into resonance vibration-rotation
transitions within the excited state the population transfer changed

the fluorescence intensity. The LMR resonances were detected by chopping the IR beam, and demodulating the fluorescence signal with a lock-
in amplifier.

The ability to use double resonance to assign J values has proved
invaluable in a recent study of the $\tilde{B}$ $^2\Pi$ state of NCO (Dixon et. al.
1987). Extensive perturbations in this state had prevented any detailed
analysis in an earlier flash photolysis study of the $\tilde{B}$ $^2\Pi$ - $\tilde{X}^2\Pi$ band
system (Dixon 1960). The $\tilde{B}$ $^2\Pi$ - $\tilde{A}^2\Sigma^+$ - $\tilde{X}^2\Pi$ double resonance was
achieved using two dye lasers pumped by a pulsed Nd:YAG laser; a blue
laser to excite to the $\tilde{A}$ state, and a red laser Raman shifted in H_2 to
the near infrared to further excite to the $\tilde{B}$ state. By this means the
vibrational origin band has been completely assigned and quantitatively
fitted to a model which takes account of perturbations by no fewer than
six high-lying vibronic levels of the $\tilde{A}$ state.

The above examples have all involved long-lived states leading to
fluorescence from the upper excited state. In contrast, the fluorescence depletion method described above has been used to resolve short-
lived levels of the E $^2\Sigma^+$ state of NO with line-widths as great as
10 cm^{-1} (Ashfold et. al. 1986). In this case A $^2\Sigma^+$ - X $^2\Pi$ fluorescence
was excited by a two-photon transition with a blue dye laser, and
$E^2\Sigma^+$ - $A^2\Sigma^+$ resonances were probed with a red dye laser. Even with
these broad resonances it was possible to produce fluorescence dips of
50% - indeed care had to be taken to avoid power broadening because of
the very large transition moment of the E-A inter-Rydberg transition.

5.2. Competitive excitation

The competitive optical-optical double resonance scheme of figure
13b has been used to give the usual great simplification over a simple
fluorescence excitation spectrum in the case of Na_2 (Klein, 1977). The
476.5 nm line of an argon ion laser is coincident with the P(28) line
in the 6-0 band of the B $^1\Pi_u$ - $X^1\Sigma_g^+$ system. A cw dye laser was tuned
across bands of the A $^1\Sigma_u^+$ - X $^1\Sigma_g^+$ system near 615 nm, with detection of
the cross-modulation of the A→X fluorescence induced on resonance by
the chopping of the argon ion pump beam. Figure 15 shows the considerable
reduction in line density achieved in this way.

This technique of population labelling was used by Johnson et. al
(1981) to simplify the crowded spectrum of the BaI $C^2\Pi(a)$ - $X^2\Sigma^+$ band
system, where many of the lines overlap within their Doppler widths.
The pump and probe lasers were tuned to the two spin components of this
transition near 561 nm and 538 nm. Both beams were chopped at different
frequencies, and a beam splitter used with two different filter-photo-
multiplier combinations to detect each spin-orbit component separately.
In this case OODR signals of one phase were detected via the depleted
common levels, with signals of the opposite phase resulting from over-
population of other lower levels after optical pumping plus fluorescence decay. These oppositely phased signals for related lines in
different branches were a great aid to assignment.

In all these double resonance experiments the pump and probe beams
excited transitions well separated in frequency. It is equally possible
for both lasers to excite the same close group of levels. For example,

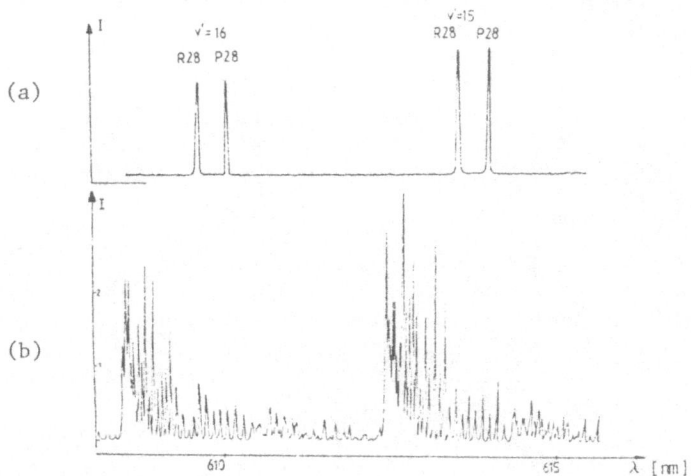

Fig.15   (a) Optical-optical double resonance signals in the spectrum
          of Na₂. The pump laser was stabilised onto a transition in
          the B–X system at 476.5 nm, while the probe laser was scanned
          across band of the A–X system. (b) LIF excitation spectrum
          of the A–X bands. (After Klein, 1977).

Demtröder et. al. (1980) used two slightly tunable argon ion lasers
operating on $\lambda=488$ nm to pump and probe the hyperfine structure of NO_2
in a molecular beam. By this means hyperfine components with a common
lower level were labelled through the OODR signal.

5.3. Stimulated emission pumping

The folded OODR scheme of figure 13c is particularly useful for
reaching levels which are not directly accessible from the molecular
ground state, or which give rise to direct absorption spectra which are
highly congested or severely overlapped. Field and coworkers have used
this technique to investigate high levels of the ground states of H_2CO
(Reisner et. al. 1984) and C_2H_2 (Abramson et. al. 1984), and have found
that the level patterns are strongly affected by Coriolis and anharmonic
coupling. Ashfold et. al. (1987) have used this same approach for a
very different problem – namely the study of line-widths from levels in
the $\tilde{A}\ ^1A''$ states of NH_3 and ND_3. We have already commented in section
4.1 on the detection of $\tilde{C}'\ ^1A_1' \rightarrow \tilde{A}\ ^1A_2''$ fluorescence following two-
photon excitation to the $\tilde{C}'$ state. The use of stimulated emission
pumping gives much greater sensitivity and resolution, the gross simplification of the band structure making it possible to resolve lines with
widths as great as 50 cm^{-1}. This work is completely complementary to
that of Ziegler discussed in section 4.4
    This chapter has discussed a wide variety of techniques which have
found applications in laser-based molecular spectroscopy. New developments in instrumentation and in human ingenuity are continually

306

extending this range, giving increasing insight into the structure and
dynamics of molecules.

REFERENCES

Abramson, E., Field, R.W., Imre, D., Innes, K.K. & Kinsey, J.L. 1984.
    J. Chem. Phys. 80, 2298
Amano, T., Kawaguchi, K., Kakimoto, M., Saito, S. & Hirota, E. 1982.
    J. Chem. Phys. 77, 159
Ashfold, M.N.R., Bayley, J.M. & Dixon, R.N. 1984. Chem. Phys. 84, 35
Ashfold, M.N.R., Bennett, C.L. & Dixon, R.N. 1987. Farad. Disc.Roy.Soc.
    Chem. 82, in press
Ashfold, M.N.R., Dixon, R.N., Prince, J.D., Tutcher, B. & Western, C.M.
    1986. Farad. Trans. 2, Roy. Soc. Chem. 82, 1257
Ashfold, M.N.R., Dixon, R.N., Rosser, K.N., Stickland, R.J. and
    Western, C.M. 1986. Chem. Phys. 101, 467
Beaudet, R.A., Weyer, K.G. & Walther, H. 1979. Chem. Phys. Lett. 60,486
Bridge, N.J., Haner, D.A. & Dows, D.A. 1968. J. Chem. Phys. 48, 4916
Brown, J.M., Chalkley, S.W. & Wayne, F.D. 1979. Mol. Phys. 38, 1521
Cariou, J. & Luc, P. 1980, Atlas du Spectre d'Absorption de la Molecule
    de Tellure, C.N.R.S., Orsay
Danyluk, M.D. & King, G.W. 1977. Chem. Phys. 22, 59
Davies, P.B., Russell, D.K. & Thrush, B.A. 1976, Chem. Phys. Lett. 37,43
Davies, P.B., Russell, D.K., Thrush, B.A. & Radford, H.E. 1977. Proc.
    R. Soc. London A353, 299
Demtröder, W., Eisel, D., Foth, H.J., Höning, G., Raab, M., Vedder,H.J.
    & Zergolis, D. 1980. J. Mol. Struct. 59, 291
Dixon, R.N. 1960. Canad. J. Phys. 38, 10.
Dixon, R.N., Field, D. & Noble, M. 1977. Chem. Phys. Lett. 50,1
Dixon, R.N. & Gunson, M.R. 1983. J. Mol. Spectrosc. 101, 369
Dixon, R.N. & Noble, M. 1980. Chem. Phys. 50, 331
Dixon, R.N., Noble, M., Taylor, C.A. & Delhoume, M. 1981. Farad. Disc.
    Roy. Soc. Chem. 71, 125
Dixon, R.N., Trenouth, M.J. & Western, C.M. 1987. Mol. Phys. 60, 779
Dixon, R.N. & Wright, N.G. 1983. Chem. Phys. Lett. 100, 311
Dixon, R.N. & Wright, N.G. 1985. Chem. Phys. Lett. 117, 280
Docker, M.P., Hodgson, A. & Simons, J.P. 1986. Mol. Phys. 57, 129
Duxbury, G. 1985. Int. Rev. Phys. Chem. 4, 237
Evenson, K.M. 1981. Farad. Disc. Roy. Soc. Chem. 71,7
Field, R.W. 1981. Farad. Disc. Roy. Soc. Chem. 71, 111
Gerstenkorn, S. & Luc, P. 1979. Atlas du Spectre d'Absorption de la
    Molecule d'Iode, C.N.R.S., Orsay: Rev. Phys. Appl. 14,791
Gottscho, R.A., Field, R.W., Bacis, R., & Silvers, S.J. 1980a. J. Chem.
    Phys. 73, 599
Gottscho, R.A., Weiss, P.S., Field, R.W. & Pruett, J.G. 1980b. J. Mol.
    Spectrosc. 82, 283
Greene, C.H. & Zare, R.N. 1982, Ann. Rev. Phys. Chem. 33, 119
Greene, C.H. & Zare, R.N. 1983. J. Chem. Phys. 78, 674
Hirota, E. 1985. High-Resolution Spectroscopy of Transient Molecules.
    Springer-Verlag
Johns, J.W.C. 1964. Canad. J. Phys. 42, 1004

Johns, J.W.C. & McKellar, A.R.W. 1973. J. Mol. Spectrosc. 48, 354

Johns, J.W.C. & McKellar, A.R.W. 1977. J. Chem. Phys. 66, 1217

Johnson, M.A., Webster, C.R. & Zare, R.N. 1981. J. Chem. Phys.75,5575

King, G.W., Littlewood, I.M. & Littlewood, N.T. 1981 b. Chem. Phys. Lett. 80, 215

King, G.W., Littlewood, I.M., McFadden, R.C. & Robins, J.R. 1979, Chem. Phys. 41, 379

King, G.W., Littlewood, I.M. & Robins, J.R. 1981a Chem. Phys. 56, 145

Klein, F.K. 1977. Diplom thesis, Physics Department, University of Kaiserslautern

Radford, H.E. & Litvak, M.M. 1975. Chem. Phys. Lett. 34, 561

Reisner, D.E., Field, R.W., Kinsey, J.L. & Dai, H.L. 1984. J. Chem.Phys. 80, 5968

Riedle, E., Moder, R. & Neusser, H.J. 1982. Opt. Commun. 43, 388

Stickland, R.J. 1987. Ph.D. thesis, University of Bristol

Urban, W. & Herrmann, W., 1978. Appl. Phys. 17, 325

Vasudev, R., Zare, R.N. & Dixon, R.N. 1984. J. Chem.Phys. 80, 4863

Webster, C.R., Brucat, P.J. & Zare, R.N. 1982. J. Mol. Spectrosc. 92, 184

Webster, C.R., May, R.D. & Gunson, M.R. 1985. Chem. Phys. Lett. 121, 429

Yamada, C., Endo, Y. & Hirota, E. 1983. J. Chem. Phys. 78, 4379

Yamada, C., Hirota, E. & Kawaguchi, K. 1981. J. Chem. Phys. 75, 5256

Yamada, C., Kawaguchi, K. & Hirota, E. 1978. J. Chem. Phys. 69, 1942

Ziegler, L.D. 1986. J. Chem. Phys. 84, 6013

Ziegler, L.D. 1987. J. Chem. Phys. 86, 1703

MULTIPHOTON IONIZATION SPECTROSCOPY WITH
PHOTOELECTRON AND MASS SPECTRAL ANALYSIS

Steven D. Colson
Sterling Chemistry Laboratory
Yale University
New Haven, CT 06511

ABSTRACT. Photoelectron and mass spectral analyses have become
important adjuncts to the recently developed field of multiphoton
ionization spectroscopy. Each of the methods are discussed and the
significant enhancement in capability provided by their joint
application is illustrated by their use in the spectroscopic study of
free radicals, molecular ions and transient excited states of molecules.

I. INTRODUCTION

A major advance in the utility of laser spectroscopy came as a
result of the development of multiphoton ionization MPI as a means of
detection of multiphoton absorption by molecules [1]. The resonance
encountered as the n-photon energy of a scanning laser becomes
coincident with that of a molecular excited state is evidenced by a
large increase in ionization rate. Since single ionization events can
be detected with near unit efficiency, this results in a very sensitive
means of detecting weak multiphoton absorption. MPI is a more widely
applicable method than laser induced fluorescence since it can be used
for non-emitting states.

Single and multiphoton ionization MPI processes will be compared
and the various approaches used to obtain the photoelectron energy
spectrum PES with laser ionization will be reviewed to compare their
advantages and disadvantages. The importance of resonant intermediate
states in the determination of the MPI-PES will be discussed to
illustrate the use of PES in providing quantum state labels for the
resonant excited state levels. Molecular Rydberg levels are
particularly clear examples of this application. In the limit wherein
the outer (Rydberg) electron is far removed from the ion core, it can be
ionized without changing the rovibrational energy of the core, except
for the requirement for conservation of angular momentum, i.e. the
Franck-Condon factors strongly favor $\Delta v=0$ for the ionization. Thus, the
vibronic quantum numbers of an unassigned Rydberg level can be
established from the PES [2]. Applications of this state labeling
technique will be illustrated by its use in the deconvolution and

A. C. P. Alves et al. (eds.), Frontiers of Laser Spectroscopy of Gases, 309–324.

analysis of complex spectra.

Deviations from the above simple model will be used to show the further application of laser PES to the study of state perturbations and non-radiative processes. If a selected resonant intermediate level is significiantly mixed with another level, the Franck-Condon factors for both "isolated" levels will contribute to the vibrational structure in the PES [3]. Variations in the PES as the laser is scanned over rotation band structure can be used to determine the rotational dependence of the interstate mixing matrix elements. State-to-state relaxation can also be seen by the time dependence of the PES. As the time between the state preparation and its subsequent ionization is increased, the non-radiative flow of energy amongst the coupled levels will be illustrated by the changes in the PES [4].

The various means of obtaining mass spectroscopy with laser ionization will be discussed. The production of ions at a point in space (and during a short time interval) by pulsed laser ionization is particularly well suited to time-of-flight TOF, as compared to quadrupole, mass analysis. The increased resolution of reflection TOF spectrometers is comparable with that attainable with quadrupole devices and yet maintains the advantage that the entire mass spectrum is obtained with each laser pulse [5].

The advantages of having mass labeled optical spectra are substantial. Many researchers have conducted exhaustive studies of "new transition", using static gas MPI or some other means, only to discover later that the results were due to trace impurities. By obtaining the excitation spectrum for the production of a particular ionic mass it is often possible to unequivocally establish the species responsible for the optical spectrum.

Because of the mass resolution, one is able to obtain simultaneously the resolved spectra of different naturally occurring isotopes. This ability to mass resolve overlapping spectra can also be applied to the study of plasmas or other complex mixtures of species wherein the analysis is greatly aided by obtaining the spectra as a function of both mass and wavelength.

A variety of means has been developed recently for the preparation of molecular beams containing sufficient concentrations of free radicals for spectroscopic analysis by MPI-MS or LIF detection. For larger species, it is particularly important to prepare them in such a manner that they can also be cooled to low temperatures by supersonic expansion to relieve the considerable spectral congestion typical of hot radical sources. To accomplish this, several devices have been developed which combine radical production with supersonic nozzle sources.

Discharge [6,7] and other high energy [8] radical sources are difficult to control. Without careful selection of the precursor species and the discharge conditions, one can generally produce a wide variety of species in addition to the one of interest. It is expected to be particularly difficult to produce energetically metastable species in this manner.

Laser photodissociation sources [9-12] can also be problematical. Because of strong absorption, often characteristic of excited molecules, it is difficult to control the degree of excitation and hence the extent of fragmentation produced by pulsed laser sources. Furthermore, the photochemical products themselves will often be photochemically labile, producing a variety of secondary products even during the time of a single nsec or psec laser pulse.

Pulsed pyrolysis is another source of radicals that has recently been developed [13]. The conditions are sufficiently mild that one is often able to limit the pyrolysis to the dissociation of a single bond. Then, by the selection of the correct precursor molecule, one is able to prepare relatively clean sources of radicals in an inert carrier gas for expansion cooling and spectroscopic analysis.

The main objective of these lectures is to focus on the experimental methods. Thus, the experimental results used to illustrate the applications of these methods will only be discussed briefly.

II. MULTIPHOTON IONIZATION PHOTOELECTRON SPECTROSCOPY

Conventional, one photon PES [14] are obtained by ionization at wavelengths sufficiently short to create electronically excited ions. The rovibronic energy of the ion is determined by measuring the kinetic energy of the ejected electron $E(e^-)$ and then applying the simple formula

$$E(\text{Internal}) = h\nu - \text{I.P.} - E(e^-)$$

Thereby, the energy distribution of $E(e^-)$ can be used to measure the distribution of ionic states produced by photoionization. Typical one photon PES are of sufficient resolution to reveal the vibrational structure, giving information on the electronic and vibrational structure of the ion. The vibronic intensities are usually determined by the Franck-Condon principle, and can often be used to learn about the change in the molecular structure upon ionization. Since the ground state structure is usually well determined by other methods, one photon PES is a valuable tool for the study of the electronic, vibrational and geometrical structure of molecular ions.

Resonantly enhanced n-photon PES is often best described as the preparation of an excited neutral state by (n-1)-photons, followed by one-photon ionization. Thus, the PES will contain the same information as conventional one-photon PES except that the "initial" state in the ionization event will be an electronic excited state rather than the ground state. Thus, the structure of the n-photon PES will reflect the difference between the excited neutral state and the states of the ion. The use of pulsed laser ionization creates the electrons at a well defined point in time and space. This makes it possible to design relatively simple spectrometers which measure the time-of-flight TOF for the electron to travel from the point of ionization to the detector. This has an advantage over conventional, one-photon ionization methods which use rare gas resonance light sources. These are continuous sources and are used without focusing so that they irradiate a larger sample volume. However, there is a disadvantage to creating the electrons in short bursts. As the first electrons fly away from the ionization region, they leave behind a charged cloud which will retard the departure of additional electrons. This degrades the resolution unless one limits the number of ionization events per laser shot. Thus, high resolution pulsed laser PES spectra are expected to require the averaging of many laser shots at relatively low power. This effect has been clearly demonstrated by Meek, et al. [15]

The three spectrometer designs that have been used to obtain Laser-PES are based upon hemispherical, time-of-flight, or "magnetic bottle" energy analysis respectively. The hemisperical analyzer controls the electron energy bandpass by the voltage difference between two hemispherical surfaces. It has the potential for very high resolution. However, while it has been used very effectively with laser ionization [3,16-18], up to 20 mev resolution, it will be difficult to obtain very high resolution with this device which only gives one electron energy measurement per laser shot.

The TOF energy analyzer takes full advantage of the high spatial and temporal definition of the ionization produced by pulsed laser ionization. In these devices, the ionization occurs at zero electric and magnetic fields and $E(e^-)$ and is measured by the time required for the electron to reach a detector ≈ 50 cm away. By using fast detectors and digitizers, it is possible to record the entire PES with each laser shot, greatly reducing the time required to obtain a PE spectrum with good signal-to-noise. Resolution of 20 mev has been achieved with TOF analyzers [19]. To achieve higher resolution will require close attention to magnetic shielding and other design criteria including the speed of the detector and recording electronics and/or the effective flight time which can be increased by retarding grids.

The major disadvantage of the TOF energy analyzer is the low solid angle for electron collection. To overcome this problem, Kruit and Read [20] developed a "magnetic bottle" analyzer. A shaped magnetic field is used to form an "electron mirror" such that if the electrons are created at the correct point in space (the mouth of the magnetic bottle) they are guided to the detector (at the bottom of the bottle) by the magnetic field lines. In this manner, half of the electrons of a given energy are collected, regardless of their initial trajectories. This gives a thousand-fold increase in the collection efficiency while maintaining the "whole-spectrum-per-shot" advantage of the field free TOF analyzer. Resolution of 15 mev is advertised by the manufacturer, Applied Laser Technology (The Netherlands). The disadvantages of this method derive from effects of the magnetic field which may perturb the molecular eigenstates, and from the confined space required to establish the 1 T magnetic field at the laser focus. This confinement makes it difficult to obtain large gas throughput (limiting supersonic beam applications) and may be the source of patch electrical fields which effect the measured energies of the electrons. This latter problem can be especially difficult when working with UV lasers which can easily cause surface charge build-up if allowed to illuminate the surfaces. Thus, repeated calibrations are essential with this type of energy analyzer when working with UV-light.

Calibration is essential for any PES device and is generally achieved by obtaining reference spectra of xenon, NO, or Fe, obtained from photodissociation of iron pentacarbonyl [21].

Some of the most interesting applications of MPI-PES come from its use in the characterization of electronic excited states of the neutral molecule. The simplest visualization of this application derives when the final step in the multiphoton ionization is from a high Rydberg state. Often, the Rydberg electron is only weakly coupled to the ion core such that the core structure is essentially identical to that of the ion. In this case, the electron can be removed without changing the rovibronic energy of the core except for the requirements imposed by selection rules. Thus, the quantum labeling of the Rydberg level can be achieved by determining the final quantum state of the ion from the PES. Viswanathan, et al. have demonstrated the rotational selection rules for the process in NO [19], and Pratt et al. have demonstrated the simultaneous preservation of the core vibrational and electronic configuration for ionization of unperturbed Rydberg levels of N_2 [22].

As will be discussed in Section V, this quantum state labeling can be very valuable in the assignment of newly discovered excited states in MPI spectra. This simple model will not apply to cases wherein.

a) The geometry of the excited state is different from that of the ion. In this case one expects extensive vibronic structure in the PES. This can be turned to advantage when it is possible to activate different sets of vibronic levels of the ion in the PES by starting with different levels of the intermediate neutral state. In this way, a very rich vibronic spectrum of the phenol ion has been obtained and analyzed [21].

b) The intermediate level is perturbed by mixing with a rovibronic level of another state. This can give rise to quite complex PES as both of the mixed states contribute to the overall ionization cross section. Such PES spectra can be very useful in the identification and characterization of such perturbations [3,22].

c) The intermediate level is ionized by a two-photon absorption process. Frequently the two-photon ionization will be enhanced by resonances or near-resonances at the energy of the first photon above the intermediate level. These resonances can give very unpredictable PES [18].

d) More than one intermediate level absorbs at the same excitation wavelength such that it is difficult to prepare a "pure" intermediate state even though the two levels are not mixed. In this case, the two pumped levels can give independent PES which can be used to deconvolute the overlapping bands. By obtaining the wavelength excitation spectra for separate PES peaks (e.g., those characteristic of the two intermediates states), two independent spectra can be generated as the wavelength of the state preparation laser is varied [23]. This application will also be illustrated by the results discussed in Section V.

e) The final photon excites the molecule to a bound level above the ionization threshold. Such bound levels are Rydberg states whose core corresponds to an excited state of the ion. This core excitation can be rotational, vibrational and/or electronic. Such levels are metastable and can autoionize by conversion of excess rovibronic energy into kinetic energy of the ejected electron. Thus, the PES will be

determined by the autoionization process rather than by direct
ionization [24].

III. MULTIPHOTON IONIZATION MASS SPECTROSCOPY

The techniques used to obtain mass spectra with multiphoton
ionization will only be discussed briefly as they will be treated
extensively in the work presented by Dr. Edward Schlag at this meeting.
Pulsed laser ionization provides the same advantages for MS as it does
for PES. The creation of ions at a well defined time and point in space
has led to the "rebirth" of the simple and relatively inexpensive linear
time-of-flight mass spectrometer. In the simplest TOF instruments, the
ions are all accelerated to a common energy (e.g. 1000 v), allowed to
drift at constant potential (~1m) and then detected with an electron
multiplier or similar device. Since the ion kinetic energy is
proportional to the mass and the velocity squared, ions of different
masses will each have different velocities and therefore different
flight times. Thus, by using fast recording devices, the entire TOF
mass spectrum can be recorded with each laser shot. Also, since the
ions are created with very little kinetic energy, they can be collected
with high efficiency by the ion acceleration fields. These factors,
combined with efficient ion detectors, make TOF mass analyzers highly
sensitive and well suited for use with laser ionization.
The main disadvantage to linear TOF analysis is resolution. This
problem can be alleviated by the use of a reflection (ion mirror), a
technique which will be discussed in detail by Dr. Schlag who has
developed it extensively in recent years [5].
Laser ionization mass spectrosocpy has aided greatly in the
understanding of multiphoton ionization and in its applications. The
ability to obtain mass resolved wavelength excitation spectra has been
particularly valuable in the identification of new spectra by knowing
the mass of the species responsible for a given optical transition. A
variety of means have been developed to prepare transient species in the
ionization region of the mass spectrometer (vide infra). The MPI
excitation spectrum is then obtained by determining the dependence of
the mass spectrum on the laser wavelength.
Laser-MS has also been used to resolve optical spectra of naturally
occurring isotopic species. In this application, a peak in the MS due
to an ion containing the isotope of interest (e.g. C-13 in benzene [25]
or other molecules [26]) is selected. The intensity of the mass peak is
then monitored as a function of the laser, producing the excitation
spectrum of the isotopically labeled molecule.
Specific applications of laser-MS to obtain mass labeled spectra of
transient species and to the study of multiphoton dissociation processes
will be discussed in Section V.

IV. PREPARATION OF FREE RADICALS IN MOLECULAR BEAMS

The molecular beam environment is ideally suited to the study of

reactive species such as free radicals. This is particularly true if
they can be prepared and formed into a beam by supersonic expansion.
The resulting temperature reduction is especially important for spectral
simplication of radicals that are usually prepared by some high energy
process which can result in nascent temperatures well in excess of
1000K. All of the standard means of radical production (chemical,
discharge, photolysis and pyrolysis) have been investigated as means of
producing radical beams of sufficient intensity for spectroscopic work.
While no attempt will be made to review this extensive field, a few
examples will be discussed to illustrate the more successful means that
have been developed.

The production of free radicals by chemical reactions is not well
suited to supersonic expansion conditions which would require the
reactions to be carried out at relatively high pressure for effective
cooling. Nevertheless, they provide direct information on the chemical
reactions themselves and have resulted in a rich variety of species,
even though the spectra that are produced are generally quite congested
due to the high rotational and vibrational temperatures involved.
Hudgens and his co-workers [27] have been very successful in the use of
fluorine atom abstraction reactions to produce free radicals in beams.
The F-atoms are produced by a microwave discharge in a flow tube and
the output is crossed with a gas stream containing the reagent species
to produce the molecular beam. By utilizing reactions such as

$$F + HM \rightarrow HF + M$$

they have obtained the spectrum of a variety of free radicals including,
SiF, ClO, BrO, CH_3, CH_3O, CH_2OH C_3H_5 and C_4H_7.

A novel discharge device for producing rotationally cold free
radicals has been developed by Carrick, et al. [6] In this device, a
fine, high voltage electrode is inserted just behind (on the high
pressure side) the orifice of a glass supersonic jet source. A corona
discharge is developed through the orifice to the walls of the vacuum
chamber. Radical species formed from seed molecules as they pass
through the orifice are cooled by the subsequent expansion. Since this
is a high energy fragmentation source, some radicals are formed in
excited electronic states and can be observed directly in emission.
Ground state species have been observed by LIF but this source has yet
to be combined with MPI-MS detection. It has been successfully used to
obtain the electronic spectra of methylnitrene [6] which has not be
obtained in supersonic beams by any other technique.

Photochemical production of free radicals with supersonic expansion
was first reported by the Smalley group [9] but has been most
effectively exploited by Miller et al. [10] They generally produce the
fragmentation with a UV laser near the output of a supersonic nozzle
where the pressure is sufficiently high to produce significant cooling
by the subsequent expansion. By using downstream, LIF detection, these
authors have obtained spectra of a large variety of free radicals.
These results are discussed in detail by Dr. Terry Miller at this
meeting.

The extent of cooling increases as the distance between the points of photofragmentation and LIF-detection is lengthened. This can be very useful in rotational analyses, for instance. The simple low temperature rotational structure can form a basis for the analysis of more complex spectra at higher temperatures, providing information on high-J levels not seen at low temperature.

We have recently developed a pulsed nozzle source which successfully combines supersonic cooling with the pyrolytic production of radicals [13]. In this source, a 1 mm I.D. tube is attached to the output of a pulsed gas source. The tube is heated with a wire coil to 1000-2000K to induce the thermal decomposition of precursor molecules selected to give desired free radicals. Bimolecular and wall reactions can be effectively suppressed by using high dilutions in rare gases. By loading the heated tube to high pressures, the gas expansion at the end of pyrolysis can result in rapid cooling to low temperatures (~40 K). In this way, the previously unresolved rotational structure in a Rydberg transition in CH_3, was found to be comprised of individual rovibronic lines with line widths ~7cm^{-1} [13].

V. APPLICATIONS OF LASER IONIZATION MASS AND PHOTOELECTRON SPECTROSCOPY

A. Photochemistry with Intense, Pulsed UV Laser Excitation

Intense, pulsed UV lasers provide a very different photochemical excitation source. Because of rate of absorption can sometimes exceed the rate of photochemical processes, some systems can be driven to high energies, giving products not otherwise expected. Laser ionization mass spectrometry is ideally suited to the study of this new type of photochemistry. Scanning the laser wavelength can produce the MPI optical spectrum of the products and these spectra can be identified with specific mass fragments from the mass spectrum as described in Section III.

Molecular photofragmentation with pulsed UV lasers is now a well established phenomena. Often, even fairly large polyatomic molecules are converted to atomic and small polyatomic fragments, a process that can require the absorption of five or more photons. It is interesting to contemplate the mechanism by which this much energy (≈25 ev) is deposited in a molecular system having numerous lower energy decay channels. The total energy absorbed is often more than four times the lowest energy dissociation channel and twice the lowest ionization potential.

One might envision two limiting cases; 1) absorption of n-photons by the parent followed by a concerted or a sequential series of dissociations into metastable fragments, or 2) a series of sequential photodissociation steps wherein the daughter species subsequently absorb at the same frequency as the parent to form smaller and smaller fragments.

The MPI-MPD of butadiene provides an excellent example of case-1 multiphoton photodissociation [28]. This is a two laser experiment where one laser is used to prepare the ion in its ground state and the other is used to study its fragmentation. The sample gas is cooled by supersonic expansion into a vacuum containing a time-of-flight mass spectrometer. The gas stream is intersected by two, 10 nsec counterpropagating laser beams, with the fragmentation laser arriving 12 ns after the ionization pulse.

The wavelength of the ionization laser is selected to pump the origin band of a Rydberg state by a two photon absorption. The absorption of one more photon results in photoionization of this Rydberg level. This is confirmed by obtaining the MPI-PES at this wavelength which shows that the ion is formed in its vibrationless ground state. At sufficiently low laser powers, the butadiene ion can be prepared with little subsequent fragmentation by the first laser.

The fragmentation laser wavelength is selected to correspond to the first absorption band in the ion. This absorption is observed to be very broad due to a rapid rate($>10^{13}$ s^{-1}) of internal conversion to the ground state of the ion. The energy of these ground state ions is insufficient to cause rapid fragmentation (10^5 s^{-1}). However, the hot ions can absorb another photon by the same electronic transition as before at rates (10^9 s^{-1}) faster than for dissociation. This will produce vibrationally and electronically excited ions which (because of the increased density-of-states at this energy) are expected to undergo internal conversion to the ground state at an even faster rate than before. This process can be repeated again and again, resulting in the cyclical pumping of vibrational excitation into the ground electronic state of the ion in 20,000 cm^{-1} increments.

The ion will continue to gain vibrational energy until the dissociation rate exceeds the absorption rate. By assuming a statistical distribution of the energy in the hot ions, RRKM theory calculations were made of the most rapid dissociation rates (to form CH) following absorption of one, two and three photons; 1.0×10^5, 3.0×10^9, and 2.0×10^{11} s^{-1} respectively. These rates and the known ion breakdown curves [29] were used to predict the fragmentation pattern in excellent agreement with the observed mass spectrum [28]. In the above case, at least half of the ions had received over 60,000 cm^{-1} of vibrational excitation before dissociation, nearly 43,000 cm^{-1} above the dissociation threshold. This mechanism is expected to be important in systems characterized by reasonably strong optical pumping ($>10^{26}$ photons s^{-1} cm^{-2}) of allowed transitions with rapid internal conversion back to the initial state. To prevent rapid dissociation above the first dissociation threshold, the excess energy must be distributed over a large number of vibronic states. Thus, larger molecules will be more susceptible to this type of pumping.

If all fragments which show secondary daughter species absorb strongly at the same frequency as the parent molecule, the main difference between cases 1 and 2 is kinetic. Case-2 presumes that dissociation is rapid compared to absorption. Thus, time resolved experiments are expected to provide a better understanding of the mechanism. Indeed, recent psec [12,30] and fsec [31] studies are proving to be very important in this regard.

The MPI-MPD of ketene is an excellent example of the combined use of a) MPI-MS to identify the photoproducts [32], b) MPI-PES to aid in the analysis of their spectra [23,32], and c) time resolved spectroscopy

[12] to show the existence of case-2 processes. The one-photon absorption band in ketene (H_2CCO) is only weakly allowed and is therefore not likely to result in cyclical pumping. While there may be a pumping cycle between two excited states, the molecule is small and rigid such that its RRKM dissociation rate will be high (compared to the optical absorption rate for this relatively weak transition) even at the first dissociation threshold. This suggests that dissociation will occur before sufficient energy is absorbed to produce all of the observed fragments, i.e. case-2. The time resolved experiments of Chen et. al [12] are consistent with this expectation. In this experiment, a sync-pumped dye laser is tuned into resonance with a two-photon atomic carbon transition to detect the C-fragment from the multiphoton dissociation of ketene. This same wavelength also overlaps the lowest electronic absorption band in ketene. The carbon ion signal (due to 2+1 MPI of neutral carbon) increases as the pulse width of the laser is increased from 5 to 50 psec. This suggests that some molecular processes must precede the photoabsorption event that produces the carbon atoms. This is confirmed by a two-pulse experiment where the psec laser is split into two beams of nearly equal intensity and recombined in the sample with a variable time delay. The efficiency of carbon atom production is greatly enhanced by the time delay, showing a dark channel for its production with a time constrast of 300 psec. From this it can be seen that the energy deposited by the first laser pulse drives a relatively slow dissociation process which precedes the production of C-atoms. When using longer pulsed lasers (10 nsec) the entire process occurs on times short compared to the pulse width and one observes copious production of carbon with each laser pulse.

B. The Optical Spectrum of CH

The analysis of the MPI spectrum of CH provides an excellent illustration of the use of MPI-MS and PES spectra in unscrambling of complex spectra. Figure 1 shows the intensity of the CH^+ MS-signal as a function of wavelength. In this case, one laser is being used to effect the multiphoton fragmentation of the parent molecule (ketene) and to cause the multiphoton ionization of the CH fragment radical. Ketene absorbs broadly enough in the region to minimize any wavelength dependence to the production of CH.

The portion of the CH spectrum shown in Figure 1 contains numerous, complex and sometimes overlapping bands, only one of which had been seen previously [33]. Thus, the "quantum labeling" capability of laser-PES proved to be invaluable in the identification and analysis of the bands. Furthermore, the CH^+ observed in the mass spectrum could result either from the MPI of CH or from the photofragmentation of a larger ion. Thus, an independent means is needed to prove that CH^+ excitation spectrum is (or is not) due to optical absorption by CH. By looking directly at the ionization channel, the PE spectrum indicates which species is being ionized, i.e., if the PES is characteristic of CH then the signal in the CH^+ channel must also be due to the MPI of CH.

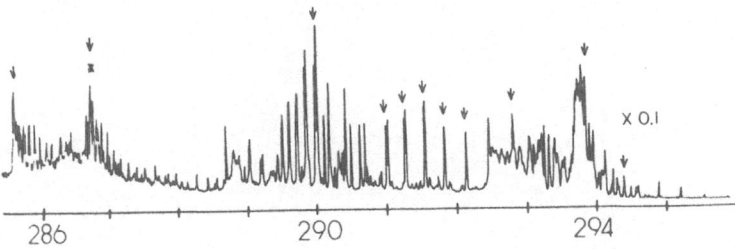

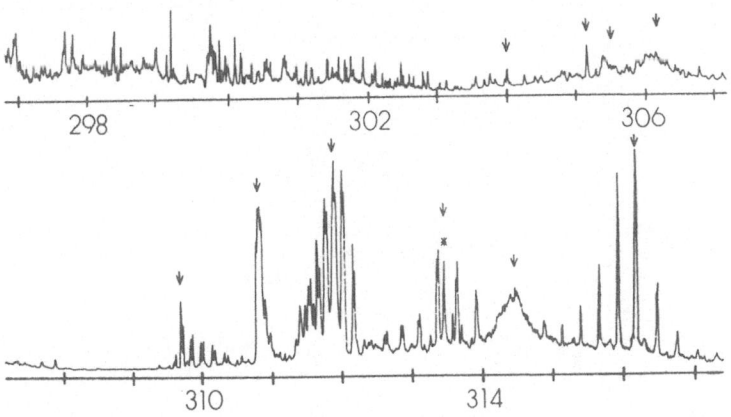

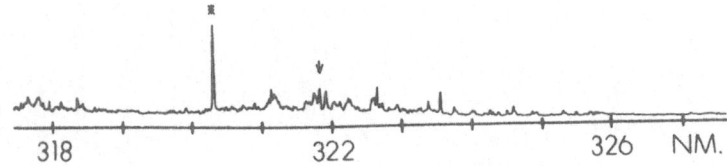

Figure 1. MPI-MS wavelength scans for m/e=13 by resonant 2 + 1 MPI of
CH produced by photodissociateion of ketene. Arrows mark wavelengths at
which photoelectron spectra have been taken. Asterisks mark peaks due
to atomic carbon transitions appearing in the CH channel.[32]

320

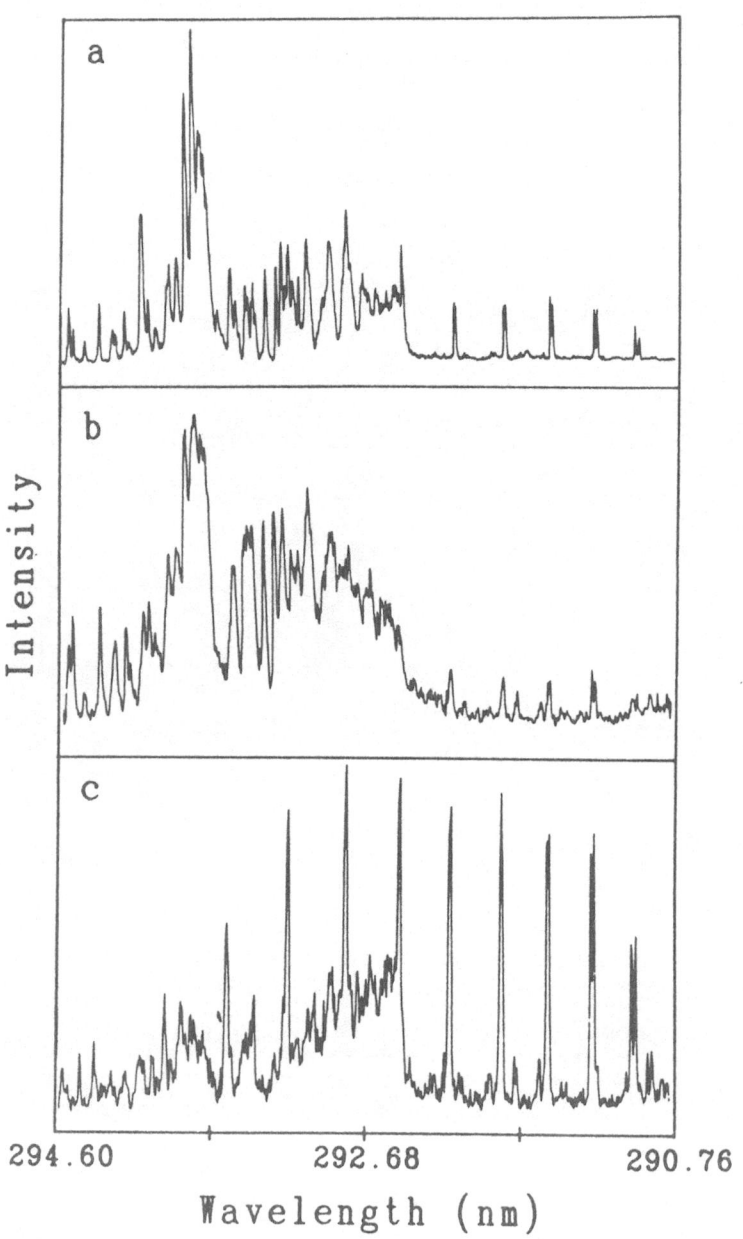

294.60   292.68   290.76

Wavelength (nm)

Figure 2. (a) Mass resolved (m/e=13) wavelength scan of the production
of neutral CH from the photodissociation of ketene. (b) and (c)
Photoelectron energy resolved wavelength scans which monitor the
intensity of the photoelectron peak corresponding to formation of the
ground state of the CH ion in the v=1 and v=2 vibrational levels,
respectively.[12]

Photoelectron spectra were obtained at each of the wavelengths indicated by an arrow in Figure 1 [32]. The three lines marked with an (*) were found to be due to C-13. (Much stronger transitions at these energies were seen in the C-12 mass channel.) All of the remaining features are due to ionization of CH. These excited states of CH have essentially the same structure as the ion. Thus, they give mainly single line PE spectra corresponding to formation of the ion in the same vibrational state as the neutral. Therefore, the PES not only identify the species being ionized (CH) but they also provide vibrational labels for each of the peaks in the MPI spectrum. From this information, it was possible to identify four different electronic transitions contributing to the two-photon absorption in this region [32]. Some of the vibronic transitions assigned to different states overlapped. However, when the overlapping states had distinct PES, the spectra could be resolved by obtaining separately the photoexcitation spectra for the production of each characteristic peak in the PES as illustrated in Figure 2 [23].

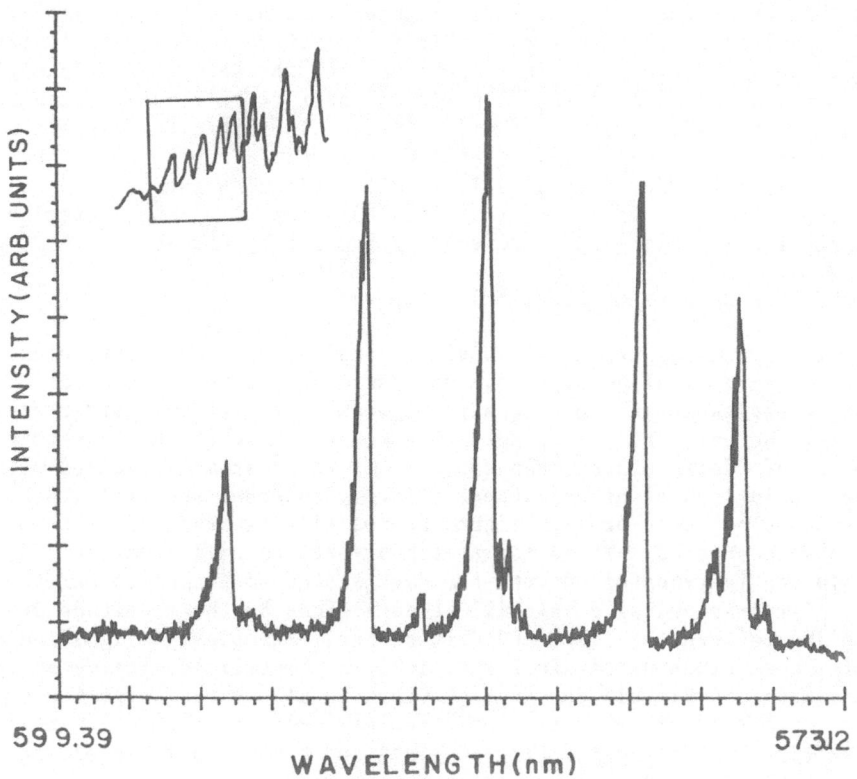

Figure 3. Photodissociation spectrum of CH_3I^+. Insert shows a section of the spectrum reported by Gross et al. [36]. The enclosed section corresponds to our spectrum shown in the lower figure.[35]

C. The Photodissociation Excitation Spectrum of State Prepared CH_3I^+

The observation of single peak, MPI-PE spectra can be used to further advantage by noting that all ions formed in this manner are in the same vibronic state. For instance, this provides a means of producing ions in specific states for studies of ion-molecule reaction cross sections as a function of vibrational energy [34]. Likewise, optical spectra of laser prepared ions can be much simpler than those obtained from electron bombardment or other high energy processes typically used as ion sources. An example of this latter application is in the CH_3I^+ photodissociation excitation spectrum. This is a two-laser experiment. The first laser is chosen to excite a Rydberg level which is selected to have a particular vibrational excitation (v) in the neutral molecule. The PES is obtained at this wavelength to confirm that the subsequent ionization of this Rydberg level produces ions without changing v. The second laser is then tuned through absorption bands above the ion dissociation threshold and the intensity of the CH_3^+ signal (due to dissociation of CH_3I^+ by the second laser) is monitored by the mass spectrometer.

The resulting spectrum [35] (Figure 3) is dramatically simplified by comparison to that obtained following electron bombardment ionization [36]. The vibrational excitation in the ion can be varied by changing the wavelength of the first, ionization laser to be resonant with a intermediate Rydberg level of differing v. This results in spectra with different vibrational activity and these results combined with the same spectra obtained for isotopically labeled molecules can be used to provide a detailed vibrational analyis [37] for this band system which was previously only subject to qualitative interpretation.

D. MPI-PES Studies of Dynamical Processes

The dynamical process we wish to consider is the radiationless decay of an optically prepared, non-stationary state. The usual model assumes the exsistence of two sets of weakly coupled zero order excited states, one set that can be optically connected to the ground state and one set of "dark" states that cannot. If the system is excited with a short optical pulse of sufficiently large, coherent spectral width, only the uncoupled, zero order, "light" states will be prepared by the optical absorption. These non-stationary states will decay as population is transferred into the dark states of the same energy. The rate of radiationless decay will be determined by the magnitude of the coupling between the light and dark states. Examples of light and dark states commonly studied are respectively a) an excited singlet state and the manifold of levels due to high vibrational excitation in the ground electronic state wherein the radiationless decay is termed internal conversion, b) an excited singlet state and the vibronic levels of an excited triplet state giving intersystem crossing, and c) an excited electronic state and levels due to a very different molecular geometry resulting in rearrangements or photochemistry.

It is often quite difficult to know which dark states are populated by radiationless decay because they cannot be "seen" by optical emission to the ground state. On the other hand, the rules governing ionization can be quite different from those restricting the emission. For example, ionization of an excited singlet or triplet level can occur with equal propability to produce an ion in its ground doublet state. Thus, while the emission can be quenched by radiationless singlet-to-triplet decay, the ionization efficiency can remain constant. However, as the vibronic character of the state is strongly altered by the relaxation, the vibrational structure in the PES will change in time. The MPI-PES of s-triazine has been used to measure the singlet-triplet decay rates for this molecule in a molecular beam [4]. The MPI-PES is obtained at a wavelength where the lowest excited singlet state absorbs. When a 2 nsec excitation pulse is used, the PES is due primarily to triplet state ionization, i.e., most of the singlet levels can decay to triplets in less than 2 nsec. When the pulse width is narrowed to 5 psec, about half of the ionization is from excited singlet levels. When the data is analyzed it is found that the coupling of the singlet and triplet levels is very inhomogeneous with singlet lifetimes ranging from nsec to sub-psec. Similar measurements have been made to show which excited states are involved in the predissociation of ammonia [38].

VI. SUMMARY

These lectures have discussed the production of free radicals for spectroscopic analysis and have focused on
   a) The use of mass analysis to identify the species responsible for observed multiphoton absorption spectra,
   b) The use of photoelectron energy analysis to provide vibronic quantum numbers for newly discovered excited states,
   c) The use of time resolved Laser-MS data to study the mechanism for multiphoton dissociation processes, and
   d) The application of time resolved Laser-PES to directly follow nonradiative decay of optically prepared, non-stationary states in molecules.

REFERENCES

1.   P.M. Johnson and G.E. Otis, Ann. Rev. Phys. Chem., **32**, 139 (1981).
2.   A. Sur, C.V. Ramana, W.A. Chupka and S.D. Colson, J. Chem. Phys., **84**, 69 (1986).
3.   M.G. White, M. Seaver, W.A. Chupka and S.D. Colson, Phys. Rev. Letters, **49**, 28 (1982).
4.   J.B. Pallix and S.D. Colson, Chem. Phys. Letters, **119**, 38 (1985).
5.   U. Boesl, H.J. Neusser, R. Welnkauf and E.W. Schlag, J. Phys. Chem., **86**, 4857 (1982).
6.   P.G. Carrick and R.C. Engelking, J. Chem. Phys. **81**, 1661 (1984); J. Chem. Phys., **83**, 1995 (1985).
7.   S. Sharpe and P.M. Johnson, Chem. Phys. Letters, **107**, 35 (1984).
8.   K.C. Smyth and W.G. Mallard, J. Chem. Phys., **77**, 1779 (1982).
9.   D.E. Powers, J.B. Hopkins and R.E. Smalley, J. Phys., Chem. **85**, 2711 (1981).

10. T.A. Miller, Science, **223**, 545 (1984).
11. R.L. Whetten, K.J. Fu, R.S. Tapper and E.R. Grant, J. Phys. Chem., **87**, 1484 (1983).
12. P. Chen, J.B. Pallix, W.A. Chupka and S.D. Colson, J. Chem. Phys. **84**, 527 (1986).
13. P. Chen, S.D. Colson, W.A. Chupka and J.A. Berson, J. Phys. Chem., **90**, 2319 (1986).
14. Molecular Photoelectron Spectroscopy, D.W. Turner, C. Baker, A.D. Baker and C.R. Brundle (Wiley, London, 1970).
15. J. Meek, S. Long and J.P. Reilly, J. Phys. Chem., **86**, 2809 (1982).
16. J.H. Glowina, S.J. Riley, S.D. Colson, J.C. Miller and R.M. Compton, J. Chem. Phys., **77**, 68 (1982).
17. S.T. Pratt, E.D. Poliaroff, P.M. Dehmer and J.L. Dehmer, J. Chem. Phys., **78**, 65 (1983).
18. M.G. White, W.A. Chupka, M. Seaver, A. Woodward and S.D. Colson, J. Chem. Phys., **80**, 678 (1984).
19. K.S. Viswanathan, E. Sekreta, E.R. Davidson and J.P. Reilly, J. Phys. Chem., **90**, 5078 (1986).
20. P. Kruit and F.H. Read, J. Phys. E. Sci. Instr., **16**, 313 (1983).
21. S.L. Anderson, L. Goodman, K. Krogh-Jesperson, A.G. Ozkabak, R.N. Zare and C.F. Zheng, J. Chem. Phys., **82**, 5329 (1985).
22. S.T. Pratt, P.M. Dehmer and J.L. Dehmer, J. Chem. Phys., **80**, 1706 (1984); J. Chem. Phys., **80**, 3444 (1984).
23. J.B. Pallix, P. Chen, W.A. Chupka and S.D. Colson, J. Chem. Phys., **84**, 5208 (1986).
24. Y. Achiba, K. Satu and K. Kimara, J. Chem. Phys., **82**, 3959 (1985).
25. U. Buesl, H.J. Neusser and E.W. Scllag, In Laser Spectroscopy IV; H. Walther and W.K. Rothe, Eds. (Springer, Berlin, 1979).
26. J.C. Miller, Anal. Chem., **58**, 1702 (1986), and references therein.
27. J.W. Hudgens, in Adv. in Multiphoton Processes and Spectroscopy, S.H. Lin, ed. (World Scientific, Singapore, 1987).
28. A.M. Woodward, W.A. Chupka and S.D. Colson, J. Phys. Chem., **88**, 4567 (1984).
29. J. Dannaucher, J. Flamme, J. Stadelmann and J. Vogt, Chem. Phys., **51**, 189 (1980).
30. D.A. Gobeli, J.D. Simon and M.A. El-Sayed, J. Phys. Chem., **88**, 178 (1984).
31. J.L. Knee, L.R. Khundkat and A.H. Zewail, J. Chem. Phys., **83**, 1966 (1986).
32. P. Chen, J.B. Pallix, W.A. Chupka and S.D. Colson, J. Chem. Phys., **86**, 516 (1987).
33. G. Herzberg and J.W.C. Johns, Astrophys. J., **158**, 399 (1969).
34. T. Ebata and R.N. Zare, Chem. Phys. Lett., **130**, 467 (1986).
35. W.A. Chupka, S.D. Colson, M.S. Seaver and A.M. Woodward, Chem. Phys. Lett., **95**, 171 (1983).
36. S.P. Goss, D.C. McGilvery, J.D. Morrison and D.L. Smith, J. Chem. Phys., **75** 1820 (1981); S.P. Goss, J.D. Morrison and D.L. Smith, J. Chem. Phys., **75**, 757 (1981).
37. A.M. Woodward, W.A. Chupka, S.D. Colson and M.G. White, J. Phys. Chem., **90**, 274 (1986).
38. J.B. Pallix and S.D. Colson, J. Phys. Chem., **90**, 1479 (1986).

MULTIPHOTON IONIZATION MASS SPECTROMETRY OF BIOMOLECULES

J. Grotemeyer   and E.W. Schlag
Institut für Physikalische und Theoretische Chemie
Technische Universität München
Lichtenbergstr.4
D-8046 Garching, Germany

ABSTRACT. Details of the novel method of laser evaporation of intact neutral molecules (LEIM) with a low powered IR-laser and the multiphoton ionization (MUPI) combined with a high-resolution Reflectron-Time-of-Flight (RETOF) mass spectrometer are explained. Some features of the method are discussed. Mass Spectra of biomolecules obtained with this method are displayed and their differences to other mass spectrometric techniques are discussed. It is shown, that Multiphoton Ionization is a general activation method for forming ions in a mass spectrometer, with additional features for an easy deducing of structures and intrinsic properties of biomolecules in contrast to other mass spectrometric ionization methods.

1. INTRODUCTION

In recent years the interest in the technique of Time-of-Flight mass separation has found a dramatic revival in mass spectrometry. TOF-mass spectrometers have several advantages in contrast to other MS-techniques One is a theoretically unlimited mass range, or more precisely the mass range is limited **only** by the sensitivity of the detector system/1/. Other advantages are fast total mass spectrum recording, high transmission of ions, ease of use and others, but the Time-of-Flight mass spectrometer also has a well known disadvantage, the low mass resolving power.

In the last years several techniques have been developed to improve the mass resolution in TOF-mass spectrometers such as the use of a reflecting field/2/ in the ion beam path or the method of velocity compaction/3/. Especially by using a Reflectron-Time-of-Flight (RETOF) instrument/4/ some shortcomings of the Time-of-Flight technique can be overcome like the limited mass resolution. The necessity of pulsed ion sources in Time-of-Flight mass spectrometry makes this technique excellent for pulsed ionization methods like laser ionization or secondary ion emission.

One common problem of all mass spectrometric methods, is the difficulty of the desorption of intact molecules of larger organic molecules and biomolecules with very low vapor pressures or thermally labile

325

A. C. P. Alves et al. (eds.), Frontiers of Laser Spectroscopy of Gases, 325–344.

structures. For the mass spectrometric investigation of these molecules, several new desorption and ionization methods have been developed in the last two decades such as field desorption/5/, direct chemical ionization/6/, secondary ion emission/7/, fast atom bombardment/8/, ^{252}Cf-plasma desorption/9/ and laser desorption of ions/10/.

The experimental conditions of these methods differ largely in their intrinsic details, but the main point, common to all of them, is the simultaneous process of vaporization and ionization of the substances. This formation of ions from the probe has, as we think, a fundamental defect. Since the energy for the ion formation is linked directly to the evaporation process, control of the softness in the ionization procedure is not possible. In many cases, this leads to a high amount of fragmentation products and very low yields of molecular ions. Another disadvantage of the simultaneous desorption and ionization process is the formation of molecular adduct ions like $M+H^+$ or $M+Na^+$ etc. and not the pure molecular ion. In contrast multiphoton ionization (MUPI)/11/ delivers only molecular ions and their fragmentation products, hence multiphoton ionization should be compared with electron ionization.

In this paper we give an overview of our mass spectrometric method, which avoids the above mentioned disadvantages. Our method has the following features:

1.) Separation of the vaporization and ionization process
2.) Laser evaporization of intact neutral molecules (LEIM)
3.) Cooling of the evaporated neutral molecules
4.) Ionization by multiphoton ionization (MUPI)
5.) Detection by a high mass resolution Reflectron-Time-of-Flight mass spectrometer (RETOF-MS).

2. METHOD

2.1. The Experimental Setup

The experimental setup consists of three different units as shown in Figure 1.

The first stage of the instrument is the laser desorption chamber. The samples to be measured are deposited onto a brass made probe tip in solution. After evaporization of the solvent, the probe tip is mounted through a vacuum lock close to the nozzle of a pulsed valve. A low powered CO_2-laser desorbs ions as well as neutral molecules, which will suffer many collisions in the nearby supersonic beam. So the desorbed molecules are cooled in their internal degrees of freedom and transported to the skimmer. A clean molecular beam of a noble gas and the desorbed molecules enters the ionization region, because due to the construction of the ion source the desorbed ions are rejected. The ionization takes place by multiphoton absorption within the focus of a frequency-doubled dye laser. The ions formed are accelerated into the field-free drift region by a constant extraction field, corrected for energy differences in an ion-reflector and detected by a tandem-channelplate detector. The signal is digitized by a transient recorder and stored by

LASER   LASER   MASS

DESORPTION   IONIZATION   SELECTIVE

                                                  DETECTION

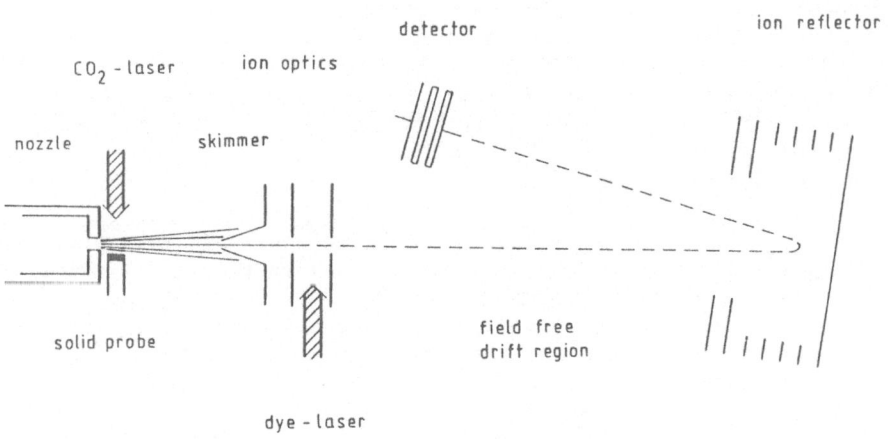

Figure 1. Scheme of the mass spectrometer system.

a computer system for further data handling.

2.2. Laser Desorption of Neutral Molecules

Most desorption techniques are used as an ion source for involatile molecules. However, it is well known that in these techniques neutrals are emitted in greater abundance than ions as demonstrated by Cotter et al./12/. In our technique we use the emission of neutrals by laser desorption/13/ and thus are able to separate desorption and ionization processes.

We use for the desorption process a small pulsed CO_2-laser (LP 30, Pulse Systems Inc.) capable of delivering 27 µs pulses of 100 mJ. The IR-laser beam is focussed through a ZnSe lens onto the solid involatile sample.

One major advantage of this setup is the possibility of cooling of translational, rotational and vibrational degrees of freedom of the desorbed molecules before ionization.

2.3. Cooling of the Desorbed Neutral Molecules in a Supersonic Beam

A very efficient cooling of the internal degrees of freedom of a molecule is the cooling in a supersonic molecular beam/14/. In our experimental setup, the supersonic beam is produced by a modified Bosch fuel injection valve, onto which a 100 μm i.d. nozzle is mounted.

A collimated beam of neutral molecules is formed by a skimmer. The skimmer is mounted onto a flange, which separates the desorption from the ionization chamber. By using a turbo molecular pump with 150 Ls^{-1} capacity at the desorption chamber, it is possible to run the system with jet pulses of 200 ms and repetition rates of 2 – 10 Hz and keep a vacuum of 10^{-4} Torr during measurements in the chamber without any influence in the supersonic beam.

It is well known that the best cooling can be reached for the translational degrees of freedom in comparison to rotational and vibrational motions/15/. Translational temperatures of 10 K and better can be reached in our molecular beam/16/.

The advantages of cooling are:

1.) low rotational and vibrational excitation leads very often to simplified structured optical spectra and enables state-selective and species-selective excitation by MUPI.
2.) low translational temperatures including narrow initial velocity distributions make high mass resolution in a Time-of-Flight mass spectrometer possible.

A common problem to all Time-of-Flight mass spectrometers is the so called "turn around time" that is due to different initial velocities of the neutrals in the ion source. Even in a RETOF these kinetic energies can destroy the achievable mass resolution of the instrument. The low translational cooling, resulting from the supersonic beam expansion, yields small kinetic energy distributions of the neutral molecules /17/.

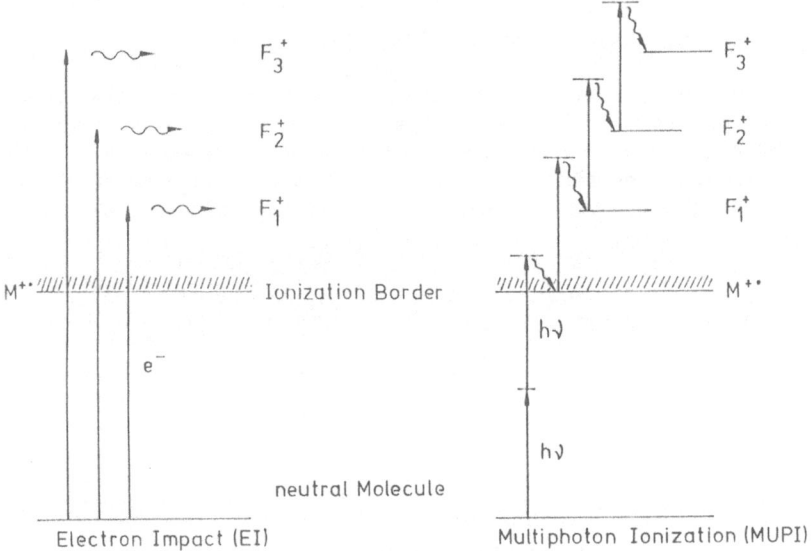

Figure 2. Difference between EI (left) and MUPI(right).

2.4. Multiphoton Ionization (MUPI)

The third step after desorption and cooling is the multiphoton ionization. As demonstrated later, in most cases the ionization takes place
by a resonance enhancement of an intermediate excited state but in several
cases the ionization can occur via a virtual intermediate state.

Since MUPI is producing molecular radical cations upon the ionization procedure, MUPI should be compared for mass spectrometric reasons
to electron ionization. But even between these two methods there is a
fundamental difference.

In electron ionization, the energy necessary for ionization and
fragmentation is imparted to the molecule by one electron, as shown in
Figure 2.

In contrast, multiphoton ionization is a multistep process. The
absorption of photons leads to the ionized molecule/18/. By choosing
low laser intensities an absorption of further photons in the ionized
molecule can be avoided. Therefore only the molecular ion is present in
the mass spectrum as a result of a real soft ionization.

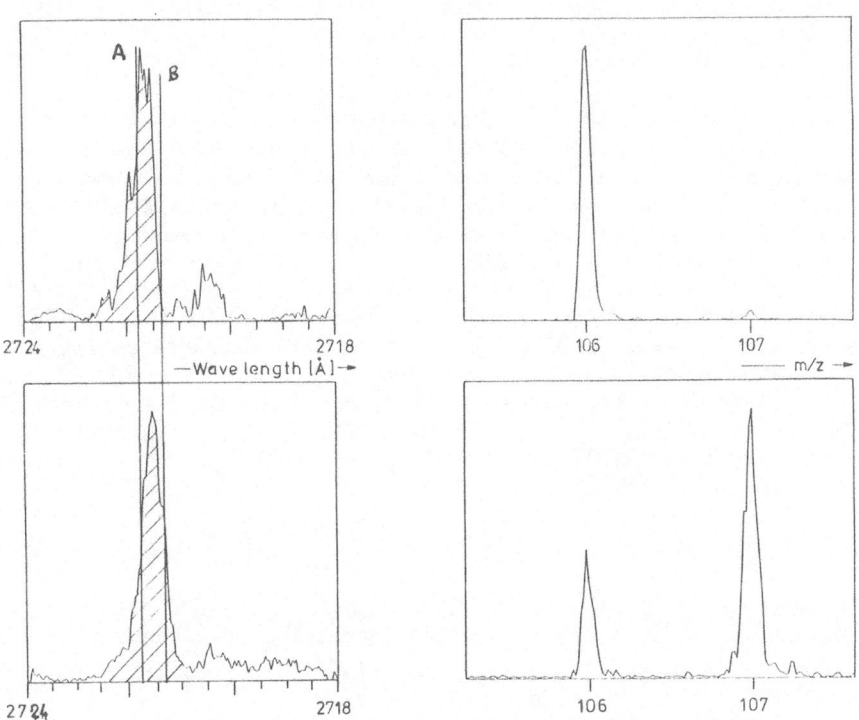

Figure 3.   Absorption mass spectra from p-Xylene. Left inserts show
the absorption spectra of m/z 106 (upper) and m/z 107 (lower). Right
inserts show the resulting mass spectra.

However, by further increasing the laser intensity, absorption by the ions and subsequent fragmentation can be induced and controlled. This absorption-fragmentation mechanism, also called the "ladder-switching" model/19/, induces the fragmentation products. As a result of this mechanism, a drastic change of the mass spectra occurs by changing the laser intensity. By increasing the laser intensity the amount of fragmentation of the molecular ions can be adjusted. At the highest laser intensity pure carbon ions are formed. Further examples of soft and hard ionization are shown by the mass spectra in the next section. Besides this feature of an adjustment of the fragmentation amount, MUPI has some further fascinating advantages especially for mass spectrometric investigations:

- by using the resonance enhancement condition fairly low laser powers are necessary for reaching high ion yields. Without resonance enhancement ions as well as molecular ions can still be detected.
- ionization processes with state selective as well as species selective conditions are possible.

Figure 3 demonstrates the feature of species selective ionization. On the left side of this figure the optical absorption-mass spectra of the molecular ion of p-Xylene and its ^{13}C-satellite are shown. The upper spectrum shows the absorption band for mass m/z 106, the molecular ion, while the lower spectrum indicates the absorption band for mass m/z 107, the ^{13}C-satellite. By choosing a wavelength corresponding to the bar indicated with "A", nearly a total discarding of the ^{13}C substituted p-xylene can be reached. On the other hand by choosing the wavelength at "B", a large discrimination of the ^{12}C-isotope and a huge intensity increase of the ^{13}C-isotopic molecule from the natural mixture is the result, as shown in the lower right mass spectrum.

In our experimental setup the dye laser (Lambda Physics FL 2001) is pumped by a XeCl-excimer laser (Lambda Physics EMG 102) and produces pulses with 2 mJ of energy within 7 ns. By frequency-doubling the dye laser, tuneable UV light in the wavelength region between 240 and 300 nm with 150 nJ energy within 5 ns is produced. The laser beam is focused by a spherical quartz lense with a focal length of 15 cm into the ion source, thus producing ions from the crossing with the supersonic molecular beam.

2.5. Detection by a High Resolution Reflectron-TOF-Mass Spectrometer

The ideal mass spectrometer to be combined with a MUPI ion source is a TOF instrument. Both are pulse techniques. The advantages of a TOF are an almost umlimited mass range, high transmission and registration of all ions formed with one laser pulse. The advantages of a laser ion source are, besides those mentioned above, short time and tight spatial characteristics. With a Reflectron-TOF considerable mass resolution can be achieved. Several authors have shown that it is possible to reach mass resolving powers of 10 000 and more with a RETOF-MS/20/. By optimizing the laser pulse width, as well as some intrinsic details of our instrument and the ionization procedure/17/, we have reached a mass resolving power of 10 250 with an accumulation of 100 laser shots. The mass resolving power of a single spectrum is better than 12 000/21/.

The field-free drift region in our instrument has a length of 80 cm, while the ion reflector is 12 cm long. This results in a total time-of-flight of about 130 µs for ions of 700 eV and m/z 500. As an ion detector, channel plates in tandem configuration are used. The flight tube is pumped by a 350 Ls^{-1} turbomolecular pump.

The mass spectra were recorded on a LeCroy TR8828 200 MHz transient recorder equipped with two MM8103 32 Kbyte memories yielding, in total, 64 Kbytes of memory. The recorder was linked over a GPIB interface to a Force 68000-VMEbus computer.

3. RESULTS

By running the laser mass spectrometer with a medium mass resolving power (3000 at mass 106), we succeeded in efficient and soft ionization as well as partial fragmentation of several involatile molecules of biological and medical interest like amino acids/22/, chlorophylls/23/, porphyrins, tripeptides /24/ and the decapeptide angiotensin I /25/.

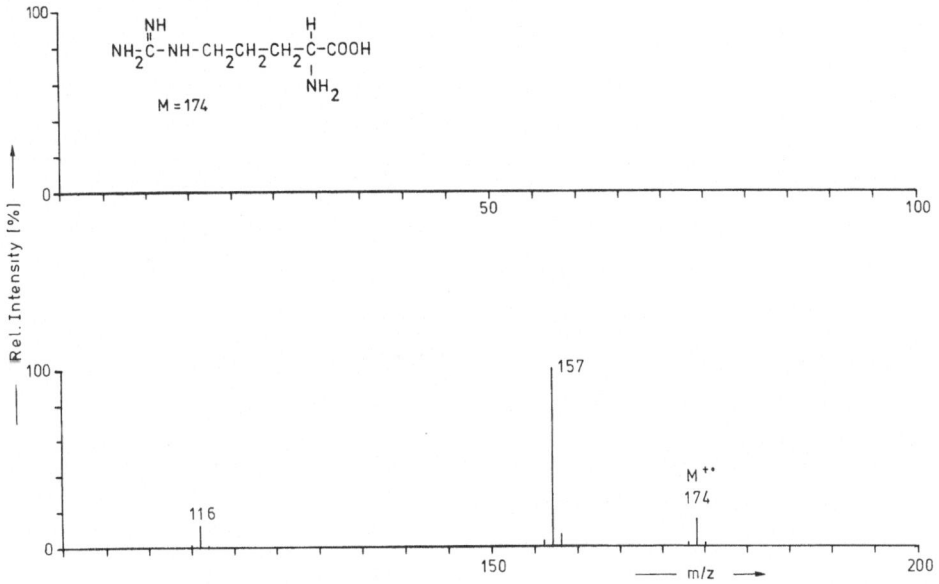

Figure 4. Non resonant multiphoton ionization mass spectrum of L-Arginine.

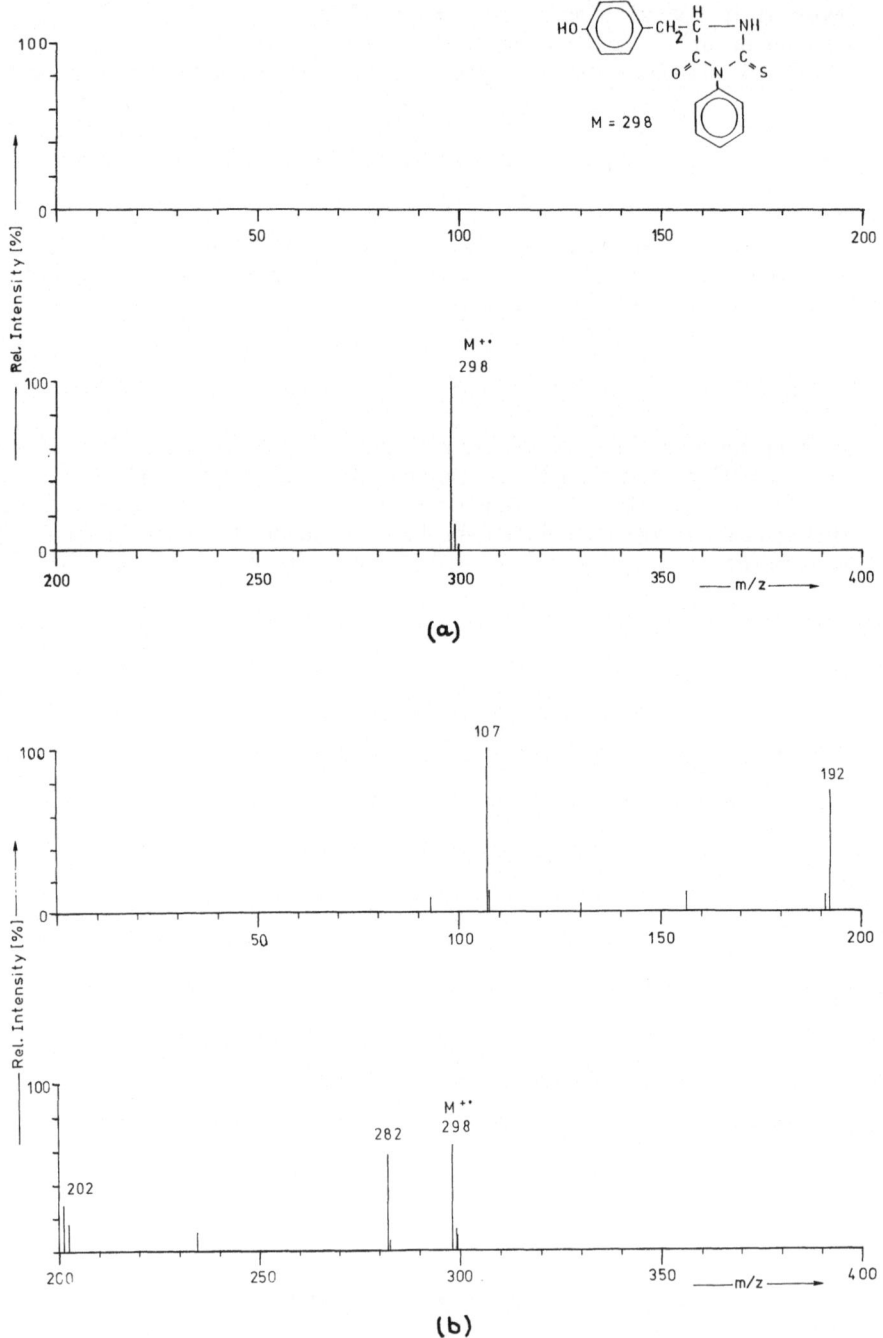

Figure 5. Soft(5a) and partial hard ionization(5b) mass spectra of the phenylthiohydantoine-tyrosine

In the following part we display and discuss some mass spectra of the above mentioned molecules as an example of the superior behaviour of multiphoton excitation in biomolecules.

3.1 Small Molecules

As an example of the multiphoton mass spectra of small molecules of biological interest, the mass spectrum of the amino acid arginine is displayed in the figure 4.

Arginine is known to be a delicate thermosensitive sample. Mass spectra of the intact molecule neither upon electron impact nor upon chemical ionization are known, since this molecule undergoes a neutral decompositon reaction upon heating. A field desorption mass spectrum of the protonated molecule has been measured/26/.

The non resonant two photon ionization mass spectrum of 1-arginine displays with a strong intensity of 15% the molecular ion. This is the first time that the radical cation of arginine is detected with a mass spectrometric technique/22/. Base peak of the spectrum is the loss of ammonia from the molecular ion, yielding the signal at m/z 157. The mass signal at mass 116 stems from the loss of the guanidino group from the molecular ion.

As a further example of the possibilities of MUPI the soft and partial hard ionization mass spectra of the phenylthiohydatoin-(PTH)-tyrosine are displayed in figure 5. Mass spectra of the thiohydantoin derivatives of amino acids are well known from literature/27/, since these derivatives are the final products in the Edman degradation of proteins and peptides. The spectrum, obtained in the soft ionization mode (figure 5a.), dis-playes only the molecular ion of the compound at mass 298. No other signal of a fragmentation reaction is detected. The partial hart ioni-zation mass spectrum shows still the molecular ion as a prominent signal, but several other ions are detected, too. The base peak of the mass spectrum at mass 107 stems from the loss of the aromatic moiety of the amino acid (hydroxytropylium ion). The intense signal at mass 192 is due to a hydrogen rearrangement and the subsequent loss of the side group from the amino acid. This signal is unique to all PTH-amino deri-vatives/22/. This ion is also detected in EI-mass spectra, but in MUPI with smaller intensities. Recently Zare and coworkers/28/ demonstrated also that it is possible to ionize such compounds under MUPI-conditions. But their mass spectra are not resolved by mass, since their instrument has only a mass resolving power of 80.

Figure 6 displays the mass spectra of a further small bioorganic molecule, the quercetin. This pentahydroxy-flavone is a highly polar substance and therefore difficult to detect in mass spectrometric inves-tigations. Figure 6a shows the soft ionization mass spectrum. Again only the molecular ion and no fragment ions are seen. In contrary, the partial hart ionization mode shows a large number of different fragmen-tations products. Main fragmentation products are the loss of the di-hydroxy-chromone moiety yielding in the signal at mass 109 and the signal at mass 137. This fragmentation product is due to a hydrogen rearrangement in the pyrone-ring and a partial loss of this ring system as indicated in figure 6b. The intense signals in the lower mass range

334

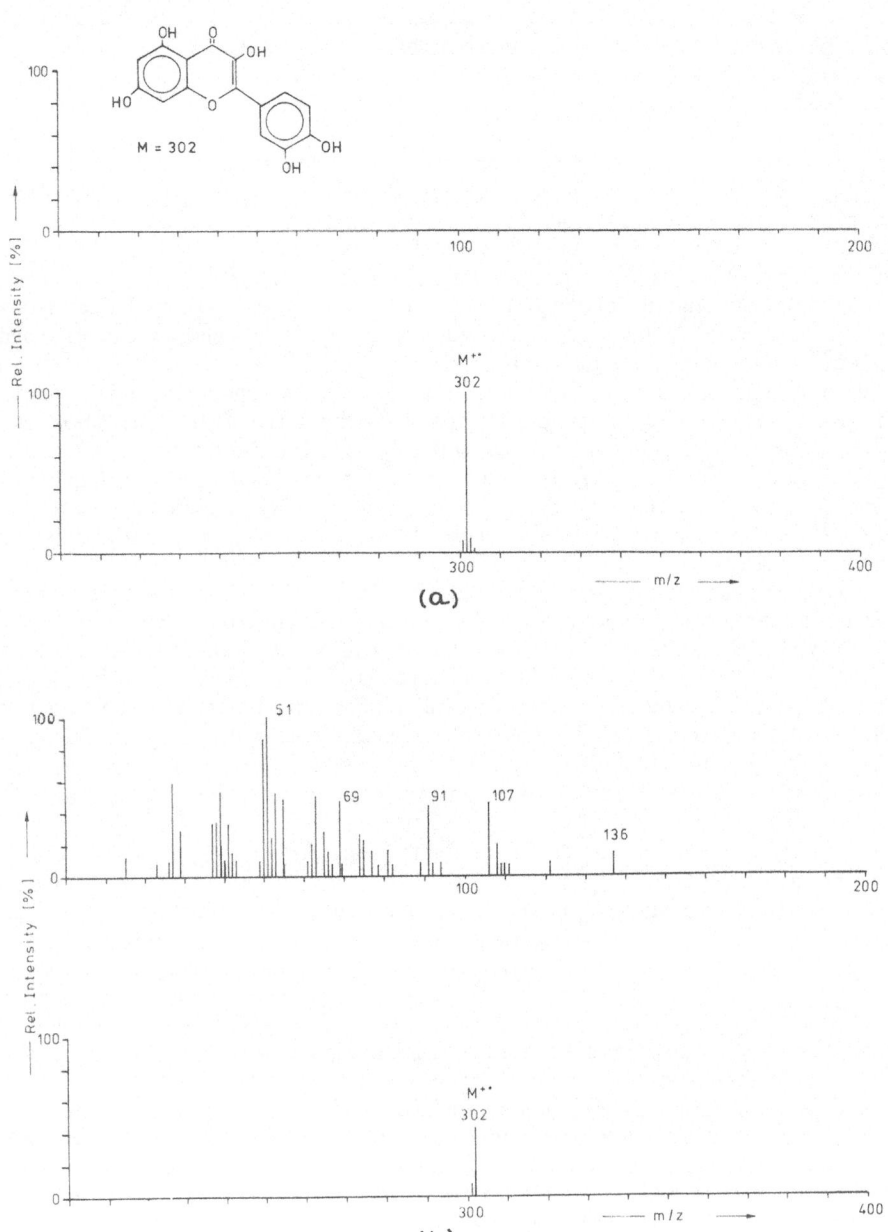

Figure 6. Soft(6a) and partial hard ionization(6b) mass spectra of quercetin.

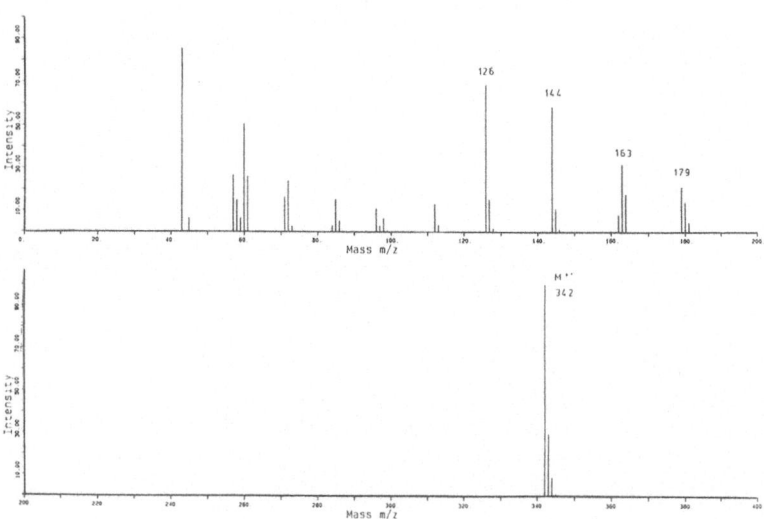

Figure 7. Non resonant multiphoton ionization mass spectrum and breakdown pattern of sucrose.

are due to the well known 'aromatic' mass series.

As a further example for small bioorganic molecules, the mass spectra of sucrose are shown in figure 7. Again this molecule has no chromophoric group suitable for the ionization. Therefore a non resonant two photon ionization takes place at a wavelength of 2500 A. The mass spectra display the molecular ion of the disaccharide in the soft ionization regime as well as in the partial hard ionization mode. As in the case of the arginine this molecule does not show a molecular ion upon electron impact ionization. It has been controversial in the literature/29/ whether a molecular ion of sucrose and other saccharides exists upon field desorption as well as electron impact ionization. From our investigation it is obvious that the radical cation of sucrose exists in the gasphase and can be detected in the mass spectrometer. The breakdown pattern of the suggar is well known from pyrolysis FI–MS investigations/30/. The main breakdown reaction consists of the bond rupture of the glycosidic bond under a hydrogen rearrangement forming a levoglucosane at mass 162. This central intermediate can be formed also by a loss of

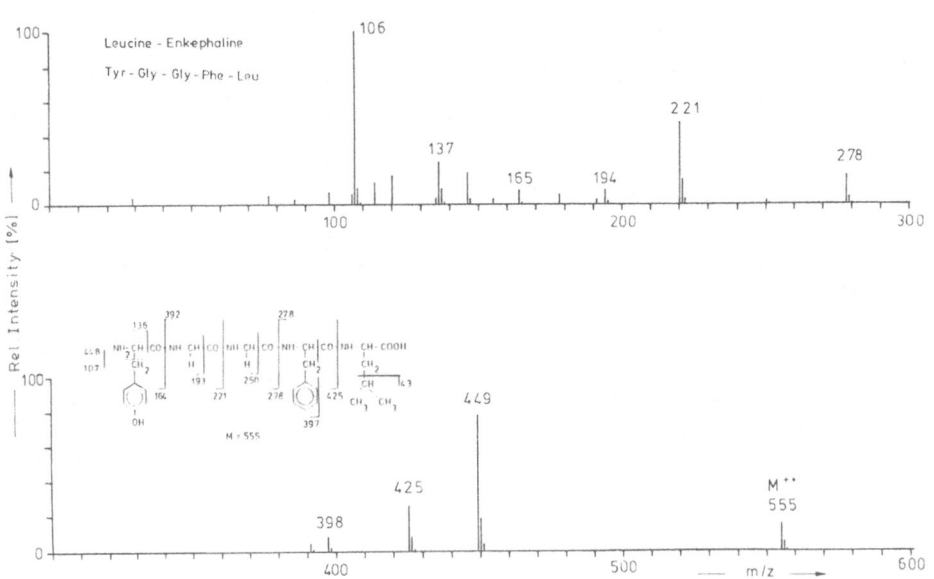

Figure 8. Partial hard ionization mass spectrum of leucine–enkephaline.

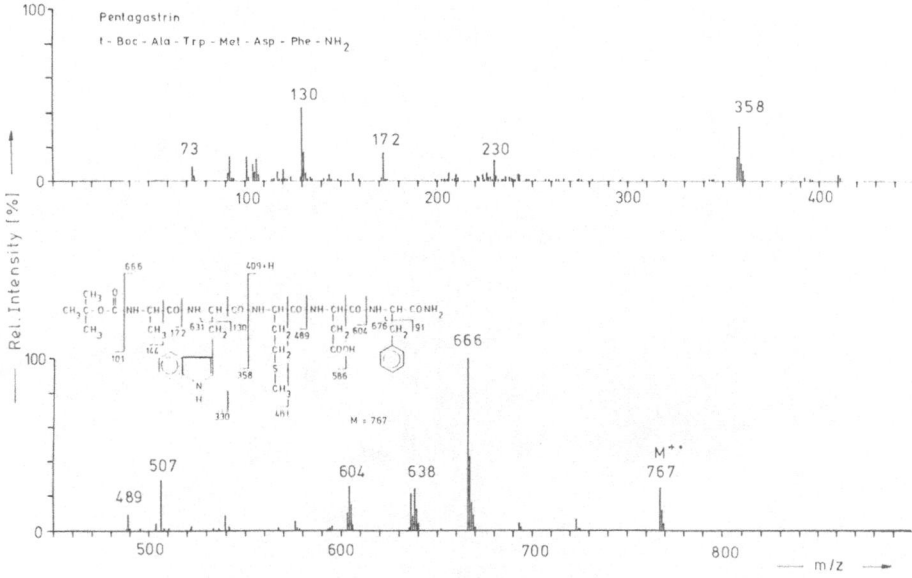

Figure 9. Partial hard ionization mass spectrum of pentagastrin.

a $C_6H_{11}O_5$- or of a $C_6H_{11}O_6$-radical yielding the signals at m/z 179 and 163, respectively, followed by either a loss of a hydroxy- or a hydrogen-radical. The further degradation reactions yield the signal at m/z 144 and at m/z 126 (levoglucosenone).

3.2 Intermediate Molecules

To demonstrate the usefullness of LEIM-MUPI mass spectrometric investigations in the intermediate mass range, e.g. between 500 - 1000 Daltons, we show the mass spectra of two smaller peptides, the leucine-enkephaline and the pentagastrine as well as the mass spectra of chlorophyll obtained from a biological sample.

The mass spectra of the leucine-enkephaline and the pentagastrine are shown in figure 8 and 9.

Both partial hard ionization mass spectra of these pentapeptides allow to detect the molecular ions as intense signals. It should be noted that in case of the leucine-enkephaline neither the C-terminal nor the N-terminal end of the molecule is blocked. In case of the pentagastrine both ends are protected. Besides these molecular ion peaks several signals due to the breakdown of the peptide bond are detectable.

As indicated in the mass spectrum of leucine-enkephaline the most intense signal, leading to the base peak, is due to a side chain reaction

338

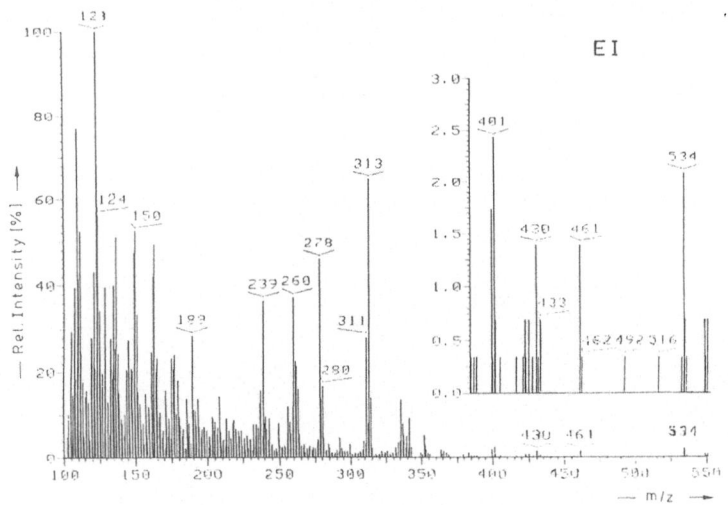

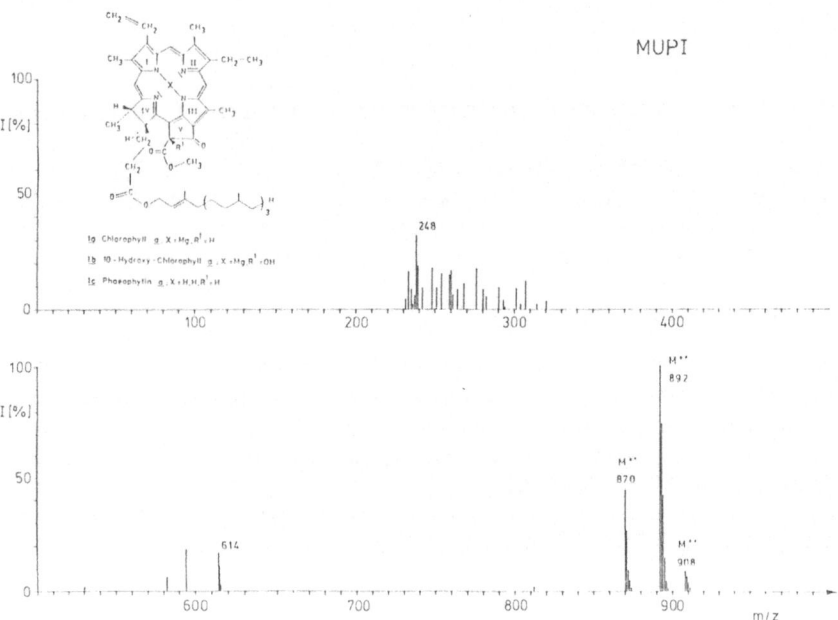

Figure 10. Comparison of the EI and the MUPI mass spectra of a biological
sample of chlorophyll **a**. (see text).

in the tyrosine amino acid, the loss of the hydroxy-benzyl-radical. The further intense signals, at mass 164,221 and 425 stem from the peptide bond rupture close to the aromatic amino acids tyrosine(164), tyrosine-glycine(221) and phenylalanine(425). These acylium ions are accompanied by satellite signals 28 mass units lower than these acylium-ions, indicating the formation of acyl-iminium ions. It should be noted that these signals from a C-terminal degradation together with the peaks stemming from the N-terminal breakdown at m/z 391 and 278 allows to deduce the primary structure of the peptide chain.

The same result can be drawn from the partial hard ionization mass spectrum of the pentagastrin. Again the molecular of this peptide can be detected at mass 767. The important signals due to the peptide chain are found at the masses 604, 358 and 172. Again these signals are accompanied by peaks due to the corresponding acyl-iminium ions. It should be noted that also the other signals due to the peptide rupture are found, but with less intensity. The intense peaks at m/z 665 and m/z 720 stem from the fragmentation of the tertiary-butyloxycarbonyl-group at the N-terminal end of the peptide chain. In the mass spectrum again also signals from the N-terminal degradation are found at m/z 596, 410 and 164.

As an example for the soft ionization possibility of MUPI in contrast to electron impact, both the EI and the MUPI mass spectra of a biological sample are displayed in figure 10. Here the mass spectra of native chlorophyll **a**, obtained by methanolic extraction of the cyano-bacterium Spirulina geitlerie without further purification, are shown. The soft ionization mass spectrum shows three different molecular ions, the chlorophyll **a** at mass 892, 10-hydroxy-chlorophyll **a** at mass 908 and phaeophytin **a** at mass 870. This mass spectrum again demonstrates the abilities of laser ionization, since by choosing the right wavelength only the porphin structures in the mixture are ionized.

3.3 Large Molecules

Figure 11 displays the soft and the hard ionization mass spectra of the decapeptide angiotensin **I**. Again the soft ionization mass spectrum shows exclusively the molecular ion of this compound. The partial hard ionization mass spectrum shows mass patterns that permit the deduction of the peptide sequence. The main signals are due to the bond breaking around the carbonyl group of the peptide chain. The breakdown pattern of this compound is shown schematically in Figure 12. Only minor fragmentation occurs in the side chain.

Figure 13 shows a mass spectrum of insuline obtained by laser desorption – multiphoton ionization. Again this mass spectrum shows as prominent signal the molecular ion of the protein at mass 5927. Besides this signal group peaks for the A-chain at mass 2334 and for the B-chain at mass 3895 appear. Especially to this mass spectrum, it should be noted, that this spectrum is drawn as measured. Neither a background correction nor a mass cut off in the lower mass range has been done. Therefore even small fragmentation products as the loss of a methyl group can be detected in the mass spectrum.

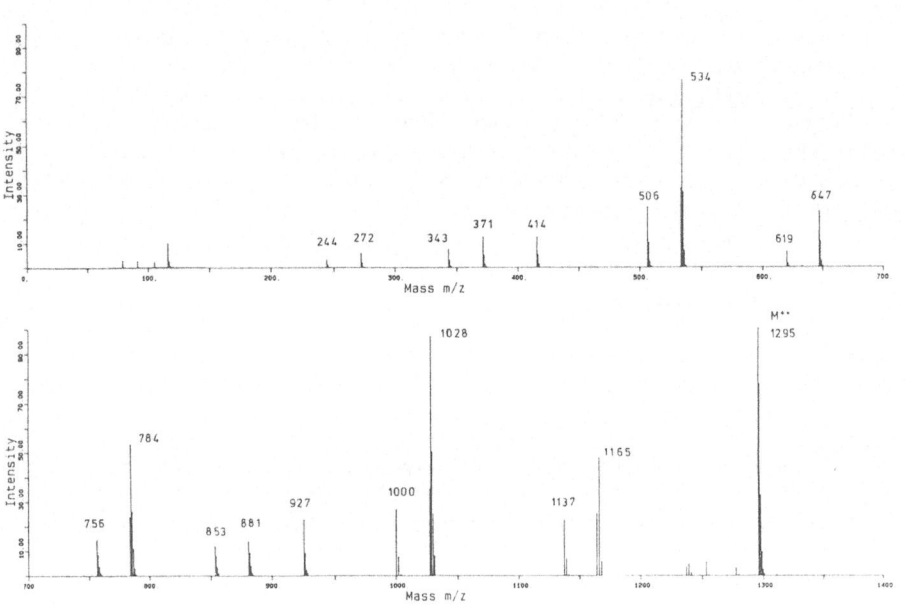

Figure 11. Breakdown pattern and partial hard ionization mass spectrum angiotensin **I.**

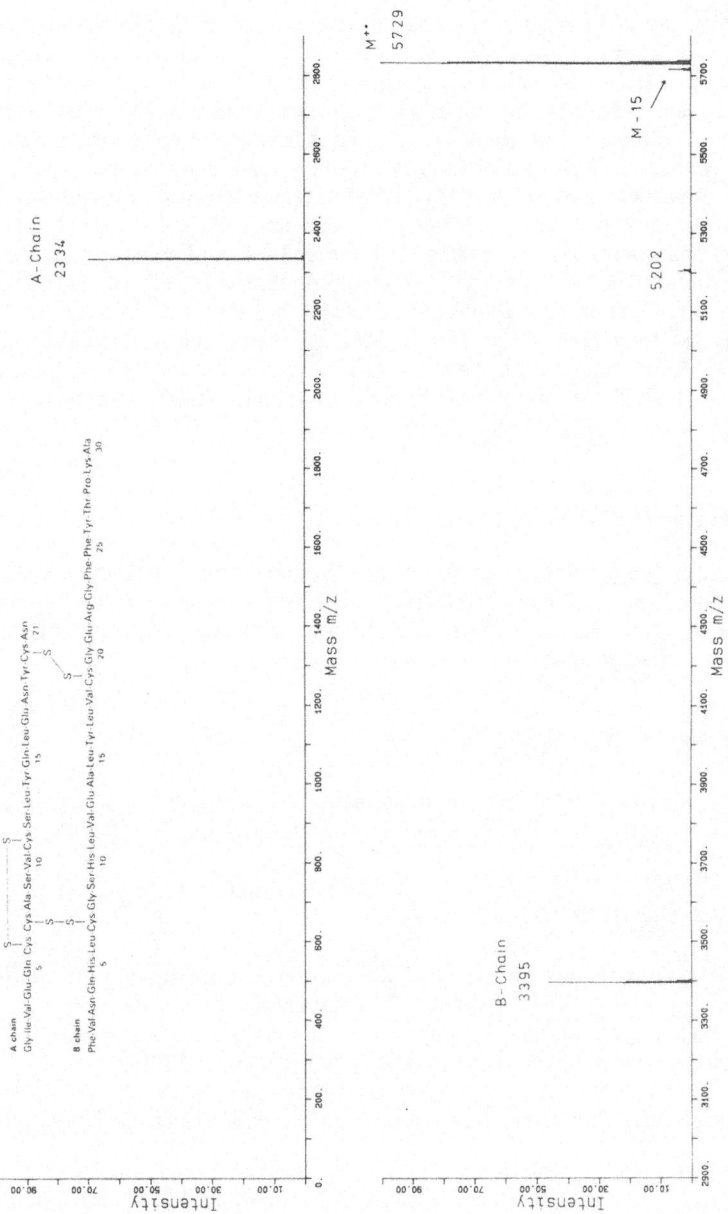

Figure 12. Multiphoton ionization mass spectrum of insulin.

4. CONCLUSION

In conclusion we will summarize the advantages of MUPI–RETOF–Mass Spectrometry using a laser desorption source for free neutral molecules.

The mass range is theoretically unlimited and with Reflectron–TOF instruments mass resolving powers of better than 10,000 can be reached. As mentioned above, the mass range is limited at present only by the detection system. The experimental setup produces mass spectra with intensive pure molecular ion signals without transfer, abstraction or addition of species like hydrogen, sodium etc. Soft, as well as species selective ionization is possible and tunable fragmentation promises new insights concerning structural information as well as ion kinetics.

We are sure that the sensitivity, which is at the moment in the fmol range, can be improved by a few orders of magnitude. A great advantage is the very low background level, also called "chemical noise". Finally, we should mention easy sample preparation, MS/MS capability/4/ and fast total mass spectrum recording.

5. ACKNOWLEDGEMENTS

This work has been supported by a grant from the Bundesministerium fur Forschung und Technologie (13N5307). The authors thank Dr. U.Boesl and Mr. K.Walter for their assistance during the development of the mass spectrometer system and the measurements.

6. REFERENCES

/1/ B.Sundqvist, A.Hedin, P.Haakanson, I.Kamensky, M.Salehpour and G.Säwe; Int. J. Mass Spectrom. Ion Proc., **65**,69 (1985).

/2/ B.A.Mamyrin, V.I.Karataev, D.V.Shmikk and V.A.Zagulin; Sov. Phys. JETP **37**, 45 (1973).

/3/ M.L.Muga; presented at the 32th Annual Conference on Mass Spectrometry and Allied Topics, San Antonio, TX, 1984. M.L.Muga; presented at the 34th Annual Conference on Mass Spectrometry and Allied Topics, Cincinnati, OH, 1986.

/4/ U.Boesl, H.J.Neusser, R.Weinkauf and E.W.Schlag; J. Phys. Chem. **86**, 4857 (1982).

/5/ H.D.Beckey; "Field Ionization Mass Spectrometry", Pergamon Press, Oxford (1971).

/6/ M.A.Baldwin and F.W.McLafferty; Org. Mass Spectrom. **7**, 1353 (1973).

/7/ A.Benninghoven and W.Sichtermann; Anal. Chem. **50**, 1180 (1978).

/8/   M.Barber, R.S. Bordoli, G.J.Elliot, R.D.Sedgwick and A.N.Tyler; Anal. Chem. **54**, 645A (1982).

/9/   B.Sundquist and R.D.Macfarlane; Mass Spectrom. Rev. **4**, 421 (1985).

/10/   M.A.Posthumus, P.G.Kistemaker and H.L.C.Meuzelaar; Anal. Chem. **50**, 985 (1978).

/11/   U.Boesl, H.J.Neusser and E.W.Schlag; Z. Naturforsch. **33A**, 1546 (1978).
L.Zandee, R.D.Bernstein and D.A.Lichtin; J. Chem. Phys. **69**, 3427 (1978).

/12/   R.B.van Breemen, M.Snow and R.Cotter; Int. J. Mass Spectrom. Ion Proc. **49**, 35 (1983).

/13/   H.v.Weyssenhoff, H.L.Selzle, E.W.Schlag; Z.Naturforsch. **40A**, 674 (1985).
W.E.Henke, H.v.Weyssenhoff, H.L.Selzle, E.W.Schlag; Verh. Dtsch. Phys. Ges. **3**, 139 (1983).

/14/   J.M.Hayes and G.J.Small; Anal. Chem., **55**, 565A (1983).

/15/   D.H.Levy, Science **214**, 263 (1981).

/16/   K.Walter, Diplom Thesis, TU München, FRG (1985).

/17/   for a complete discussion see :
U.Boesl,J.Grotemeyer,K.Walter and E.W.Schlag; Anal.Instrum., 1987, in press.

/18/   U.Boesl, H.J.Neusser and E.W.Schlag; Chem. Phys. **55**, 193 (1981).

/19/   U.Boesl, H.J.Neusser and E.W.Schlag; Chem. Phys. Lett. **87**, 1 (1982).

/20/   E.Niehuis, T.Heller, H.Feld and A.Benninghoven; J. Vac. Sci. Technol.,(1987), to be published.
E.Niehuis, T.Heller, H.Feld and A.Benninghoven; presented at the 34th Annual Conference on Mass Spectrometry and Allied Topics, Cincinnatti, OH, 1986.
M.Yang, J.R.Millard and J.P.Reilly; Opt. Communications, **55**, 41 (1985).
M.Yang and J.P.Reilly; presented at the 34th Annual Conference on Mass Spectrometry and Allied Topics, Cincinnatti, OH, 1986.

/21/   K.Walter, U.Boesl and E.W.Schlag; Int. J. Mass Spectrom. Ion Proc., **71**, 309 (1986).

/22/ J.Grotemeyer, K.Walter, U.Boesl and E.W.Schlag; Int. J. Mass Spectrom. Ion Proc., in press.

/23/ J.Grotemeyer, U.Boesl, K.Walter and E.W.Schlag; J. Am. Chem. Soc. **106**,4233 (1986).

/24/ J.Grotemeyer, U.Boesl, K.Walter and E.W.Schlag; Org. Mass Spectrom. **21**, 645 (1986).

/25/ J.Grotemeyer, U.Boesl, K.Walter and E.W.Schlag; Org. Mass Spectrom. **21**, 595 (1986).

/26/ H.U.Winkler and H.D. Beckey; Org. Mass Spectrom. **6**,655 (1972). H.J.Heinen, U.Giessmann and F.W.Röllgen; Org. Mass Spectrom. **12**,710 (1977).

/27/ H.Hagenmaier, W.Ebbinghausen, G.Nicholsen and W.Völtsch; Z. Naturforsch., **25b**,681 (1970).

/28/ F.Engelke, J.H.Hahn, W.Henke and R.N.Zare; Anal.Chem., **59**, 909 (1987).

/29/ H.J.Veith and F.W.Röllgen; Org. Mass Spectrom., **20**, 689 (1985). D.E.C.Rogers and P.J.Derrick; Org. Mass Spectrom., **19**, 490 (1984). P.J.Derrick, N.Tang-Trong and D.E.C. Rogers; Org. Mass Spectrom., **20**, 690 (1985).

/30/ H.R.Schulten and W.Görtz; Anal. Chem. **50**, 428 (1978). F.Shafizadeh and Y.L.Fu; Carbohyd. Res. **29**, 110 (1973).

UNIMOLECULAR DECAY OF ENERGY-SELECTED POLYATOMIC MOLECULAR IONS BY RESONANCE-ENHANCED MULTIPHOTON IONIZATION IN A REFLECTRON MASS SPECTROMETER

A. Kiermeier, H. Kühlewind, H. J. Neusser, E. W. Schlag
Institut für Physikalische und Theoretische Chemie der
Technischen Universität München
Lichtenbergstr. 4, D-8046 Garching, Germany

ABSTRACT.   It is shown by a photoelectron energy analysis that resonantly enhanced two-photon ionization leads to vibrational state-selected benzene ($C_6H_6^+$, $C_6D_5H^+$ and $C_6D_6^+$) cations. These are further excited with a second laser pulse of variable frequency beyond the dissociation threshold for the four decay channels of lowest energy. The excitation results in a metastable decay of the isotopic benzene cations with typical decay rate constants of some 10^6 s^{-1}. Decay rate measurements of internal energy-selected ions were performed in a reflectron time-of-flight mass spectrometer. We found the results to be in good agreement with our RRKM calculations. This confirms the statistical character of the benzene ion dissociation.

1. INTRODUCTION

It has been demonstrated during the last decade that multi-photon ionization (MPI) is a novel source of ions in a mass spectrometer /1-5/. This technique has particular features not present in conventional ion sources like electron impact, etc. For instance, at low light intensities ($I < 10^7$ W/cm^2) soft ionization, i.e., the almost exclusive production of parent ions, is possible /1/. At high light intensities ($I > 10^8$ W/cm^2) however, a rich and highly structure-specific fragmentation pattern is obtained /2,3,6,7/. It was shown in previous work /2/ that the mechanism responsible for soft ionization as well as for hard fragmentation is "ladder switching" from the neutral molecules to the parent ions and successively to fragment ions of decreasing size. The ladder switching mechanism causes highly structure-specific fragmentation patterns even for those isomeric cations which normally cannot be distinguished by electron impact or other conventional ion sources /6/. Another advantage of resonance-enhanced MPI is the high efficiency which leads to nearly complete ionization of all neutral molecules within the laser focus /8/, a precondition for sensitive trace analysis.

    Resonance-enhanced MPI is extremely selective due to the intermediate state spectrum which is sharp even in many polyatomic

345

A. C. P. Alves et al. (eds.), Frontiers of Laser Spectroscopy of Gases, 345–352.
© 1988 by Kluwer Academic Publishers.

molecules. This fact does not only allow the selection of particular molecules from a mixture /9/ - a feature of great practical interest for mixture and trace analysis - but also permits the selection of molecules in a particular vibrational state, even though many states might be thermally populated. In this paper it will be shown that resonance-enhanced two-photon ionization via suitable intermediate vibronic states leads to the production of vibrational state- and therefore energy-selected polyatomic molecular ions. State- and energy-selected ions are of great interest for the precise investigation of dissociation kinetics. We will present an example for the decay time measurements of highly excited internal energy-selected molecular ions.

2. PHOTOELECTRON KINETIC ENERGY ANAYSIS

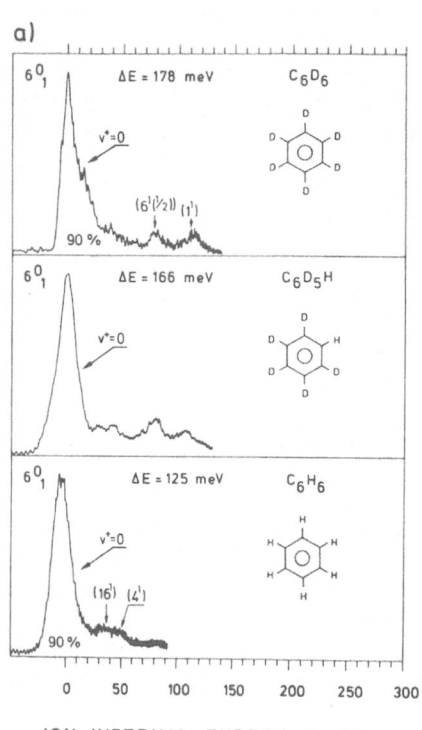 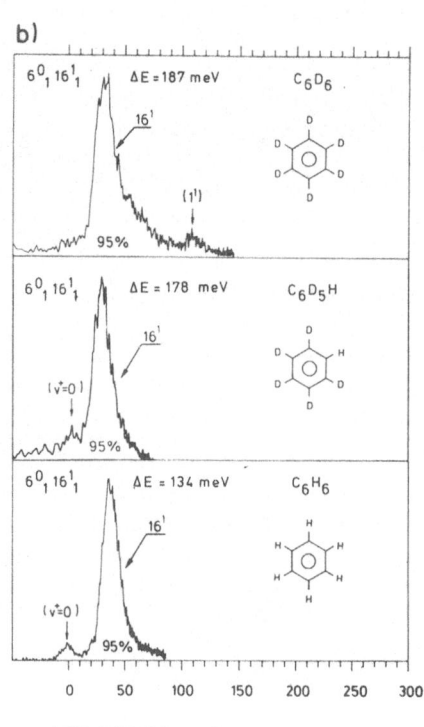

Figure 1: Photoelectron spectra of benzene isotopes $C_6H_6^+$, $C_6D_5H^+$ and $C_6D_6^+$ after resonance-enhanced two-photon ionization
a) via the 6^0_1 band of the $S_1 \leftarrow S_0$ transition of the neutral benzene
b) via the $6^0_1 16^1_1$ band of the $S_1 \leftarrow S_0$ transition of the neutral benzene

Even though the excitation of the resonant intermediate state guarantees state-selection in the neutral molecules, the vibrational state-selection of the ion is not automatically achieved after the second absorption step to the ionization continuum /10/. This is due to the excess energy above the ionization continuum present in a one-laser experiment where the laser frequency is determined by the first resonant absorption to the intermediate state. The resulting excess energy can be distributed between the kinetic energy of the produced photoelectrons and internal energy of the ions. In order to check the internal energy distribution of the produced ions, an analysis of photoelectron energy was performed with a time-of-flight photoelectron analyzer /11/. In Fig. 1a the resulting photoelectron spectra of three benzene isotopes $C_6H_6^+$, $C_6D_5H^+$ and $C_6D_6^+$ are shown. Ionization took place via the 6^0_1 transition at 37481.6 cm^{-1}, 37673.1 cm^{-1} and 37712.0 cm^{-1}, respectively, which leads to the vibrationless S_1 state as an intermediate state. The corresponding excess energies ΔE above the ionization threshold are indicated as an inset. It is clearly seen that for this chosen intermediate state more than 90% of the isotopic benzene ions are produced in a single vibrational state, i.e., the vibrationless ground state.

Alternatively, when the laser frequency is tuned to the $6^0_1 16^1_1$ sequence band at 37321.1 cm^{-1}, 37471.6 cm^{-1} and 37595 cm^{-1}, respectively, the 16^1 vibrational state acts as an intermediate state in S_1. As shown in Fig. 1b for this particular transition again state-selected ions are produced, however, now almost exclusively in the 16^1 state with a well defined internal energy of some 40 meV, again preserving the vibrational character of the excited intermediate state into the ion.

3. UNIMOLECULAR ION DECAY IN A REFLECTRON TIME-OF-FLIGHT MASS SPECTROMETER

The excitation scheme for our kinetic studies of unimolecular decay is shown in fig. 2.

Benzene ions are produced in an effusive molecular beam inside the acceleration field of a reflectron time-of-flight mass spectrometer /12/. (see Fig. 3). Laser 1 is tuned to the frequency of the 6^0_1 or $6^0_1 16^1_1$ band and produces, via a resonance-enhanced two photon absorption, state- and energy-selected benzene cations.

200 ns later, after the ion cloud has left the molecular beam region, a second laser pulse of variable photon energy (5.07 – 5.52 eV) further excites the ions to an energy level slightly above the thresholds for the several different dissociation channels with low threshold energy E_0 (see Fig. 2). This leads to the production of $C_6H_5^+$, $C_6H_4^+$, $C_4H_4^+$ and $C_3H_3^+$ ionic fragments /13/ and their deuterated analogues, respectively. The internal energy of the ions is sufficient to induce metastable decay on a µs timescale, i.e., a decay of the ions on their way from the ion source to the reflecting field. The reflecting field, working in the energy correcting mode, substantially improves the mass resolution (M/ΔM = 4000) compared to a linear time-of-flight mass spectrometer, a precondition for the exact measurement of metastable signals. Since the reflecting field can also act as an kinetic energy

348

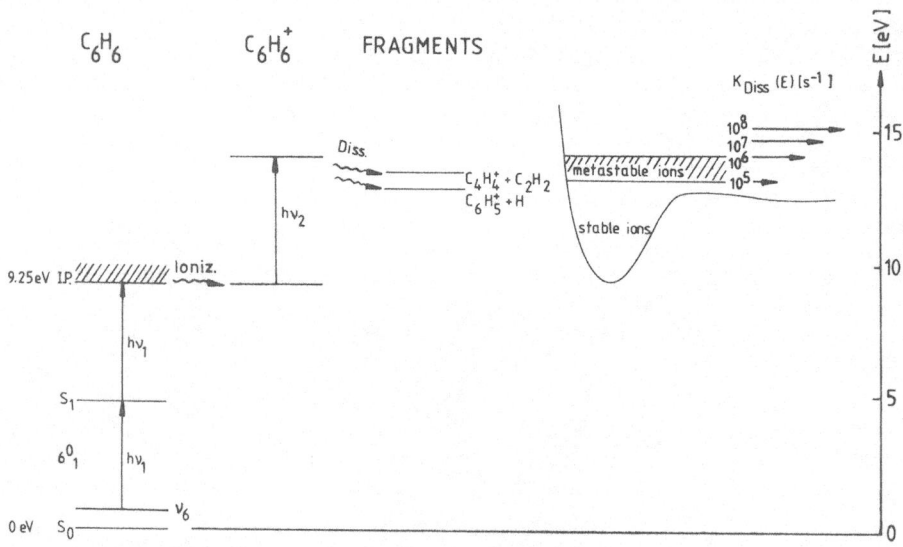

Figure 2: Excitation scheme for the two laser pump-pump experiment. State-selected benzene cations are produced in a resonantly enhanced two-photon ionization process. A second laser pulse of variable photon energy excites the ions to a well defined energy level above the dissociation threshold and metastable dissociation takes place (taken from ref. /16/).

analyzer, daughter ions resulting from a metastable decay are separated from interfering stable ions /14/. Signals of metastable ions produced in the acceleration region are well separated in the mass spectrum from those resulting from a decay in the field free drift region due to their different kinetic energies and the consequently resulting time-of-flight differences. From the ratio of these two signals the unimolecular decay rate constant is determined according to the technique described in our previous work /15/. Since all four decay channels of low energy, independent of their H-loss or C-loss character, are competing and originate from one electronic state /15,16/ the measured decay rate constant represents the total decay rate constant. The total decay rate constant is given by the sum of the individual rate constants of all competing decay channels. The experimental determined variation of the total decay rate constant with internal energy of the ions is shown in Fig. 4 for the three benzene isotopes.

In separate experiments ions in the vibrationless state and, alternatively, ions in the 16^1 state were each excited into the metastable energy range. In that way total decay rate constants for ions of the same internal energy but resulting from different excitation pathways could be compared directly.

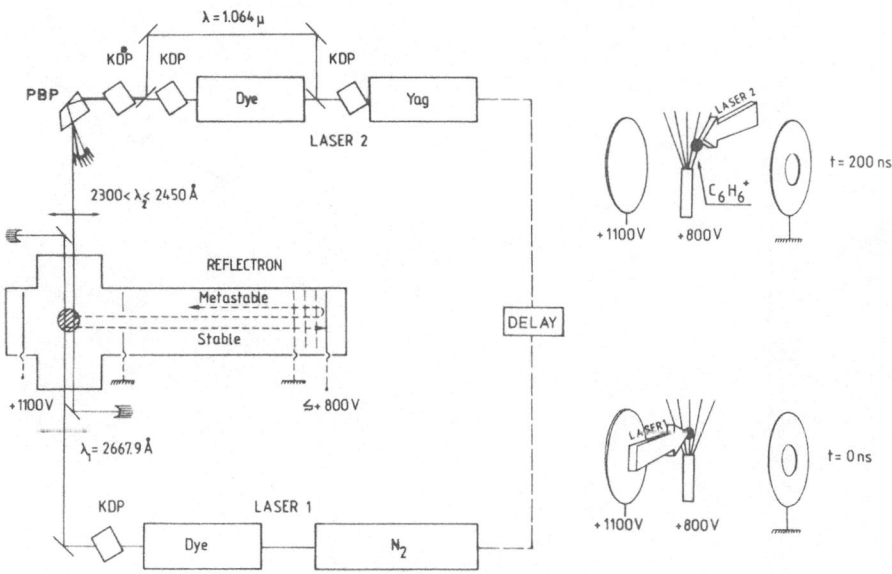

Figure 3: Scheme of a two laser pump-pump experiment for the production of internal energy-selected molecular ions in a reflectron time-of-flight mass spectrometer (from ref. /15/). Laser 1 produces state-selected molecular ions and 200 ns later laser 2 excites these ions to a well defined internal energy level above dissociation threshold. The dissociation rate constants of the energy-selected ions are measured by the technique of detection and energy analysis of metastable ions.

Several points of interest are found:

i)   there is a smooth tenfold increase of the total decay rate constant in the observed internal energy range;

ii) no difference in the total decay rate constant is found for ions of the same internal energy but resulting from different pathways of production. Ions produced via the vibrationless S_1 state (●) and, on the other hand, ions produced via the 16^1 state (▨) display the same decay rate constant if excited to the same internal energy;

iii) a pronounced kinetic isotope effect for the three different benzene isotopes in the observed energy range is found.

The results point to a dissociation which proceeds according to a statistical model of unimolecular dissociation. In particular, no vibrational specificity of the decay rate constant is found when excitation proceeds via the two different excitation pathways described in ii).

350

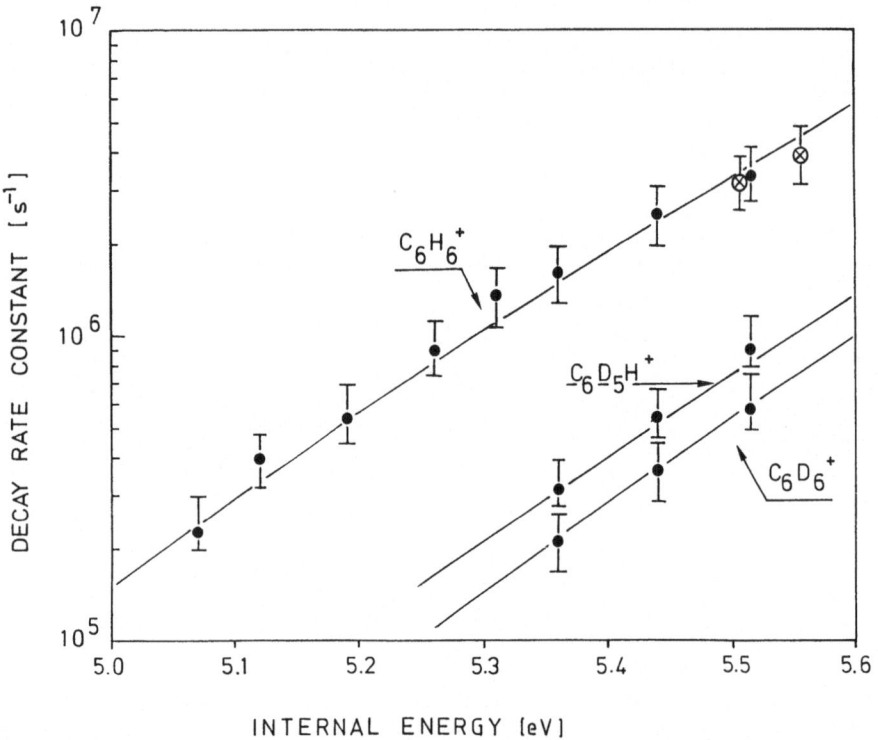

Figure 4: Total decay rate constants of internal energy-selected benzene cations $C_6H_6^+$, $C_6D_5H^+$ and $C_6D_6^+$ as a function of their internal energy. The solid lines represent the result of RRKM calculations fitted to the experimental data. Vibrationless benzene cations are produced via the 6^0_1 transition (●). Benzene cations in the 16^1 vibrational state are produced via the $6^0_1 16^1_1$ transition (⊗). Subsequently both ion species are excited with the second laser pulse to the internal energy indicated on the abscissa.

For an interpretation of our results we performed statistical RRKM calculations of the individual decay rate constants of all four competing decay channels at low threshold energy. The latter have been experimentally extracted from the directly measured total decay rate constant (see Fig. 4) and the simultaneously measured branching ratios of the relevant fragment ions /16/. For different isotopically labelled species a good simulation of experimental results is obtained with a single set of parameters for the determination of the frequencies of the activated complex (solid line in Fig.4). Isotope shifts of the vibrational frequencies were obtained by use of the Teller-Redlich product rule. This points to a high reliability of the set of parameters used and yields detailed information on the structure of the activated complex for the four decay channels under consideration /16/. In particular, the looseness of the activated complex with respect to the complete set of frequencies as well as the threshold energies E_0 and activation entropies $\Delta S^{\#}_0$ of the decay channels under consideration are deduced from the experimental data. Furthermore the measured intramolecular and intermolecular isotope effects yield a microscopic probe of the looseness or tightness of C-H bonds in the activated complexes and give detailed insight into the character of the reaction coordinate /17/.

4. CONCLUSION

In conclusion, we have shown that resonantly enhanced two-photon ionization is a versatile method for the production of state- and energy-selected polyatomic molecular ions. This was explicitly demonstrated by an analysis of the kinetic energy distribution of the ejected photoelectrons. In a reflectron time-of-flight mass spectrometer the total decay rate constants and individual decay rate constants of internal energy-selected molecular ions have been measured for various well defined internal energies. From our experimental results detailed information about the statistical character of the dissociation mechanism and the structure of the activated complex is obtained.

In the future it seems possible to study the ergodic character of the energy redistribution prior to dissociation for even larger polyatomic molecular ions.

REFERENCES

/1/ U. Boesl, H.J. Neusser, E.W. Schlag, Z. Naturforsch, **33A**, 1546 (1978)

/2/ U. Boesl, H.J. Neusser, E.W. Schlag, J. Chem. Phys., **72**, 4327 (1980)

/3/ L. Zandee, R.B. Bernstein, D.A. Lichtin, J. Chem. Phys., **69**, 3427 (1978)

/4/ V.S. Antonov, I.N. Knyazev, V.S. Letokhov, V.M. Matiuk, V.G. Morshev, V.K. Potapov, Opt. Lett., **3**, 37 (1978)

/5/ E.W. Schlag, H.J. Neusser, Acc. Chem. Res., **16**, 355 (1983)

/6/ H. Kühlewind, H.J. Neusser, E.W. Schlag, J. Phys. Chem., **89**, 5600 (1985)

/7/ T. Carney, T. Baer, J. Chem. Phys., **75**, 477 (1981)

/8/ U. Boesl, H.J. Neusser, E.W. Schlag, Chem. Phys., **55**, 193 (1980)

/9/ U. Boesl, H.J. Neusser, E.W. Schlag, J. Amer. Chem. Soc., **103**, 5058 (1981)

/10/ S.R. Long, J.T. Meek, J.P. Reilly, J. Chem. Phys., **79**, 3206 (1983)

/11/ A. Kiermeier, H. Kühlewind, H.J. Neusser, E.W. Schlag, to be published

/12/ U. Boesl, H.J. Neusser, R.Weinkauf, E.W. Schlag, J. Phys. Chem., **86**, 4857 (1982)

/13/ H.M. Rosenstock, J.T. Larkins, J.A. Walker, Int. J. Mass Spectrom. Ion Phys., **11**, 309 (1973)

/14/ H. Kühlewind, H.J. Neusser, E.W. Schlag, J. Chem. Phys., **82**, 5482

352

(1985)

/15/ H. Kühlewind, H.J. Neusser, E.W. Schlag, <u>J. Phys. Chem.</u>, **88**, 6104 (1984)

/16/ H. Kühlewind, A. Kiermeier, H.J. Neusser, E.W. Schlag, <u>J. Chem. Phys.</u>, **85**, 4427 (1986)

/17/ H. Kühlewind, A. Kiermeier, H.J. Neusser, E.W. Schlag, to be published

INFRARED SPECTROSCOPY OF MOLECULAR IONS

Takeshi Oka

Department of Chemistry and
Department of Astronomy and Astrophysics
The University of Chicago
Chicago, Illinois 60637 U.S.A.

ABSTRACT

In this course of three one-hour lectures, I cover the three main features of molecular ion spectroscopy using tunable infrared sources and discharges, (I) production of ions, (II) spectroscopy of molecular ions, and (III) analysis of infrared ion spectra. There are four subject matters that play major roles in this course: (1) Molecular ions with its tremendous activity and challenge (2) Infrared laser spectroscopy with its high sensitivity and resolution. (3) Plasmas full of ions and mystery, and (4) Interstellar space with a high degree of ionization and inexhaustible richness.

I. PRODUCTION OF MOLECULAR IONS

1.1. Introduction

Let us take the hydrogen molecule H_2 as the starting point, a stable, neutral system with two protons and two electrons. We can produce charged species, i.e., molecular ions, by subtracting or adding an electron or a proton. The simplest way to make an ion from H_2 is to knock off an electron to produce H_2^+. This occurs in the laboratory plasma through electron bombardment and in space through cosmic ray ionization and photoionization due to star radiation. Radiofrequency spectroscopy of H_2^+ by Dehmelt and Jefferts[1] and infrared spectroscopy of HD^+ by Wing, Ruff, Lamb and Spezeski[2] are the classic work on the ionic species. We can also attach an electron to H_2; we then obtain H_2^-. This anion is known to exist in spite of the negative electron affinity of H_2.[3]

A. C. P. Alves et al. (eds.), Frontiers of Laser Spectroscopy of Gases, 353–377.

354

Instead of subtracting and adding an electron to produce the
cation and anion, we can do it also by adding and subtracting protons.
We then obtain H_3^+ and H^- from H_2. The whole situation can be
summarized in a diagram shown in Figure 1.

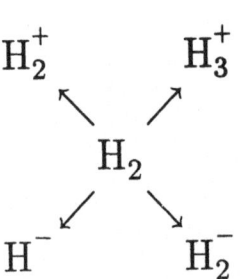

Fig. 1. Four
"stable" ions
from the hydrogen
molecule.

While H_2^+ and H_2^- produced by subtracting or adding an electron
are paramagnetic radical ions and chemically very active, H_3^+ and H^-
are isoelectronic to H_2 and therefore diamagnetic and quite stable.
They are called protonated and deprotonated ions. *These ions have
been known to mass spectroscopists over many years but their spectroscopy has not been conducted successfully until very recently.[5]

Some of you may think that these ionic species are unstable and
are very rare exotic species in the nature. They are rare in the
terrestrial atmosphere. However our terrestrial atmosphere is an
exception in nature. The universe as a whole is much more highly
ionized; atomic and molecular ions are abundant and play important
roles in various aspects of astrophysics. For example, since the work
of Bethe[6] and Hylleraas,[7] stability of the hydrogen atom anion H^- is
well-known. The H^- ions exist abundantly around the sun and its
absorption of the solar radiation is the major cause for the sun's
opacity as initially noted by Wildt[8] and shown conclusively by a
series of papers by Chandrasekhar and his colleagues.[9]

The stability of protonated hydrogen H_3^+ has been well known
since its discovery by J.J. Thomson.[10,11] Although it is yet to be
discovered in space,[12] it is the most important ingredient in our
understanding of ion-molecule reaction scheme of dense molecular
clouds.[13-16] Ab-initio calculations show the energy relation

---

*It is none other than J.J. Thomson who discovered H_3^+ in 1912. For
the long history of H_3^+ see Reference 4.

$$H_3^+ = H_2 + p - \Delta H_e$$

with $\Delta H_e = 4.59$ eV.[17,18] The large value of ΔH_e (proton affinity) shows the stability of this ion.

In general protons attach to any stable molecule or atom to produce stable protonated ions, in other words the proton affinities of stable molecules are always positive. A convenient list of proton affinity is found in Ref. 19. Some typical values are: H_2 4.4 eV, He 1.9 eV, CH_4 5.4 eV, NH_3 8.8 eV, H_2O 7.3 eV, CO 6.1 eV, N_2 4.9 eV, HCN 7.5 eV, C_2H_2 6.5 eV. On the contrary electron affinities of most stable molecules are negative. Values of electron affinity are found in Reference 20.

Because of the large proton affinities protonated ions are quite stable. For example, if we use a thermodynamical argument the exothermicity of

$$H_2O + p \rightleftharpoons H_3O^+$$

reaction is so large at room temperature ($\Delta H = 7.3$ eV ~ 290 kT), that the ratio of H^+ to H_2O^+ in solution is 10^{-130}. Hence Sidgwick's remark,[21] "in 10^{70} universes filled with a normal acid solution there would be one unsolvated hydrogen ion". Because of this stability, protonated ions have been well-known to mass spectroscopists. However their spectra had never been observed in any spectral region until 1970 when radio astronomers Snyder and Buhl[22] observed the microwave emission of unknown species and named it X-ogen. This caught laboratory spectroscopists by surprise and (although Klemperer correctly conjectured[23] the carrier of the emission line to be protonated carbon monoxide HCO^+), it took five years until in 1975 the spectral line was identified in the laboratory of Woods and his group.[24] As is often the case, the message came from the sky.

1.2. Molecular Ions

We can extend the manoeuvre shown in Fig. 1 and obtain a great many fundamental molecular ions as shown below. Fig. 2 summarizes molecular ions containing hydrogen atoms and one heavy atom; we limit ourselves to carbon, nitrogen and oxygen which are highly abundant in nature. The most well studied ones are of course the stable species CH_4, NH_3 and H_2O. They are the most fundamental examples of spherical top, symmetric top, and asymmetric top, respectively, and much of our knowledge on spectroscopy has been obtained from the enormous amount of studies on these molecules.[25,26] From each of these stable species we obtain free radicals CH_3, CH_2, CH; NH_2, NH; OH which have also been studied spectroscopically.[26,27] Now the subjects of this course are the charged species, cations and anions produced from the

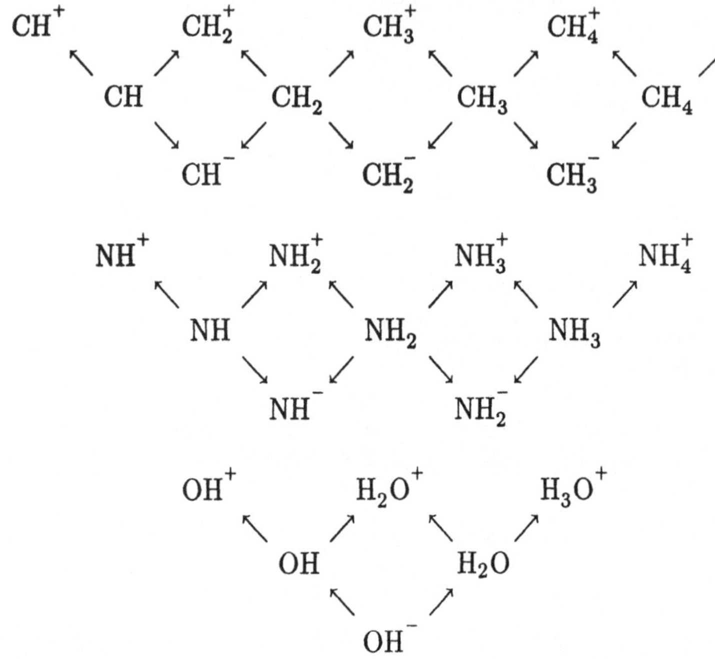

Fig. 2. Hydride molecules, radicals and ions.

neutral species. There are so many of them as seen from Fig. 2 and each gives a fascinating spectrum.

Some of these ions (diatomic cations CH^+, NH^+, OH^+ and H_2O^+) have been studied through their electronic spectra prior to the present activity of molecular ion spectroscopy. We use laser infrared spectroscopy: infrared because all these ions absorb infrared radiation strongly and laser spectroscopy because of its high sensitivity and resolution.

Figure 2 is extremely rich in its content and we can have a lot of fun looking at it in various ways. First of all if you draw vertical lines between the group of carbon, nitrogen, and oxygen compounds, we have isoelectronic species connected. The species in the groups $CH-NH^+$, $CH^- - NH - OH^+$, $CH_2 - NH_2^+$, $CH_2^- - NH_2 - H_2O^+$, $CH_3 - NH_3^+$, $CH_3^- - NH_3 - H_3O^+$, $CH_4 - NH_4^+$, have some number of nuclei and electrons and usually have the same geometrical and electronical symmetry and multiplicity. For example, $CH^- - NH - OH^+$ all have $^3\Sigma^-$ ground state (although CH^- is yet to be observed), and $CH_3^- - NH_3 - H_3O^+$ all have the C_{3v} pyramidal structure of $^1A_1$ electronic symmetry with double minimum potential and inversion (although this is yet to be experimentally observed for CH_3^-).

We can show isoelectronicity in a wider sense when we consider protonation and deprotonation. For example, CH_5^+- CH_4- CH_3^-- NH_4^+- NH_3 - NH_2^- - H_3O^+ - H_2O - OH^- are isoelectronic and have singlet ground states. The species in the group CH_4^+ - CH_3 - CH_2^- - NH_3^+ -NH_2- NH^- - H_2O' - OH^- have all doublet ground states. The series CH_3^+ - CH_2 - CH^- - NH_2^+ - NH - OH^+ have triplet ground states except for CH_2^+ which has a singlet ground state (CH_2 too has almost a singlet ground state).

For the analysis of rotational structure of the spectrum, we group the species in Fig. 2 according to the number of protons. For example we can look at CH_3^+ - CH_3 - CH_3^-, NH_3^+ - NH_3, and H_3O^+. CH_3^+ is a rigid planar molecule whereas CH_3 and NH_3^+ are planar with low frequency out of plane vibration. NH_3, H_3O^+ and CH_3^- (yet to be studied) are pyramidal with inversion splitting (NH_3 0.8 cm^{-1}, H_3O^+ 55 cm^{-1}). If we look at the CH_2^+ - CH_2 - CH_2^-, NH_2^+ - NH_2 - NH_2^-, H_2O^+ - H_2O series, they are all bent but the rigidity of bending varies from well bent molecules (H_2O, H_2O^+, NH_2^-, NH_2) to quasi-linear molecules (CH_2, NH_2^+). We are currently studying the ν_3 fundamental band of NH_2^+ but it looks like this ion is even more quasi-linear than CH_2.

We can extend Fig. 2 to more complicated ions. Figure 3 lists the similar species containing two heavy atoms. Again we can discuss various molecular properties systematically using such figures. We

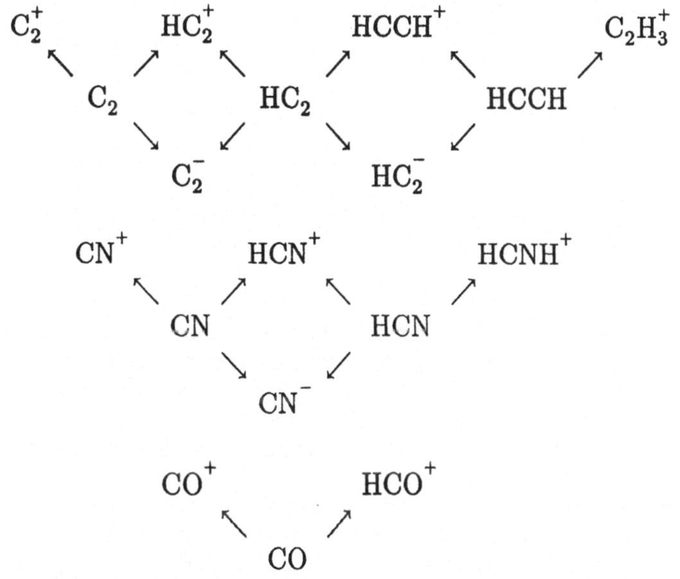

Fig. 3. Molecules, free radicals, ions containing CO, CN and CC.

can extend such figures to more complicated molecules. Among the molecular ions appearing in these figures, the following ions have been studied since 1983 through their infrared spectra: CH_3^+, NH^+, NH_2^+, NH_3^+, NH_4^+, NH^-, NH_2^-, OH^+, H_2O^+, H_3O^+, OH^-, $HCCH^+$, $C_2H_3^+$, C_2^-, $HCNH^+$, CO^+, HCO^+. Spectra of many other ions will be discovered in the near future. Spectroscopy and chemistry is much enriched through these studies.

1.3. Ion Kinetics

Langevin Rate The Langevin rate is one of the most fundamental concepts which is indispensable in the discussion of ion kinetics (Langevin, 1905).[28] In the laboratory discharge and in interstellar space an ion collides mostly with neutral molecules. If the polarizability of the neutral molecule is α, the energy of interaction between the ion and the molecule (charge-induced dipole interaction) is

$$W = -\frac{1}{2}\alpha E^2 = -\frac{\alpha e^2}{2r^4} \tag{1}$$

where $E = e/r^2$ is the Coulomb electric field of the ion at the position of the neutral molecule. The collision cross section for such an interaction is[29]

$$\sigma = \pi\sqrt{\frac{\alpha}{m}} \cdot \frac{e}{v} \sim 100 \ \overset{\circ}{A}{}^2 . \tag{2}$$

for a typical velocity v. This large cross section and the corresponding rate constant

$$k = \sigma v = \pi\sqrt{\frac{\alpha}{m}} \, e \sim 10^{-9} \ cm^3/sec \tag{3}$$

are called the Langevin cross section and the Langevin rate constant, respectively. This large rate constant applies both to many ion-molecule reaction rates and to a relaxation process such as rotational energy transfer. Thus in the laboratory where the typical concentration is $\sim 10^{17}/cm^3$, the rate is $\sim 10^8/sec$; in space where the typical concentration is $\sim 10^4/cm^3$, the rate is $\sim 10^{-5}/sec$. We note in Eq. (3) that k is independent of velocity and thus temperature. Maxwell used this potential for his calculation of the transport process for sheer mathematical convenience.[30] The Langevin rate for rotational relaxation leads to pressure broadening on the order of ~ 10 MHz/torr Vibrational relaxation without chemical reaction is several orders of magnitude slower.

<u>Ionization</u>   In the laboratory plasma, molecular ions are initially
produced through ionization by electron bombardment. Since the electron temperature of the plasma (typically 2-3 eV (20 eV $\sim$ 30,000K)[32]) is
considerably less than the ionization potential of molecules (typically 12 $\sim$15 eV) the ionization occurs through the tail of the energy
distribution of thermal electrons. If a Boltzmann distribution is
assumed the rate constant can be calculated from

$$k = \int_0^\infty \sigma(\varepsilon)vdn(\varepsilon) \tag{4}$$

where $v = \sqrt{2m\varepsilon}$ is velocity of electrons and the energy distribution of electron is

$$dn(\varepsilon) = \frac{2}{\sqrt{\pi}} \left[\left(\frac{\varepsilon}{kT} \right)^{1/2} e^{-\frac{\varepsilon}{kT}} d\left(\frac{\varepsilon}{kT} \right) \right] . \tag{5}$$

The cross section $\sigma(\varepsilon)$ as a function of electron energy is often
approximated as a linear function of ε

$$\sigma(\varepsilon) = a\varepsilon - b. \tag{6}$$

The experimental values of a and b can be found from the $\sigma(\varepsilon)$ curve in
the compilation by Kieffer.[33] The value of k is typically 2×10^{-12}
cm3/sec, that is, if we have 10^{11}/cm^3 of electron density a molecule
is ionized 0.2/sec. For a plasma with a pressure of 1 torr the ion
production is on the order of $\sim 10^{16}$/cm^3 sec, that is, 10^{16} ion pairs
are produced per cm^3 per second. Once ions are formed they react with
neutrals with the large Langevin rate, that is $\sim 10^{-8}$ sec.

<u>Recombination</u>   Because of the long range Coulomb interaction, the
electronic recombination usually occurs much faster than the Langevin
rate. The typical value of the recombination rate constant is $\sim 10^{-7}$
cm^3/sec. We can find electron recombination rates in Ref. 34 and 35.
So, for an electron density of $\sim 10^{11}$/cm^3, we find the recombination
rate to be on the order of 10^4/sec. At the Royal Society Meeting in
London (March, 1987), David Smith surprised us by announcing that the
H_3^+ recombination is at least 10^4 times slower than previously
believed.[36] This will radically change the ion kinetic argument in
hydrogen dominated plasmas both in the laboratory and in space.

In a laboratory plasma the dominating ion destruction mechanism
is often not the electron recombination in the plasma but the electron
recombination at the wall after ambipolar diffusion. Because of the
light mass and high temperature, electrons diffuse in the plasma

faster than atomic and molecules ions by say $\sim 10^3$. In order to preserve quasi-neutrality in the plasma, cations move with the electrons towards the wall.[37,38] The data for this is found in Ref. 39. In the last few years, there has been mounting evidence that ambipolar diffusion is much faster than previously believed.[40] In a recent plasma diagnotics of H_2 and He discharges,[41,42] we have used diffusion rates of $\sim 10^5$ cm/sec. Thus destruction of ions by ambipolar diffusion is more significant than that by electron recombination if the diameter of the discharge tube is on the order of a few centimeters.

1.4. Degree of Ionization

Laboratory Plasma

The ion concentration in a laboratory plasma can be estimated from the simple formula

$$\frac{I}{S} = nev. \tag{7}$$

From the measured current density I/S we can determine the electron concentration n if we assume the electron drift velocity v. The drift velocity in various discharges can be found in Ref. 32. We usually find the value like $I/S \sim 0.2$ A/cm^2, $v \sim 10^7$ cm/sec and $n \sim 10^{11}$/cm^3. From the overall neutrality this must also be the number density of cations (the densities of molecular anions probably are smaller).

This ion concentration agrees with what we expect from the ion kinetics in the previous section. Using the ion production rate of $\sim 10^{16}$/cm^3 sec and destruction rate by ambipolar diffusion of $\sim 10^5$/sec (i.e. lifetime of $\sim 10^{-5}$ sec) we obtain the steady state ion concentration of $\sim 10^{11}$/cm^3. Since the pressure of the gas is ~ 3 Torr, the fraction of ion (which we call degree of ionization) is typically $\sim 10^{-6}$. This is a typical value for a glow discharge. Depending on the main component of the gas the degree of ionization varies. For example, in an Ar discharge the electron drift velocity is smaller and thus the degree of ionization is larger by an order of magnitude. The degree of ionization is also larger in hollow cathodes and in negative glow, due to larger current density and/or small electron drift velocity.

Astronomical Plasma

The degree of ionization in various astronomical objects have wide ranging values depending on the energy of the object. Inside the star and cosmic ray, of course, the electrons are completely stripped from nuclei - the fully ionized plasmas. In the solar corona, where the temperature is $\sim 2 \times 10^6$ K, highly charged atomic ions are observed. As we go towards lower temperature, planetary nebula - diffuse cloud - dense cloud, the degree of ionization decreases.

In the dense clouds which we are most interested in, the star radiation is completely shielded and the ionization occurs through cosmic ray bombardment. The universal cosmic ray flux is fairly well established[43] to be $\zeta = 10^{-17}$/sec. This should be compared with 0.2/sec given earlier for the laboratory plasma. The degree of ionization in dense clouds can be estimated by equating the production rate and the destruction rate by electron recombination.

$$\zeta[X] = k [X^+] [e] . \tag{8}$$

We have

$$[X^+] = [e] = \sqrt{\frac{\zeta}{k} [X]} \sim 10^{-3}/cm^3 \tag{9}$$

if we use $k = 10^{-7}$ cm^3/sec and density $[X] = 10^4/cm^3$. This gives the degree of ionization $[X^+]/[X] \sim 10^{-7}$. We note that even the least ionized part of the universe has degrees of ionization comparable to that of laboratory plasma (as we mentioned earlier, planetary atmosphere is exceptional because of the high molecular concentration. If we use $[X] = 10^{18}/cm^3$ in Eq. (9) we obtain $[X^+]/[X] - 10^{-13}$).

1.5. Production of Specific Ions

As we have seen, we have a pretty good idea about how much charged species we can produce in a discharge. To produce a specific ion, however, is more difficult. Plasma chemistry is almost alchemy. There is not much logic before the execution of the experiment; we just try to see if it works. The simplest is the production of H_3^+ in the H_2 discharge, through the ion-molecule reaction[44]

$$H_2^+ + H_2 \rightarrow H_3^+ + H. \tag{10}$$

The reaction is very exothermic (1.8 eV) with a large Langevin cross section of [45], and therefore H_3^+ is the dominant ion in a pure hydrogen discharge.[46]

The simple hydride cations are produced by the hydrogen extraction reaction,

$$X^+ + H_2 \rightarrow HX^+ + H. \tag{11}$$

For the oxygen series the extraction reactions occur efficiently throughout the chain

$$O^+ \rightarrow OH^+ \rightarrow H_2O^+ \rightarrow H_3O^+ \tag{12}$$

which are all exothermic with the Langevin rates. For the nitrogen series with the chain

$$N^+ \to NH^+ \to NH_2^+ \to NH_3^+ \to NH_4^+, \tag{13}$$

all reactions are very efficient except for the last reaction $NH_3^+ \to NH_4^+$ which has a barrier and has a rate constant which is lower by a factor of 10^3. The individual reaction rates can be found conveniently in the table of ion-molecule reactions by Albritton.[47] These reactions are very important in the ion chemistry scheme in interstellar space and are used a great deal in the literature.[13,14,15] The series of work by Huntress[48,51,52] is particularly informative. Because of the chains shown in Eqs. (12) and (13), it is easy to produce much H_3O^+ and NH_4^+. They are sometimes called "end of the food chain" ions. In order to produce plasmas which have abundant intermediate ions such as OH^+, H_2O^+, NH_2^+, NH_3^+, etc. we use a large amount of He to slow down the reaction reaching the end of the chains.[49,50]

The chain of reactions for carbocations is more complicated,

$$C^+/CH^+ \to CH_2^+ \to CH_3^+/CH_4^+ \to CH_5^+. \tag{14}$$

This chain has two hangups.[51,52] As a result of these hangups, CH_3^+ is relatively easy to produce.

The diatomic carbon chain

$$C_2^+ \to C_2H^+ \to C_2H_2^+ \to C_2H_3^+/C_2H_4^+ \tag{15}$$

also has a hangup.[51] The last reaction $C_2H_2^+ \to C_2H_3^+$ goes only for vibrationally excited $C_2H_2^+$. Therefore $C_2H_2^+$ and $C_2H_3^+$ are relatively abundant in hydrocarbon dicharges.[54,55] A major difficulty in the production of carbo-ions results from efficient polymerization. Almost all reactions between carbocations and neutral hydrocarbons such as $CH_3^+ + CH_4$ (C_2H_4, C_2H_6, etc.) have the Langevin cross sections and produce polymeric hydrocarbons eventually leading to a large amount of soot deposited on the wall of the discharge tube. The method we use to cope with this difficulty is to simply dilute the discharge gas mixture with a large amount of He. Helium also increases the efficiency of ion production through Penning ionization[56] and, because of its high ionization potential, tends to increase the electron temperature in the plasma thus helping to fragment larger carbon compounds. For example, for the spectroscopy of CH_3^+ or $C_2H_3^+$ we used a gas mixture of He: H_2:X of 700:20:1 (X=CH_3, C_2H_2, C_3H_6 etc.) with a total pressure of $\sim$7 torr. A more detailed discussion of carbon chemistry appears in Refs. 54 and 57.

For the spectroscopy of carbo-ions we have found it useful to use various multiple-inlet-outlet discharge tubes. We use an air-cooled discharge tube (nicknamed "spider"), a water cooled tube ("tarantula") and a liquid nitrogen tube ("black widow") depending on purposes. The water cooled tube is schematically shown below in Fig. 4.

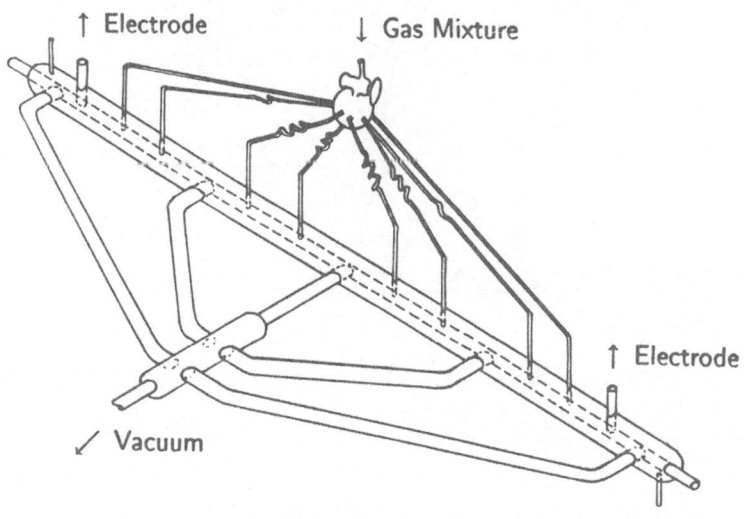

Figure 4.

A fresh mixture of gases is introduced into the discharge tube through the eight inlet ports. From the color of the discharges we note that fresh chemical reactions occur in the region of the inlet ports. In the usual single inlet-outlet cell, the fresh reaction tend to occur at electrodes, depositing a large amount of soot in that area. The reaction chain of Eq. (14) plays an important role in interstellar chemistry as seen from Fig. 5 provided by A. Dalgarno.[58,59]

There are many other reactions which form specific molecular ions. For example the universal reaction

$$X + H_3^+ \rightarrow HX^+ + H_2 \tag{16}$$

occurs very efficiently in a hollow cathode discharge. Amano has been extremely successful in using this reaction in a hollow cathode discharge tube with X = CO_2, N_2O, OCS, CH_3CN to produce $HOCO^+$ (Ref. 60), HN_2O^+ (Ref. 61), $HOCS^+$ (Ref. 62), CH_3CNH^+ (Ref. 63) and $HCCCNH^+$ (Ref. 63). The reaction of Eq. (16) is also universal in interstellar

364

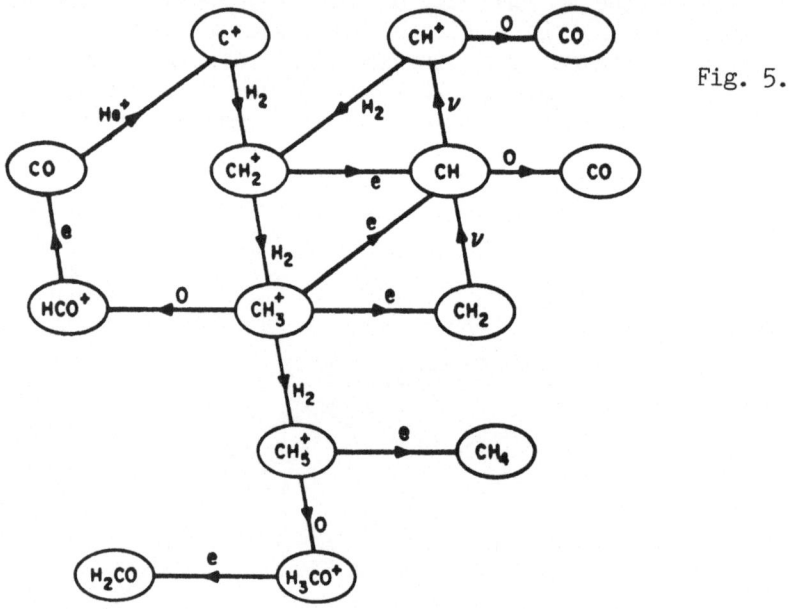

Fig. 5.

space and leads to HCO^+, HN_2^+ and numerous other molecular ions observed by radio astronomers. In Suzuki's treatment of chemical reactions in molecular clouds[15] as many as 70 reactions of this type have been considered.

Negative ions can also be produced in abundance in discharges as has been demonstrated recently by the detection of OH^- and NH_2^- infrared spectrum in discharges.[64,65] The production of negative ions is optimized under different discharge conditions from that for cations. For example, metal coating the wall of a discharge cell increases the concentration of OH^- at least an order of magnitude.[66] Recently M. Polak, M. Gruebele and R.J. Saykally have been extremely successful in producing polyatomic negative ions by using the general reaction

$$XO + NH_2^- \rightarrow XN^- + H_2O. \tag{17}$$

Using N_2O, CO_2 and OCS, they have obtained spectra of N_3^- (Ref. 67), NCO^- (Ref. 68) and NCS^- (Ref. 69).

Obviously there are a great many more ways to produce other ions.

II. INFRARED SPECTROSCOPY OF MOLECULAR IONS USING FREQUENCY TUNABLE LASER SOURCES

Infrared spectroscopy is the most universal spectroscopic method to study the simple fundamental ions discussed so far. There are cases such as H_3^+ for which no stable electronic excited state exists. Even if there are excited states, they are often predissociated and do not give discrete spectra. Thus few hydride ions would show discrete electronic spectra. On the other hand, almost all molecular ions (except for some homonuclear diatomics) absorb infrared strongly.

In discussing the experimental aspect of spectroscopy there are two major factors - sensitivity and resolution. Comparing the two neighboring spectral ranges, i.e., microwave and visible, the traditional grating infrared spectroscopy had been inferior in both of the two attributes. The availability of monochromatic sources and the relative ease of electronic controls of various elements has made microwave spectroscopy an extremely sensitive ($\Delta I/I < 10^{-7}$) and high resolution technique. In the visible-ultraviolet region, fluorescence spectroscopy and the photographic method has made the spectroscopy also extremely sensitive. Many large grating visible and UV spectrometers had near Doppler limited resolution of $\Delta \nu / \nu \sim 10^{-6}$. In the grating infrared spectrometer the sensitivity has been typically $\Delta I/I \sim 10^{-2}$ and the resolution $\Delta \nu / \nu \sim 10^{-5}$. This situation has been changed radically with the advent of laser spectroscopy and Fourier transform spectroscopy (I shall not touch on the latter here). The sensitivity of laser infrared spectroscopy by using a suitable modulation is comparable to that of microwave, and the resolution using sub-Doppler technique is $\sim 10^{-9}$. This increase of a factor of $\sim 10^4$ both in sensitivity and resolution has made the infrared spectroscopy much more powerful than before. A great many things are now possible. Ion spectroscopy is only a small part of it. We have not really understood the impact of this yet.

2.1. Frequency Tunable Laser Infrared Sources

Although it is already more than a quarter century since Maiman first operated his laser,[70] it is only in the last ten years or so that frequency tunable infrared sources have become available to practicing spectroscopists. A review by Pine[71] contains useful information on various sources and spectroscopy using them. Following are brief sketches of various sources in relation to molecular ion spectroscopy.

Difference Frequency Laser System

This non-linear optical method of generating frequency tunable infrared radiation was initially operated by Boyd and Ashkin,[72] and developed by Pine[73] into a very powerful spectroscopic tool. Radiation from a dye laser (ν_D) and radiation from a single-mode Ar ion laser (ν_A) are mixed in a LiNbO$_3$ crystal. Their planes of polarization are perpendicular to each other. The infrared difference frequency (ν_{IR}) is generated when the energy balance

$$\nu_{IR} = \nu_A - \nu_D, \tag{18}$$

and the momentum balance condition

$$\underset{\sim}{k}_{IR} = \underset{\sim}{k}_A - \underset{\sim}{k}_D. \tag{19}$$

are satisfied. The latter (phase matching) is achieved for a special value of temperature for collinear $\underset{\sim}{k}_A$ and $\underset{\sim}{k}_D$ through the temperature dependence of the dielectric constants of the birefringent LiNbO$_3$ crystal. (Note $k = h\nu/C(T) = h\nu/cn(T)$ where $n(T)$ is the temperature dependent refractive index). The frequency of infrared is tuned by tuning the dye laser frequency. The monochromatic ($\Delta\nu \sim 2$ MHz) radiation obtained has power of $1 \sim 500\mu$ Watt which is sufficient for linear spectroscopy (Note that the noise equivalent powers of high sensitivity detectors are typically $10^{-10} \sim 10^{-11}$ Watts).

The advantage of this source is its wide and continuous coverage (2.2-4.2μm) of the important infrared region. Practically all hydrogen stretch vibrations fall in this region. This radiation source was indispensable for my first work on H$_3^+$ (Ref. 5) and still is the most powerful tool in our laboratory. The spectrometer used for the H$_3^+$ spectroscopy is shown below in Fig. 6.

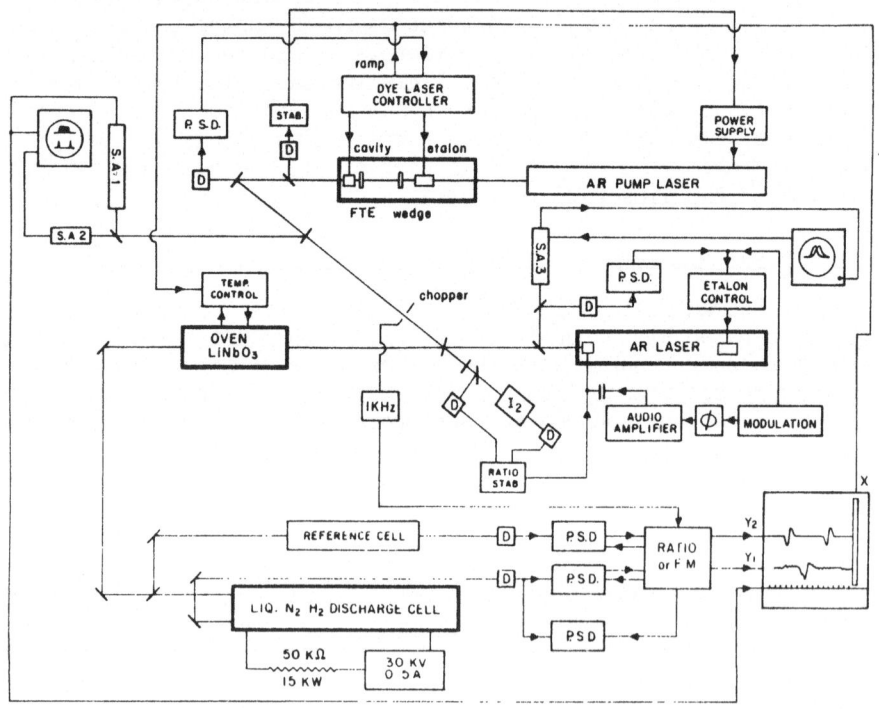

This type of spectrometer is used now by Amano in Ottawa and in Chicago for molecular ion spectroscopy. An extension of the frequency coverage is being attempted using $LiIO_3$. It will then cover a wide wavelength region of $2 \sim 5.6 \mu m$ albeit with less power.

Color Center Laser

The color center laser was invented by Mollenauer and Olson[74] in 1974. It has become well-known among high resolution spectroscopists through the beautiful and energetic work of Saykally and his colleagues. It is the infrared version of the dye laser. Pumped by a Kr ion laser, it generates tunable infrared radiation over a wide range of $2.3 \sim 3.45 \mu m$ with high power (2.5 - 100 mWatts). This radiation source covers the N-H and O-H stretching regions and much of the C-H stretch also. The available high power makes this source particularly attractive for non-linear experiments such as sub-Doppler spectroscopy[76] and double resonance spectroscopy.[77] Color center lasers are used for ion spectroscopy in Saykally's laboratory in Berkeley[75] and in Urban's laboratory in Bonn.[78] The extensive electronic and computer control of this system by the latter group[79] and by Curl and his colleagues[80] is noteworthy.

Diode Lasers

The semiconductor lasers are commercially available from 3 to $30 \mu m$. The advantage of diode lasers is that they are relatively low priced, easy to operate, they generate a fair amount of power $(10 \mu W \sim 1 mW)$ of low amplitude noise, and their frequency is easily modulated. They are the only frequency tunable infrared sources which cover more than three octaves (if a sufficient number of diodes are available). This infrared source was indispensable for the detailed studies of the ν_2 vibration-inversion mode of H_3O^+ in which three bands between $10-30 \mu m$ had to be studied.[81] The disadvantage of this source on the other hand is the narrow and patchy coverage of an individual diode laser and their rather unpredictable optical quality.

Diode lasers are used for ion spectroscopy by McKellar in Ottawa, Davies in Cambridge, Destombes in Lille, Kawaguchi and Hirota in Okazaki, Sears in Brookhaven, Saykally in Berkeley, and in Chicago. Perhaps there are some more people doing this.

Microwave Modulation Sidebands on CO_2 Laser Lines

The $10 \mu m$ radiation from a CO_2 laser (ν_c) and microwave radiation (ν_m) are mixed in a CdTe crystal to generate sidebands at $\nu_c \pm \nu_m$. This technique, developed by Bonek and Magerl,[82] provides a relatively simple way to produce tunable infrared radiation with high spectral purity and sufficient power $(1 \sim 3$ mWatts) for sub-Doppler spectroscopy. While this source is not very suitable for the search of new ion spectra, it allows very high resolution observation

($\Delta\nu/\nu$ 10^{-9}) of a known spectrum. The experimental set up is shown in Figure 7.

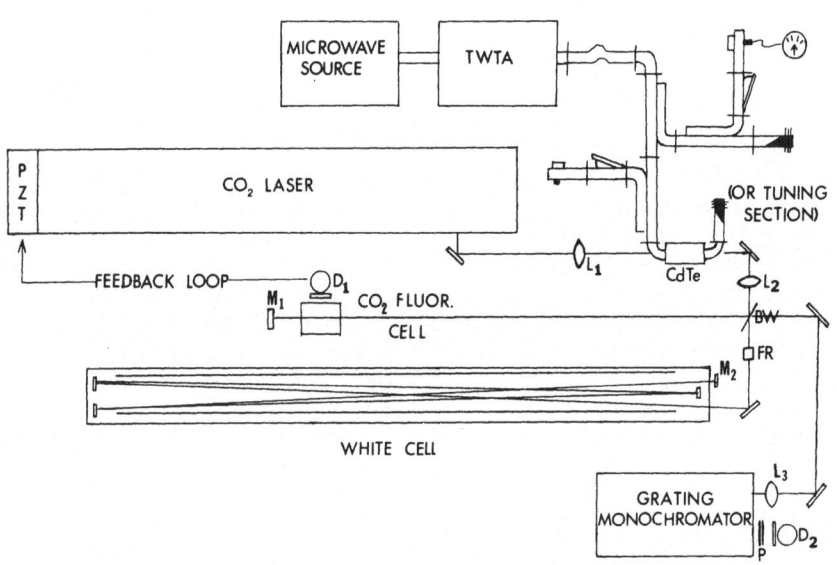

Figure 7. CO_2 microwave sideband sub-Doppler spectrometer with multiple path.

This approach has recently been applied to the H_3O^+ ion spectroscopy.

2.2. Increasing the Sensitivity

We saw earlier that the number concentration of ions in typically $\sim 10^{11}/cm^3$. Since the infrared absorption coefficient for the hydrogen stretching vibration of simple molecules (such as the ν_3 band of CH_4) is on the order of ~ 0.1 cm^{-1}/Torr, we are dealing with absorption coefficient on the order of $\sim 3 \times 10^{-6}$ cm^{-1}. If a path length of 1m is used the absorption is $\sim 3 \times 10^{-4}$. This is too small for the traditional grating spectroscopy but is sufficiently large for laser spectroscopy. In fact the existence of charge usually makes the transition moment larger and ion absorptions are often more than the above value. We have seen $\sim 20\%$ absorption due to ArH^+ using a 2m path length.[1] In favorable cases we can detect ions with number concentrations on the order of $\sim 10^8/cm^3$.

One complication in infrared spectroscopy is that, unlike electronic spectroscopy, many bands appear overlapped in the a same wavelength region. Thus, for example, the spectra of the C-H stretching vibrations of CH_3^+, $HCCH^+$, and $C_2H_3^+$ appear intermixed with those of stable neutral hydrocarbons CH_4, $HCCH$, C_2H_4, C_2H_6 ... and a great many others which are always produced abundantly in plasmas regardless of

which hydrocarbon discharge is used. Since the ions are a very small
fraction of neutrals this causes a serious problem of detection and
assignment. The ingenious method of velocity modulation invented by
Gudeman, Saykally and others[75,81] is a very powerful tool to avoid
this difficulty and, at the same time, to increase sensitivity. Let
us first start from the drift velocity of ions in plasmas.

Ion Drift Velocity

In plasmas ions are always under acceleration by the discharge
electric field E which is typically 10 V/cm in a positive column.
At the same time ions are decelerated upon collisions with neutrals.
So on the average ions move from anode to cathode with a constant
speed. We write the ion drift velocity v_d phenomenologically as

$$v_d = KE \tag{20}$$

where the proportionality constant K is called mobility. Equating the
drift velocity to the diffusion velocity formula $v_d = (D/N)(dN/dz)$,
we obtain the Einstein relation

$$K = D \frac{e}{kT} \tag{21}$$

between the mobility K and the diffusion constant D.[56] Using the
Langevin cross section of Eq. (2) we obtain

$$K = \frac{3}{16N} \left(\frac{3}{2\pi\mu\alpha} \right)^{1/2} \tag{22}$$

from the Chapman-Enskog theory [38,85] of diffusion.

In order to have a more microscopic picture, it is instructive
to consider a hypothetical one-dimensional model. The acceleration of
a single ion is

$$m\ddot{z} = eE \quad \text{or} \quad \dot{z} = v_o + \frac{eE}{m} t \tag{23}$$

An ion with zero initial velocity will have a velocity $\varphi\Delta t$ after a
time interval Δt ($\varphi \equiv eE/m$). Suppose it collides after this time
interval with a neutral molecule with zero velocity and with mass m_o.
The velocity of an ion after the collision will be $\rho\varphi\Delta t$ from momentum
conservation where $\rho \equiv m/(m+m_o) < 1$. The ion starts from this initial
velocity now and gets accelerated. After another time Δt, it has the
velocity $(1+\rho) \varphi\Delta t$ and collides with a neutral. If we continue this
argument we find that after many collisions, the ion speed approaches

$$v \doteq (1 + \rho + \rho^2 + \dots) \varphi\Delta t = \frac{\varphi}{1-\rho} \Delta t \tag{24}$$

just before the final collision. If we substitute the expressions
of φ and ρ given earlier and $\Delta t = (n\sigma v)^{-1}$ with the Langevin cross

370

section of Eq. (2) we obtain

$$v = \frac{1}{2\pi N} \frac{1}{\sqrt{\mu\alpha}} E \qquad (25)$$

The value of mobility thus obtained has the same μ, α dependence as that in Eq. (22); its value is larger than that in Eq. (22) by about 20%. Eq. (25) gives the value just before the collision, the average velocity is

$$\bar{v} = \frac{2m+m_0}{2(m+m_0)} v = KE \qquad (26)$$

While this simplistic picture requires much extension taking into account the three dimension and the stochastic nature of collision (see the simplified one-dimensional picture and its extension in Chandrasekhar's article on Brownian motion[86]), we can obtain from it a basic picture and orders of magnitude of the microscopic motion. We can calculate approximate values of velocity from Eqs. (25) and (26) for various combinations of molecular ions and neutrals. We see from the discussion leading to Eq. (24) that, if $\rho \sim 1$ ($m \gg m_0$) it takes many collisions for the ion to reach the steady state while if $\rho \ll 1$ ($m \ll m_0$) the ion is slowed down by each collision but reaches the ultimate speed by the time the next collision occurs.

Spectroscopic Measurements of Ion Drift Velocity

The traditional method of measuring ion drift velocity has been to use time-of-flight drift tubes employing a mass spectrometer as the detector.[39] The high sensitivity ion spectroscopy has introduced an additional method to measure this quantity. The advantage of this new method compared with the traditional one is that it is in situ and non intrusive and it can be carried out with high spatial resolution. The basic setup is shown in Figure 8. The radiation from a diode laser is

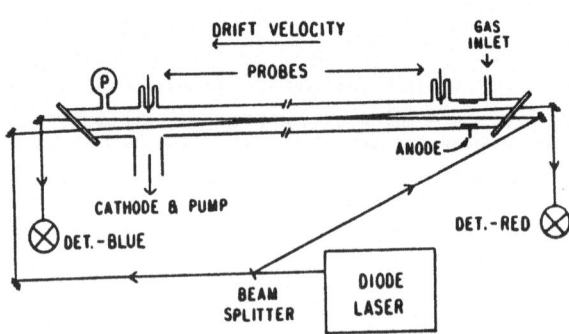

Fig. 8. Doppler-shift discharge cell and optical arrangement.

split into two counter propagating beams with a beam splitter
detected with two separate Hg CdTe detectors. For ions moving from
with velocity v , the two radiations appear blue shifted and red
shifted by ∿v/c. Thus we have Doppler shifted spectra as shown below
in Figure 9.

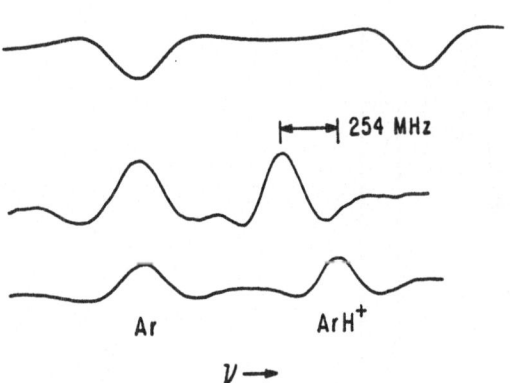

Fig. 9. Doppler-shift
spectrum. The top
trace is Ge etalon
transmission with F.S.R.
of 1420 MHz. The
middle and the bottom
traces show red-and
blue-shifted spectra,
respectively. The
unshifted line on the
left is of a neutral
Ar absorption.

The Doppler shifted line in Fig. 9 is the P(5) fundamental transition
of ArH$^+$ at 2479.411 cm^{-1}. We observed this spectrum using a He:Ar:H$_2$
mixture of ∿60:3:1 with the total pressure of ∿3 Torr, the discharge
electric field was ∿5.3 V/cm. We clearly see from Fig. 9 that ArH$^+$
ions are drifting from anode to cathode with the velocity of ∿512 m/se
and that this speed is comparable or larger than the random velocity
of the molecular ions in the discharge. Such a measurement gives ion
mobility in discharges. For more details of these results see Ref. 85.

Velocity Modulation

Early in 1983 Gudeman, Begemann, Pfaff and Saykally[83] published
a paper in which they used AC glow discharge applied to the ν_1 fundamental band of HCO$^+$. Their apparatus is shown below in Fig. 10. In
the AC discharge the molecular ions are accelerated back and forth by
the alternating electric field thus frequency modulating the absorption line through Doppler effect. This modulation is effective for
molecular ions since, as we see from Fig. 9, the Doppler shift is comparable to the Doppler linewidth due to the random velocity. Suppose
we use the velocity modulation with frequency of 10 kHz, that is, an
ion is accelerated for a time interval of 5 x 10^{-5} sec to one direction and then to the other. The ion will collide with neutrals about
∿10^4 times during the time interval (a pressure broadening of 30 MHz
is used in this estimate). The ion reaches their ultimate velocity
of Eqs. (24)-(26) after 1∿10 collisions. The velocity is on the order
of ∿500 m/sec, that is, the ion moves a few centimeters in its plasma
and changes direction. This method allows us to discriminate ion
lines from much stronger neutral absorption and at the same time

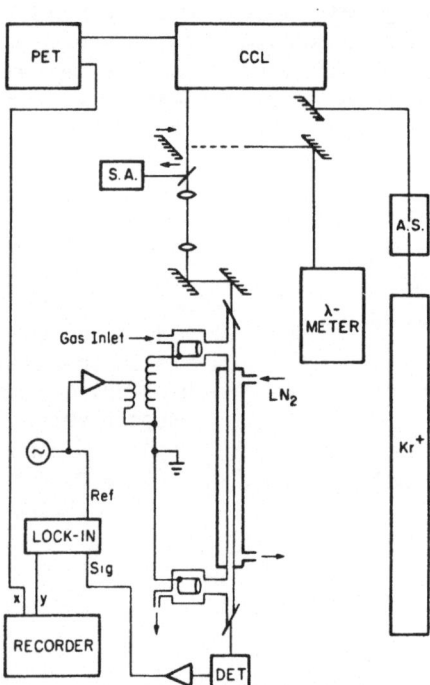

Fig. 10. Schematics of the color center laser-velocity modulation spectrometer of Saykally and coworkers.[75]

increase the sensitivity of detection. From my viewpoint this is the most important technical development for ion spectroscopy after introduction of the tunable infrared source. More details of this method can be found in the review by Gudeman and Saykally.[75]

Other Methods

Several other methods have been utilized to increase sensitivity and discrimination of molecular ion spectroscopy. Kawaguchi, Yamada, Saito and Hirota[88] developed the magnetic field modulation method exploiting the sensitivity of molecular ion absorption to a magnetic field. Since the magnetic field varies the plasma conditions in general, the neutral absorption lines are also affected by the magnetic field but the effect on ion lines is more drastic. This method is particularly useful for cases when the velocity modulation method is not applicable such as for a hollow cathode discharge used very effectively by Amano and his colleagues.[60-63]

The traditional method of noise subtraction has been revived and used very effectively for ion spectroscopy.[89] The laser beam is split into two parts, one for spectroscopy and the other for reference and detected by two matched infrared detectors. The signals from the two infrared detectors are elctronically combined with opposite phase to subtract noise. This method is particularly effective for a color center laser and a difference frequency laser system where the amplitude noise of the laser is larger (a few percent). Improvements on sensitivity of 10 ∿100 have been reported.

The method of White multiple path cell[90] is also adapted to the ion spectroscopy to increase the absorption path length. For ion spectroscopy without using velocity modulation such as that for H_3^+ or hollow cathode discharge,[5,60,88] usual White cell arrangement[91] is used and increase the path by 8-16. When the velocity modulation method is used, it is necessary to arrange a unidirectional multiple path as shown in Figure 11. Using the arrangement we typically gain sensitivity by a factor of ∿4 (Ref. 57,54).

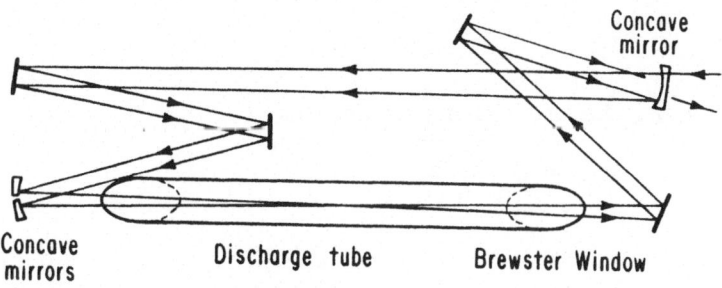

Concave mirror

Concave mirrors   Discharge tube   Brewster Window

2.3. Spectroscopy of Molecules Under Acceleration

There is a qualitative difference between translational motion of neutral molecules in normal spectroscopy and that of molecular ions in discharge spectroscopy. In normal spectroscopy a molecule is moving with a uniform velocity between collisions and if it resonates with the laser radiation the resonance is interrupted only by collision. In ion spectroscopy, however, the ions are under acceleration between collisions and, depending on the condition, the resonance may not last for the whole collision interval. This fact introduces a new source of uncertainty broadening which I call here "transit" broadening (the quotation mark is to discriminate this effect from the usual transit time broadening which is much smaller).

The calculation of this new broadening is simple. The velocity of an ion under acceleration is given as

$$v = v_o + \frac{eE}{m} t. \tag{27}$$

The variation of velocity during the collision interval Δt is $v = eE\Delta t/m$, which means the variation of resonance frequency is

$$v = \frac{eE}{m\lambda} \Delta t \tag{28}$$

where λ is the wavelength of radiation. If we compare this with the usual equation of uncertainty broadening

$$\Delta\nu = \frac{1}{2\pi\Delta t} \; , \tag{29}$$

we find that they are inverse in their Δt dependence. If Δt is short $\Delta\nu$ is determined from Eq. (29), but if Δt is long, the ion moves out of resonance after some time interval δt which is shorter than δt. I speculate that the value of δt is obtained by equating Eq. (28) and (29) for that value of δt, that is

$$\delta t = \sqrt{\frac{m\lambda}{2\pi eE}} \; , \tag{30}$$

which means the uncertainty broadening

$$\delta\nu = \sqrt{\frac{eE}{2\pi m\lambda}} \; . \tag{31}$$

Some numerical examples are given below.

ion	$\lambda(\mu)$	$E(V/cm)$	$\delta\nu(MHz)$
H_3^+	4	15	31
H_3O^+	10	20	9
OH^+	100	3	1.2
H_3O^+	1000	1	0.6

One additional subtlety; if the coordinate displacement of the ion δz during the time δt is smaller than the wavelength of the radiation, the radiation cannot sense the off-resonance (Dicke narrowing[92,93]). Thus for the last case of the above numerical examples, $\delta\nu$ will be smaller than the calculated value.

The numerical examples indicate that very low electronic field discharge is required for sub-Doppler saturation spectroscopy of molecular ions.

I have run out of the paper allotted to me and I have to cut the chapter on the analysis of ion spectra. While in principle, the analysis of ion spectra is not different from that for usual molecules, there are certain cases where unexpectedly rich and irregular spectral features are observed because of (paradoxically) simplicity of the species (H_3^+ see Refs. 5,12,94) and because of non-rigidity of the species ($C_2H_3^+$; see Ref. 54). Readers are referred to review articles[75,95] for more sources.

References

1. H.G. Dehmelt and K.B. Jefferts, Phys. Rev. 125, 1318 (1962).
   K.B. Jefferts, Phys. Rev. Lett. 20, 39 (1968) 23, 1976 (1969).
2. W.H. Wing, G.A. Ruff, W.E. Lamb and J.J. Spezeski, Phys. Rev. Lett. 36, 1488 (1976).
3. A. Dalgarno and M.R.C. McDowell, Proc. Phys. Soc. (London) A-69, 615 (1956).
4. T. Oka, Molecular Ions: Spectroscopy, Structure and Chemistry, T.A. Miller, V.E. Bondeby (Ed.) North-Holland (1983).
5. T. Oka, Phys. Rev. Lett. 45, 531 (1980).
6. H. Bethe, Z. Phys. 57, 815 (1929).
7. E.A. Hylleraas, Z. Phys. 63, 291 (1930).
8. Wildt, Ap. J. 89, 294 (1939).
9. S. Chandrasekhar, Rev. Mod. Phys. 16, 301 (1944).
10. J.J. Thomson, Phil. Mag. 24, 409 (1912).
11. See for a history of H_3^+, T. Oka, Molecular Ions: Spectroscopy, S Structure and Chemistry, T.A. Miller and V.E. Bondeby (Ed.) North-Holland (1983).
12. T. Oka, Phil. Trans., Roy. Soc. Lond. A303, 543 (1981).
13. E. Herbst and W. Klemperer, Ap. J. 185, 505 (1973).
14. W.D. Watson, Rev. Modern Phys. 48, 513 (1976).
15. H. Suzuki, Prog. Theor. Phys. 66, 936 (1979).
16. de Jong, A. Dalgarno and W. Boland, Astron. Astrophys. 91, 68 (1980).
17. G.D. Carney and R.N. Porter, J. Chem. Phys. 65, 3547 (1976).
18. C.E. Dykstra and W.C. Swope, J. Chem. Phys. 70, 1 (1979).
19. R. Walder and J.L. Franklin, Int. J. Mass. Spectrom. Ion Phys. 36, 85 (1980).
20. B.K. Janoncek and J.I. Brauman, in Gas Phase Ion Chemistry, Vol. 2, pp. 53, M.T. Bowers (Ed.) Academic Press, New York, San Francisco, London (1979).
21. N.V. Sidgwick, The Chemical Elements and Their Compounds, Vol. 1, p. 19, Oxford Press (1950).
    Quoted by P.A. Giguere, J. Chem. Education 56, 571 (1979).
22. D. Buhl and L.E. Snyder, Nature 227, 1230 (1970).
23. W. Klemperer, Nature 227, 1230 (1970).
24. R.C. Woods, T.A. Dixon, R.J. Saykally, P.G. Szanto, Phys. Rev. Lett. 35, 1269 (1975).
25. G. Herzberg, Molecular Spectra and Molecular Structure II. Infrared and Raman Spectra of Polyatomic Molecules, D. Van Nostrand Co., Inc. (1945).
26. G. Herzberg, Molecular Spectra and Molecular Structure III. Electronic Spectra and Electronic Structure of Polyatomic Molecules, D. Van Nostrand Co., Inc. (1966).
27. G. Herzberg, Molecular Spectra and Molecular Structure I. Spectra of Diatomic Molecules, D. Van Nostrand Co., Inc. (1950).
28. P. Langevin, Ann. Chim. Phys. 5, 245 (1905).
29. L.D. Landau and E.M. Lifshitz, Mechanics§18, Nauka, Moscow (1965).

30. J.C. Maxwell, Phil. Trans. R. Soc. 157, 49 (1867).
31. R.C. Woods, Molecular Ions: Spectroscopy, Structure and Chemistry pp. 11, T.A. Miller and V.E. Bondley (Ed.) North-Holland (1981).
32. A. von Engel, "Ionized Gases", Oxford Press (1965).
33. L.J. Kieffer, Atomic Data 1, 19 (1969).
34. J.N. Bardsley and M.A. Biondi, Adv. Atom. Mol. Phys. 6, 1 (1970).
35. J.W. McGowan, P.M. Mul, V.S. D'Angelo, J.B.A. Mitchell, P. Defrance and H.R. Froelich, Phys. Rev. Lett. 42, 373 (1979).
36. D. Smith, Phil. Trans. Roy. Soc. Lond., in press.
37. W. Schottky, Phys. Z. 25, 635 (1924).
38. E.M. Lifshitz and L.P. Pitaevsky, Physical Kinetics, Pergamon Press (1981).
39. E.W. McDaniel and E.A. Mason, The Mobility and Diffusion of Ions in Gases, John Wiley and Sons, New York (1973).
40. Ganguly and A. Garscadden, Appl. Phys. Lett.   (1985).
41. F.S. Pan and T. Oka, Phys. Rev., in press.
42. M.W. Crofton and T. Oka, Int. J. Mass Spectr. Ion Phys., in press.
43. L. Spitzer, Physical Processes in the Interstellar Medium, John Wiley Pub., New York (1978).
44. T.R. Hogness and E.G. Lunn, Phys. Rev. 26, 44 (1925).
45. D.P. Stevenson and D.O. Schissler, J. Chem. Phys. 29, 282 (1958).
46. M. Saporoschenko, Phys. Rev. 139A, 349 (1965).
47. D.L. Albritton, Atomic Data and Nuclear Data Tables 21, 1 (1978).
48. J.K. Kim, L.P. Theard and W.T. Huntress, Jr., J. Chem. Phys. 62, 45 (1975).
49. M.W. Crofton, R.S. Altman, M.-F. Jagod and T. Oka, J. Phys. Chem. 89, 3614 (1985).
50. B. Dinelli, M.W. Crofton and T. Oka, J. Mol. Spectrosc., in press.
51. W.T. Huntress, Jr., Ap. J. Supp. 33, 495 (1977).
52. S.S. Prasad and W.T. Huntress, Jr., Ap. J. Supp. 43, 1 (1980).
53. M.W. Crofton, W.A. Kreiner, M.-F. Jagod, B.D. Rehfuss and T. Oka, J. Chem. Phys. 83, 3702 (1985).
54. T. Oka, Phil. Trans. Roy. Soc. London, in press.
55. M.W. Crofton, M.-F. Jagod, B.D. Rehfuss and T. Oka, J. Chem. Phys. 86, 3755 (1987).
56. E.W. McDaniel, Collision Phenomena in Ionized Gases, John Wiley & Sons, Inc. (1964).
57. M.W. Crofton, M.-F. Jagod, B.D. Rehfuss, W.A. Kreiner and T. Oka, J. Chem. Phys., in press.
58. J.H. Black and A. Dalgarno, Ap. J. Supp. 34, 405 (1977).
59. A. Dalgarno, Physics of Ion-Ion and Electron-Ion Collisions, F. Brouilland and J.W. McGowan (Ed.) Plenum Pub. Co. (1983).
60. T. Amano and K. Tanaka, J. Chem. Phys. 83, 3721 (1985).
61. T. Amano, Chem. Phys. Lett. 167, 101 (1986).
62. T. Nakamaga and T. Amano, Mol. Phys., in press.
63. T. Amano, private communication.
64. J.C. Owrutsky, N.H. Rosenbaum, L.M. Tack and R.J. Saykally, J. Chem. Phys. 84, 605 (1985).

65. L.M. Tack, N.H. Rosenbaum, J.C. Owrutsky and R.J. Saykally, J. Chem. Phys. 85, 4222 (1986).
66. D.J. Liu and T. Oka, J. Chem. Phys. 84, 2426 (1986).
67. M. Polak, M. Gruebele and R.J. Saykally, J. Am. Chem. Soc. 103, 2884 (1987).
68. M. Gruebele, M. Polak and R.J. Saykally, J. Chem. Phys. 86, 6631 (1987).
69. R.J. Saykally, Phil. Trans. Roy. Soc., in press.
70. T.H. Maiman, Nature 187, 493 (1960).
71. A.S. Pine, Phil. Trans. R. Soc. Lond. A307, 481 (1982).
72. G.D. Boyd and A. Ashkin, Phys. Rev. 146, 187 (1966).
73. A.S. Pine, J. Opt. Soc. Am. 64, 1683 (1974) 66, 97 (1976).
74. L.F. Mollenauer and D.H. Olson, Appl. Phys. Lett. 24, 386 (1974).
75. C.S. Gudeman and R.J. Saykally, Ann. Rev. Phys. Chem. 35, 387 (1984).
76. J.L. Hall, T. Baer, L. Hollberg and H.G. Robinson, Laser Spectroscopy V, pp. 15, Springer-Verlag (1981).
77. R.L. DeLeon, P.M. Jones and J.S. Muenter, Appl. Opt. 20, 525 (1981).
78. A. Stahn, H. Solka, H. Adams and W. Urban, Mol. Phys. 60, 121 (1987).
79. H. Adams, R. Bruggemann, P. Dietrich, D. Kirsten, H. Solka and W. Urban, J. Opt. Soc. Am. B2, 815 (1985).
80. P.G. Carrick, A.J. Merer and R.F. Curl, Jr., J. Chem. Phys. 78, 3652 (1983).
81. D.J. Liu and T. Oka, Phys. Rev. Lett. 54, 1787 (1985).
82. G. Magerl and E. Bonek, J. Appl. Phys. 47, 4901 (1976).
83. C.S. Gudeman, M.H. Begemann, J. Pfaff and R.J. Saykally, Phys. Rev. Lett. 50, 727 (1983).
84. A. Einstein, Ann. d. Physik. 17, 549 (1905) 19, 371 (1906).
85. S. Chapman and T.G. Cowling, The Mathematical Theory of Non-Uniform Gases, Cambridge University Press (1970).
86. S. Chandrasekhar, Rev. Mod. Phys. 15, 1 (1943).
87. N.N. Haese, F.-S. Pan and T. Oka, Phys. Rev. Lett. 50, 1575 (1983).
88. K. Kawaguchi, C. Yamada, S. Saito and E. Hirota, J. Chem. Phys. 82, 1750 (1985).
89. D. J. Nesbitt, H. Petek, C.S. Gudeman, C.B. Moore and R.J. Saykally, J. Chem. Phys. 81, 5281 (1984).
90. J.U. White, J. Opt. Soc. An. 32, 285 (1942).
91. H.J. Bernstein and G. Herzberg, J. Chem. Phys. 16, 30 (1948).
92. R.H. Dicke, Phys. Rev. 89, 472 (1953).
93. D.R. Rao and T. Oka, J. Mol. Spectrosc. 122, 16 (1987).
94. J.K.G. Watson, S.C. Foster, A.R.W. McKellar, P. Bernath, T. Amano, F.S. Pan, M.W. Crofton, R.S. Altman and T. Oka, Can. J. Phys. 62, 1875 (1984).
95. T.J. Sears, J. Chem. Soc. Faraday Tran. 2, 83, 111 (1987).

DYNAMICS OF THE PHOTODISSOCIATION OF SMALL MOLECULES

Peter Andresen
Max-Planck-Institut für Strömungsforschung
Bunsenstrasse 10
D-3400 Göttingen
West Germany

ABSTRACT.   The dynamics of the photodissociation of small molecules is discussed in view of a few recent experiments to demonstrate the state of the art in this field. The highly improved understanding of photodissociation processes is essentially due to modern   experiments in which the product motion is analysed in great detail by state selective detection methods. In this contribution an effort is made to understand the origin for this motion in terms of the strength of final state interaction. Using the photodissociation of water in the first absorption band as an example, it is shown that in the case of weak final state interaction product motion arises essentially from parent motion. In the case of strong final state interaction additional product motion is generated during the fragmentation by forces that originate from anisotropies of the excited state potential surface. Particular efforts are made to understand the origin for the selective population of Λ-doublet states.

1. INTRODUCTION

Photodissociation is the break up of a molecule due to the absorption of light:

$$ABC + h\nu \longrightarrow AB + C \qquad (1)$$

It is the first step in the whole field of photochemistry, that generates radicals and initiates subsequent reactions, and this is why photodissociation has been studied actively first in this field.[1] However, the primary interest in photochemistry has been the question which radicals can be formed from which parent molecules at which wavelength and what subsequent reactions occur. Detailed studies about the dynamics of photodissociation have been performed only more recently, mainly because they became only possible with modern laser technology. Today, most experimental work is devoted to a detailed characterisation of process (1) itself, with the intention to understand the motion of both nuclei and electrons during fragmentation.

A. C. P. Alves et al. (eds.), Frontiers of Laser Spectroscopy of Gases, 379–419.

One of the main tools that lead to the highly improved understanding of photodissociation dynamics in the recent years is the <u>analysis of the characteristic motion of the products</u>. The analysis of the characteristic motion in the products leads to a microscopic understanding of photodissociation processes, because it contains very detailed information about the interaction. It is the same method that is used to understand other dynamical processes like chemical reactions, inelastic collisions or surface scattering. In the case of reactive collisions it led to a microscopic understanding of chemical reactions and to the Nobel prize in chemistry for Lee, Herschbach and Polanyi in 1986.

The aim of the present paper is to demonstrate that the field of photodissociation dynamics is in a state where the origin of product motion is at least qualitatively understood. The origin of product motion is either initial motion in the parent molecule, or it is generated in the fragmentation step by forces that are introduced by anisotropies in the excited state potential surfaces. In a few fortunate cases, photodissociation processes can be understood almost quantitatively. One of these cases is the photodissociation of water in the first absorption band in the VUV, which will be used here as a model system to discuss features that are important in other fragmentation processes as well. The whole discussion in this paper is kept on a qualitative level and several simplifications are made to stress the basic ideas and not the finer details.

2. EXPERIMENTAL TECHNIQUES TO ANALYSE PRODUCT MOTION

In the photodissociation of a triatomic molecule ABC

$$ABC + h\nu \longrightarrow AB(f) + C$$

the product AB is formed only in well defined quantum states f. The probability for the formation of AB in different quantum states f is called "product state distribution". This product state distribution contains the information about the characteristic motion.

Nascent product state distributions have to be measured for a single photodissociation process, i.e., the products have to be analysed before they collide again with the residual gas. Otherwise the characteristic motion that emerges from the fragmentation process itself is changed by secondary collisions and the direct information about the fragmentation process is lost.

Today two important techniques are used to avoid secondary collisions. The first technique is the molecular beam method, in which the density is kept so low that no collisions occur before the products reach the detector (for example a mass spectrometer). The second technique, that became more popular in the recent years, is based on time resolution. In this "pump and probe" technique, two short laser pulses

are used to study dissociation processes: the first laser dissociates
the parent molecule ABC and the second laser (delayed from the first)
analyses the formation of the product AB for the different quantum
states. If pressures in the range 100mTorr and delays less than
100nsec are used between dissociation and analysis the collision free
product state distribution is usually obtained.

2.1. Direct detection of product state distributions

The product state distribution can be measured directly, for example
by laser induced fluorescence (LIF) or resonance enhanced multiphoton
ionisation (REMPI). Both of these techniques yield the quantum specific density of AB molecules that are created in the dissociation
process. However, these methods can be applied only to a limited
number of molecules. Whereas LIF is essentially restricted to a few
diatomic molecules, REMPI allows in a few favoured cases also the
state selective detection of larger molecules.
Vibrational and rotational product state distributions of diatomic
fragments can be determined usually by just scanning the laser and
recording fluorescence intensity as a function of frequency. In many
cases the spectra are so clearly resolved that the fluorescence intensity at a given frequency allows the determination of the population in single rovibrational states.   For many photodissociation
processes the products are radicals with additional electronic fine
structure degrees of freedom, as for example Λ-doublets and spin. With
laser techniques, these distributions are obtained as easily as the
vibrational and rotational state distributions.   Different spin states
are usually resolved by frequency. Λ-doublet states are usually too
close in energy to be separated by frequency with standard pulsed
lasers. In this case optical selection rules help to determine the
population in Λ-doublet states separately. For example in the LIF of
OH using the $^2\Sigma - {}^2\Pi$ absorption band of OH, the parity selection rule
implies that the $\Pi^-\Lambda$-doublet is probed by Q-lines whereas the Π^+
Λ-doublet is probed by R-lines.
With these techniques, an almost complete (exception: hyperfine and
magnetic substates) analysis of the population in different quantum
states is often possible.

2.2. Indirect detection of product state distributions

The product state distribution can also be obtained by analysing the
relative translational motion of AB versus C. The excess energy is
distributed over internal energy of AB and translational energy of AB
versus C. Because AB is formed only in well defined quantum states f,
only certain well defined translational energies E_f are possible.
This implies that the C-atoms are formed only with certain well
defined velocities v_f after fragmentation. The number of C-atoms with
velocity v_f will reflect the number of AB molecules formed in quantum
state f.
The velocity of the C-atoms can be measured by different methods. A
few examples are given below.

First, a mass spectrometer might be used for detection. In this case the atoms C arrive at the detector at different times after the dissociation laser pulse, because of their different velocities. The peak intensities corresponding to different arrival times yield the information about the formation of the AB product in different states.[2]

In the second method, a laser is used to ionize the atomic fragment C at the same position where the dissociation laser is fired.[3] Because no impulse is transferred to the atom during ionisation, the velocity of the ion is identical to that of the atom. The ions drift away from the point of dissociation with the velocity of the atoms and arrive at different times at the detector. As before, the peak intensities at different arrival times yield the information about the formation of AB in different quantum states. For the photodissociation of water along the B state the application of this technique revealed, for example, that the OH fragment is formed rotationally very hot.[3] It should be emphasized that a direct laser detection of such hot OH radicals by LIF or REMPI is impossible.

Third, the velocity of atom C can be measured using the Doppler effect. If the atom C is excited with a laser, the absorption frequency depends on its velocity component relative to the direction of the laser. If LIF is used to excite the atom, the fluorescence intensity at different frequencies will contain the information about the formation of AB in different quantum states. If the atom is detected at the same time and the same position that is used to dissociate the parent molecule, Doppler broadened lines are obtained.[4] The information about the formation of the AB product in different states is in this case contained in the edge of the Doppler broadened line. Usually it is not easy to obtain information from a noisy edge of a Doppler broadened line.
An improved Doppler method has been used recently. In this method the probe laser is delayed relative to the dissociation laser so that only the products that fly parallel or antiparallel to the detection laser are excited.[5] This yields much better state resolution and a deconvolution of complicated Doppler line shapes is no longer necessary.

The preceding remarks show that an analysis of the translational motion of atom C gives information about the states in which the partner fragment AB is formed. For the photodissociation of larger molecules, this yields an interesting perspective. If a diatomic fragment from a larger molecule is probed by LIF or REMPI, information might be obtained in a similar way about the partner fragment. Although the diatomic fragment is detected only in one quantum state f, the partner fragment can still be formed in many different quantum states. One of the interesting problems in photodissociation is the question to what extent is the formation of products in different quantum states correlated, i.e., the question, in which states the partner fragment is formed, if the diatomic fragment is formed in one particular state f. If the diatomic fragment is probed in a well defined quantum state, the translational motion will again reflect the energy content in the partner fragment. This yields simultaneous knowledge of the quantum state in which one fragment is formed and the corresponding probability to form the partner fragment in different states. This corresponds to a coincidence experiment, where features of both fragments can be correlated to each other.

2.3. Analysis of product alignment

The analysis of translational motion can not only yield information about the internal state distributions of the fragments, but also about directional properties of the relative motion. These directional properties can be analysed by polarisation techniques.

The use of a polarised laser for dissociation implies that only a certain subset of molecules is selected for dissociation. For a given electronic transition of a triatomic molecule, the absorption dipole (or "transition moment"), has usually a well defined orientation relative to the plane of the nuclei. As an example, we consider the photodissociation of water in the first absorption band.[6] Because the absorption dipole is in this case perpendicular to the molecular plane, only the component of the polarised dissociation laser along the absorption dipole leads to excitation of the molecule. If the electric vector of the dissociation laser is along the z-axis in the laboratory frame, the molecules in the (x,y)-plane are excited with highest efficiency. For a planar dissociation, the OH products will consequently rotate in the (x,y) plane.

The detection of OH is done by LIF via the $^2\Sigma$-$^2\Pi$ absorption band. For Q-line detection in the high-J-limit, the absorption dipole for this transition is perpendicular to the OH rotation plane.[6] Because the OH molecules rotate preferentially in the (x,y)-plane the absorption dipoles are essentially along the z axis. If the detection is now done with a polarised laser, the excitation efficiency - and therefore the resulting fluorescence signal - depends on the direction of the electric vector of the probe laser. For the photodissociation of water in the first absorption band this yields not only information about the planarity of the dissociation process, but also information about the electronic structure of the OH radical. Of particular importance was the result that the orientation of the unpaired electron in the Λ-doublet states was opposite to all earlier assumptions. Because of these erroneous earlier assumptions all interpretations for astronomical OH masers turned out to be wrong.

Polarisation techniques have been applied to a number of dissociation processes. The directional properties of the light, together with the directional properties of the particular transitions in parent and product, can be used to measure the spatial distribution of fragments relative to parent molecule. They yield a powerful tool to understand the dynamics of photodissociation.

A more recent development is based on combination of Doppler techniques with polarisation techniques.[7] For a particular frequency of a Doppler broadened line, only species with a well defined velocity component along the direction of the probe laser are excited. The excitation efficiency of the molecules flying in a particular direction depends again on the direction of the absorption dipole relative to the direction of the electric vector of the probe laser. The resulting shape of the Doppler broadened line depends in this case on the relative direction of the fragment velocity to the rotation plane of the probed fragment. An analysis of such experiments yields information about the direction of the vectors for the product angular momentum relative to the velocity vector ("vector correlation").

Some of these techniques have been applied to the photodissociation of H_2O and will be discussed in more detail in the following.

3.   PHOTODISSOCIATION OF H_2O IN THE FIRST ABSORPTION BAND

3.1. The nature of the electronic transition

The photodissociation of H_2O in the first absorption band

$$H_2O(\tilde{X}^1A_1) + h\nu \longrightarrow H_2O\ (\tilde{A}^1B_1) \longrightarrow OH(^2\Pi) + H \qquad (2)$$

is a model system for simple fragmentation processes, essentially because a very rare and clear physical situation is met in this case.[8] This is explained schematically by the potential diagram in Fig. 1.

H_2O is excited from the ground state $^1A_1$ to the first excited state, in the example at 157nm. At this wavelength there is only absorption to the first electronically excited state $^1B_1$, because higher states of H_2O are not within reach of the photon energy. The first absorption, belonging to this state, is shown at the upper right side. The $^1B_1$ state is strongly repulsive, there are no crossings with other states, so that the dissociation is fast and direct, i.e. predissociation is impossible. The only possible products are OH and H, both in their electronic ground states.

The nature of the electronic transition is well known, because in the first absorption band the weakest bound electron is excited to the next higher lying orbital. The orbitals in H_2O are known from quantum chemistry. The lower left side shows the weakest bound electrons in the so called $1b_1$ orbital, which is an almost undisturbed p_z orbital of the O-atom. This orbital contains two electrons. The next higher orbital is the so called $4a_1{}^*$ orbital, which is very strong antibonding and symmetric to the nuclear plane. This implies that in the photodissociation of H_2O in the first absorption band one of the two electrons in the $1b_1$ orbital is promoted to the unoccupied, strongly repulsive $4a_1{}^*$ orbital, leaving an unpaired $1b_1$ electron perpendicular to the nuclear plane behind.

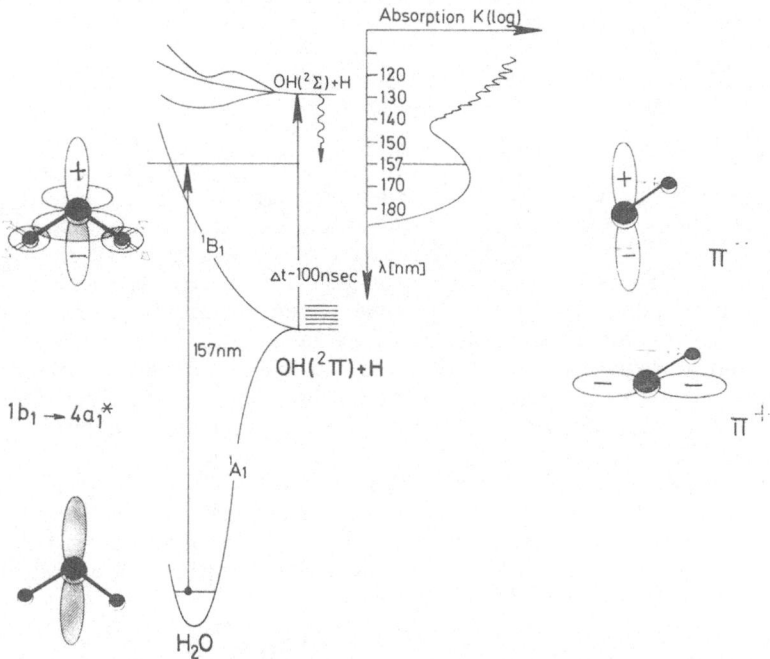

Figure 1. Correlation diagram for the photodissociation of water in the first absorption band with the first electronically excited states of H_2O and the corresponding absorption (upper right side). The left side illustrates the nature of the elctronic transition. The right side shows the important difference between the Λ-doublet states in OH, i.e., the orientation of the unpaired $p\pi$-electron relative to the OH rotation plane in the high-J-limit.

The upper left side shows the resulting shape of the corresponding excited complex, with the symmetric $4a_1^*$ orbital and the now <u>singly occupied</u> $1b_1$ electron. Its structure is very well characterized, because both the electronic structure and the nuclear configuration are known. A very interesting feature of the excited complex is its electronic symmetry with respect to a reflection at the nuclear plane. Because of the singly occupied $p\pi$ lobe perpendicular to the molecular plane (the $1b_1$ electron), the electronic wavefunction is antisymmetric to the plane.

On the right side the two different electronic states in the OH product are visualized for a fast rotating molecule. The OH has an unpaired electron in a $p\pi$ orbital, which can be oriented either perpendicular to the rotation plane, or in the rotation plane, as shown in the figure. In the first case the electronic wavefunction is antisymmetric, in the second case symmetric relative to the rotation plane. These two different electronic configurations represent the almost degenerate Λ-doublet states of OH, which will play an important role in the following.

3.2. Selective population of Λ-doublets

The correlation of the potential surfaces to the product OH seems obvious. The asymmetric excited state $^1B_1$ correlates to the asymmetric Λ-doublet, whereas the symmetric ground state $^1A_1$ correlates to the symmetric Λ-doublet. Conservation of electronic symmetry in the fragmentation predicts the formation of OH exclusively in the asymmetric Λ-doublet state. Conservation of electronic symmetry implies that the motion proceeds along the same potential surface and that transitions to other potential surfaces are negligible. This so called "adiabatic behaviour" is expected to hold for most molecular processes. It seems immediately clear that the origin for the selective population of Λ-doublet states is due to adiabatic behavior.

Selective population of Λ-doublet states is a rather general phenomenon, which is found in chemical reactions, inelastic collisions, photodissociation and surface scattering. In most cases the selective population of Λ-doublet states is indeed explained by adiabatic behaviour, i.e. by the "adiabatic rule"

$$
\begin{array}{ccc}
8 \rightarrow 8 & & oo \rightarrow oo \\
& \text{or} & \\
as \rightarrow as & & s \rightarrow s
\end{array}
\qquad (3)
$$

This rule expresses conservation of the lobe orientation (or conservation of electronic symmetry) in the break up of an excited complex in the sense that antisymmetric transition states yield the antisymmetric Λ-doublet state, whereas symmetric transition states yield the symmetric Λ-doublet state. The symbols "8" and "oo" are used here and in the following to represent the lobe orientation in OH perpendicular and parallel to the rotation plane.

This rule, if correct, is very useful. If for a given molecular process selective population of the 8 (or oo) Λ-doublet is found in an experiment, we conclude that the process was proceeding along an antisymmetric (or symmetric) potential surface. This implies a memory of the electronic symmetry in the transition state.

A nice example of this behaviour is the selective population of Λ-doublet states of NO in the photodissociation of $(CH_3)_2NNO$ at two different wavelengths.[9] In this case, it was found that the antisymmetric Λ-doublet was dominating at one wavelength, whereas the symmetric Λ-doublet was dominating at the other wavelength. It was concluded that the excitation leads to two different electronically excited states.

It will be demonstrated in this paper that the adiabatic rule is qualitatively correct, but quantitatively wrong. The real origin for selective population of Λ-doublet states is considerably more complex.

The experimental results in Fig. 2 are selected to demonstrate that the rule is qualitatively, but not quantitatively, correct. Shown is the relative population of Λ-doublet states for different rotational states of OH for the photodissociation of H_2O at different temperatures. There is a strong preference for the asymmetric Λ-doublet, in

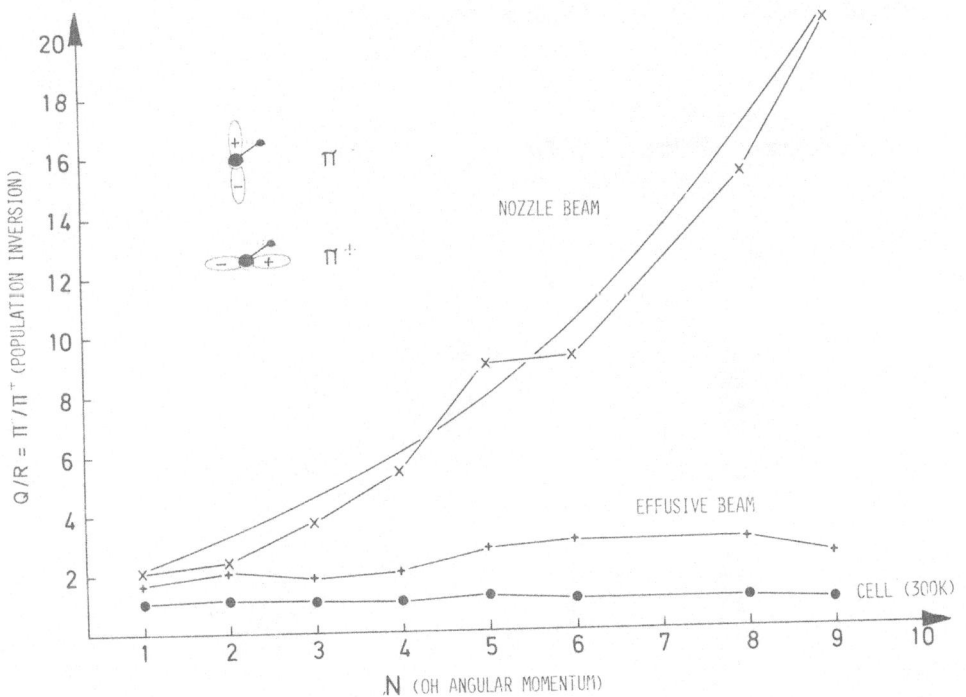

Figure 2. Relative population of Λ-doublet states of OH $(^2\Pi_{3/2})$ after 157nm photodissociation of H_2O in a cell, in a slightly expanded beam and in a nozzle beam. The dominance of the 8-Λ-doublet is obvious, in particular for the nozzle beam H_2O at higher OH angular momentum N.

particular for the photodissociation of jet cooled H_2O at high rotational states of OH. The preference is however obviously not 100%. It depends both on the final rotational states of OH and on the temperature of the parent molecule H_2O. The adiabatic argument explains qualitatively but not quantitatively the dominance of the asymmetric Λ-doublet state. Obviously there are non adiabatic transitions, in paticular at low N and for warmer H_2O.

It will be one of the major topics of this article to discuss the nature of these non adiabatic transitions and to clarify the real origin for the selective population of Λ-doublet states. That the selective population of Λ-doublet states is much more involved was found in a more recent full state to state experiment[10], which is described in the following.

3.3. State to state photodissociation of H_2O

The experimental idea for the study of state to state photodissociation is described in the figure below.

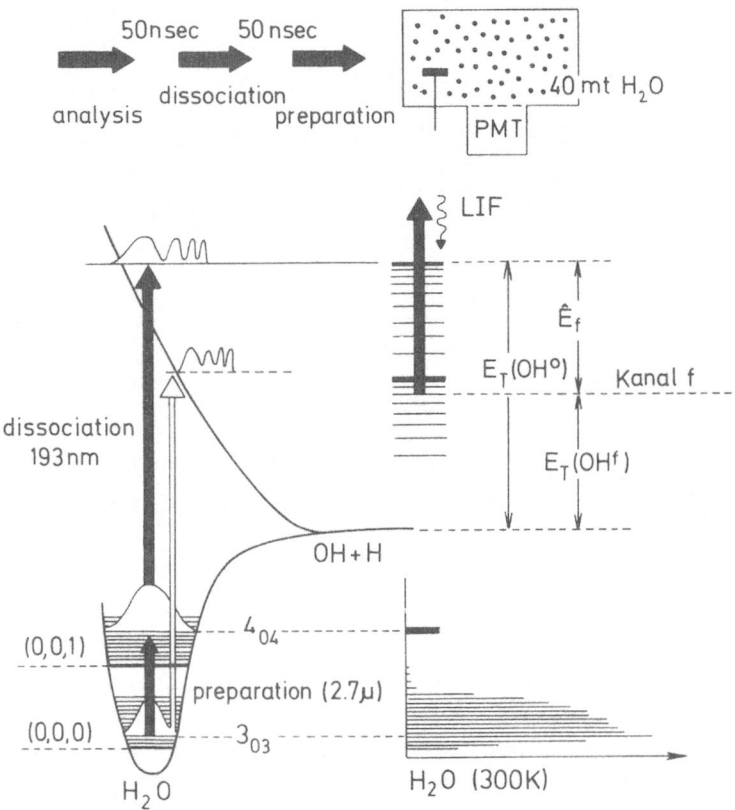

Figure 3. State to state photodissociation of H_2O.

The upper part shows an H_2O containing cell with a microphone and a phototube, which is irradiated by three different laser pulses to 1) prepare a single rotational state by IR excitation 2) to dissociate the molecule from this state and 3) to analyse the OH product. The process is explained in detail in the correlation diagram below. The lower right part gives a rough idea about the population of quantum states of H_2O after IR excitation.

Only very few real state to state experiments have been done in photo-dissociation. The reason for that is very simple. The main problem of a real state to state experiment is to achieve product formation from single rotational states. In a gas at room temperature, many rotational states are populated and contribute to product formation. Even in most nozzle beams the state preparation is often incomplete, as will be demonstrated below. The main trick of the present state to state experiment is the preparation of single rotational states in vibrationally excited H_2O and the subsequent dissociation at a wavelength where photolysis out of the vibrational ground state is impossible.

The idea of the experiment is illustrated in Fig.3. The upper part of the figure shows a simple cell filled with H_2O. A slow gas flow is maintained at a constant pressure of about 40mTorr. The cell is irradiated by three different laser pulses, which follow each other within less than 50nsec. First an infrared laser at 2.7μm is fired to vibrationally excite H_2O and to prepare a well defined rotational state in the asymmetric stretch mode.[11] The second laser, an excimer laser at 193nm, dissociates the H_2O and the third laser analyses the formation of the OH products by LIF. The infrared excitation of H_2O is monitored by the photoacoustic method with the microphone shown in the cell. At the short delays and the low pressures the products are analysed collision free.

In the lower part of Fig.3, this is explained in the potential diagram. A few rotational states are shown both for the vibrational ground state (0,0,0) and the vibrationally excited state (0,0,1) (asymmetric stretch mode). The infrared laser excites a particular rotational transition from the vibrational ground state (000) to the asymmetric stretch mode (001) in the wavelength range around 2.7μm, in the example from rotational state 3_{03} to rotational state 4_{04}. The rotational quantum numbers are J, K_a and K_c for total J and its projection on the A- and C-axis. Other rotational states can be prepared, simply by scanning the infrared laser to other rotational transitions.

Now the dissociation is done at 193nm. At this wavelength, absorption from vibrationally ground state H_2O is negligible, whereas absorption from the vibrationally excited state (0,0,1) is very strong. If the OH products can be formed only from vibrationally excited states, the products have to come from the single rotational state 4_{04} of the asymmetric stretch mode. This is the idea of the experiment.

The origin for the highly enhanced absorption for vibrationally excited state becomes obvious by a look at the nuclear wavefunctions in the electronic ground and the electronically excited state. With the additional energy of the infrared photon, the total available energy is somewhat higher if the dissociation starts from (0,0,1). Therefore the scattering wavefunction is further to the left in the excited state, giving a good overlap with the ground state. If dissociation starts from (0,0,0), the overlap of scattering and ground state wavefunctions is much worse. Because the overlap is responsible for the absorption probability, a highly enhanced absorption is obtained from vibrationally excited states.

The problem of this method is that a large enhancement is required for the experiment. The typical excitation efficiency with the infrared laser is around 10-20% and only a small fraction (1-2%) of the H_2O molecules is in the appropriate rotational state (3_{03} in the example). Because most molecules are still in the vibrational ground state, an enhancement of approximately 1000 will yield as many OH products from the many species in the vibrational ground state as from the one prepared state in the vibrationally excited state. The enhancement could be measured directly because some products were already formed from the ground state. It was found to be approximately 500[10], in good agreement with the factor 450 found in theoretical calculations by Schinke et al.[12]

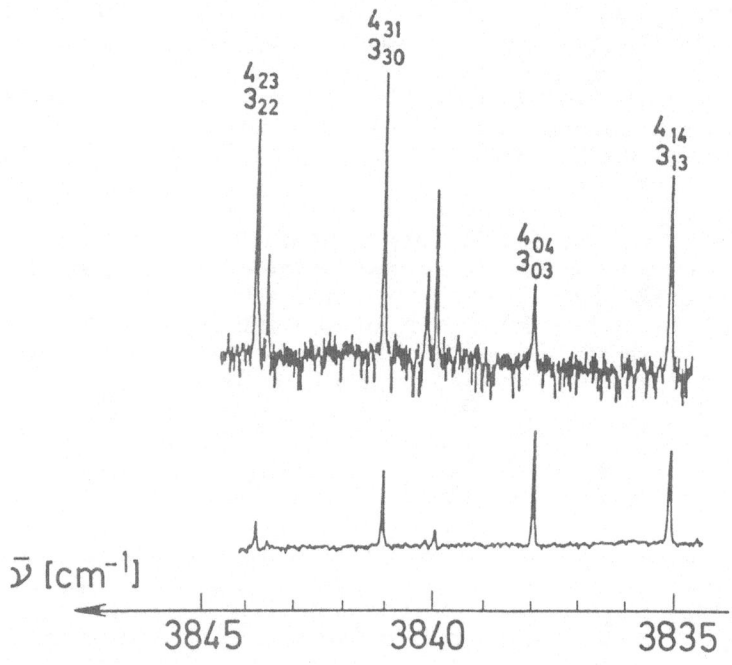

Figure 4. IR laser scans to prove the photodissociation from single rotational states. Excitation is to the asymmetric stretch mode. Above: photoacoustic signal at the microphone. The numbers specify for example the rotational transition 3_{03} to 4_{04}. Below: OH formation from the full experiment

Fig.4 proves that this idea works. It shows in the upper trace the photoacoustic signal at the microphone as a function of the infrared laser wavelength. The lines originate from vibrational excitation of

H_2O, the spectrum can be identified and a few transitions are indicated. The lower trace shows the enhanced formation of OH as a function of the infrared laser wavelength, with the OH probe laser tuned to the $R_1(0)$ transition. The comparison of the upper and the lower trace reveals that there is enhanced OH formation whenever the infrared laser excites a particular rotational transition from (0,0,0) to (0,0,1). The signals in the lower trace are due to the photodissociation of H_2O from single rotational states!

For the next experiments the infrared laser is tuned to a fixed transition to prepare a single rotational state in H_2O and the OH product state distribution originating from this one state is measured by tuning the probe laser.

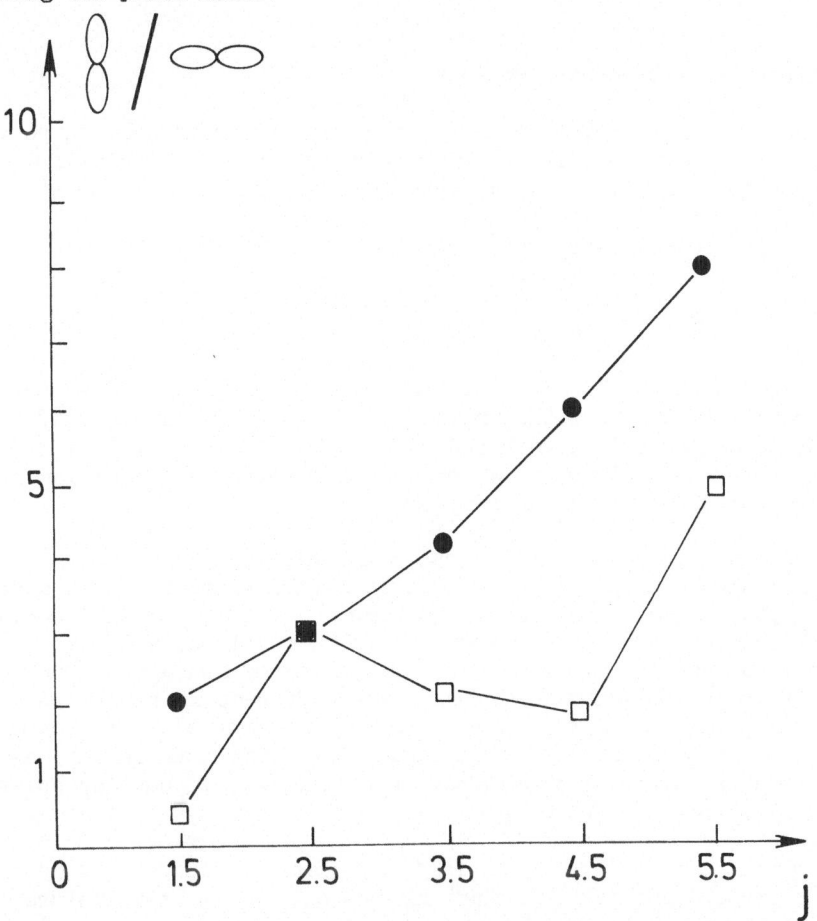

Figure 5.   Comparison of the relative population of the Λ-doublet states in OH after photodissociation of H_2O in a nozzle beam (full circles) and after photodissociation from the single rotational state 0_{00} of the (001) asymmetric stretch state (open squares).

In this section just one such experimental result is briefly discussed. Fig.5 shows the relative population of Λ-doublets originating exclusively from the lowest rotational state 0_{00} of H_2O in comparison with the distribution for jet cooled dissociation of H_2O (see Fig.2). The distributions are expected to be similar, because the rotational state 0_{00} should be most populated in a nozzle beam. However, the distributions are very different, indicating that the H_2O nozzle beam is not as cold as expected. Because there is ortho- and para-H_2O and often incomplete cooling in nozzle beams, this may not be too surprising. The really surprising feature in Fig.5 is the fact, that the Π^+ Λ-doublet is dominating at J=1.5 more than a factor of three over the Π^- Λ-doublet state, exactly opposite to the adiabatic rule. This will however only be discussed later on.

3.4 State to state cross sections

Because state to state cross sections are measured only in very few experiments, its definition and its relation to the product state distributions and the total absorption cross section are briefly discussed.

For the photodissociation of a triatomic molecule ABC

$$ABC(i) + h\nu \xrightarrow{\sigma_{if}} AB(f) + C$$

the state to state photodissociation cross section σ_{if} is essentially the probability to form the product AB in a quantum state f, if 1) the parent molecule is in a well defined initial state i and 2) the photon energy $h\nu$ is well defined. A somewhat more precise definition of σ_{if} in terms of theory is given later on.

Although the formation of products in different states is often analysed almost completely, only very few experiments yield state to state photodissociation cross sections. Usually photodissociation experiments are done with a gas at room temperature, where many rotational states are populated in the parent molecule. In this case product formation originates from many different initially populated rotational states of the parent molecule. Even in experiments, in which jet cooling is used to prepare low rotational states, usually more than one state is populated, in particular, if nuclear spins are important, like in H_2 or H_2O.

In a gas at temperature T, an initial rotational state i is populated with a certain probability, that depends only on the temperature T and is given by the Boltzmann law:

$$p_i(T) = A(2J+1)\exp(E_i/kT) \tag{4}$$

Here k is the Boltzmann constant, J the total angular momentum and E_i the energy for state i.

In general all states with appreciable population will contribute to product formation. The molecule is formed in the final state f from initial state i with the probability σ_{if}. This has to be weighted with the probability $p_i(T)$ for state i and summed over all the populated initial states i to give the probability for the formation of the product in state f:

$$n(f,T) = \sum_i p_i(T) \cdot \sigma_{if} \qquad (5)$$

This is the product state distribution that has been mentioned before and that is measured in most experiments.

The product state distribution in Eq.5 no longer depends directly upon the initial state i, but upon the temperature T instead. It is interesting to discuss in which case the product state distributions really depend upon temperature.
If the state to state cross sections σ_{if} do not depend upon the initial state i, then the product state distributions are not temperature dependent, simply because σ_{if} can be extracted from the sum and the remaining Boltzmann sum is 1. Only if the state to state cross sections do depend upon the initial state, temperature dependent product state distributions are possible.
If the product state distributions are experimentally found to be temperature dependent, it is concluded that the state to state cross sections do depend on the initial state. Temperature independent product state distributions suggest that the state to state cross sections are independent of the initial state. It will be seen later that the question whether σ_{if} depends on i is closely related to the strength of final state interaction.

The total absorption cross section represents essentially the probability of forming the product AB at all. It is obtained from state to state cross sections by summing over all initial states populated in the parent molecule - weighted with their corresponding probabilities - and by summing also over all final states in which the AB product is formed

$$\sigma_{tot} = \sum_f \sum_i p_i(T) \cdot \sigma_{if} \qquad (6)$$

The difference from Eq.5 is the additional averaging over final states. Most experiments, in which product state distributions are measured, are done at one particular wavelength. In contrast, the total absorption cross section is typically measured as a function of wavelength. Although the absorption cross section is a higher averaged quantity, it contains important information about electronically excited states.

Not only the total absorption cross section, but all other features of a given dissociation process can be calculated if the state to state cross sections are known for each i, each f and each photon energy. The complete knowledge of the state to state cross sections implies, that the photodissociation process is quantitatively understood.

To the present time, the photodissociation of H_2O in the first absorption band seems to be the only photodissociation process that can be understood on this level of sophistication.

4.   WEAK AND STRONG FINAL STATE INTERACTION

4.1. The two different steps in photodissociation

In the following photodissociation processes are considered from a more general point of view. The photodissociation of a triatomic molecule ABC consists of two very different steps. This is seen already in the formal description

$$ABC + h\nu \xrightarrow{\ 1\ } ABC^* \xrightarrow{\ 2\ } AB + C$$

where in the first step the absorption of a photon leads to an excited complex ABC^* and where in the second step the molecule fragments to the products AB and C.

The absorption of the photon in the first step leads to electronic excitation of the ABC molecule. Because it takes only a very short time to absorb a photon, there is no time for the nuclei to move. This implies, that the nuclear frame of the stable ground state molecule is conserved in the excitation step. This step is essentially electronic motion: an electron is promoted to a higher lying orbital. Immediately after absorption the excited molecule has therefore the same nuclear configuration as the ground state, but a very different electronic shell.

The second step, the fragmentation, is of completely different nature. It is motion of heavy nuclei against each other on an excited state potential surface. Because the break up of the excited molecule to products resembles the second half of a reaction collision, photodissociation processes are sometimes called "half collisions".

Whereas the absorption of light in the first step and the resulting features of the excited molecule belong to the general field of spectroscopy, the second step belongs more to the field of collision dynamics. Photodissociation dynamics is a field in between spectroscopy and scattering and this may be the reason why it took so long to understand these processes in detail.

According to the two very different steps, a rough classification of photodissociation processes according to the origin of product motion can be given. First, motion can be generated by forces acting on the nuclei in the excited state. This happens in the second step and is called "final state interaction". If these forces are negligible, there will still be motion in the products, because initial motion in the parent molecule can be transferred to the products. Obviously these are two basically different reasons for motion in the products.

4.2 Final state interaction

Final state interaction describes the effect of the excited state potential (the final state) on the motion of the products. Immediately after absorption of the photon, the molecule finds itself in the excited state, but with the nuclear frame of the ground state. This nuclear configuration is usually not energetically favored in the excited state, i.e., other nuclear configurations may be energetically lower. This implies that forces are exerted during fragmentation.

These forces are given by derivatives of the interaction potential with respect to different coordinates. For example, a non zero derivative along the reaction coordinate will cause a radial force, which repels the fragments from each other. Other non zero derivatives generate rotational or vibrational motion in the product. Because the nuclear configuration of the electronic ground state is often highly unfavoured in the excited state, the forces can be large. The nature of these forces and their importance for product motion is demonstrated below.

In the following three figures the excitation and fragmentation steps are illustrated for a triatomic molecule. In the electronic ground state V (lower plane) the stable configuration of the parent molecule ABC is shown. The nuclear equilibrium coordinates characterize the structure of the molecule. The zero point vibrational energy in the parent molecule causes some variation of these coordinates. Immediately after absorption of the photon, the ABC molecule is on the excited state potential surface V^* with the nuclear configuration from the ground state (middle). Because this configuration is in general not favored energetically in the excited state, different forces can be exerted during fragmentation and lead to vibration or rotation in the products (upper plane). The following pictures show the importance of "bond length" and "bond angle" in the excited state. They demonstrate that a different nuclear configuration in electronic ground and excited states can yield large motion in the products.

4.2.1 Planar rotational excitation

Fig.6 shows an example for the fragmentation of an ABC molecule, in which a force located in the ABC-plane leads to the excitation of planar rotational motion in the AB product. Planar forces are expected for example from antibonding orbitals that are symmetric to the nuclear plane.

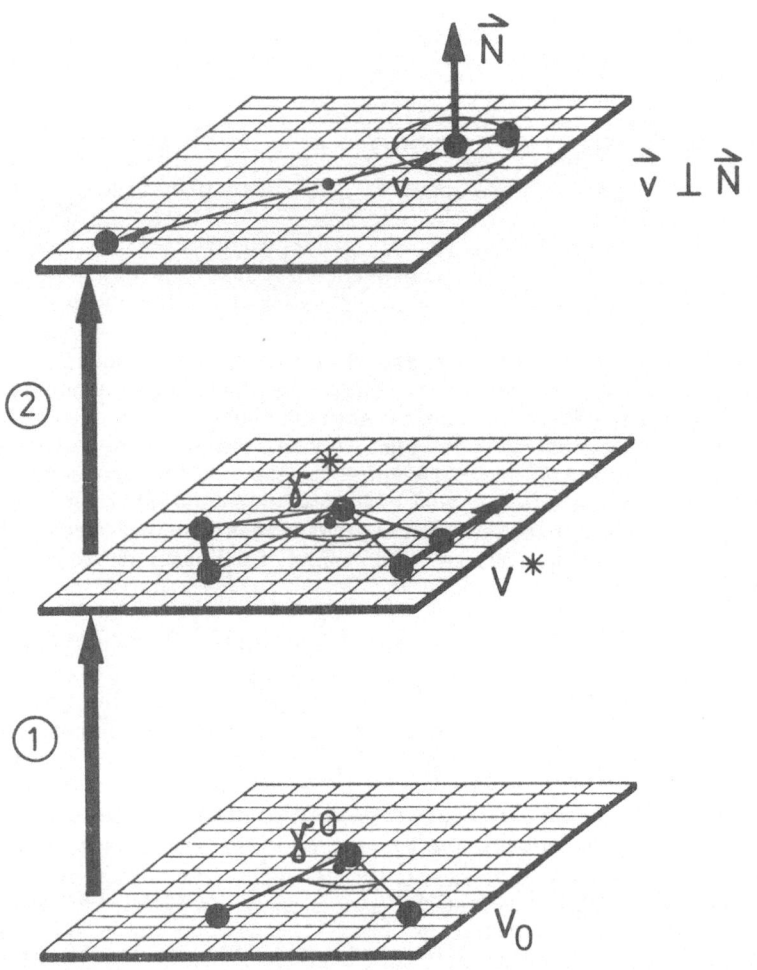

Figure 6: Planar rotational excitation of a triatomic molecule ABC

The three planes show the situation before the absorption of the photon (below), directly after absorption with still the nuclear configuration from the ground state but with another favored geometry (middle) and after fragmentation (above). The direction of the force is indicated by the arrows. Because the force is assumed to be located in the plane, the products will rotate in the same plane.

In the ground state the bond angle is assumed to be γ^o and this will be the angle of the molecule in the excited state immediately after absorption of the photon. However, in the excited state another bond angle γ^* may be energetically more favoured. In this case, the excited state potential V^* does change with the angle γ and the force

$$F = \frac{\partial V^*}{\partial \gamma} , \tag{7}$$

is exerted on the departing fragments. Depending on the magnitude of this force there will be large or small rotational excitation in the products.

The largest forces are expected to be exerted in the early part of the fragmentation simply because potentials are usually steeper for smaller internuclear distances. This implies, that most of the torque results immediately after absorption, i.e., from nuclear configurations where the ground state wavefunction is non zero. The amount of rotational excitation in the fragmentation step is determined by the variation of the excited state potential surface with γ in that particular range.

If the bond angle is conserved in an electronic transition, i.e., if the excited state potential as a function of γ has a minimum in the same place, little rotational excitation is expected. If the bond angle changes dramatically, like in a linear-bent transition, large rotational excitation will result.

An impressive example of large rotational excitation is the photodissociation of water in the second absorption band (the B-state), where OH rotational states are populated up to N=45.[4] In contrast, in the first absorption band of the same molecule, very little rotational excitation is found in the OH product, indicating an extremely small anisotropy in the excited state potential surface. This demonstrates, that the rotational state distribution in the products is very sensitive to the features of the excited state potential surface, in this case to its anisotropy with respect to γ. The large difference in the rotational distributions for the same molecule demonstrates also that dynamics and not kinematical constraints are responsible for this effect.

Another feature, that is typical for the case of planar rotational excitation, is the relative orientation of the velocity vector v to the rotational angular momentum vector N. In contrast to the case of torsional excitation, v is perpendicular to N (compare figure 8).

4.2.2 Vibrational excitation

Fig.7 shows an example for vibrational excitation in the products. It is assumed in this case that another bond length r^* (= distance between the two atoms of the diatomic fragment) is energetically favored in the excited state.

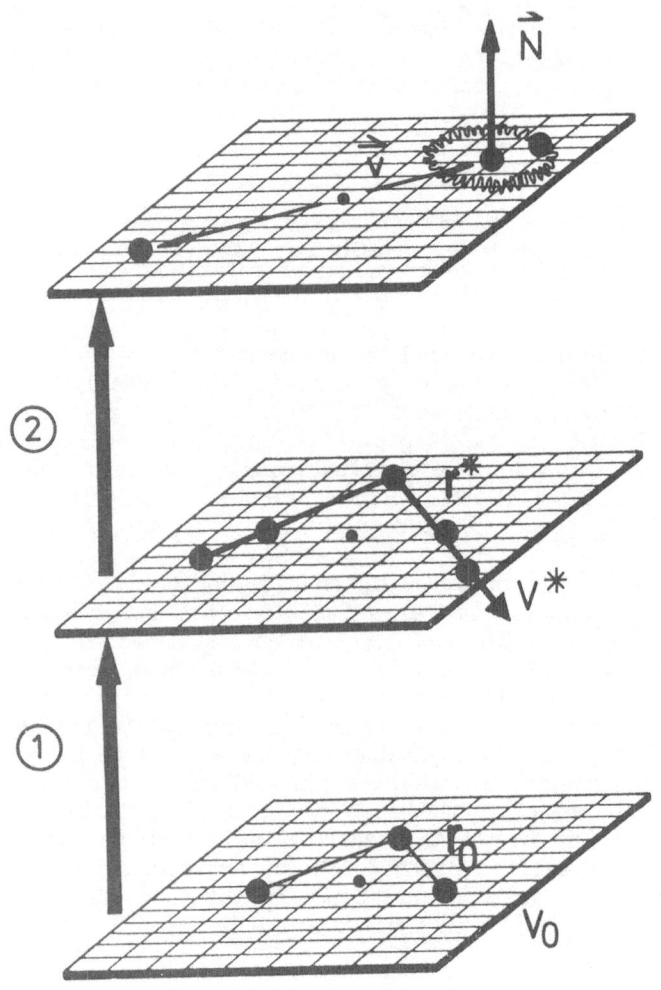

Figure 7: Vibrational excitation

A radial force along the bond causes vibrational motion in the products.

This leads to a force

$$F = \frac{\partial V^*}{\partial r} \tag{8}$$

that is exerted on the departing fragments in the excited state.
Depending on this force there may be large or small vibrational energy
contents in the product. An example of this behaviour is the photo-

dissociation of water in the first absorption band at 157nm[8], where the interaction potential has a strong anisotropy in the OH coordinate r.

Fig.8 shows a somewhat more interesting case of the excitation of torsional motion in the fragmentation of a four atomic molecule A_2B_2, like for example H_2O_2. It is assumed, that the torsional angle is different in the excited state to that in the ground state.

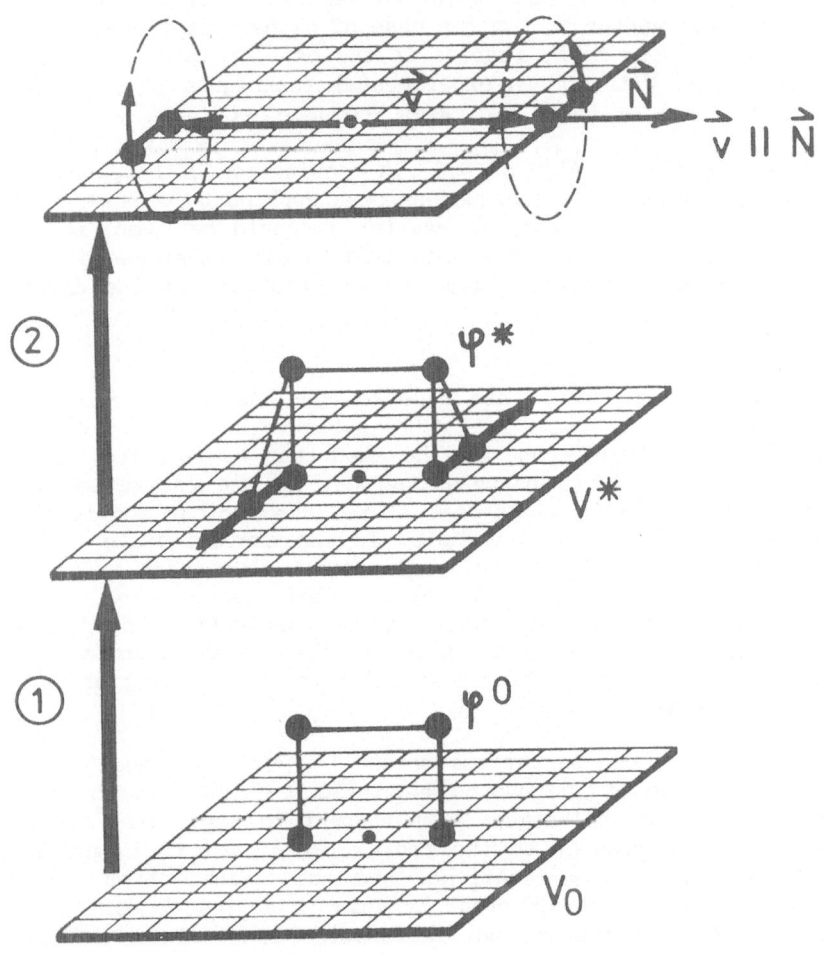

Figure 8: Excitation of torsional motion

The corresponding force

$$F = \frac{\partial V^*}{\partial \phi} \qquad (9)$$

leads consequently to torsional excitation of the complex. However, because the molecule fragments, this torsional motion shows up as rotational motion in the two AB fragments.

The obvious difference from the case of planar fragmentation is the different direction of the relative velocity vector with respect to the rotational vector N. For the case of planar rotational excitation, v is perpendicular to N, whereas in the case of torsional motion v is parallel to N. With modern experimental techniques it is possible to distinguish between these two cases (compare section 2.3).

For the photodissociation of H_2O_2, a combination of Doppler spectroscopy and polarisation technique has been applied to measure v relative to N. The effect of torsional excitation could be verified in this case.[7] The same technique has been used in some other cases and may turn out to be a powerful tool for the analysis of the dynamics of photodissociation.[14]

A few summarising remarks are made to conclude this section on final state interaction,
From the above figures one of the important origins for the nuclear motion in the products becomes obvious. In the excitation from the ground to the excited electronic state, the nuclear configuration is conserved due to the FC principle. In the excited state the nuclear configuration is often not in equilibrium with respect to the different nuclear coordinates. The forces that are exerted during the fragmentation process are given by the derivatives of the excited state potential surface with respect to the nuclear coordinates. These derivatives represent nothing else than the anisotropies of the interaction potential.

The anisotropies can be very different for different degrees of freedom. In the case of the photodissociation of water in the first absorption band, the anisotropy with respect to γ is small, whereas the anisotropy with respect to the radial coordinate is large. This explains the rather low rotational and the rather high vibrational energy contents in the OH product. This is different for the absorption of water in the second absorption band, where a very large anisotropy with respect to γ leads to an enormous rotational energy content in the OH product.

It should be emphasized that the instability of the nuclear configuration in the excited electronic state is closely related to the electronic structure of the excited state. The "change in bond length" or the "change in bond angle", i.e., the difference in the equilibrium nuclear configuration between the electronic ground and excited state yields the main active forces in photodissociation processes. Therefore the analysis of product motion from photodissociation processes yields important information about the nature of the electronic structure in excited states. Large rotational motion indicates for example a change in bond angle, a large vibrational motion a change in bond length in the excitation process.

In the quantum mechanical treatment of scattering theory couplings rather than the forces from Eqs.7-9 are used. The couplings between various degrees of freedom are responsible for the conversion between different kinds of motions. Different kinds of motion can be converted into each other only if these couplings are non-zero. The strength of the coupling is however directly related to the magnitude of the potential derivatives and therefore to the strength of the forces exerted during fragmentation. The above mentioned effects can be described in good approximation by classical mechanics, i.e., classical trajectory calculations yield results similar to a full dynamical treatment. This is in contrast to the case of weak final state interaction, which can only be described by quantum mechanics.

4.3 Weak final state interaction or FC-limit

If there is no anisotropy in the excited state potential surface, no extra motion is generated in the fragmentation step. In this case the product state distribution must already be determined after the first step. Because no motion is generated in the fragmentation step, the product motion must originate from that of the parent molecule. The motion in the parent molecule consists not only of initial rotation of the parent molecule in the space fixed frame but there is also zero point motion in the bending mode or vibrational motion in symmetric or asymmetric stretch modes. It is this parent motion that shows up in the product after the fragmentation process. In the case of zero final state interaction, which is here called the Franck-Condon-limit, the product motion originates exclusively from initial motion in the parent molecule.

The case of zero final state interaction can never be realized, because there will be no fragmentation without a force $\partial V/\partial R$ along the reaction coordinate. However, there is not only motion along the reaction coordinate. The forces that are responsible for generating rotational or vibrational motion have been shown to be $\partial V^*/\partial \gamma$ and $\partial V^*/\partial r$. Zero final state interaction can be found for a degree of freedom, if the corresponding anisotropy of the excited state potential surface is zero or close to zero. If, for example $\partial V^*/\partial \gamma$ is equal to zero, no additional rotational motion is generated in the fragmentation step and all product rotation will originate from parent motion.

The amount of energy that shows up in the products in the FC-limit depends very much upon the degree of freedom. For product rotation only zero point vibrational and initial rotational energy is important. Because only a comparably small amount of energy is contained in this motion only a limited amount of rotational energy is expected in the products. In contrast, rather large vibrational energy may be transferred from parent to products, as will be seen below.

A characteristic feature for weak final state interaction is the "memory of the initial state". Memory of the initial state means, in this case, that the state to state cross sections σ_{if} depend on the initial state i and this in turn implies that the product state distributions are different for different initial states.

For example, in the state to state experiment on the photodissociation of H_2O in the first absorption band, different initial rotational motions are prepared in the parent molecule. Because this motion is transferred to the OH product, it is clear that different initial motions in the parent molecule will lead to different motions in the product. For example, if the parent molecule rotates fast (or slow) the products are expected to rotate fast (or slow). This implies, that the state to state cross sections depend upon the initial state, i.e., σ_{if} will depend on i. According to the discussion in section 3.4, this might give temperature dependent product state distributions.

This is different for the case of strong final state interaction. For H_2O in the second absorption band, the enormous torque exerted during fragmentation on the OH product generates so much rotation that the transfer of initial rotation from the parent molecule is almost negligible. In this case, the state to state cross sections may become independent of the initial state of the parent molecule, i.e., σ_{if} will not depend on i. This results in temperature independent product state distributions.

Depending on the strength of final state interaction, the product state distributions may be temperature dependent or temperature independent. State to state experiments are only reasonable if the state to state cross sections do depend upon the initial state i. Therefore, it is worthwhile to first check the temperature dependence of the product state distributions before going to an ambitious state to state experiment. The temperature dependence of product state distributions can contain already important clues about the strength of final state interaction.

4.3.1 State to state cross sections in the FC-limit

To understand the features that are described in the Franck-Condon-
limit, we introduce the definition of the state to state photodissociation cross sections:

$$\sigma_{if} = \langle \, i \, | \, \mu \, | \, \Psi_f^*(E) \, \rangle. \qquad (10)$$

Here $|i\rangle$ is the wavefunction describing the parent molecule ABC in the
bound state i in the electronic ground state and Ψ_f is a solution of
the full scattering problem, that ends asymptotically exclusively in
channel f. These wavefunctions are already considerably complex for a
triatomic system like H_2O, because there are different degrees of
freedom. To understand the general structure of these wavefunctions,
we introduce the coordinates from Fig.9.
The relative motion is described by the vector R pointing from the
center of mass (CM) to the departing C-atom. The vector $\vec{R}$ is described
by its polar coordinates R,θ and Φ in the space fixed (x,y,z) coordinate system (not shown). The orientation of the internuclear axis of
the diatomic product is described by the vector $\vec{r}$, which points from A
to B. The motion of the internuclear axis is described by polar coordinates (r, γ, φ), which are defined relative to the vector R as
"z-axis".

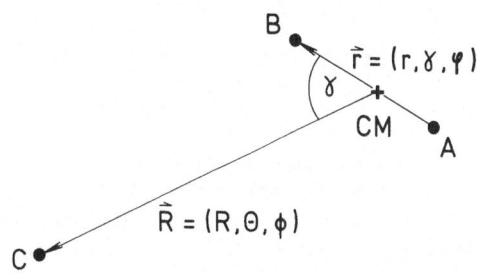

Figure 9   Coordinates for the triatomic molecule ABC

The wavefunction $|i\rangle$ in the definition of the state to state cross
section   describes the rotational, vibrational and electronic motion
of the stable parent molecule ABC in the electronic ground state.   If
this motion is decoupled we can write

$$|i\rangle = |rot\rangle \, |vib\rangle \, |el\rangle. \qquad (11)$$

The description of the scattering wavefunction is much more involved, because $\Psi_f^*(E)$ represents a full solution of the multidimensional Schrödinger equation i.e., $\Psi_f^*(E)$ solves the scattering problem on the excited state potential surface.

Here we discuss only the very simple case of zero final state interaction, which corresponds to the asymptotic scattering problem. In this case the motion of the products AB and C is completely decoupled. The corresponding wavefunctions are the wavefunctions for the separated fragments. If the internal motion of AB is described by $|AB\rangle$, the internal motion in C by $|C\rangle$ and the relative motion by $|AB-C\rangle$, the solution $|f\rangle$ for the asymptotic scattering problem becomes

$$|f\rangle = |AB-C\rangle\, |AB\rangle\, |C\rangle. \tag{12}$$

This is a special solution for the full scattering problem in the trivial case where the asymptotic scattering wavefunction is not changed at all by the interaction potential, i.e., for the case of zero final state interaction. This implies that the full scattering solution Ψ_f^* can be replaced by the asymptotic wavefunction $|f\rangle$. This yields the state to state photodissociation cross section in the FC-limit:

$$\sigma_{if} = |\langle i | \mu | f \rangle|^2. \tag{13}$$

A further simplification of this is obtained under the usual assumption that the transition moment μ is constant:

$$\sigma_{if} = |\langle i | f \rangle|^2 \tag{14}$$

This can be evaluated without extensive numerical calculations, because the wavefunction for both the initial state and the separated products can be written down explicitly for many cases.

It is interesting to discuss the meaning of Eq.14. In quantum mechanics any matrix element of the type $\langle i|f\rangle$ represents the probability that state f is contained in state i. The states $|f\rangle$ form a full basis set and the matrix elements $\langle i|f\rangle$ are the coordinates of $|i\rangle$ in the basis f. This may be expressed in a different way. The matrix element $\langle i|f\rangle$ represents the probability that the motion described by f is already contained in the motion described by state i. This illustrates that the FC-limit describes the transfer of motion from parents to products.

The FC-limit holds only in the case of zero final state interaction. However, as discussed above, there may be zero final state interaction for some degrees of freedom and strong final state interaction for other degrees of freedom. It will be seen, that equation 14 can be correct for some degrees of freedom and wrong for others.

4.3.2 Transfer of motion from H_2O to OH and H

For the specific case of H_2O, the rotational motion is described by the quantum numbers J for the total angular momentum, K_a for the projection of J on the A-axis and K_c for the projection of J on the C-axis. The set of quantum numbers $|JK_aK_c\rangle$ describes the motion of the nuclear plane in a space fixed (x,y,z) coordinate system and this motion depends only on the angles θ, Φ and ϕ. The vibrational motion in H_2O may be separated into the stretch modes $|v_1v_3\rangle$ and the bending mode by $|v_2\rangle$. The stretch modes $|v_1v_3\rangle$ depend on the coordinates r and R, whereas the bending mode $|v_2\rangle$ depends only on γ. To describe the electronic motion in the ground state we use the symbol $|e_0\rangle$. We assume that vibrational, rotational and electronic motion is de-coupled and obtain

$$|i\rangle = |JK_aK_c\rangle \; |v_2\rangle \; |v_1v_3\rangle \; |e_0\rangle \qquad (15)$$

The quantum number i used above corresponds therefore to the whole set of quantum numbers $(J, K_a, K_c, v_1, v_2, v_3$ and $e_0)$.

For the products OH and H the situation is a little more complicated. The vibrational wavefunction for OH is denoted by $|v\rangle$ and depends on the coordinate r only. The rotational motion, given by $|N,m_N\rangle$, is described relative to $\vec{R}$ and depends therefore only on the angles γ and ϕ. The symbol $|e^*\rangle$ is used to describe the electronic motion in both OH and H. The relative motion consists of a radial part $|Rel\rangle$ depending only on the reactive coordinate R and an angular part $|1,m_1\rangle$, depending on the angles θ, Φ.

$$|f\rangle = |Rel\rangle \; |1m_1\rangle \; |v\rangle \; |Nm_N\rangle \; |e^*\rangle \qquad (16)$$

The quantum number f for the final state consists consequently of a whole set of quantum numbers $(N, m_N, v, ,e^*, 1, m_L)$ and by $|Rel\rangle$ for the relative motion.

For the fragments OH and H from H_2O, the internal motion of OH is not only rotation and vibration, but also electronic motion. In addition, rotational and electronic motions are coupled in OH with the coupling being intermediate between Hund's cases a and b. Although the other product is an atom, the spin of the electron of the H-atom has to be coupled to the partner product OH. This is why in most cases, (16) does not describe the asymptotic motion of the products correctly. Instead, linear combinations of such wavefunctions have to be used in the general case. Nevertheless we first discuss the simplified case of decoupled motion.

To discuss the different effects that are described in the FC-limit, we insert the wavefunctions from Eqs. 15,16 in the matrix element $\langle i|f\rangle$ and separate the terms that depend on different coordinates

$$\langle i | f \rangle = \langle v_1 v_3 | v \rangle | R \rangle \quad \langle JK_a K_c | \langle v_2 | \mu | Nm_N \rangle | 1m_\varrho \rangle \quad \langle e_0 | \mu | e^* \rangle \quad (17)$$

It should be repeated that this assumes decoupling of motion in the different degrees of freedom. This is often not true. The discussion here is simplified in order to show that there are three very different types of overlap matrix elements that describe different effects for the transfer of parent motion to products.

The first term is a vibrational overlap, the second term a rotational overlap and the third term an electronic overlap. The first term is responsible for the vibrational distribution in the products, the second term for the rotational distribution and the last term for the electronic fine structure distribution.

The "vibrational overlap matrix element" $\underline{V}$

$$\underline{V} = \langle v_1 v_3 | \; v \rangle \; | R \rangle. \quad (18)$$

depends only on the radial coordinates r, R. It transfers the stretch motion $| v_1 v_3 \rangle$ in the parent molecule (depends on both r and R) to the vibrational motion $| v \rangle$ in AB (depends only on r) and to the translation $| \mathrm{Rel} \rangle$ of AB versus C.

This matrix element describes for example the effect of a different A-B bond length in the isolated AB molecule and the parent molecule ABC. To illustrate that, we assume that the motion along r is given by $| v_3 \rangle$ and the motion along R by $| v_1 \rangle$ and that these motions are decoupled:

$$\underline{V} = \langle v_1 | v \rangle \; \langle v_3 | \mathrm{Rel} \rangle. \quad (19)$$

Then the matrix element $\langle v_1 | v \rangle$ determines the vibrational state distribution in AB, whereas the matrix element $\langle v_3 | \mathrm{Rel} \rangle$ is important for translational motion.

Fig. 10 shows the effect of different bond lengths on the vibrational product state distribution. In the stable ABC molecule the AB bond length (r_0) is assumed to be smaller than in the isolated AB molecule (r^*). The wavefunction $| v_1 \rangle$, describing the vibration of AB in the stable ABC molecule, is shown in the upper part. The wavefunctions $| v \rangle$, describing the free motion of AB, are plotted in the lower part for different v. The overlap $\langle v_1 | v \rangle$ for different v represents the probability for the formation of AB in different vibrational states v. Because there is only overlap with states for v larger than 2, high vibrational excitation results in AB.

This may be considered from a somewhat different point of view. If atom C is eliminated from the ABC molecule without final state inter-action, the motion of AB, that was originally hindered by the presence of the C-atom, becomes free and leads to vibrational excitation in AB. This is the physical content described by the matrix element $\langle v_1 | v \rangle$.

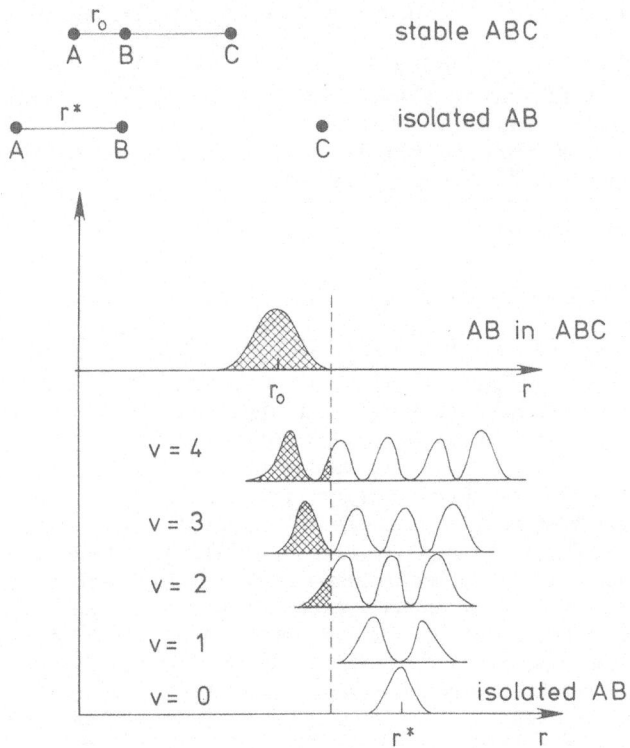

Figure 10   The effect of a difference in bond length on the products vibrational distribution.

It is interesting to note that, in this case, a force is exerted in the excited state which causes the relative motion of A versus B to change. Nevertheless, this has nothing to do with final state interac-tion. Here, it is explicitly assumed that there are no anisotropies in the interaction potential and, therefore, the couplings are zero. However, the interaction potential is the full ABC potential minus the asymptotic potential and the asymptotic diatomic potential is not zero. The asymptotic potential is typically a Morse type potential that depends still on r. The initial non-equilibrium configuration of

the nuclei in the parent molecule relative to this asymptotic potential is responsible for the vibrational excitation in the AB product.

This point is important because this type of effect can be explained
in the FC-limit without extensive numerical close coupling calculations. An example for this behaviour is the photodissociation of
H_3C-I, where the relative distance of the plane of the H-atoms to the
C-atom is very different in H_3C-I and in the isolated CH_3. CH_3 is
planar, whereas the H_3C group in CH_3I is decidedly pyramidal. The
dominant part of the vibrational excitation in CH_3 can be explained in
the FC-limit.

The second type of matrix element may be called a "rotational overlap
matrix element", because it depends only on the angular coordinates,
and because it is responsible for the transfer of rotational motion
from the parent ABC to the products AB and C.

The rotational matrix element is given by

$$\underline{R} = \langle JK_aK_c| \ \langle v_2| \ \mu \ |Nm_N\rangle \ |1m_\varrho\rangle \qquad (20)$$

The initial rotation of H_2O $|JK_aK_c\rangle$ depends on the angles θ, Φ and ϕ
that define the location of the H_2O molecule in space. The bending
mode $|v_2\rangle$ depends only on the bond angle γ. The OH rotation $|Nm_N\rangle$
depends on the coordinates γ and ϕ, whereas the angular part of the
relative motion of AB versus C depends on θ and Φ.
The transfer of rotational motion from parent to products may be
illustrated by the simple example of a planar fragmentation of initially non rotating H_2O to OH and H. Although there is no initial
rotation in the parent molecule there is bending zero point motion in
the coordinate γ. If one H-atom were not present, the other
H-atom could rotate freely around the O-atom. However, due to the
presence of the H-atom, this rotational motion is hindered. If the H
is eliminated without any force in the exit channel, the rotation becomes free and shows up in the products. This is the type of effect
described by the matrix element $\langle v_2|Nm_N\rangle$.

In the real case of H_2O the evaluation of the matrix element requires
considerable numerical effort. The rotational distribution that
results from this matrix element, explicitly evaluated for the case
of H_2O, is shown in Fig. 11.

Obviously the theoretical result does not describe the experiment at
all. The bad agreement is however not due to a failure of the FC-
limit, it is due to coupled rotational and electronic motion in OH.
From an experimental point of view, the bad agreement was not too
surprising, since the distributions in the two Λ-doublet states are
already very different, and it is not clear what has to be compared.
The only meaningful comparison would be with a distribution that is
averaged over the Λ-doublet states. However, this also leads to bad
agreement.

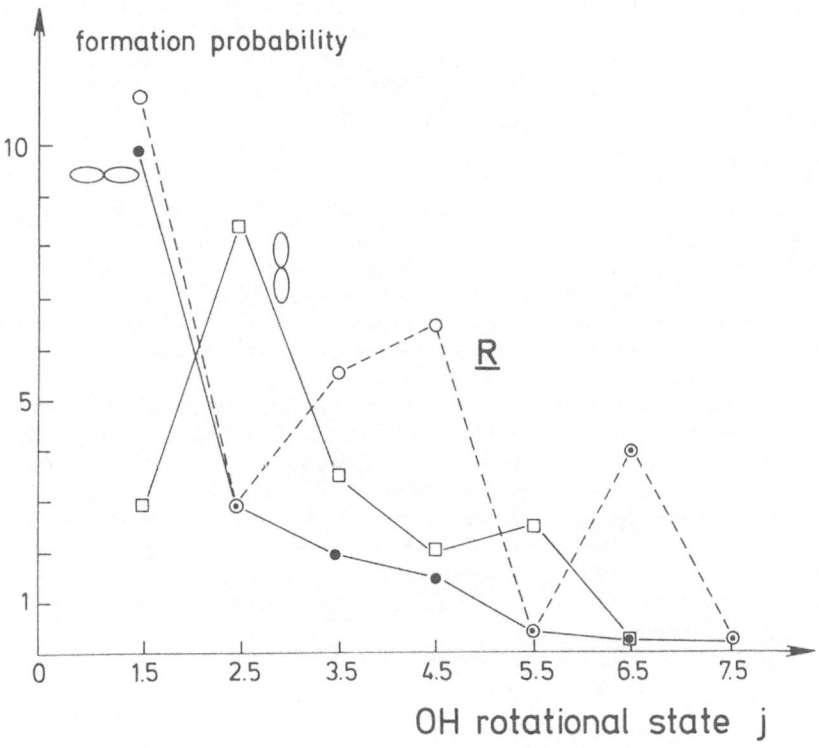

Figure 11 Comparison of the theoretical and experimental rotational state distribution for the photodissociation of H_2O from the 0_{00} state using Eq.20

The third type of matrix element may be called "electronic overlap matrix element" $\underline{E}$

$$\underline{E} = \langle e_0 | \mu | e^* \rangle \tag{21}$$

It describes the electronic structure of H_2O in the ground state on the left side. $|e_0\rangle$ is determined only by the active electrons in the excitation process, i.e., the $^1b_1$ and the $4a_1^*$ orbital. The active electrons in the products are the unpaired electron of OH and the 1s electron on the H-atom. We discuss only the case where the unpaired electron in OH and the unpaired $p\pi$ electron perpendicular to the nuclear plane in the excited H_2O are taken into account.

If we use the perfectly oriented lobes from Fig. 1 for both OH and H_2O, the overlap matrix element can be evaluated simply as

$$\underline{E} = \langle 8 | 8 \rangle = 1 \text{ and } \underline{E} = \langle 8 | 00 \rangle = 0 \tag{22}$$

This implies formation of OH exclusively in the Π^--Λ-doublet state, as would be expected from the simple argument of adiabatic behaviour.

The actual electron density distribution in OH is however different, because there is a coupling of rotational and electronic motion. Due to this coupling, the unpaired electron density is distributed over both lobes.[14] The distribution of the electron density over both lobes is very different for different molecules. The orientation is closely related to the Hund's coupling cases. In Hund's case a the electronic density is the same for both Λ-doublet states, giving cylindrical symmetry around the internuclear axis. Only in Hund's case b is the lobe perfectly oriented as in Fig.1. This is why in most cases where selective population of Λ-doublet states has been observed, it is for molecules in the Hund's case b limit.

We can use the J-dependent electron densities instead of the perfectly oriented lobes and form the overlap with the unpaired $1b_1$ electron in the parent molecule H_2O again. In this overlap only the probability for the lobe out of plane of the OH is projected, because the corresponding lobe in the parent H_2O is perfectly oriented perpendicular to the nuclear plane. The probability for the lobe out of plane is c_j^2 for the π^- state and $1-c_j^2$ for the π^+ state of OH.[14] This yields the probability for the formation of OH in the different Λ-doublet states:

$$\pi^- = c_j^2$$
$$\pi^+ = 1-c_j^2 \tag{23}$$

and the relative populations of the Λ-doublet states is

$$\frac{\pi^-}{\pi^+} = \frac{c_j^2}{1 - c_j^2} \tag{24}$$

Eq.24 predicts the relative populations of Λ-doublet states and is called "one electron overlap" model. The results of this model have been shown already in Fig.3 as the solid smooth line. It works very well for the formation of OH from the jet cooled water. However, for internally warm H_2O, and in particular for the full state to state experiment this model breaks down completely.
The breakdown can be seen by going back to Fig.5, where the results for the photodissociation of the 0_{00} state have been compared with those for the photodissociation of jet cooled water. Because the predictions of Eq.24 agree almost quantitatively with the data for jet cooled H_2O (compare Fig.3), Fig.5 yields also the comparison of the predictions of the "one electron overlap model" with the state to state photodissociation from the 0_{00} state. Obviously there is bad agreement between experiment and the predictions of this "one electron overlap" model on the state to state level. This is true for other

initial states of the parent H_2O as well. As before, the breakdown is not due to a failure of the FC-limit, but to coupling of electronic and rotational motion in the OH product.

Obviously the separate description of the transfer of rotational and electronic motion by the matrix elements E and R, yields the wrong result both for the rotational and the electronic fine structure degrees of freedom.

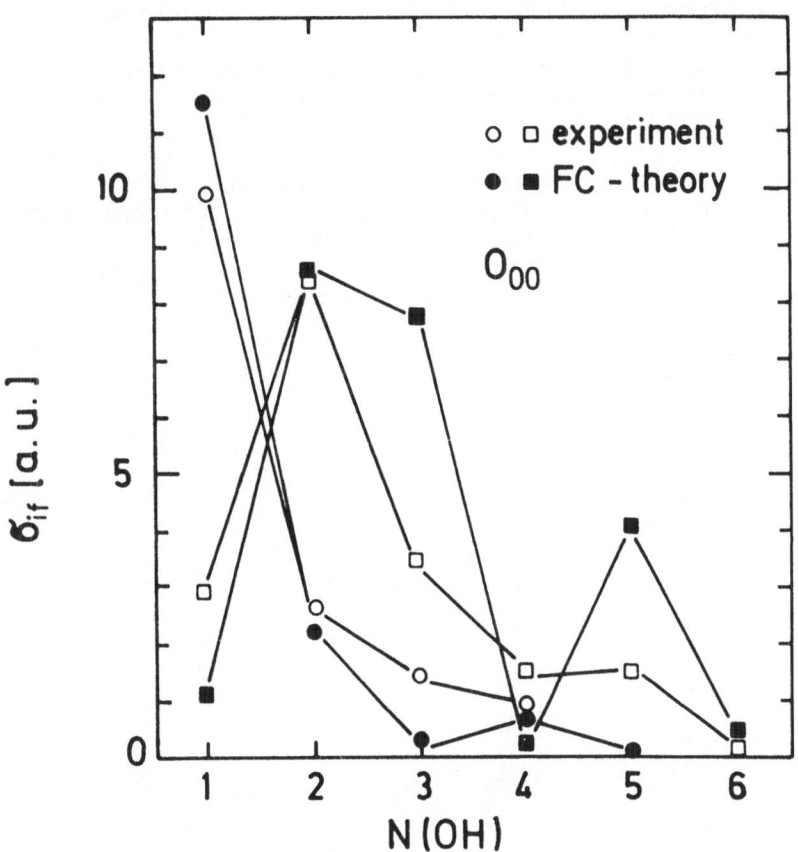

Figure 12   OH formation in different quantum states from the photo-dissociation of H_2O in the 0_{00} state, comparsion of experiment and theory. Squares: OH formation in the π^--Λ-doublet state. Circles: OH formation in the π^+-Λ-doublet state.

The inclusion of the coupling of electronic and nuclear motion in the FC-limit has been given by G. Balint-Kurti.[15] A complete description of this theory is way beyond the scope of this qualitative discussion. However, an important point is that due to the coupling the electronic overlap matrix element E and the rotational overlap matrix element R can not be considered separately. Only a sum of products of rotational and electronic overlap matrix elements

$$\sum_n \alpha_n\, E_n\, \underline{R}_n \qquad\qquad (25)$$

describes the simultaneous transfer of electronic and rotational motion to the OH product. Hidden in the coefficients α_n are all the details about the coupling of electronic and nuclear motion.

If this type of theory is adopted, the comparison of experiment and theory becomes almost quantitative. This is shown in the following two figures 12 and 13.

Fig.12 gives this comparison for the photodissociation of rotational ground state water, i.e., from H_2O the 0_{00} rotational state in the asymmetric (001) vibrational mode. The rotational distributions are shown for both Λ-doublet states. The agreement is reasonable for the π^--Λ-doublet and good for the π^+-Λ-doublet.

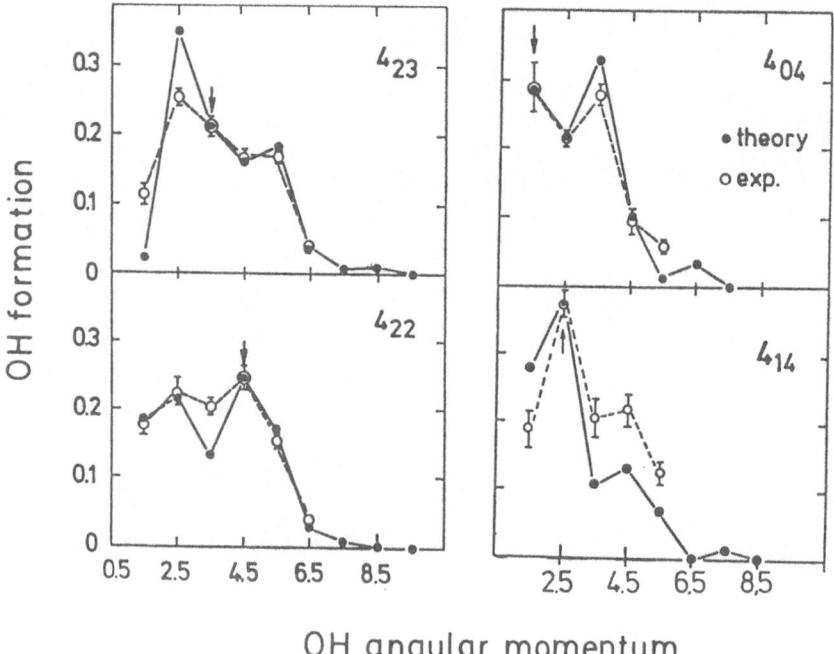

Figure 13   OH formation in different rotational states of $^2\Pi_{3/2}$, (π^+-Λ-doublet) from the photodissociation of H_2O in various initial states.

The OH product state distributions in Fig.13 are obviously different for different initial rotational states of H_2O. This is qualitatively understood by the transfer of motion from parent to product: different motion in the parent has to lead to different motion in the products. In addition there is a pronounced quantum structure in these distributions. These and other distributions show, at least in the beginning, an oscillation with $\Delta J=1$. This oscillation is essentially caused by the different parity and shows up also for different Λ-doublet states.

Both Fig.12 and Fig.13 show impressively the almost quantitative agreement of experiment and theory. It should be emphasized that an agreement with such highly structured distributions is certainly not accidental. Obviously the rotational and electronic fine structure distributions predicted by the FC theory reproduce almost quantitatively the experimental results.

According to the discussion above, the FC-limit can only be found in an experiment if the anisotropies in the interaction potential are close to zero. This is why we can conclude that there is neither anisotropy in the rotational nor in the electronic degrees of freedom. In particular we can conclude directly from this experimental result that the excited state potential has to prefer the same angle (104.5 degrees) that water has in its electronic ground state. In addition, we conclude that couplings among the electronic degrees of freedom, like spin-orbit effects, are negligible in the fragmentation step.

This demonstrates that from the validity of the FC-limit direct information is obtained about the excited state potential surface. The validity of the FC-limit for the rotational degrees of freedom implies a conservation of the bond angle in the excitation step and the validity of the FC-limit for the vibrational degree of freedom implies that the bond length is conserved.

The conservation of the bond angle for excitation of H_2O in the first absorption band is in agreement with the quantum chemistry calculations by Staemmler et al.[16] In the range of coordinates, where the nuclear wavefunction in the electronic ground state is different from zero, the anisotropy with respect to γ is almost zero, independent of the distance of the departing H-atom.

It is interesting to note that the "ab initio" potential surface is at the same time strongly anisotropic in the radial coordinate r of the interaction potential, leading to the rather high vibrational excitation in the OH product observed experimentally at 157nm. The FC model predicts zero vibrational excitation for this case simply because the bond length in the free OH is almost the same as in the parent H_2O (compare Fig.10). The FC-theory can hold for one degree of freedom and can be completely wrong for another degree of freedom in the very same process.

The results of the FC-limit are by no means trivial and yield detailed information about the transfer of parent motion to products, including electronic degrees of freedom. It may be discouraging that the results look so complex, that an intuitive physical picture seems no longer possible.

It turns out that most of the oscillatory structure in the OH product state distributions is due to parity. It is clear that the total parity has to be the conserved. A closer look at the experimental results reveals that the formation probability of OH in different quantum states depends strongly upon the parity of the final state. In the rotational distributions for one Λ-doublet state, the intensity alternates with $\Delta J=1$, i.e., with parity. The alternations are found in both Λ-doublet states and are opposite to each other, i.e., the rotational distribution decreases in the one Λ-doublet, if it increases in the other. This is again an oscillation with parity. The oscillations are also opposite in the multiplet states. This implies that most of the complicated structure is due to parity.

In the last part of this paper, it is shown that it is possible to come back to a simpler view for the origin of the selective population of Λ-doublet states. This simpler view is obtained by averaging over states of different parity.

If the OH product state distributions are, for example, averaged over the Λ-doublet states, the rotational state distributions become smooth Boltzmann-like distributions, reflecting the rotational energy content in the parent molecule.

Another procedure is to average over initial H_2O states of different parity. Fig.14 shows relative populations of Λ-doublet states for different initial rotational states of H_2O, averaged over two initial rotational states of H_2O. For J=6, for example, the contributions from the states 6_{06} and 6_{16} are summed, modelling an in-plane rotation (upper panel). If the contributions from 6_{60} and 6_{61} are summed, an out-of-plane rotation is modelled (lower panel). The words "in-plane " and "out-of-plane" have to be used with care, because H_2O is an asymmetric rotor. Nevertheless, the essential part of the rotation will be in-plane for the upper and out-of-plane for the lower part of Fig.14.

For the in-plane rotation of H_2O, the data look very similar to the data from Fig. 3. The π^- Λ-doublet state is preferred, and the preference grows with increasing J. This is still in agreement with the conservation of electronic symmetry.

For initial out-of-plane rotations the situation is different. If there is a preference at all, it is for the π^+-Λ-doublet state! The simple adiabatic argument becomes completely wrong, because it predicts always a preference of the π^--Λ-doublet state, independent of the motion of the nuclei.

This implies that if the parent rotation is in-plane the lobe stays in-plane, but if the parent rotation is out-of-plane, the lobe goes out-of-plane. Obviously the lobe follows the rotation of the parent molecule!

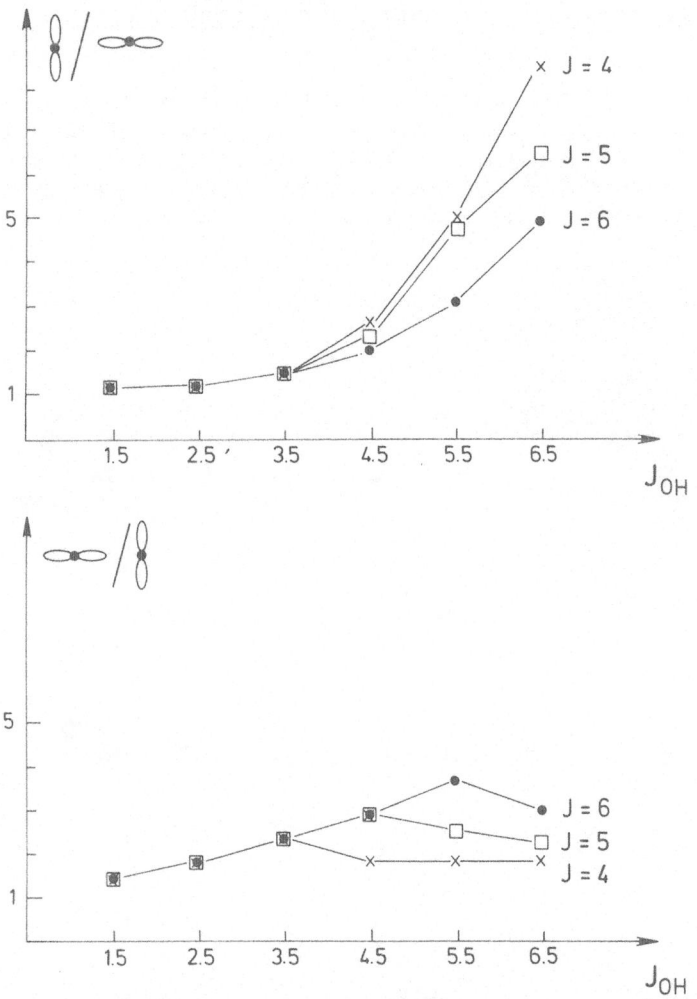

Figure 14   Relative population of Λ-doublet states for different initial parent motion. Above: "in-plane" rotations with different initial total J. Below: "out-of-plane" rotations with different initial total J. The results are obtained by averaging over two states of different parity in the parent molecule.

The implications of this behaviour are illustrated in Fig.15. It shows a triatomic molecule that is rotating in-plane (above) and out-of-plane (below). For the initial parent rotation, the lobe is always perpendicular to the nuclear plane. In the break up of the parent molecule the lobe continues to rotate in the same direction as before in the parent molecule.

416

This demonstrates that the real origin of Λ-doublet selectivity is not the adiabatic rule (3). It is the conservation of the lobe orientation relative to the initial rotational motion of the excited complex.

It should be noticed, that this intuitive picture results only if the parity is eliminated by averaging. If no averaging effects are involved, like for example in a full state to state experiment, the parity has a very important effect on the selective population of Λ-doublet states.

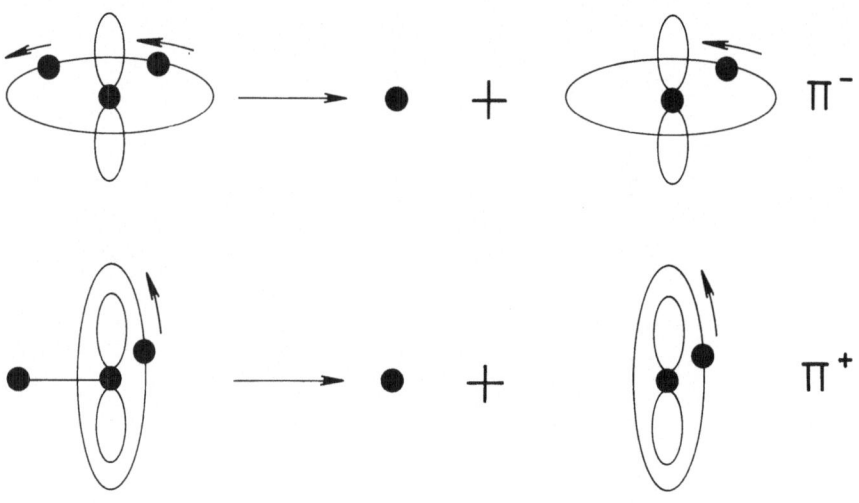

Fig.15   Conservation of the lobe orientation relative to the rotation in the excited complex.

Conclusions

In most realistic photodissociation processes there will be some anisotropy of the excited state potential surface. The first step will always lead to some product motion and the second step will alter the motion originating from the first step. In the general case, however, both initial motion in the parent molecule and forces in the excited state will contribute to product motion.

The first case of very weak final state interaction yields the FC-limit in which the product motion originates exclusively from the parent molecule. In the second case of extremely strong final state interaction, the motion may be so strongly dominated by the forces in the exit channel, that the initially transferred motion is negligible. For example, an extremely strong torque exerted in the exit channel, as in the photodissociation of H_2O along the B-state, gives OH in such high rotational states that OH rotation from the initial parent motion step is negligible compared to that generated in the second step.

The main conclusion is that the rotational and electronic fine structure distributions can be understood quantitatively in terms of a parameter free FC-theory. This is the first case in which the selective population of Λ-doublet states could be understood quantitatively. As a matter of fact, it is the only case in which the FC-theory with electronic fine structure has been applied to a triatomic system.

1) According to the above discussion, the validity of the FC-theory for the rotational and electronic fine structure distributions implies that there is zero final state interaction not only for the rotational degree of freedom, but also for the electrons. Although the FC theory does not depend on the interaction potential at all, the validity of the FC-limit yields important information about the excited state potential surface, i.e., that the anisotropies are zero for the rotational and electronic degree of freedom.

2) The selective population of Λ-doublet states is exclusively determined by the parent motion and the absorption step. The origin for the selectivity is conservation of the unpaired pπ-lobe relative to the direction of initial nuclear rotation. In contrast to the adiabatic picture, it is not simply electronic symmetry that is conserved. The selectivity depends sensitively on the motion of the parent molecule (initial in-or out-of-plane rotations) as well as on the coupling of the electronic motion in the products. Each rotatonal state of the parent molecule leads to another Λ-doublet distribution.

It is interesting to speculate about possible implications of these results.

In most cases of photodissociation, the products have open shells, simply because one usually starts with a stable molecule and a chemical bond is broken in the process. The electronic structure in the products can be very different depending on the particular dissociation process. For example in NO_2 photodissociation, the product NO is also formed in a $^2\Pi$ state, but the O atom is formed either in the ^3P or the ^1D state. This is already different because, in the detailed FC-theory, the spin of the other fragment is explicitly coupled in a different way to the angular momenta. For each dissociation process the FC-theory is expected to give different results, and the appropriate theories have to be worked out.

The selective population of electronic fine structure states is observed in many other molecular processes, like chemical reactions, inelastic collisions and surface scattering. The basic origin for the selectivity is the same in all these cases. It requires some orientation of the unpaired lobe in the product and a well defined rotational motion during the break up of the complex. If too many different initial rotations are present, the Λ-doublet selectivity is smeared out and a statistical population is obtained. In most cases, the Λ-doublet selectivity is indeed considerably lower than expected from the degree of electron alignment. This suggests, in the view of these results, that still considerable out-of-plane motion is present in the excited complex.

Acknowledgement
The author thanks all collaborators in this work and in particular Dr. R. Schinke and Prof. E.W.Rothe for many, many stimulating discussions.

References

[1] H.Okabe, "Photochemistry of small Molecules" Wiley Interscience, New York (1978)

[2] S.J.Riley and K.R.Wilson, Faraday Disc. of the Chem.Soc. 53, 132, (1972)

[3] H.J.Krautwald, L.Schnieder, K.H.Welge and M.N.R.Ashfold Faraday Discuss.Chem.Soc. 82, paper 7 (1986)

[4] R.Schmiedl,H.Dugan,W.Meier and K.H.Welge, Z.Phys.A 304, 137 (1982)

[5] Z.Xu, B.Koplitz,S.Buelow,D.Baugh and C.Wittig Chem.Phys.Lett. 127, 534 (1986)

[6] P.Andresen and E.W.Rothe J.Chem.Phys. 78, 989 (1983)

[7] G.E.Hall,N.Sivakumar,R.Ogorzalek,G.Chavla, H.P.Haerri and P.L. Houston Farad. Disc. Chem.Soc. 82, paper 4 (1986)
G.E.Hall, N.Sivakumar and P.L.Houston Phys.Rev.Lett. 56, 1671 (1986)
K.H.Gericke, S.Klee,F.J.Comes and R.N.Dixon J.Chem.Phys. 85, 4463 (1986)
M.Dubs,U.Brühlmann and J.R.Huber J.Chem.Phys. 84, 3106 (1986)
M.P.Docker, A.Hodgson and J.P.Simons Chem.Phys.Lett. 128, 264 (1986)

[8]) P.Andresen, G.S.Ondrey,B.Titze and E.W.Rothe, J.Chem.Phys. $\underline{80}$, 2548 (1984)

[9]) R.Lavi, I.Bar and S.Rosenwaaks   J.Chem.Phys. to be published

[10])   R.Schinke,V.Engel,P.Andresen,D.Häusler   and   G.G.Balint-Kurti, Phys.Rev.Lett. $\underline{55}$, 1180, (1985)
P.Andresen,V.Beushausen,D.Häusler,H.W.Lülf   and   E.W.Rothe J.Chem.Phys. $\underline{83}$, 1429 (1985)
D.Häusler,P.Andresen and R.Schinke J.Chem.Phys. accepted 1987

[11]) C.Camy-Peyret,J.M.Flaud,G.Guelachvili and C.Amiot, Mol. Phys. $\underline{26}$, 825 (1973)
J.Y.Mandin,C.Camy-Peyret,J.M.Flaud and G.Guelachvili, Can.J.Phys. $\underline{60}$, 94 (1982)
A.R.H.Cole, "Tables of wavenumbers for the calibration of infrared spectrometers" Pergamon Press, Oxford (1977)
J.M.Flaud,C.Camy-Peyret and R.A.Toth, "Water vapour line parameters from microwave to medium infrared" Pergamon Press, Oxford (1981)

[12]) R.Schinke and V.Engel to be published

[13])   P.L.Houston, Conference on "Dynamical Stereochemistry", Proceedings Jerusalem 1986

[14]) P.Andresen and E.W.Rothe, J.Chem.Phys. $\underline{82}$, 3634 (1985)

[15]) G.G.Balint-Kurti J.Chem.Phys. $\underline{84}$, 4443 (1986)

[16]) V. Staemmler and A. Palma, Chem. Phys. $\underline{93}$, 63 (1985)

LASER SPECTROSCOPY OF CHEMICAL INTERMEDIATES IN SUPERSONIC FREE JET EXPANSIONS

Stephen C. Foster, Richard A. Kennedy,* and Terry A. Miller, Laser Spectroscopy Facility, Department of Chemistry, The Ohio State University, 120 West 18th Avenue, Columbus, Ohio 43210 U.S.A.

I. Introduction

Lasers have contributed enormously to various areas of spectroscopy. Indeed, different forms of spectroscopy have utilized nearly all of the laser's unique characteristics, i.e. coherence, high power, high resolution, etc. In this respect, the laser spectroscopy of chemical intermediates is exemplary. Most often the experimental environment is dictated by the stringent demands of the production and preservation of chemical intermediates, rather than by spectroscopic considerations. The fact that laser induced fluorescence spectroscopy requires only the passage of a photon into and out of a "remote," easily perturbed, medium makes it in a number of cases the only spectroscopic tool available for probing this environment. Beyond this advantage, numerous experiments on chemical intermediates have utilized the unique power and resolution capabilities of lasers.

In the experiments described herein, the environment chosen for the chemical intermediates is a supersonic free jet expansion. We have found the jet in many ways an ideal device for studying the spectroscopy of a variety of chemical intermediates, including isolated molecular ions, ionic clusters, and both small and moderately large organic and inorganic neutral free radicals.

The fundamentals of the experiment are illustrated in Fig. 1 which shows the characteristics of the jet expansion. Chemical intermediates

A. C. P. Alves et al. (eds.), Frontiers of Laser Spectroscopy of Gases, 421–449.

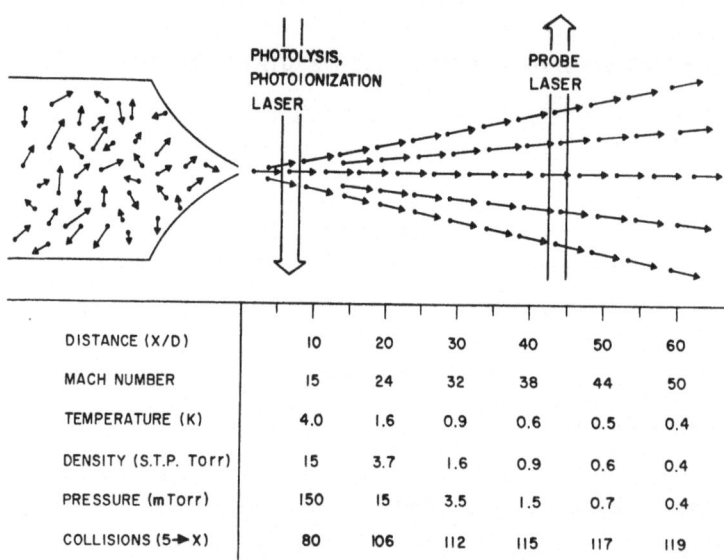

DISTANCE (X/D)	10	20	30	40	50	60
MACH NUMBER	15	24	32	38	44	50
TEMPERATURE (K)	4.0	1.6	0.9	0.6	0.5	0.4
DENSITY (S.T.P. Torr)	15	3.7	1.6	0.9	0.6	0.4
PRESSURE (mTorr)	150	15	3.5	1.5	0.7	0.4
COLLISIONS (5→X)	80	106	112	115	117	119

Fig. 1   Schematic diagram of a supersonic free jet expansion. Shown
below the diagram is a scale indicating distance in units of
nozzle diameters, and the characteristics of the expansion at
various points downstream.   For this illustration, it is
assumed that the reservoir pressure is 10 atmospheres of He
at 300°K; for the last row a collision cross section of 50 Å²
is assumed.

cannot be seeded into the inert gas reservoir of the expansion,
because, generally speaking, they will react with the walls,
impurities, or each other, before they escape the nozzle. Instead, the
reactive chemical intermediates must be made *in situ* in the expansion
from precursor molecules seeded into the reservoir. Fig. 1 shows a
photolysis or photoionization beam, typically from an ArF or KrF laser,
crossing the expansion near its throat, and thereby creating ions or
radicals from the seeded, precursor molecules.

As indicated at the bottom of Fig. 1, in the region of the excimer
laser crossing, a number of gas collisions take place. By carefully
shifting the laser position relative to the nozzle and/or varying the
reservoir pressure, it is possible to produce chemical intermediates
either isolated or complexed, particularly with inert gas atoms, for
spectroscopic studies by the downstream laser.

The remainder of this paper is divided in the following manner. The next section discusses in some detail the supersonic free jet expansion with an emphasis upon those properties particularly important for spectroscopy. This is followed by sections which discuss the spectroscopy of ions and ionic clusters and finally free radicals.

II. Supersonic Free Jet Expansions

A. Properties of the Expansion

Supersonic free jet expansions are by now a relatively common laboratory tool. Nonetheless, some of their more fundamental properties are not widely understood. Perhaps this is in part due to the fact that much of the information is widely scattered in the literature. This section attempts to summarize many of the pertinent facts about expansions relevant to their use for spectroscopic purposes. Several other articles[1-5] may also be useful, for supplementing the material contained herein.

The local temperature in an expanding jet can be calculated from the isentropic equation of state for an ideal gas:[2,5,6]

$$\frac{T}{T_R} = \frac{1}{1 + \frac{1}{2}(\gamma-1)M^2} \tag{1}$$

where T and T_R are the jet and reservoir temperatures; γ is the heat capacity ratio, C_p/C_v; and M is the Mach number. This latter quantity is the ratio of the flow velocity of the jet to the local speed of sound. The Mach number can be calculated with the approximate relationship:[2,5,7,8]

$$M = A(X/D)^{\gamma-1} = A X_D^{\gamma-1} \tag{2}$$

Here A is a constant which depends on γ, and $X_D \equiv X/D$ is the ratio of the downstream position to the nozzle diameter. Eqn. 2 should not be used when X_D is very small or very large. For small distances a more complete expression is given by Anderson and Fenn.[7] At large distances the treatment of the expanding gas as a continuum becomes increasingly inaccurate; the actual rarefaction of the gas reduces the frequency of collisions and results in a maximum Mach number (and hence minimum temperature) which has been estimated to be,[2,5,7]

$$M_T = 133(P_R D)^{0.4} \tag{3}$$

where P_R, is expressed in atmospheres and D in cm. This latter equation is valid for all monatomic gas expansions.

Translational temperatures within the expanding jet can be calculated from equations 1 and 2. For any monatomic gas (such as He or Ar), for which $\gamma = 5/3$ and $A = 3.26$, we can rewrite Eq. 1 as

$$T = \frac{T_R}{1 + 3.54\, X_D^{1.33}} \tag{4}$$

This simple relationship implies that the temperature attained in an expanding free jet is independent of the backing pressure. It should be noted that the <u>minimum</u> temperature achieved is pressure dependent, as a consequence of the rarefaction discussed above. Molecules seeded into the jet rapidly equilibrate their rotational motion to the low translational temperature of the carrier gas. Vibrational energy is less efficiently transferred to the cold jet and, as a result, higher vibrational temperatures are sometimes seen.

The photolysis or photoionization of a stable molecule in a free jet will, in general, deposit large amounts of energy into the vibrational and rotational degrees of freedom of the molecular products. This energy can be transferred to the cold carrier gas via two-body collisions, but the rate of this secondary cooling is clearly strongly dependent upon the collision rate. The temperature of these fragments can conveniently be controlled by varying the number of collisions they undergo with the cold carrier gas. Changing the distance between the nozzle and the photolysis region, the distance between the photolysis and probe lasers, or the stagnation pressure of the reservoir will achieve this by varying the number of collisions.

The collision frequency, z, at various points downstream was elucidated in a paper by Lubman, et al.[9] One can model the expansion as having, superimposed upon the stream velocity, a Maxwellian distribution of velocities determined by some local, reduced temperature. Within this assumption of local equilibrium the collision frequency for a given molecule is just

$$z = 2^{\frac{1}{2}} n\sigma\bar{v} \tag{5}$$

where

    n = local number density

    $\bar{v}$ = local mean velocity

    σ = collision cross section

The local mean velocity is related to the mean velocity $\bar{v}_R$ in the reservoir by

$$\bar{v} = \bar{v}_R \left(\frac{T}{T_R} \right)^{\frac{1}{2}} = \bar{v}_R [1 + \tfrac{1}{2} (\gamma-1) M^2]^{-\frac{1}{2}} \tag{6}$$

where the last equality follows from Eq. (1).

The local density can be expressed[2] in terms of the reservoir density, n_R,

$$n = n_R [1 + \tfrac{1}{2} (\gamma-1) M^2]^{-1/(\gamma-1)} \tag{7}$$

Combining equations 5, 6 and 7 yields

$$z = 2^{\frac{1}{2}} n_R \sigma \bar{v}_R [1 + \tfrac{1}{2} (\gamma-1) M^2]^{-\frac{(\gamma+1)}{2(\gamma-1)}} \tag{8}$$

or under the conditions where Eq. 2 is valid

$$z = 2^{\frac{1}{2}} n_R \sigma \bar{v}_R [1 + \tfrac{1}{2} (\gamma-1) A^2 X_D^{2(\gamma-1)}]^{\frac{-(\gamma+1)}{2(\gamma-1)}} \tag{9}$$

For a monatomic gas Eq. 9 reduces to

$$z = 2^{\frac{1}{2}} n_R \sigma \bar{v}_R [1 + 3.54 X_D^{1.33}]^{-2} \tag{10}$$

Integration of Eq. (10) from zero to a particular point X_D' yields the approximate, total number of collisions that a molecule suffers in the expansion to the point X_D'. In this way, the last row of Fig. 1 was calculated, assuming σ was independent of X_D, i.e. independent of the local collision velocity (temperature).

The formation of clusters in the free jet can be considered both a problem and a useful feature of this source. Cluster formation releases the heat of condensation, which increases the temperature of the jet. Additional problems arise because cluster features may obscure portions of the spectrum of the uncomplexed molecule.

An alternative view would be that clusters provide an interesting way of studying the transition between gas-phase and condensed phase properties, and hence should be maximized. Either viewpoint makes it important to understand the conditions necessary to inhibit or promote cluster formation. As noted above, fragments are cooled by two body collisions, whereas cluster formation requires, at a minimum, a three body process. The number of higher order cluster-forming collisions in a jet scale[5] as $P_R{}^2D$, whereas the two-body collision rate is proportional to $P_R D.[4]$ These scaling laws show that cluster formation is promoted by small nozzles and high reservoir pressures. Conversely, maximum cooling without condensation is promoted by large nozzles at lower backing pressures.

Two other quantities must be considered when designing this type of experiment: the stream velocity of the carrier gas, and the location of the Mach disk or shock front. The stream velocity, v_s, defines the flight time for molecules in the jet and thus must be known in order to correctly time the firing of the production and probe lasers. This quantity can conveniently be calculated from the Mach number, which can be expressed as:

$$M = v_s/C_o , \tag{11}$$

where C_o is the local speed of sound:

$$C_o = (\gamma kT/m)^{\frac{1}{2}} . \tag{12}$$

In equation 12, k is the Boltzman constant and m is the atomic mass. The stream velocity is not a constant quantity for any given gas, but varies by less than 5% for Mach numbers in the range 6 - ∞. For He, v_s is found to equal 1.78×10^5 cm s^{-1}. Stream velocities calculated from these simple relationships are found experimentally to be accurate to better than 1%.[7]

The location of the Mach disk must be considered when designing a continuous nozzle expansion. When the expanding gas meets this shock front, the turbulence destroys the desirable properties of the jet.

Ashkenhas and Sherman have found experimentally that for any gas the location of the Mach disk, X_D^M, is given by:[8]

$$X_D^M = 0.67 \left[\frac{P_R}{P_C} \right]^{\frac{1}{2}} \tag{13}$$

At first sight this equation might be taken to imply that the Mach disk can be moved further from the nozzle simply by increasing the backing pressure. However, if we express the chamber pressure, P_c, in terms of the system pumping speed, S;[10]

$$P_c = \frac{1}{4} P_R \pi \left(\frac{D}{2} \right)^2 \bar{v}_R \cdot \frac{1}{S} , \tag{14}$$

we can write:

$$X_D^M = \frac{2.68}{D} \left[\frac{S}{\pi \bar{v}_R} \right]^{\frac{1}{2}} . \tag{15}$$

Here $\bar{v}_R$ is as before the mean velocity of the gas in the reservoir,[10]
$$\bar{v}_R = [8\, kT_R/\pi m]^{\frac{1}{2}}$$
For He at 300K, $\bar{v}_R$ is equal to 1.26×10^5 cm s^{-1}. Equation 14 implicitly assumes that the system pumping speed is independent of the chamber pressure. Within this relatively good approximation, we see that the Mach disk location is independent of the nozzle diameter and the backing pressure. Only the type of gas and the pumping speed affect the shock-front location.

As a numerical example, consider a free jet pumped by a small Roots system ($S = 70$ ℓs^{-1}). The Mach disk in a helium expansion from a 300K reservoir is located ~11 mm from the nozzle tip - more than 70 nozzle diameters for a typical 150 μm orifice.

In most spectroscopic applications, it is only necessary to have a free jet expansion. Occasionally, however, skimming the jet to produce a molecular beam is important. Campargue[11] determined an empirical relationship for the optimum nozzle skimmer separation, X_D^S,
$$X_D^S = 0.125\, [(D/\lambda_R)(P_R/P_C)]^{1/3} \tag{16}$$
where the mean free path in the reservoir, λ_R, is defined by
$$\lambda_R = [2^{\frac{1}{2}} \sigma_R n_R]^{-1} \tag{17}$$

with

σ_R = reservoir collision cross section.

Combining Eqs. 16 and 17 yields

$$x_D^S = 0.125 \ (D \ 2^{\frac{1}{2}} \ \sigma_R \ n_R)^{1/3} \ (P_R/P_C)^{1/3} \tag{18}$$

Combining Eq. 18 with Eq. 14 yields

$$x_D^S = 0.241 \left[\frac{\sigma_R n_R}{\bar{v}_R} \right]^{1/3} \left[\frac{S}{D} \right]^{1/3} \tag{19}$$

where S is again the pumping speed.

Consider the free jet example discussed earlier (S = 70 ℓs^{-1}, D = 150 μm). Taking the collision cross section for He as 13 Å^2, Eq. 19 predicts that the optimum nozzle-skimmer distance for a 10 atm., 300 K jet is 9 mm.

B.   Photolysis and Photoionization in Jet

As described in the Introduction, free-jet spectra of radicals and ions can be produced by the photolysis or photoionization of a stable precursor seeded into a rare-gas expansion. The stable molecule is seeded into the jet by passing the high pressure carrier gas over a liquid or solid sample maintained at a fixed temperature. The percentage mixed into the carrier can be simply varied by changing the temperature of the precursor reservoir and hence the vapor pressure of the sample. The seeded rare gas is then expanded through a pinhole to form a cw expansion as shown schematically in Fig. 2.

Reactive intermediates are prepared by photoionization or photolysis with a high-powered laser, typically an ArF or KrF excimer laser. This production laser is focussed with a long focal length lens (often 1m for the excimer lines) to form a stripe which "cuts" the expanding jet a small distance above the pinhole. The products of this first laser are probed further downstream with a tunable dye laser. The timing of the lasers is obviously crucial to this experiment. The delay before firing the dye laser must be precisely controlled to detect the ions or radicals as they fly through the region probed by the dye laser. This overall timing can be easily calculated from the stream velocity discussed in the previous section.

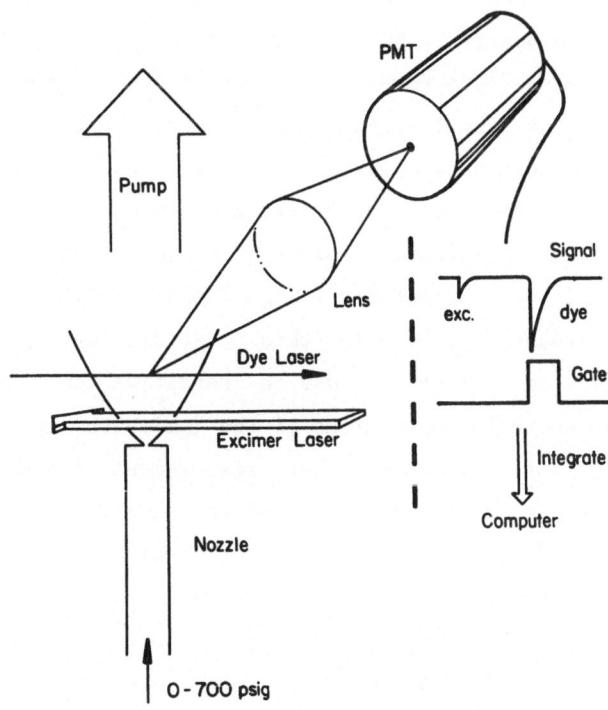

Fig. 2   Schematic diagram of free jet apparatus showing the positions
           of the lasers and the fluorescence detector.   On the right
           side of the diagram, the timing sequence is illustrated for
           the excimer and dye laser as well as for the integrator gate.

The distance between the two pulsed lasers, and hence the overall
timing, is not chosen arbitrarily. The photolysis process, in general,
will leave a large amount of excess energy in the fragments. This
energy will be partitioned among the various vibrational and rotational
degrees of freedom of the molecules, resulting in a very "hot" product.
By allowing these hot products to travel downstream in the jet and to
suffer collisions with the cold carrier gas, the fragments are rapidly
cooled. Clearly, if the two lasers are focussed to points well
separated in the jet, the resultant spectra will be vibrationally and
rotationally cold. As we move the lasers towards each other, the
spectra will become progressively warmer. This is in fact a convenient
way to record and analyze spectra. A cold spectrum containing only a
small number of lines is recorded initially, and is usually readily
assigned.   The spectrum is then progressively warmed, to add extra
lines and hence to provide a more accurate set of molecular constants.

The fluorescence produced by the detection laser is collected with an f/1 lens and spatially filtered to minimize the scattered excimer light reaching the detector. Laser excited fluorescence spectra can be recorded in the usual fashion by observing the change in total fluorescence with a photomultiplier tube as the dye laser is scanned. Alternatively, the dye laser wavelength can be set to be in resonance with a molecular rovibronic transition and the fluorescence dispersed with a small monochromator giving detailed information about the electronic ground state. In this latter arrangement, it is particularly convenient to use a sensitive optical multichannel analyzer to record large sections of the "emission" spectrum after each laser shot. Not only does that arrangement speed the recording process, but it also removes problems caused by shot-to-shot power fluctuations and by the decreasing power level of the excimer laser over its fill lifetime.

III. Spectroscopy of Ions

A. Isolated Ions

The general techniques described in the previous section have direct application to the spectroscopic study of isolated molecular ions. Beyond the fact that the free jet expansion represents a hospitable environment for the production and maintenance of reasonably high ion concentrations, the principal advantage of the jet for spectroscopy is the very low temperatures attained by the ion's internal degrees of freedom in the expansion. While these low temperatures are most useful in the study of large ions, they are most easily demonstrated and monitored in the case of small ions.

Figure 3 shows the LIF excitation spectrum of the CO^+ ion. The ion is produced[12] from neutral CO seeded into the expansion by resonantly enhanced multi-photon ionization in a so-called 2+1 process. Molecules in the $X^1\Sigma^+$ ground state of CO can be pumped to the $B^1\Sigma^+$ state by a 2-photon transition requiring radiation of ≈ 230 nm. This light is generated by a frequency-doubled dye laser whose output is summed with the fundamental of a Nd:YAG laser. The excitation spectrum

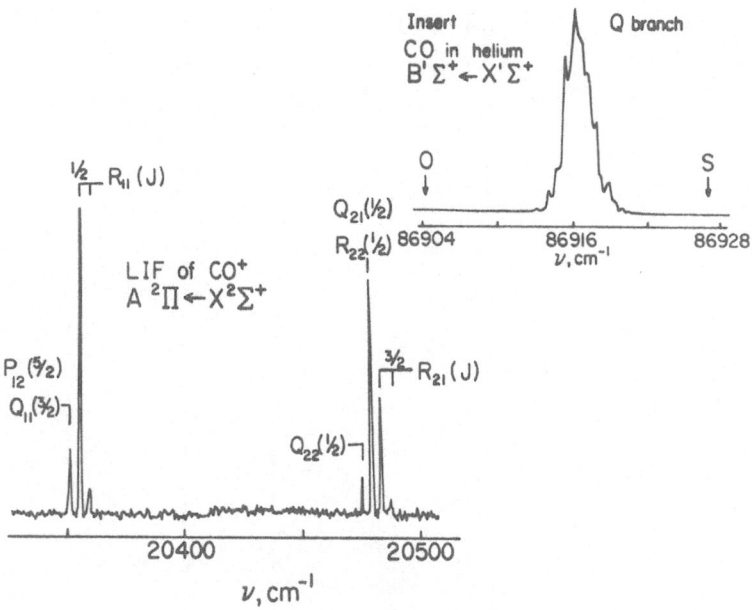

Fig. 3   Rotationally resolved LIF excitation spectrum of the 0-0 $A^2\Pi$ ← $X^2\Sigma^+$ transition of CO^+ in a He free jet expansion. The upper right insert is the 2-photon excitation spectrum of the 0-0 $B^1\Sigma^+$ ← $X^1\Sigma^+$ transition of CO, which is monitored by its fluorescence to the $A^1\Pi$ state.

for this process is shown in the upper right-hand insert in Fig. 3.   To produce CO^+ the excited $B^1\Sigma^+$ CO absorbs a third 230 nm photon.

As can be seen from the excitation spectrum of CO^+ in Fig. 3,   the ion is very cold.   Only the lowest rotational levels are populated corresponding to a temperature of $\lesssim 3$ K.   This temperature is primarily determined by the fact that the CO neutral has reached a very low temperature in the jet before ionization.   To a good approximation the temperature of the neutral is mapped upon the ion, because neither the photon nor the exiting electron can carry significant angular momentum. This conservation consideration requires approximately equivalent angular momentum content in the neutral CO and its ion, CO^+.

Although quantitatively more difficult to measure, the effect of the jet's low temperature on larger molecular ions appears even more dramatically in their spectra.   Fig. 4 shows the laser excitation

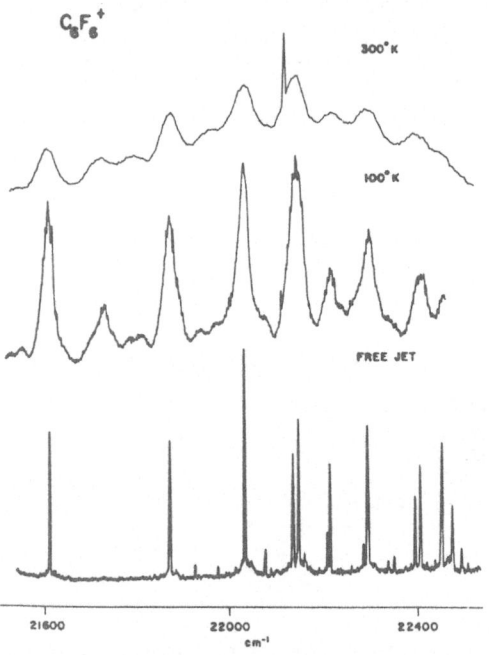

Fig. 4   Laser excitation spectra of $C_6F_6^+$ under three different temperature conditions.

spectrum of the hexafluorobenzene cation, $C_6F_6^+$. This ion is produced by a non-resonant or quasi-resonant 2-photon ionization of neutral C_6F_6 seeded into the He reservoir by passing the inert gas over a sample of liquid C_6F_6 contained in a temperature controlled bath.   The ionizing photons are from an excimer laser operating on the ArF transition, whose wavelength is sufficiently short so that 2 photons contain enough energy to produce $C_6F_6^+$.

    In the top trace of Fig. 4, taken at room temperature, congestion from a number of vibronic transitions with significant rotational contours limits the effective spectral resolution to ≈ 40 cm^{-1}. In the middle trace cooling to $\approx 100°$K reduces the number of rotational and vibrational levels populated so that the spectral resolution has improved to ≈ 20 cm^{-1}. In the lowest trace, taken in the jet, rotational and vibrational temperatures of the ion are reduced to such an extent that the spectral lines are now $\lesssim 1$ cm^{-1}, and indeed no longer limited by congestion, but in this trace by the laser bandwidth.

Studies now in progress[13] on $C_6F_6^+$ with a much higher resolution laser indicate that its rotational temperature is comparable to that indicated above for the CO^+ ion.

The symmetrically (3- or 6-fold axis) halogenated benzene cations are an extremely interesting group of ions for spectroscopic study.[14-19] Members of this category for which laser excitation spectra have been obtained include $C_6F_6^+$, and the 1,3,5-substituted $C_6H_6F_3^+$, $C_6H_3Cl_3^+$, $C_6F_3Cl_3^+$, and $C_6F_3Br_3^+$. A variety of spectroscopic techniques have been employed to elucidate the Jahn-Teller effect in these molecules. These techniques include LIF in a discharge flow reactor (the source of top two traces in Fig. 4), traditional grating spectroscopy of the emission from a discharge through the parent halobenzenes and laser spectroscopy of matrix isolated ions. While much of this work laid the foundation for the Jahn-Teller analysis, it was the jet work exemplified by Fig. 4 that resolved various ambiguities and controversies involving the earlier work.[20-22]

In Fig. 5, we reproduce the experimentally determined energy level position for the Jahn-Teller active vibronic energy levels of the

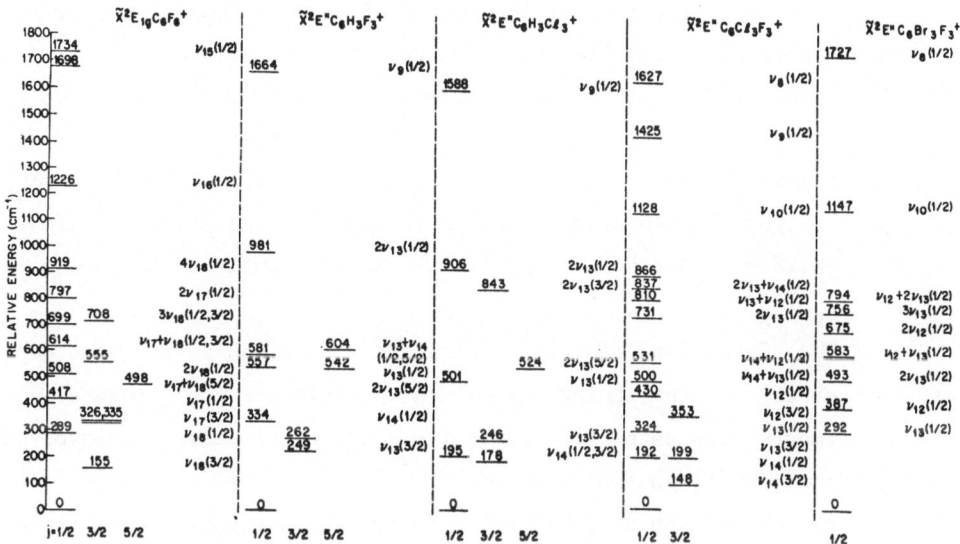

Fig. 5   Experimentally determined positions of the Jahn-Teller active vibronic energy levels of $C_6F_6^+$, $C_6H_3F_3^+$, $C_6H_3Cl_3^+$, $C_6F_3Cl_3^+$, and $C_6F_3Br_3^+$.

ground states of $C_6F_6^+$, $C_6H_3F_3^+$, $C_6H_3C\ell_3^+$, $C_6F_3C\ell_3^+$, and $C_6F_3Br_3^+$. These energy levels have been subject to exhaustive Jahn-Teller analyses which determined for the ions the unperturbed oscillator frequency, ω_i, Jahn-Teller stabilization energy, ϵ_i, the linear, D_i, and quadratic, K_i, Jahn-Teller interaction parameters, and the geometric distortion, ρ_{im}, of each molecule at the minimum of the potential energy surface. These results are contained in Table I.

B. Ionic Clusters

As we have seen, the free jet expansion technique offers a powerful means for probing the details of the electronic structure of even relatively large organic ions. Another species equally or more interesting is the ionic cluster, i.e. an ion complexed by one or more neutral atoms or molecules.

The success of the spectroscopic approach to the investigation of intermolecular forces is amply demonstrated by the many studies of neutral clusters.[23,24] Although ion-molecule complexes have attracted considerable experimental attention,[25] detailed spectroscopic information concerning them is extremely sparse. Schwartz[26] has published low resolution IR spectra of charged H_2O and NH_3 clusters, while Okumura et al. have observed vibrational predissociation spectra of H_5^+, H_7^+ and H_9^+[27] and of $H_7O_3^+\cdot H_2$ and $H_9O_4^+\cdot H_2$.[28] Miller et al.[29] have reported the red shift of the origin of the $\tilde{B}-\tilde{X}$ band of $C_6F_6^+$ on complexation by one or more He, Ne or Ar atoms. Kennedy and Miller[30] first reported vibrational progressions in the cluster modes of $C_6F_6^+\cdot X$, $X = $ He and Ne. Kung et al.[31] have recently extended the work on $C_6F_6^+\cdot X_n$.

The formation of ionic clusters in the jet environment depends critically on the point at which photoionization occurs. Our experiments indicate that typically a number of steps are involved in the formation of an ion cluster:

i)   Formation of a neutral cluster, which occurs very close to the nozzle (within 10 nozzle diameters).

TABLE I

Summary of Jahn-Teller Parameters for Benzenoid Cations. The units are as follows: i, D_i, K_i-dimensionless; ϵ_i, ω_i-cm^{-1}; ρ'_{im}-a.m.u.$^{\frac{1}{2}}$ Å. The column headings for the modes are only qualitatively descriptive. The total stabilization energy $\sum_i \epsilon_i$ is given beside each ion.

	C-C Stretch	C-F (Cl) Stretch	C-C-C Bend	C-F Bend	C-Cl Bend
$C_6F_6^+(\sum_i\epsilon_i=821)$					
i	15	16	17	18	---
ω	1610	1215	425	265	---
D	0.23	0.05	0.68	0.38	---
K	---	---	0.006	---	---
ϵ	370	61	289	101	---
ρ'_{im}	0.0098	0.053	0.328	0.310	---
$C_6H_3F_6^+(\sum_i\epsilon_i=922)$					
i	9	12	13	14	---
ω	1570	960	480	335	---
D	0.35	0.01	0.73	0.03	---
K	---	---	0.0007	---	---
ϵ	550	12	350	10	---
ρ'_{im}	0.122	0.029	0.319	0.078	---
$C_6H_3C\ell_3^+(\sum_i\epsilon_i=563)$					
i	9	11	13	---	14
ω	1540	1060	420	---	190
D	0.18	0.02	0.62	---	0.03
K	---	---	≤0.005	---	---
ϵ	277	21	260	---	5
ρ'_{im}	0.089	0.036	0.315	---	0.094
$C_6C\ell_3F_3^+(\sum_i\epsilon_i=575)$					
i	8	---	12	13	14
ω	1550	---	390	310	185
D	0.15	---	0.60	0.32	0.05
K	---	---	---	---	---
ϵ	233	---	234	99	9
ρ'_{im}	0.081	---	0.321	0.263	0.135
$C_6Br_3F_3^+(\sum_i\epsilon_i=660)$					
i	8	10	12	13	---
ω	1560	1100	385	260	---
D	0.15	0.10	0.33	0.75	---
K	---	---	---	---	---
ϵ	234	110	127	195	---
ρ'_{im}	0.080	0.078	0.240	0.440	---

ii) Photoionization of the neutral cluster to leave a bare ion. Fragmentation of the neutral cluster during photolysis disposes of the excess energy deposited in the organic molecule, thereby lessening fragmentation of the parent molecule during or after photoionization.

iii) Attachment of one or more neutral species to the bare ion in three body collisions. Note that direct photoionization of a neutral cluster to form an ionic cluster does not appear to be a significant process in most of our experiments.

In section III.A, we discussed in detail the spectroscopy of isolated halobenzene cations. These cations are particularly well suited as nucleation sites for ionic clusters for spectroscopic studies for several reasons. Firstly, as indicated above, the spectroscopy of the ion is well understood. Further, the experimental production of these ions is straightforward. Finally we would argue that at the present state of knowledge with regards to the spectra of ionic clusters, the study of one charged species compared to another is of little importance. The key differences are between neutral van der Waals molecules and those complexes containing a charge. The very different, and stronger, electrostatic binding in the charged complexes is the critical issue. For this reason we have spent considerable time studying[29-31] a series of halobenzene cation inert gas atom complexes. The clusters most thoroughly studied include $C_6F_6^+ \cdot X_n$ where X = He, Ne, and Ar.

We have found that the optimum conditions for producing a simple cluster like $C_6F_6^+ \cdot He$ are as follows. A He reservoir pressure of ~15 atm is appropriate. $C_6F_6^+$ is produced as before via the 2-photon ionization by an ArF excimer laser of C_6F_6 seeded into He expansion. The optimum nozzle-excimer distance is about 10 nozzle diameters (~1.5 mm). Moving the nozzle closer to the excimer leads to the appearance of a variety of photolysis fragmentation products including C_2, while increasing the nozzle-excimer distance does not allow the bare ions to undergo the 3-body collisions necessary to form clusters. If larger clusters, $C_6F_6^+ \cdot He_n$ (n>1) are desired, one simply increases the He reservoir pressure maintaining the other conditions constant.

The best approach to generating $C_6F_6^+ \cdot Y$ (Y = Ne or Ar) is first to produce good spectra of $C_6F_6^+ \cdot He$, and then add to the carrier gas a small amount (at most a few percent) of the species, Y, to be attached to $C_6F_6^+$. The formation of $C_6F_6^+ \cdot Y$ may well proceed by displacement of the He atom:

$$C_6F_6^+ \cdot He + Y \longrightarrow C_6F_6^+ \cdot Y + He$$

Addition of larger quantities of Y leads to the formation of larger clusters $C_6F_6^+ \cdot Y_n$, but also quenches the generation of $C_6F_6^+$ by excimer laser photoionization.

Spectra for the cluster species $C_6F_6^+ \cdot Ne$ and $C_6F_6^+ \cdot He$ are well known, especially in the region near the origin. Indeed all members of the vibrational progressions, save the origins, appear as doublets. This vibrational structure gives us a detailed experimental probe into the nature of the intermolecular potential. The position of these transitions has been predicted[30,31] semi-quantitatively by calculation of the vibrational energy levels based upon a model potential which consists of a Lennard-Jones 6-12 van der Waals interaction as well as a charge induced dipole contribution. Importantly, there are no adjustable parameters in this potential. This potential gives a well depth of 136 cm^{-1} for the He complex and ~210 cm^{-1} for the Ne complex. These are binding energies for the excited electronic state of the ion. Combining these results with the measured red shifts of the cluster spectra yields ground state binding energies of 100 and 150 cm^{-1} for $C_6F_6^+ \cdot He$ and $C_6F_6^+ \cdot Ne$ respectively. The match between experiment and theory is appealing, suggesting that the model potential constructed from considerations of the purely physical interactions within the complex provides a useful description of the system.

$C_6F_6^+ \cdot Ar_n$ has also been studied extensively.[31] The most remarkable feature about this species is our ability to generate in the gas-phase clusters $C_6F_6^+ \cdot Ar_n$ with $n \approx \infty$, i.e. the condensed phase limit, and record their spectra. We know this limit has been obtained because i) the complex spectra no longer change as the reservoir pressure is increased and ii) the spectral shift of this cluster is, practically speaking, identical to that previously reported[32] for $C_6F_6^+$ isolated in an Ar matrix.

Perhaps the most interesting fact revealed by the spectroscopic work on the ionic clusters is that they do not approach the condensed phase spectra in a simple monotonic and asymtotic manner. Indeed in all these cluster species $C_6F_6 \cdot X_n$, X=He,Ne and Ar, the spectral "solvent" shifts for the clusters with n = 2 exceed the shifts of the condensed phase species. In the series $C_6F_6^+ \cdot Ar_n$, it appears that this spectral shift undergoes several oscillations as n increases. At present there is no theory that accounts for these strange effects. However, these shifts are clearly intimately connected to the electronic and geometric structure of these clusters. It is equally clear that LIF spectroscopy of ionic clusters gives us a powerful probe for understanding the structure of these complex species, which are the prototypes for the species found in ionic solutions, where so much interesting chemistry takes place.

IV.   Neutral Free Radicals

A.   The Cooling of Radicals

In our discussion of chemical intermediates, we have to this point described exclusively the spectra of ions produced from parent molecules by the ejection of an electron. It was argued that the conservation of angular momentum and the similarity of the force fields of the ions and neutrals insured that the ion would be produced "cold" if the precursor neutral was cold.

If we consider the production of radicals via laser radiation, the details of the photophysics change drastically. For example, a typical photolysis producing a free radical can be written

$$R-R' + h\nu \rightarrow R\cdot + R'\cdot$$

Because of the existence of two heavy particles $R\cdot$ and $R'\cdot$, no general statement about the angular momentum content of the individual radicals is possible even if $R-R'$ is cold. Similarly the complete change in the potential surfaces between the radicals, $R\cdot$ and $R'\cdot$, and the precursor $R-R'$, forbids any general statement about the vibrational energy content of the radicals, except that it will often be considerable because of the breaking of a chemical bond.

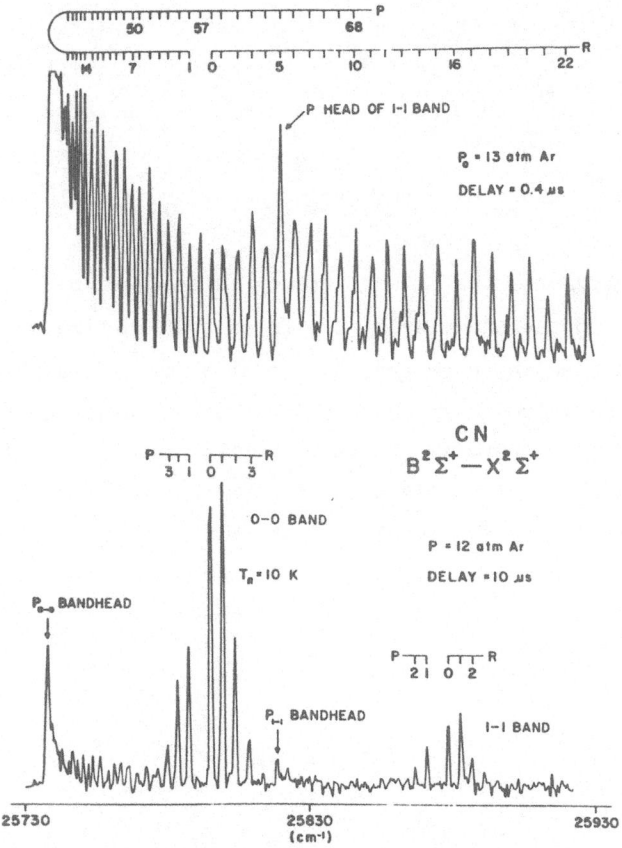

Fig. 6   LIF spectra of CN produced in an Ar free jet expansion by ArF
         laser photolysis of BrCN. In the top trace the photolysis
         and probe laser are closely overlapped in space and time and
         the detected CN distribution is nearly nascent. In the lower
         trace the probe laser is translated downstream in space and
         delayed in time sufficiently so that the CN rotations have
         partially come into equilibrium with the jet's translational
         temperature.

The question is then whether it is possible to   retain   the   very
considerable advantages of the low internal temperature that ions in
jets display, for free radicals produced in jets.   Fortunately the
answer to this question is affirmative.   Fig. 6 shows the LIF
excitation of the CN radical produced[33] from the excimer laser
photolysis of BrCN. In the top trace the photolysis and probe lasers

are closely overlapped in time and space and the spectrum therefore represents CN with its internal states populated nearly nascently. This rotational distribution is not quite Boltzmann but it is nearly so, and would be best represented by a temperature of about 4000 K. Only the lowest two vibrational levels are observed in this spectrum, but a Boltzmann temperature of $\approx$2000 K would be required to reproduce their populations.

In contrast, the lower trace is produced by altering only the probe laser in the experiment. It is delayed in time by $\approx$10 μsec, with a compensatory downstream translation of $\approx$6 mm. Thus the lower trace is representative of CN radicals which have suffered numerous equilibrating collisions with the cold inert gas atoms of the expansion. The CN radicals in the lower rotational levels have thus equilibrated to a near Boltzmann distribution with the temperature near 10 K.

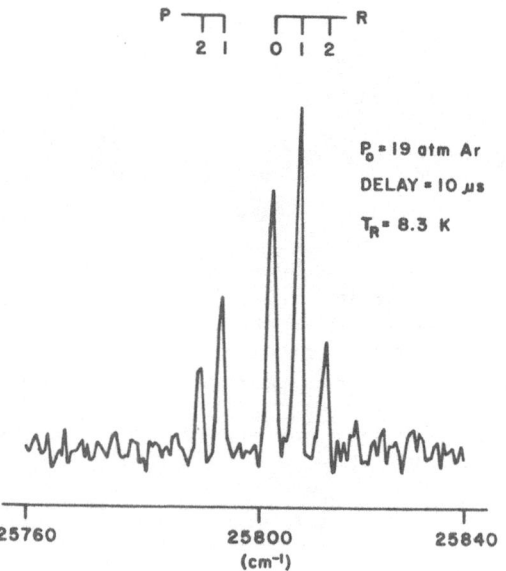

Fig. 7   Spectrum of the CN radical under conditions similar to Fig. 6 except that the Ar reservoir pressure has been increased to 19 atmospheres causing almost complete equilibrium of the translational and rotational degrees of freedom of the radical.

There remains some non-Boltzmann distribution in the higher rotational levels but, as shown in Fig. 7, if the pressure in the reservoir is increased to nearly 20 atmospheres, only the lowest few rotational levels are significantly populated and have a temperature of ≈8 K.

If Ar is used as the expansion gas the v = 0 and 1 vibrational populations are little changed by collisions in the jet, so internal cooling is not complete in this case. On the other hand, the non-equilibrium population of higher vibrational levels is often useful to the spectroscopist. In addition, we have found that the CN population in v > 0 can be eliminated by substituting He as the expansion gas, although with He the rotational cooling is not as efficient as with Ar.

B. Organic Free Radicals

The above example with the CN radical illustrates in a quantitative manner that it is possible to produce quite cold radicals in a jet, even though their nascent distribution immediately following photolysis is anything but cold. This cooling can be most important for the study of large organic radicals whose spectral congestion at room temperatures is enormous.

The principal objective of LIF studies of jet-cooled radicals is to obtain high resolution revealing rotational fine structure. From an analysis of the rotational structure one can, in principle, determine the geometric structure of the radicals. From the fine, and perhaps hyperfine, structure one can learn the details of the electronic distribution within the radical. There have now been a number of organic free radicals for which more or less completely resolved rotational fine structure has been observed. These include methoxy,[34] CH_3O, ethoxy,[35] C_2H_5O, iso-propoxy,[35] C_3H_7O, vinoxy,[36] C_2H_3O, mono-methyl sulfide,[37] CH_3S, cyclopentadienyl,[38] C_5H_5, and benzyl,[38] $C_6H_5CH_2$.

To illustrate how LIF spectroscopy of free radicals in jets can be useful, let's consider several closely related alkoxy radicals, CH_3O, C_2H_3O, C_2H_5O, and $i\text{-}C_3H_7O$. These species are shown schematically in Fig. 8. As can be seen from Fig. 8, each radical has a pair of π molecular orbitals more or less localized on the O atom. In the ground state of the radical, there are three electrons available to populate

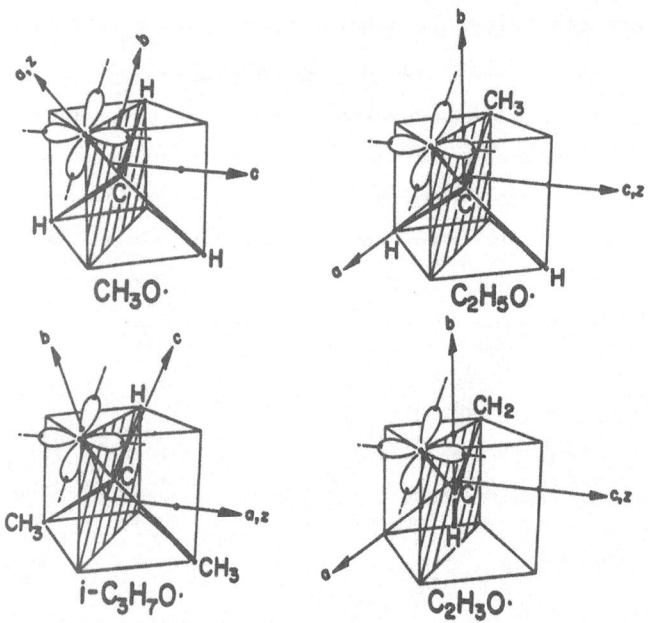

Fig. 8   Schematic representation of the geometries of the alkoxy
          radicals, methoxy, ethoxy, vinoxy, and i-propoxy. The
          highest occupied π-orbitals, approximately localized on the O
          atom, are indicated for each molecule.

those molecular orbitals. The electronic transition observed in the
LIF experiment consists, roughly speaking, of the promotion of an
electron from the $p\sigma$ bonding orbital localized along the CO bond, to
completely fill the above-mentioned π orbitals. This transition
results in a significant decrease in the C-O bond strength as evidenced
by a large decrease in the frequency of the C-O stretch in the excited
state.   In all the molecules the electronic transition lies in the UV
region of the spectrum at an energy of order 30000 cm^{-1}.   (The
situation is slightly more complicated in vinoxy as the existence of
the C=C double bond gives rise to another electronic configuration of
the same symmetry as that described above for the ground state
configuration. Significant configuration interaction causes mixing and
repulsion of the resulting states. It is believed that the electronic
transition observed in LIF takes place between these two states in
vinoxy, but is characterized by approximately the same transition

frequency and a similar decrease in CO bonding character in the excited state as for the other radicals.)

As is well known, CH_3O has a three-fold axis of symmetry with a doubly degenerate ^2E electronic ground state. Using Fig. 8 as a reference, the two components (or linear combinations thereof) of this degenerate ground state can be thought of as arising from placing two electrons in the in-plane π-orbital and one in the out-of-plane π-orbital or *vice versa*.

Turning to C_2H_5O, C_2H_3O, and i-C_3H_7O, one sees that the degeneracy of the ground state is resolved into two components, A′ and A″, due to the removal of the 3-fold axis by substitution for one or more of the H-atoms of methoxy. The A′ and A″ states corresponding respectively to putting the odd electron in the in-plane orbital or the out-of-plane orbital.

It is by no means intuitively obvious which of the A′ or A″ states will be lowest in energy. However, if we make the assumption that the methyl or methylene substitution is a small perturbation, while it breaks the degeneracy of the ground state, it is unlikely to alter the ordering of the much more widely separated excited states. Ergo, it is reasonable to assume that the excited state in all the electronic transitions is the same, which is $^2A_1$ in CH_3O and correlates to A′ in the other molecules. (Because of the configuration interaction in vinoxy which was mentioned earlier, it seems well established that the excited state in the transition for vinoxy alone is an A″ state which does not exist for the other species.)

Knowing the symmetry of the excited states gives one a means of determining the symmetry of the non-degenerate ground state of the molecules lacking a 3-fold axis. One simply must resolve the rotational structure of the electronic transition sufficiently to determine whether it is consistent with A-type, C-type, A-B hybrid, etc. selection rules.[39] Since the selection rules correlate with whether the electronic state is A′ or A″, this experimental observation answers the question of the nature of the ground state.

Figs. 9-12 show respectively the LIF spectra obtained for CH_3O, C_2H_3O, C_2H_5O, and i-C_3H_7O in the jet. As these figures show, in all

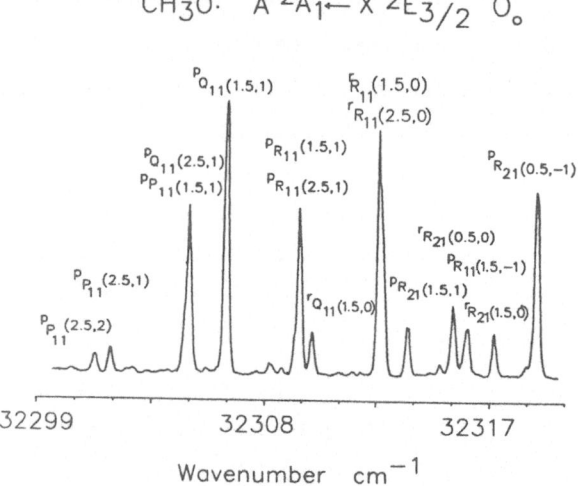

$$CH_3O. \quad \tilde{A}\ ^2A_1 \leftarrow \tilde{X}\ ^2E_{3/2} \quad O_o^o$$

Fig. 9   Laser excitation spectrum arising from $\tilde{X}\ ^2E_{3/2}$ CH_3O showing resolved rotational structure, with the assignments as indicated.

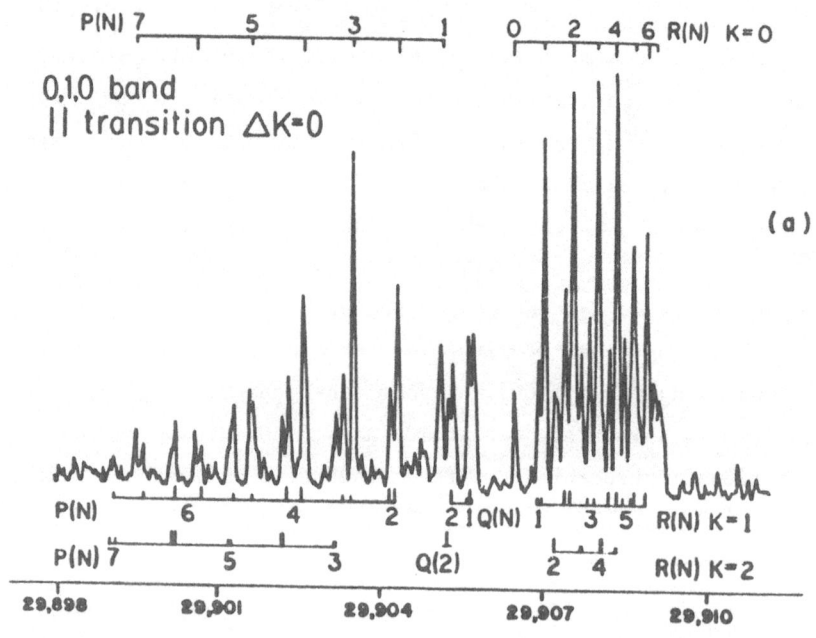

Fig. 10   Laser excitation spectrum of the vinoxy radical showing resolved rotational structure, with the assignments as indicated.

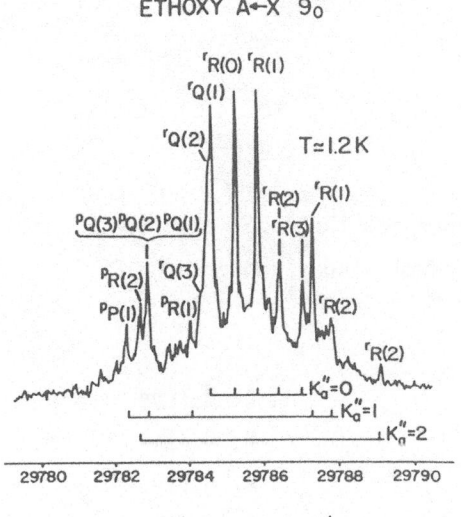

Fig. 11 Laser excitation spectrum of the ethoxy radical showing rotational structure with the indicated assignments.

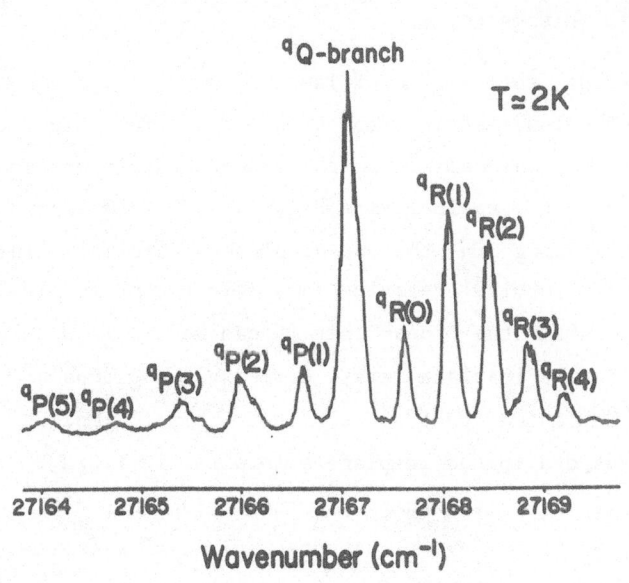

Fig. 12 Laser excitation spectrum of i-C_3H_7O showing rotational structure with the indicated assignments.

cases the rotational structure is sufficiently resolved to determine the selection rules. For the details of this determination, it is best to consult references 35 and 39. The results of this analysis are given in Table II.

TABLE II

Observed band types and the lower electronic state symmetry derived therefrom for several alkoxy radicals.

Molecule	Band Type	Lower State[a]
CH_3O	B, C $\perp$	^2E
C_2H_3O	A, B hybrid	$^2A''$
C_2H_5O	C $\perp$	$^2A''$
i-C_3H_7O	B, C hybrid	$^2A'$

[a] The lower state symmetry is determined assuming the upper state is A', (A_1 for CH_3O) in all cases except vinoxy where it is assumed A'', consistent with numerous calculations and configuration interaction unique to this molecule.

Comparing the results of Table II with Fig. 8 shows that in C_2H_5O and C_2H_3O the lower state, from which the transition originates and which can be assumed to be the ground state because of the very low temperature of the radical (see Figs. 9-12) corresponds to the odd electron residing in the out-of-plane a'' orbital. This implies that the in-plane π-orbital carrying two electrons is lowest in energy consistent with the idea that it can be involved, albeit weakly, in bonding with the in-plane methyl or methylene groups.

Somewhat in contrast, we see that Table II implies that the odd electron resides in the in-plane a' orbital in i-C_3H_7O. However, if we refer to Fig. 8, we see that this result is nonetheless consistent with our previous thinking. The out-of-plane a'' π-orbital containing two electrons can now be slightly stabilized by bonding, so-called hyperconjugation, to the now out-of-plane methyl groups in i-C_3H_7O.

While the complete analysis of the rotational structure of the alkoxy radicals is still underway in an effort to obtain precise geometric structural parameters for the radicals, it seems clear that much about the electronic structure of these radicals has already been gleaned. LIF experiments on other cold organic radicals should yield similar information about those molecules in the near future.

V. SUMMARY

For many years, spectroscopists have recognized that one of the most exciting challenges remaining was to obtain spectra of short-lived, highly reactive chemical intermediates. It has now been demonstrated that the combination of laser spectroscopy and the supersonic free jet expansion constitutes a powerful tool in this effort. Spectra of ions, ionic complexes, and neutral free radicals have been recorded. In addition, the very low temperature of the ions and radicals in the jet allow the analysis of the spectra of even moderately complex organic species. These efforts promise to push even further the spectroscopist's efforts to monitor and understand the very heart of chemistry.

Acknowledgment: This work was supported by the National Science Foundation under grant CHE-8507537.

REFERENCES

*Present address: Department of Chemistry, University of Birmingham, P.O. Box 363, Birmingham, B15 2TT U.K.

1. T. A. Miller, Science 223, 545 (1984).

2. R. E. Smalley, L. Wharton, and D. H. Levy, Acc. Chem. Res. 10, 139 (1977).

3. L. Wharton, D. Auerbach, D. H. Levy, and R. Smalley, Adv. Laser Chem., A. H. Zewail, ed. (Springer, Verlag, New York, 1978).

4. D. H. Levy, L. Wharton, R. E. Smalley, Chemical and Biochemical Applications of Lasers, C. B. Moore, ed. (Academic Press, New York, 1977).

5. D. H. Levy, *Ann. Rev. Phys. Chem.* **31**, 197 (1980) and D. H. Levy, *Science* **214**, 263 (1981).

6. H. W. Liepmann and A. Roshko, *Elements of Gas Dynamics*, p. 40, Wiley, New York, 1957.

7. J. B. Anderson and J. B. Fenn, *Physics of Fluids* **8**, 780 (1965).

8. H. Ashkenhas and F. S. Sherman, *Rarefied Gas Dynamics,*, 4th Symposium, ed. J. H. deLeeuw, vol. 2, p. 84 (Academic, New York, 1966).

9. D. M. Lubman, C. T. Rettner, and R. N. Zare, *J. Phys. Chem.* **86**, 1129 (1982).

10. The simple kinetic theory used to derive this equation can be found in many texts. See, for example, *Physical Chemistry*, P. W. Atkins, pp. 799-804, W. H. Freeman and Co., San Francisco (1978) and explicitly *Chemical Applications of Molecular Beam Scattering*, M. A. D. Fluendy and K. P. Lawley, p. 67, Chapman and Hall, London (1973).

11. R. Campargue, *Entropie* **30**, 15 (1969).

12. L. DiMauro and T. A. Miller, *Chem. Phys. Lett.*, in press.

13. S. C. Foster, L. Yu, J. M. Williamson, and T. A. Miller, to be published.

14. T. Sears, T. A. Miller, and V. E. Bondybey, *J. Am. Chem. Soc.* **103**, 326 (1981).

15. T. J. Sears, T. A. Miller, and V. E. Bondybey, *J. Chem. Phys.* **74**, 3240 (1981).

16. T. J. Sears, T. A. Miller, and V. E. Bondybey, *Discuss. Faraday Soc.* **71**, 175 and 341 (1981).

17. V. E. Bondybey, T. J. Sears, T. A. Miller, C. Vaughan, J. H. English, and R. H. Shiley, *Chem. Phys.* **61**, 9 (1981).

18. "The Jahn-Teller Effect in Benzenoid Cations: Theory and Experiment," T. A. Miller and V. E. Bondybey, "Molecular Ions: Spectroscopy, Structure, and Chemistry," T. A. Miller and V. E. Bondybey, eds., (North-Holland, 1983).

19. C. Cossart-Magos, D. Cossart, and S. Leach, *Chem. Phys.* **41**, 345 and 363 (1979).

20. S. Leach and C. Cossart-Magos, *Discuss. Faraday Soc.* **71**, 336 (1981).

21. T. A. Miller, *Discuss. Faraday Soc.* **71**, 341 (1981).

22. C. Cossart-Magos, D. Cossart, S. Leach, J. P. Maier and L. Misev, *J. Chem. Phys.* **78**, 3673 (1983).

23. D. H. Levy, *Adv. Chem. Phys.* **47** (1981) Part 1, 323.

24. J. A. Beswick and J. Jortner, *Adv. Chem. Phys.* **47** (1981) Part 1, 363.

25. T. D. Märk and A. W. Castleman, Jr., *Adv. At. Mol. Phys.* **20** (1985) 66.

26. H. A. Schwartz, *J. Chem. Phys.* **67** (1977) 5525 and **72** (1980) 284.

27. K. Okumura, L. I. Yeh and Y. T. Lee, *J. Chem. Phys.* **83** (1985) 3705.

28. M.Okumura,L.I.Yeh,J.D.Myers and Y.T.Lee,*J.Chem.Phys.* **85** (1986) 2328.

29. L. F. DiMauro, M. Heaven and T. A. Miller, *Chem. Phys. Lett.* **104** (1984) 526.

30. R. A. Kennedy and T. A. Miller, *J. Chem. Phys.* **85**, 2326 (1986).

31. C.-Y. Kung, T. A. Miller, and R. A. Kennedy, *Proc. Roy. Soc.* (Lond), accepted for publication.

32. V. E. Bondybey, J. H. English, and T. A. Miller, *J. Am. Chem. Soc.* **100**, 5251 (1978).

33. M. Heaven, T. A. Miller, and V. E. Bondybey, *Chem. Phys. Lett.* **84**, 1 (1981).

34. D. E. Powers, J. B. Hopkins, and R. E. Smalley, *J. Phys. Chem.* **85**, 2711 (1981).

35. S. C. Foster, Y.-C. Hsu, C. P. Damo, X. Liu, C.-Y. Kung, and T. A. Miller, *J. Phys. Chem.* **90**, 6766 (1986).

36. L. F. DiMauro, M. Heaven, and T. A. Miller, *J. Chem. Phys.* **81**, 2339 (1984).

37. X. Liu, Y.-C. Hsu, and T. A. Miller, to be published.

38. M. Heaven, L. DiMauro, T. A. Miller, *Chem. Phys. Lett.* **95**, 347 (1983).

39. G. Herzberg, *Molecular Spectra and Molecular Structure*, D. Van Nostrand Co., Princeton, NJ, 1966.

THE INFRARED SPECTROSCOPY OF THE H_3^+ AND HD$^+$ MOLECULAR IONS AT THEIR DISSOCIATION LIMITS

By Alan Carrington, Iain R. McNab and Christine A. Montgomerie
Department of Physical Chemistry
South Parks Road
Oxford OX1 3QZ

ABSTRACT. Previous theoretical and experimental work on the ions H_3^+ and HD$^+$ is reviewed. Earlier work using ion beam techniques to study the uppermost bound levels and quasibound levels of H_3^+ is discussed, and a suitable model for the molecule in these levels is proposed. Earlier work using ion beam techniques to study the vibration-rotation levels of HD$^+$ is reviewed.

The nuclear hyperfine and spin-rotation structure of HD$^+$ is described, and recent observations of the vibration-rotation satellite lines which will yield absolute values of the deuterium hyperfine constants are presented.

1. THE SPECTROSCOPY OF H_3^+

1.1. Introduction

The H_3^+ molecular ion is the simplest polyatomic molecule, and was discovered by J.J. Thompson in 1911 (1). Although its chemistry has been studied extensively using mass spectrometric methods, its spectrum has only recently been observed. The first spectroscopic studies were described by Oka (2) for H_3^+, and by Shy, Farley, Lamb and Wing (3) for D_3^+ and H_2D^+. These studies were confined to the first few vibration-rotation levels of the molecules and confirmed the essential correctness of the theoretical descriptions of the molecule in these low energy states.

The equilibrium geometry of H_3^+ in its ground electronic state ($^1A_1'$) has been the subject of many theoretical calculations, the unanimous conclusion being that this geometry is that of an equilateral triangle; this was confirmed experimentally in 1978 by Coulomb explosion experiments (4). Our work suggests that this description is not appropriate for the highest lying bound and quasibound levels of the electronic ground state, at which energies the molecule may more profitably be considered as a hydrogen molecule/proton complex.

1.2 Experimental Methods

The experiment has been described in detail elsewhere (5), and only an outline is presented here. The instrument used is a modified Vacuum Generators ZAB 1F tandem mass spectrometer (Fig 1). Molecular ions are produced in a conventional source by electron bombardment of hydrogen gas. A beam of these ions is produced by applying an accelerating potential to the source, which may

451

A. C. P. Alves et al. (eds.), Frontiers of Laser Spectroscopy of Gases, 451–460.

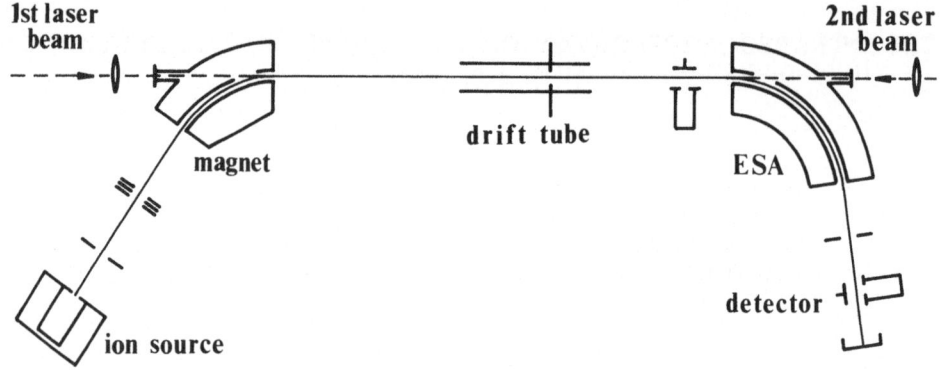

Figure 1   modified ZAB 1F Tandem Mass Spectrometer

be floated between 1 and 10kV above earth potential; these ions then pass into a 55° magnetic sector which acts as a mass analyser. The single focus mass spectrum may be observed by sweeping either the magnetic field or the source potential, and monitoring the ion beam on the first off-axis electron multiplier.

The ion beam is brought to a focus at the intermediate slit, which is positioned 3/4 of the way through a cylindrical metal tube hereafter refered to as the drift tube. The beam then enters an 81.5° electrostatic analyser (ESA) which may be set to transmit either the parent ions, or fragment ions produced by dissociation of the parent beam. Those ions transmitted by the ESA may be detected on the second off-axis electron multiplier, or on a Faraday cup.

The major modification to the spectrometer is the addition of windows which allow a laser beam to be brought into collinear coincidence with the ion beam, in either a parallel or an anti-parallel orientation. External lenses are used to bring the laser beam to a focus at the intermediate slit. All the work described has been carried out using cw carbon dioxide lasers (Edinburgh Instruments PL3 and PL4) operated with either $^{12}CO_2$ or $^{13}CO_2$ as the lasing medium. The CO_2 lasers are line tunable over the frequency range 880-1100cm^{-1}, but full frequency coverage for light ions is achieved by utilising the Doppler effect: the source potential (and hence the ion velocity) is swept, and the ions experience a laser frequency given by the relativistic doppler expression

$$\nu_{res} = \nu_{laser} \times \left[\frac{1 \mp (k/c)}{1 \pm (k/c)}\right]^{\frac{1}{2}}$$

where k is the velocity of the ions, and the upper and lower signs apply to parallel and antiparallel alignment of the laser.

The drift tube has several uses. It may be held at a potential of up to ±500V relative to earth, and this allows us to discriminate in favour of fragment ions produced at this potential (within the drift tube) by means of the ESA. A square wave potential of up to ±10V may also be applied to the drift tube; this velocity modulates the ion beam and hence, through the Doppler effect, is equivalent to frequency modulation, thus allowing phase sensitive detection to be employed.

Photodissociation spectra are recorded by scanning the accelerating potential and monitoring the number of fragment ions produced by photodissociation in the drift tube on the second off-axis electron multiplier. Total fragment ion signal is then plotted against the effective frequency. The Doppler width of transitions is extremely small (less than 3MHz), because of an effect called kinematic compression (6), which is due to the velocity spread of the molecules being compressed relative to the high velocities to which the ions are accelerated. A further feature of the instrument is that the ESA has sufficient resolution to allow us to record laboratory kinetic energy spectra of the fragment ions. This enables the centre of mass kinetic energy release to be determined upon applying a suitable transformation.

1.3   The Infrared Predissociation Spectrum of H_3^+

The H_3^+ molecule is easily formed by electron bombardment of H_2, being formed by the ion-molecule reaction

$$H_2 + H_2^+ \rightarrow H_3^+ + H$$

which is exothermic by about 1.7eV. In our spectrometer we can achieve ion fluxes greater than 10^{12} ions s^{-1}. We observe strong infrared photodissociation to $H_2 + H^+$, which implies that states close to or above the lowest dissociation limit are significantly populated.

Scanning the source potential and monitoring H^+ ions, reveals an extensive spectrum, most features of which have been described in detail by Carrington and Kennedy (5). These features are now reviewed, and new work described.

1.3.1. <u>Density of Lines.</u>   Over the frequency region 872 - 1094cm^{-1}, nearly 27000 lines were observed at an average laser power of 7W cw using velocity (frequency) modulation and phase sensitive detection.

1.3.2. <u>Linewidths.</u>   The Doppler width is typically 3MHz (see above), but most of the lines measured in the initial recording of the spectrum had widths between 3 and 60MHz. This does not necessarily reflect the broad lifetime range of the states observed, as the velocity modulation technique discriminates in favour of the narrower lines. There are certainly other lines which are undermodulated due to greater width, and hence undetected. To some extent this problem may be overcome by using amplitude modulation of the laser beam, which then allows linewidths of several hundred MHz to be observed. The large variation in linewidths strongly suggests that they are determined by the lifetimes of predissociating states. Studies with two lasers prove that the spectrum does not arise by direct photodissociation following a bound to bound transition.

1.3.3. <u>Power Dependence of Lines.</u>   Detailed studies of the power dependence of different sections of the spectrum using a wide range of laser powers show no evidence for two photon, or multiphoton processes. It is seen that for the strongest lines saturation may occur at laser powers of 5W, and the spectrum is therefore considerably simplified at low laser powers.

1.3.4. <u>Kinetic Energy Measurements.</u>   The centre of mass kinetic energy release of photofragment protons for any line may be made using the ESA. Most lines are associated with energy releases of 0-500cm^{-1}, but some show energy releases of 3000cm^{-1} or more. As the infrared photons used have a maximum energy of 1100cm^{-1}, such transitions must involve two levels, both of which lie at least 1900cm^{-1} above the lowest dissociation limit.

454

1.3.5. <u>Lifetimes of the Predissociating Levels.</u>   The experimental geometry defines a "lifetime window" on those states between which we may observe transitions. Only those transitions which involve a lower level of sufficiently long lifetime to reach the detection region (drift tube) with a significant population may be observed. At a beam potential of 5kV this implies the initial state lifetime must be $3\mu s$ or greater. Conversely, the final state lifetime must be sufficiently short that the molecule has time to dissociate within the detection region; at a beam potential of 5kV this implies that the final lifetime must be less than $0.7\mu s$. The existence of this experimental "lifetime window" has important consequences for a possible assignment of the spectrum which are discussed later.

Absolute lifetimes of levels involved in transitions may sometimes be determined if a transition is observable using two different laser lines at different potentials. As the ion takes longer to reach the detection region at a low beam potential, it is posssible that the population of an intial state will be depleted relative to that reaching the detection region at a high beam potential. For this case, the relative intensity of an observed line due to a transition involving such a state will be decreased in a spectrum recorded at a low beam potential, compared with a spectrum recorded at a high beam potential. Many examples of this have been observed, and one is presented (Fig 2). The strongest line in the high potential scan is smaller by a factor of four in the low potential scan. The difference in flight times is $\approx 2\mu s$, and we conclude that for this line the initial state has a lifetime of ca. $1\mu s$.

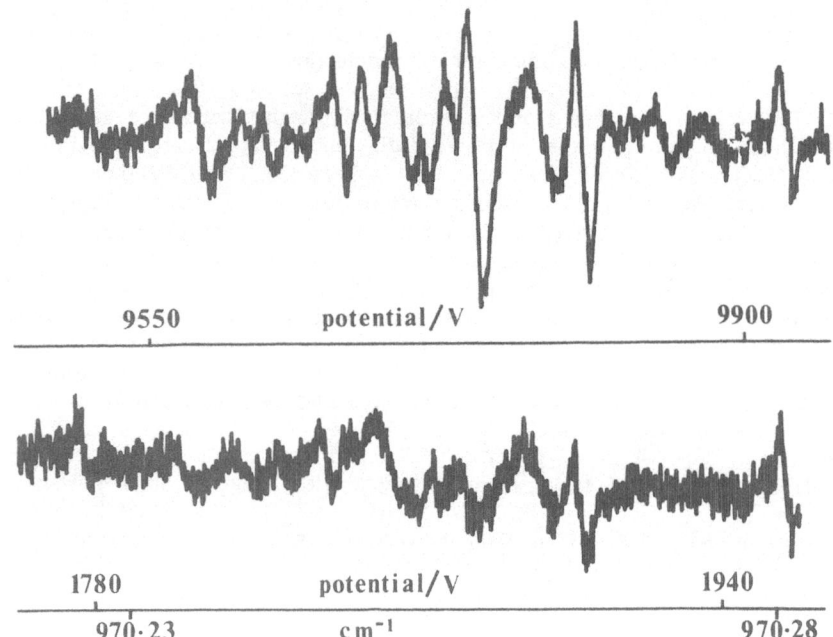

Figure 2.   Section of the infrared predissociation spectrum of H_3^+ recorded at high and low beam potentials.

1.3.6. <u>Isotope Studies.</u>   The isotopomers H_2D^+, D_2H^+ and D_3^+ also show predissociation spectra which are in general similar to that of H_3^+. In the case of D_2H^+, the spectrum may be recorded by monitoring either the D^+ or H^+ photofragment ions. It is found that the spectra obtained by monitoring H^+ ions are different from those obtained by monitoring D^+ ions over the same frequency region (Fig 3). It is also found that the D^+ spectra are much weaker than the H^+ spectra, despite the increased collection efficiency for D^+ ions; the two spectra shown were recorded at 15s/point for D^+, and 9s/point for H^+ fragments. The intensities of the H^+ spectra from D_2H^+ are generally lower than in the H_3^+ spectrum, but the density of lines is even greater. The spectra of D_3^+ and H_2D^+ are also weaker and also exhibit a greater density of lines than the spectrum of H_3^+.

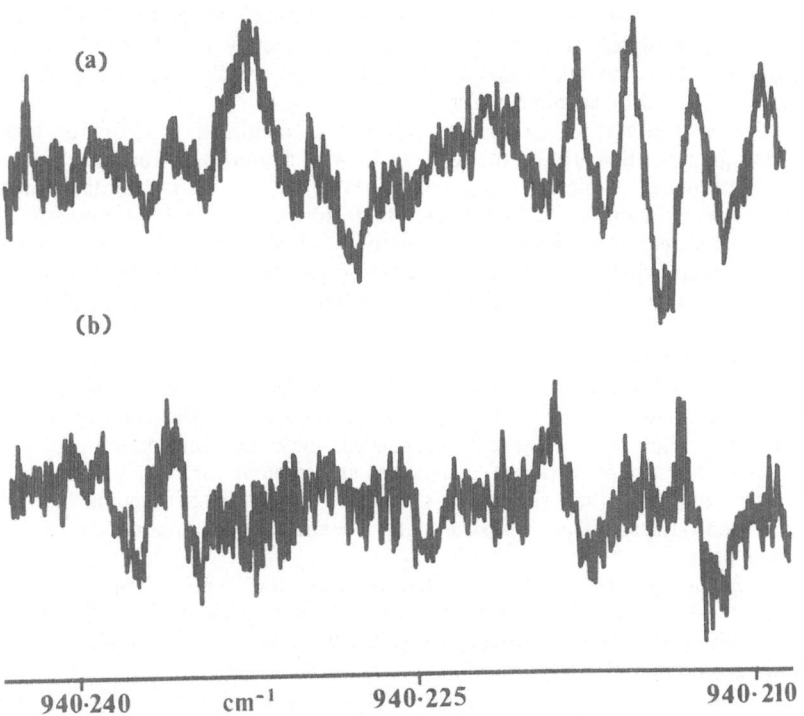

Figure 3.   Section of the predissociation spectrum of the D_2H^+ predissociation spectrum recorded by monitoring H^+ (a) and D^+ (b) ions.

1.4. Assigning the Spectrum
    The H_3^+ infrared predissociation spectrum is so extensive that it is unlikely that a full assignment could ever be made; the most that we can currently hope to achieve is an understanding of how such a spectrum could originate. An obvious first step to such an understanding would be to take a low resolution spectrum of the same region, in order to see whether any recognizable structure could be seen. Unfortunately our experiment has an inherently high resolution which prevents us

from doing this. A pseudo low resolution convolution of the high resolution data has been made (5) which showed four clear peaks. These peaks were found to lie at the Δj=3-5 transition frequencies of the hydrogen molecule in the first four vibrational levels of the ground electronic state.

The correlation of the pseudo low resolution spectrum with transitions in the hydrogen molecule suggests that close to (or above) its dissociation limit, the H_3^+ molecule is perhaps best considered as a hydrogen-proton complex. A van der Waals type model may therefore be the best description for understanding the structure and intramolecular dynamics, although it is in no way suggested that the bond is anything other than chemical in nature. Recent semiclassical analyses by Child (7), and Pfeiffer and Child (8) support this approach. Using a high angular momentum H_2/H^+ complex model, Pfeiffer and Child were able to account for a sufficient density of states to give rise to the observed density of lines. Simulated weak laser field spectra showed features between 700 and 900cm^{-1} due to Δl=1 transitions, where l is the quantum number associated with rotation of the H_2---H^+ complex. Transitions at higher frequency could only be accounted for by assuming that the spectrum was saturated, in which case the observation of 27000 lines was readily accounted for.

The lifetimes of predissociating states calculated by Pfeiffer and Child varied from greater than 10^{-6}s to 10^{-12}s. It was found that for any given total angular momentum J, typically four states lay in the experimental "lifetime window" discussed previously. It is believed that this number might increase to ten with an expansion of the basis. The relatively few states for a given total angular momentum which could be observed suggests that the possibility of seeing any "normal" spectroscopic progressions within the H_3^+ predissociation spectrum is remote.

A classical analysis of the H_3^+ system has also recently been performed by Berblinger, Pollack and Schlier (9) which suggests that low angular momentum complexes are almost certainly *not* involved in the observed spectrum. They conclude that the observed resonances may be interpreted as molecular shape resonances, whose classical analogues are regular or chaotic bound states in the continuum, and that sufficient states are available to account for the density of observed lines. The classical and semiclassical treatments are therefore in agreement.

Berblinger *et al.* further believe that the type of spectrum found for H_3^+ should be a general feature of many systems with sufficiently deep wells in the interaction region. We intend to conduct a search for similar spectra from such systems in the near future.

One aspect of the spectrum which yields hope for future analysis is the enormous simplification that is possible on re-recording the spectrum at lower laser powers. We have re-recorded one 20cm^{-1} segment at a laser power of 2W using amplitude modulation, and have seen a density of lines less than twenty times that of the spectrum recorded at 7W. If theoretical predictions of which transitions are most strongly electric dipole allowed can be made, it may prove possible to assign the lines observable at low laser powers.

2. THE SPECTROSCOPY OF HD$^+$

2.1. Introduction

The hydrogen molecular ion has presented a great challenge to both experimentalists and theoreticians for many years. For theoreticians it is an example of the three-body problem in quantum mechanics, and with only one electron it has the advantage that no electron interactions are present. Calculations of energy levels have been carried out to increasingly higher levels of approximation. The first analytical solution of the problem within the Born-Oppenheimer approximation was made by Hylleraas in 1931 (10). A complete tabulation of all bound vibration-rotation levels within the adiabatic approximation was published by Hunter, Yau and Pritchard in 1974 (11), and the best calculations to date including non-adiabatic, radiative and relativistic corrections have just been presented by Wolniewicz and Poll (12). A detailed review of both the theory and spectroscopy has been published by Carrington and Kennedy (13).

Spectroscopy on the hydrogen molecular ion is difficult. Early results came from photoelectron (14), photoionization (15) and Rydberg spectra (16). Dehmelt and Jefferts (17) succeeded in measuring the radiofrequency spectra of H_2^+ and characterizing the hyperfine structure in the vibrational levels v=4-8.

The first ion beam studies on HD^+ were carried out by Wing et al. (18), using Doppler tuning to bring vibration-rotation transitions into resonance with an infrared carbon monoxide laser, and detecting resonances by monitoring the change in the cross section for charge exchange with a collision gas. Results were obtained on the vibrational levels v=0-3.

2.2. Experimental Methods

Most of our work on HD^+ has been carried out on the apparatus described in section 1. The important feature which allows HD^+ to be studied in this way is that on electron impact ionization all bound vibrational levels are populated, up to v=21. This is due to the displacement of the potential minimum of HD^+ relative to HD, which results in significant overlap of the ground state wavefunction with all vibrational levels of HD^+. Molecules in vibrational states v=17 and above can be photodissociated by an infrared photon, which excites molecules from the ground electronic state into the first excited state which is repulsive at all internuclear distances (save for a van der Waals minimum).

A two photon experiment may therefore be carried out; transitions between levels involving a state which can be photodissociated may be monitored by changes in the photodissociation yield. This can be achieved using either one laser (and hence one frequency) to supply both photons, or by using two lasers - one at low power to drive transitions with minimum power broadening, and one at high power to drive the photodissociation process. The two laser experiment is clearly capable of achieving better resolution.

Recently we have constructed a new ion beam machine, similar to the machine with which Carrington, Buttenshaw and Roberts first observed the vibration-rotation transitions in HD^+ (19). The ions are produced in an electron impact source from a 50/50 mixture of H_2/D_2 which is passed over a heated Palladium catalyst (10% Pd on activated carbon). The ions are accelerated to a maximum of 5kV and a 90° magnetic sector mass selects the ion of interest, in this case HD^+. The HD^+ ions are focused by an electrostatic lens across the main chamber, in which the ion beam interacts with a collinear CO_2 laser beam. This apparatus also contains a drift tube, to which a scanning potential is applied in order to achieve Doppler tuning. At the end of the drift tube there is a second electrostatic lens which focuses the ion beam into a second 90° magnetic sector, which is used to separate the photofragments from the parent ion beam. A resonance is detected by a change in the number of photofragment ions reaching the electron multiplier after the second magnet. As above, the drift tube potential is modulated, and phase sensitive detection employed.

2.3. The HD$^+$ Spectrum close to Dissociation

A feature which complicates the spectroscopy of the HD$^+$ ion is the existence of two dissociation limits which differ in energy by 28.9cm^{-1}. One may ask whether the electron distribution in the molecule is sensitive to this difference, or becomes sensitive to it as the lowest dissociation level (H$^+$+D) is approached. Carrington and co-workers have measured the vibration-rotation transitions for the HD$^+$ bands 18-16 (20), 17-14 (21), 17-15 (22) and 20-17 (22), by means of the two photon experiment outlined above.

2.4. Hyperfine Structure in HD$^+$ Spectra

In order to describe the hyperfine levels of HD$^+$ we must couple together the angular momenta due to complete rotation of the molecule (N), electron spin (S=$\frac{1}{2}$), nuclear spin of the proton (I$_H$=$\frac{1}{2}$), and nuclear spin of the deuteron (I$_D$=1). The coupling scheme suggested by the relative magnitudes of the interactions is:

$$\underset{\sim}{S} + \underset{\sim}{I}_H = \underset{\sim}{G}_1 \qquad\qquad G_1 = 1 \quad ; \quad 0$$

$$\underset{\sim}{G}_1 + \underset{\sim}{I}_D = \underset{\sim}{G}_2 \qquad\qquad G_2 = 2,1,0 \quad ; \quad 1$$

$$\underset{\sim}{G}_2 + \underset{\sim}{N} = \underset{\sim}{F}$$

The values which G$_1$ and G$_2$ are allowed to take are shown.

The effective Hamiltonian used to describe these interactions is

$$b_H\underset{\sim}{I}_H\cdot\underset{\sim}{S} + b_D\underset{\sim}{I}_D\cdot\underset{\sim}{S} + t_D(2I_{Dz}S_z - I_{Dx}S_x - I_{Dy}S_y) + \gamma\underset{\sim}{S}\cdot\underset{\sim}{N} +$$
$$+ t_H(2I_{Hz}S_z - I_{Hx}S_x - I_{Hx}S_y)$$

where b_H is the proton Fermi contact interaction
b_D is the deuteron Fermi contact interaction
t_H is the axial component of the proton dipolar interaction
t_D is the axial component of the deuteron diploar interaction
γ is the spin rotation interaction.

Explicit expressions for relevant matrix elements evaluated in this basis have been given by Carrington and Kennedy (22).

The electric dipole allowed vibration-rotation transitions obey the selection rules: $\Delta N=\pm1$, $\Delta G_1=0$, $\Delta G_2=0$, $\Delta F=0,\pm1$.

At low laser power a hyperfine splitting is observed for each transition corresponding to transitions between levels of different values of G$_1$. This splitting only depends on *differences* in the hyperfine constants for the upper and lower states. The splitting is particularly interesting in the 20-17 band; unambiguous evidence was found from lineshapes and splittings due to nuclear hyperfine interactions that the electron distribution was indeed sensitive to the lower dissociation limit (D+H$^+$) and became asymmetric in favour of it. Unfortunately the observed electric dipole transitions were diagonal in G$_1$ and G$_2$, and hence could not yield absolute values of the hyperfine constants.

At high laser powers (>2W) the central vibration-rotation resonance is broadened into one line. With our new machine we have observed weaker satellite peaks on either side of the central resonance in the 16-18 band (23); these have been identified with the forbidden transitions $\Delta G_2\neq0$, although they are predicted to be three orders of magnitude weaker than the central resonance. One such spectrum is shown (Fig 4).

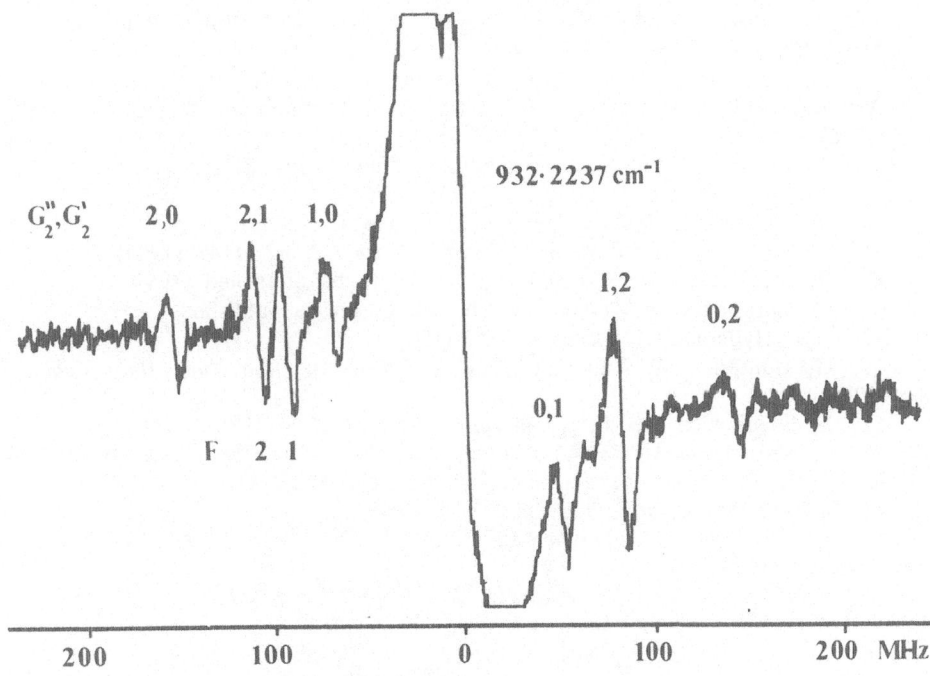

Figure 4. 16,1-18,2 vibration-rotation transition in HD⁺ showing satellite lines due to forbidden transitions.

These results will be able to give unambiguous determinations of the absolute values of the deuteron hyperfine constants in the upper and lower states. As a first approximation we have ignored the smaller dipole-dipole and spin-rotation constant terms in the Hamiltonian, and the differences between different rotational levels. This gives an estimate for the deuteron Fermi contact parameter b_D in v=16 of 110MHz and in v=18 of 106MHz. These are close to the values calculated by Carrington and Kennedy (22).

Our immediate aim is to increase the data on the satellite peaks of the 16-18 band. It will then be possible to improve the fit of the results to the effective Hamiltonian, giving absolute values of the nuclear hyperfine constants for the vibration rotation levels of HD⁺ as it approaches its lowest dissociation limit.

A. C. thanks the Royal Society for a Research Professorship, whilst I. R. M. and C. A. M. thank the S.E.R.C. and The British Petroleum Company P.L.C. respectively for Post-graduate Research Studentships. We are also grateful to the S.E.R.C. for their financial support in the purchase and construction of equipment.

REFERENCES

1   J.J. Thompson, *Philos. Mag.* 21, 225 (1911)

2   T. Oka, *Phys. Rev. Lett.* 45, 531 (1980)

3   J.-T. Shy, J.W. Farley, W.E. Lamb, Jr., and W.H. Wing, *Phys. Rev. Lett.* 45, 753 (1980)

4   M.J. Gaillard, D.S. Gemmell, G. Goldring, I. Levine, W.J. Pietsch, J.C. Poizat, A.J. Ratowski, J. Remillieux, Z. Vager, and B.J. Zabranski, *Phys. Rev.* A 17, 1797 (1978)

5   A. Carrington and R.A. Kennedy, *J. Chem. Phys.* 81, 91 (1984)

6   S.L. Kaufman, *Opt. Comm.* 17, 309 (1976)

7   M.S. Child, *J. Phys. Chem.* 90, 3595 (1986)
      J. Chem. Soc. Faraday Trans. 2, 82, 1143 (1986)

8   R. Pfeiffer and M.S. Child, *Mol. Phys.* to be published (1987)

9   M. Berblinger, E. Pollak and Ch. Schlier, to be published

10   E.A. Hylleraas, *Z. Phys.* 71, 739 (1931)

11   G. Hunter, A.W. Yau and H.O. Pritchard, *At. Data Nucl. Data Tables* 14, 11 (1974)

12   L. Wolniewicz and J.D. Poll, *Mol. Phys.* 59, 953 (1986)

13   A. Carrington and R.A. Kennedy, in *Gas Phase Ion Chemistry*, vol. 3, *Ions and Light*, (ed. M.T. Bowers), pp. 393. London: Academic Press, Inc.

14   L. Åsbrink, *Chem. Phys. Lett.* 7, 549 (1970)
      J.E. Pollard, D.J. Trevor, J.E. Reutt, Y.T. Lee and D.A. Shirley, *J. Chem. Phys.* 77, 34 (1982)

15   W.A. Chupka and J. Berkowitz, *J. Chem. Phys.* 51, 4244 (1969)
      W.B. Peatman, *J. Chem. Phys.* 64, 4093 (1976)

16   S. Takezawa (a) *J. Chem. Phys.* 52, 2575 (1970)
                  (b) *J. Chem. Phys.* 52, 5793 (1970)
      G. Herzberg and Ch. Jungen, *J. Mol. Spectrosc.* 41, 425 (1972)

17   H.G. Dehmelt and K.B. Jefferts, *Phys. Rev.* 125, 1318 (1962)

18   W.H. Wing, G.A. Ruff, W.E. Lamb and J.J. Spezeski, *Phys. Rev. Lett.* 36, 1488 (1976)

19   A. Carrington, J. Buttenshaw and P.G. Roberts, *Mol. Phys.* 38, 1711 (1979)

20   A. Carrington, and J. Buttenshaw, *Mol. Phys.* 44, 267 (1981)

21   A. Carrington, J. Buttenshaw and R.A. Kennedy, *Mol. Phys.* 48, 775 (1983)

22   A. Carrington and R.A. Kennedy, *Mol. Phys.* 56, 935 (1985)

23   A. Carrington, I.R. McNab and C.A. Montgomerie, *Phil. Trans. Roy. Soc.* to be published (1987)

THE INTERPRETATION OF VIBRATIONAL OVERTONE SPECTRA OBSERVED BY FOURIER TRANSFORM AND LASER PHOTOACOUSTIC SPECTROSCOPY.

by I. M. Mills
Department of Chemistry,
University of Reading,
Reading RG6 2AD, England

ABSTRACT

In this talk I shall be discussing the spectra of vibrational overtone and combination states in polyatomic molecules. I shall be mainly discussing the overtones of hydrogen stretching vibrations, with energies up to about 16 000 cm^{-1}, corresponding to about one third of the energy necessary to dissociate the bond, and I shall be considering the effect of the bond-dissociation type of anharmonicity on the vibrational wavefunctions and the pattern of the vibrational energy levels. The concept of "local modes" will be central to my talk.

I shall first discuss briefly the experimental techniques involved. I shall then review the effects of bond dissociation anharmonicity in a diatomic molecule. Next I shall introduce the idea of local modes with a simple classical model, and then extend this to a mathematically defined quantum mechanical model which I shall discuss in detail for the case of two symmetry related stretching vibrations, as in the water molecule. I shall then introduce the effects of Fermi resonance, and describe some of our recent work on the dichloromethane molecule. I shall also describe similar fits to the overtones of carbonyl stretching vibrations in metal carbonyls. Finally I shall comment briefly on the implications of this work for intramolecular vibrational relaxation (IVR) and chemical dynamics.

1.   EXPERIMENTAL TECHNIQUES

My talk will not be technique-oriented, but it is appropriate to mention briefly how high overtone spectra are observed. They are of course much weaker than the fundamental vibrations when they are observed directly by the methods of infrared or Raman spectroscopy. However they may be observed as conventional infrared absorption spectra

A. C. P. Alves et al. (eds.), Frontiers of Laser Spectroscopy of Gases, 461–489.
© 1988 by Kluwer Academic Publishers.

simply by using long pathlengths, provided an instrument is
available equipped to operate in the appropriate spectral
region. We have been using a Fourier Transform instrument
at Reading [1] (Nicolet 7199) with path lengths of up to
15 m for spectra up to 10 000 cm⁻¹; longer paths are
needed to get into the visible. We use a quartz
beamsplitter, an indium arsenide or a silicon detector, a
tungsten strip filament lamp source, and appropriate optical
filters.

An alternative technique that we have also been using at
Reading is laser photoacoustic spectroscopy [2]. A tunable
dye laser is operated with a fairly short absorption cell
(ca. 5 cm long with Brewster angle windows) inside the laser
cavity, equipped with a small microphone detector. The pump
beam is chopped and the microphone signal is detected with a
lock-in amplifier. The effective pathlength is Ql, where
Q is the quality factor of the cavity and l is the
pathlength of the cell. This effective pathlength may be
only a few metres, but the photoacoustic detection is so
sensitive that hydrogen stretching spectra at 16 000
cm⁻¹ (V = 6 - 0) are easily observed. A disadvantage of
this technique is that it is difficult to calibrate the
intensity of absorption. The method of laser photoacoustic
spectroscopy is illustrated diagrammatically in Figure 1.

Highly excited vibrational energy levels may also be
observed by other techniques, such as Raman spectroscopy
[3] or electronic spectroscopy, or the methods described by
Carrington in the preceding talk, but these methods are not
so generally applicable and have not yet been widely used.

2.   ANHARMONICITY IN A DIATOMIC MOLECULE

Figure 2 shows the potential curve and the vibrational
energy levels of a typical diatomic hydride molecule. The
curve shown in the diagram is actually a Morse curve, which
is a simple empirical function that is a good - but not a
perfect - approximation to the potential of real diatomic
molecules. The parameters in the diagram have been
approximately adjusted to fit the OH diatomic molecule (half
a water molecule, as it were). However the points to note
are general for any diatomic molecule. The spacing of the
energy levels closes up approximately arithmetically as the
vibrational energy increases and we approach dissociation.
The energy levels, measured from the minimum in the
potential, may be represented by the term formula

$$G(v) = \omega_m(v+\tfrac{1}{2}) + x_m(v+\tfrac{1}{2})^2 \qquad (1)$$

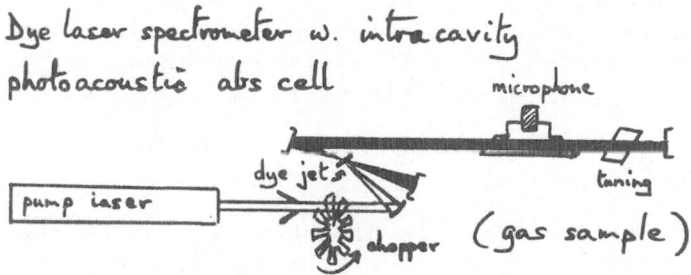

Dye laser spectrometer w. intra cavity
photoacoustic abs cell

microphone

dye jet's

pump laser

tuning

chopper   (gas sample)

Fig.1 Photoacoustic dye laser spectroscopy

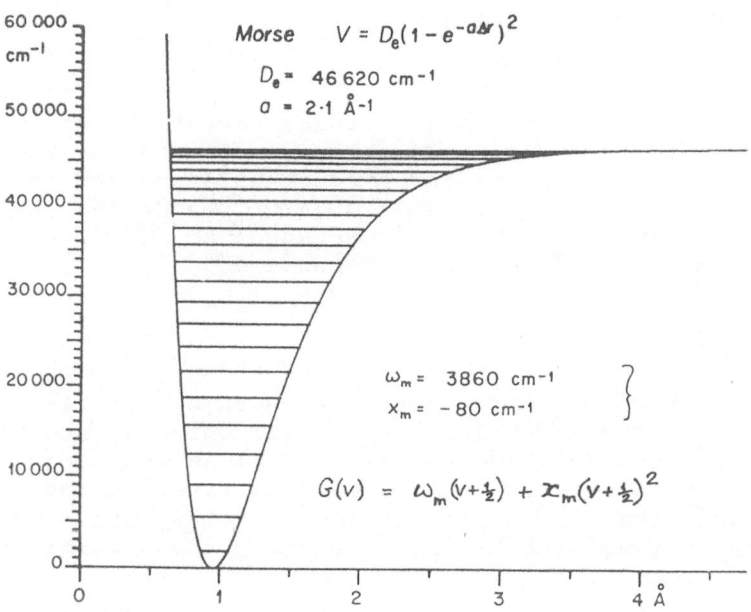

60 000
cm⁻¹

50 000

40 000

30 000

20 000

10 000

0

Morse $V = D_e(1 - e^{-a\Delta r})^2$

$D_e = 46\,620$ cm⁻¹
$a = 2.1$ Å⁻¹

$\omega_m = 3860$ cm⁻¹
$x_m = -80$ cm⁻¹

$G(v) = \omega_m(v + \tfrac{1}{2}) + x_m(v + \tfrac{1}{2})^2$

0   1   2   3   4 Å

Fig.2 The Morse function for a diatomic molecule

where the anharmonicity constant x_m is negative. For a
Morse potential this formula is exact (the subscript m is a
mnemonic for "Morse"), and it is easy to show that the
dissociation energy D_e is related to ω_m and x_m by
the formula

$$D_e = -\omega_m^2/4x_m \tag{2}$$

The vibrational frequency is directly related to the spacing
between successive energy levels, and thus the vibrational
frequency of a diatomic molecule <u>decreases as the
vibrational energy increases</u>. This important result may be
understood classically as follows: for each value of the
internuclear distance the restoring force is the gradient of
the potential curve at that point, the force being always
such as to return the bond length to its equilibrium value.
For high vibrational energy the gradient at the outer
turning point of the vibration gets less as the energy
increases, and we approach dissociation; hence the restoring
force gets less, and the vibrational motion is slow to turn
round at the outer turning point. The effect is that <u>the
frequency decreases as the energy increases</u>.

3. A SIMPLE CLASSICAL MODEL

Consider the hydrogen stretching vibrations of the
acetylene (ethyne) molecule, shown in Figure 3. The
symmetric stretch ν_1 is observed at 3374 cm^{-1} and the
antisymmetric stretch ν_3 at 3289 cm^{-1}. We define an
interbond coupling constant λ as equal to half the
splitting between the two fundamentals,

$$\nu_1 - \nu_3 = 2\lambda, \tag{3}$$

so that in this case $\lambda = 42.5$ cm^{-1}. Figure 3 also shows
the symmetric and antisymmetric stretching vibrations of the
diacetylene (butadiyne) molecule, for which ν_1 and ν_3
are both observed close to 3329 cm^{-1}, less than 1 cm^{-1} apart,
so that in this case $\lambda \approx 0$ cm^{-1}. This fits with our
intuition, that in the longer molecule the interbond
coupling between the two hydrogen stretching vibrations is
weaker.

Consider now the effect of isotopically substituting one
end of the molecule with a ^{13}C atom. It is easy to show
that in the case of the C–H diatomic the effect of ^{13}C
substitution is to lower the vibration wavenumber by about
10 cm^{-1}. For acetylene, the effect is that both ν_1 and ν_3
are lowered by 5 cm^{-1}, but they remain symmetric and
antisymmetric combinations of the two CH stretching

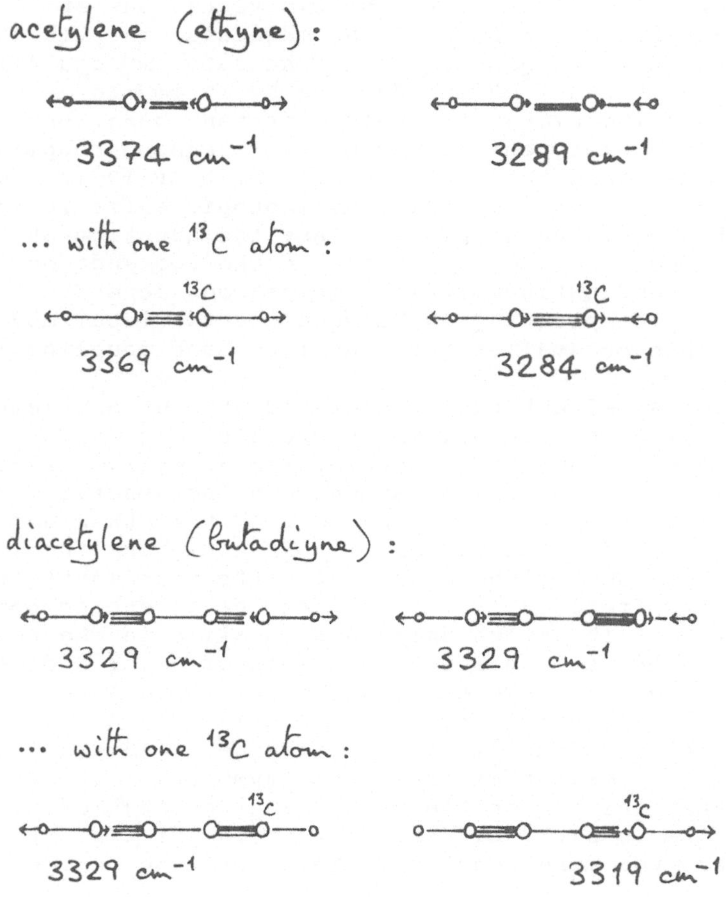
acetylene (ethyne):
3374 cm⁻¹
3289 cm⁻¹
... with one 13C atom:
13C
13C
3369 cm⁻¹
3284 cm⁻¹
diacetylene (butadiyne):
3329 cm⁻¹
3329 cm⁻¹
... with one 13C atom:
13C
13C
3329 cm⁻¹
3319 cm⁻¹

Fig.3 The effect of ^{13}C isotopic substitution on the vibration wavenumbers of acetylene and diacetylene

vibrations respectively. For diacetylene the effect is different (see Figure 3): the two vibrations split and localize at the two ends of the molecule, the vibration at the ^{12}CH end remaining unchanged at 3329 cm^{-1} and the vibration at the ^{13}CH end falling by 10 cm^{-1} to 3319 cm^{-1}. The reason for this different behaviour is that in the acetylene molecule the interbond coupling λ is much larger than the ^{13}C isotopic shift $\Delta\nu(13-12)$, whereas in the diacetylene molecule the isotopic shift is larger than the interbond coupling. Thus the question as to whether the vibrations localize in the two ends or remain as symmetric and antisymmetric combinations depends on whether the isotopic splitting $\Delta\nu$ between the two local modes is larger than or smaller than the interbond coupling λ.

A similar effect can be achieved without any isotopic substitution, by vibrationally exciting one end of the molecule only, and thus reducing its effective vibrational wavenumber through the anharmonicity (as described above for a diatomic molecule). If the bond at the other end is relatively unexcited so that it has a higher vibrational wavenumber, and if the anharmonic effect is sufficient to make the difference $\Delta\nu$ larger than the interbond coupling λ, then the vibrations split and localize in the two ends of the molecule. This is the origin of local mode effects in molecules with symmetry, where one would intuitively expect the vibrations to occur as symmetric and antisymmetric normal modes. It is the vibrational excitation itself that breaks the symmetry. It should be noted that the higher the vibrational excitation, and the bigger the possible difference between the excitation at the two ends, the more localized the vibrations become.

4. A SIMPLE QUANTUM MECHANICAL MODEL

The following quantum mechanical hamiltonian reproduces the same effects described above with a classical model. We consider a zeroth order hamiltonian defined as the sum of a set of independent harmonic oscillators in a set of symmetrically related X-H bonds:

$$H_o = \sum_i (\tfrac{1}{2}g_{rr}p_{ri}^2 + \tfrac{1}{2}f_{rr}\Delta r_i^2) \tag{4}$$

The two terms represent kinetic and potential contributions to the hamiltonian respectively, g_{rr} and f_{rr} are the appropriate diagonal g and f matrix elements [4] for each X-H vibrational coordinate Δr_i, and p_{ri} is the momentum conjugate to r_i (i.e. $p_{ri} = -i\hbar\partial/\partial r_i$). It is assumed that symmetry requires each bond coordinate to have the same g and f matrix element. We now add two

perturbing effects. The first, H_1, is the interbond
coupling due to off-diagonal g matrix and f matrix elements
coupling pairs of X-H vibrational coordinates, $g_{rr'}$ and $f_{rr'}$,

$$H_1 = \sum_{i<j} (g_{rr'}P_{ri}P_{rj} + f_{rr'}\Delta r_i \Delta r_j) \tag{5}$$

The second, H_2, is a sum of terms that represent the
effect of bond dissociation anharmonicity associated with
each bond oscillator. The effect of H_2 is to convert
the harmonic potential associated with each bond oscillator
in (4) into a Morse potential. It may be written as an
appropriate sum of cubic, quartic, and higher anharmonic
terms in the vibrational coordinates Δr_i:

$$H_2 = \sum_i [(1/6)f_{rrr}\Delta r_i^3 + (1/24)f_{rrrr}\Delta r_i^4 + \dots] \tag{6}$$

where it is to be understood that the anharmonic force
constants f_{rrr} and f_{rrrr} etc. are related to the
quadratic force constant f_{rr} by the appropriate Morse
relations [5]

$$f_{rrr} = -3af_{rr} \quad \text{and} \quad f_{rrrr} = +7a^2 f_{rr} \tag{7}$$

Here a is the parameter in the exponent of the Morse
function

$$V(\text{morse}) = D_e[1 - \exp(-ar)]^2 \tag{8}$$

The sum of the terms in (4), (5) and (6) define a simple
hamiltonian for the hydrogen stretching vibrations. Its
limitations will be considered later.

To calculate energy levels and wavefunctions it is
convenient to diagonalize the matrix of this hamiltonian in
an appropriate set of basis functions. If there are N
equivalent vibrational coordinates (where N is the number of
terms in the sum in (4); N=2 for H_2O, N=3 for NH_3,
etc.) then the basis functions must span N dimensional
coordinate space. There are two obvious choices for the
basis functions. The first, called the local mode basis, is
to take them as products of one-dimensional oscillator
functions in the individual bond coordinates Δr_i; and
the second, called the normal mode basis, is to take them as
products of one dimensional oscillator functions in symmetry
combinations of the Δr_i. Either choice gives basis
functions which are eigenfunctions of H_o in (4), but they
represent different linear combinations of the degenerate
sets corresponding to a given total vibrational excitation.
In either case off-diagonal matrix elements arise from the
perturbations in (5) and (6). For the local mode basis we
may label the basis functions with the number of quanta in

each bond oscillator,

[m, n, ...]   – local mode basis,   (9)

and for the normal mode basis with the number of quanta in
each normal mode oscillator,

$(v_1, v_2, ...)$ – normal mode basis.   (10)

In either case the eigenvalues of the resulting hamiltonian
matrix give the energy levels, and the eigenfunctions give
the wavefunctions in terms of the appropriate basis
functions.

Further approximations are made in the usual application
of this model [6,7]. If we define a total excitation
quantum number V as the sum of quanta excited in each basis
function,

$$V = m + n + ... = v_1 + v_2 + ...$$ (11)

then matrix elements off diagonal in V are neglected, so
that the hamiltonian matrix is factorised in V. Also all
matrix elements are calculated using formulae derived from
harmonic basis functions, and the bond dissociation
anharmonicity is treated in the same way as it would be for
a Morse function. The approximations involved, and the
formulae for the matrix elements, are considered more fully
by Mills and Robiette [7], where it is shown that the
effects of H_1 may be summarised in terms of a parameter
λ which we describe as the interbond coupling constant,
and the effects of H_2 may be summarized in terms of a
parameter x_m which is the usual Morse anharmonicity
constant. The interbond coupling λ is related to the off
diagonal f and g matrix elements by the equation

$$\lambda = \tfrac{1}{2}\omega_m[(f_{rr'}/f_{rr}) + (g_{rr'}/g_{rr})]$$ (12)

The energy levels obtained from this model are
illustrated in the form of a correlation diagram in Figure
4. The energy levels on the left are obtained by setting
the anharmonicity x_m to zero, i.e. they are the eigenvalues
of $H_0 + H_1$ only. The corresponding matrix would be diagonal
in the normal mode basis functions. Similarly the energy
levels on the right are obtained by setting λ to zero,
i.e. they are the eigenvalues of $H_0 + H_2$ only. The
corresponding matrix would be diagonal in the local mode
basis functions. The true energy levels are shown in the
centre of the diagram.

For low values of the total vibrational energy, and in

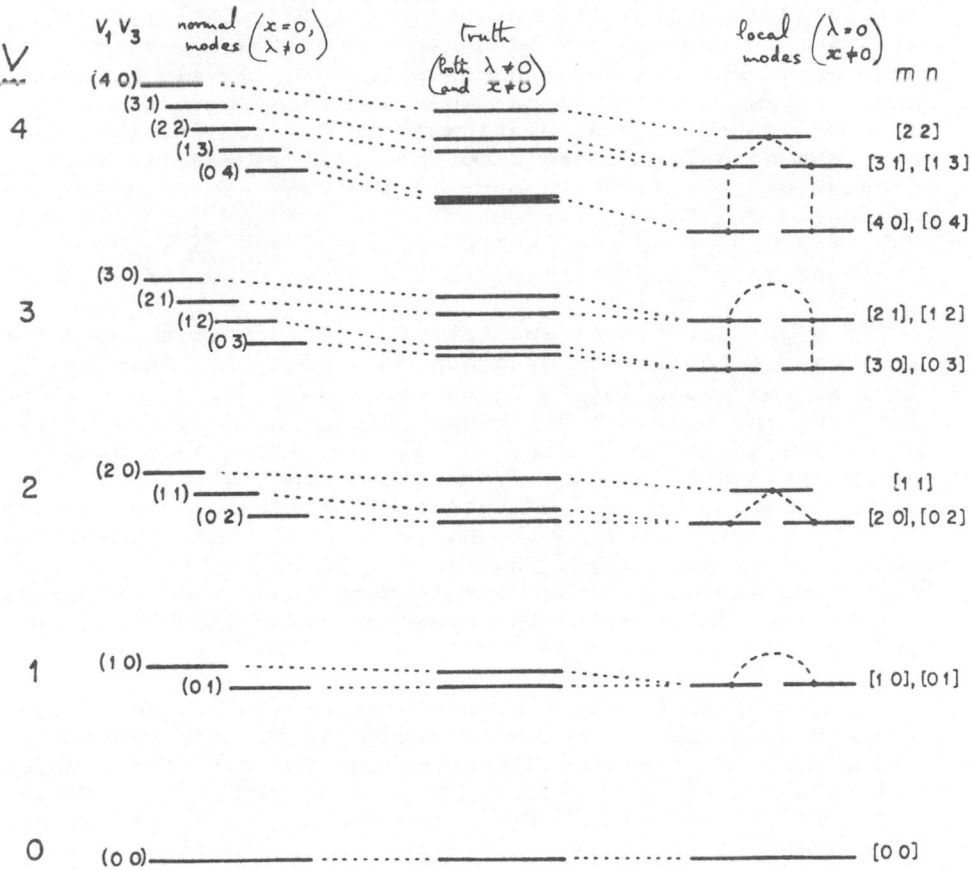

Fig.4   Correlation of the energy levels of two coupled oscillators
between the normal mode and local mode limits

particular for the fundamental vibrational excited states
usually observed in mid-infrared and Raman spectroscopy, the
pattern of vibrational energy levels is determined by the
interbond coupling effects, as on the left hand side of
Figure 4.   However for high vibrational energies the
anharmonic effects win and the pattern of levels becomes
characteristic of the local mode limit on the right hand
side of Figure 4.   Mathematically the reason for this switch
from the normal mode basis to the local mode basis with
increasing vibrational energy is that the anharmonic matrix
elements, due to H_2, increase in proportion to V^2, whereas
the interbond coupling matrix elements, due to H_1,
increase only in proportion to the first power of V.

At high vibrational excitation, for each value of V there
is always a lowest energy group of N nearly degenerate
levels, corresponding to V quanta of excitation in one bond
and none in the other N-1 bonds, represented by the local
mode wavefunctions [V,0,0, ...].   The stationary state
wavefunctions are always symmetrized combinations of this
group of N functions, but because of their near degeneracy
the individual basis functions [V,0,0, ...] are themselves
nearly stationary state functions.   These represent truly
localized excitation in a single bond, and these states play
a key role in the molecular dynamics of excited vibrational
states.

The observed OH stretching vibrational overtone levels of
the H_2O molecule are shown in Figure 5, to illustrate the
pattern of the transition from normal modes to local modes.
One should note that already at V=3 the two lowest levels
are only 13 cm⁻¹ apart, with a gap of 250 cm⁻¹ up to the next
level; at V=4 the lowest pair are only 2 cm⁻¹ apart with a
gap of 400 cm⁻¹ to the next level.   This is characteristic
of the local mode pattern, and it would already be
inappropriate to label the levels at V=3 as $3\nu_1$, $2\nu_1+\nu_3$,
$\nu_1+2\nu_3$, and $3\nu_3$ (in ascending order), because this
implies normal mode wavefunctions.   An alternative and more
appropriate labelling of the levels is illustrated in the
Figure 5.   Even this notation becomes somewhat clumsy when
extended to more complicated molecules like NH_3, CH_4,
and C_6H_6; a tidy and appropriate notation for local mode
states has still to be found.

A further comment must be made about the representation
of bond dissociation anharmonicity in the normal mode basis.
The effects of anharmonicity on the normal modes of a
polyatomic molecule are usually described by the expression

$$G(v_r, \ldots) = \sum_r \omega_r (v_r + \tfrac{1}{2}) + \sum_{r \leq s} x_{rs} (v_r + \tfrac{1}{2})(v_s + \tfrac{1}{2}) \qquad (13)$$

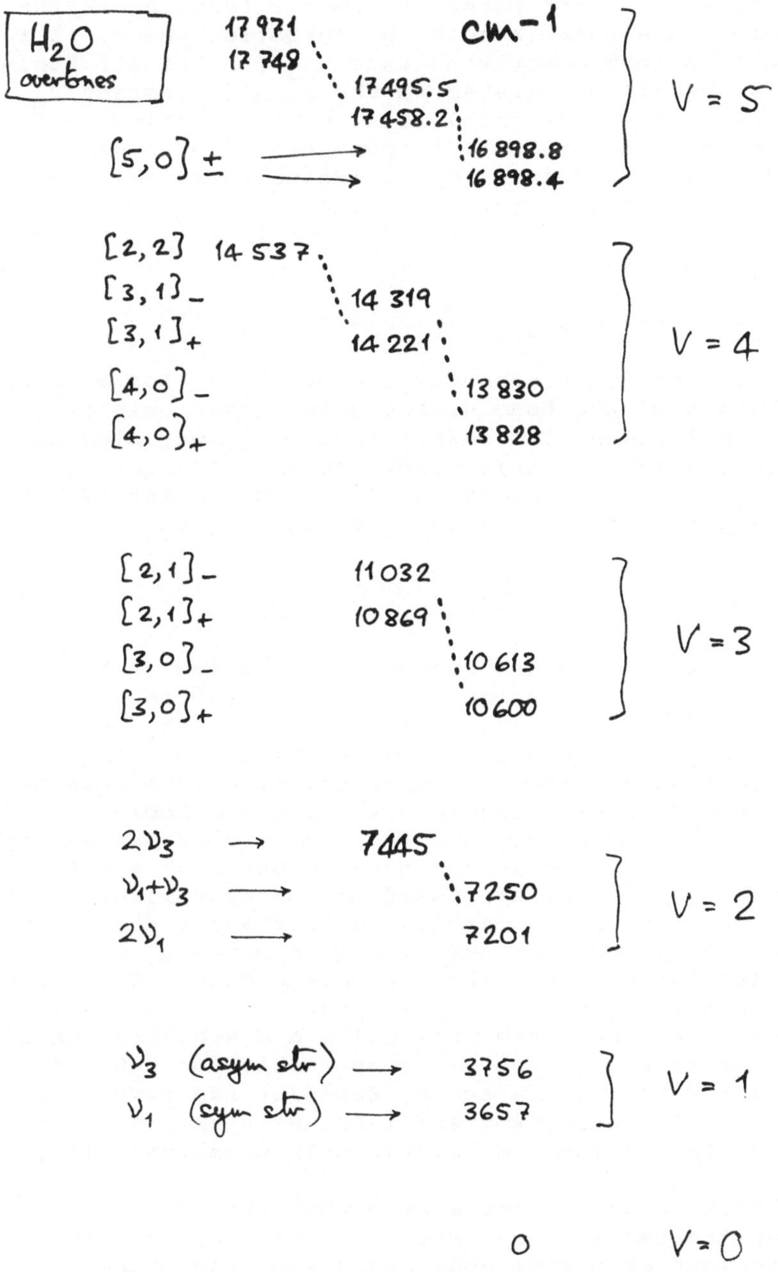

H_2O overtones

17 971
17 748
cm^{-1}
17 495.5
17 458.2
$[5,0] \pm$ ⟶ 16 898.8
⟶ 16 898.4
$\left.\right\}$ $V = 5$

$[2,2]$ 14 537
$[3,1]_-$ 14 319
$[3,1]_+$ 14 221
$[4,0]_-$ 13 830
$[4,0]_+$ 13 828
$\left.\right\}$ $V = 4$

$[2,1]_-$ 11 032
$[2,1]_+$ 10 869
$[3,0]_-$ 10 613
$[3,0]_+$ 10 600
$\left.\right\}$ $V = 3$

$2\nu_3$ ⟶ 7445
$\nu_1 + \nu_3$ ⟶ 7250
$2\nu_1$ ⟶ 7201
$\left.\right\}$ $V = 2$

ν_3 (asym str) ⟶ 3756
ν_1 (sym str) ⟶ 3657
$\left.\right\}$ $V = 1$

0    $V = 0$

Fig.5 Observed hydrogen stretching overtones with
normal mode and local mode labels for H_2O

When the effects of the Morse anharmonic terms H_2 in the present model are calculated by perturbation theory it is found that the anharmonic constants x_{rs} for the stretching vibrations can all be related to the single constant x_m appropriate to the (diatomic) local bond vibration, and it is also found necessary to introduce the effects of Darling-Dennison resonance [8] by including off-diagonal matrix elements of the type $\Delta v_r = \pm 2$, $\Delta v_s = \mp 2$,

$$<v_r+2,v_s-2|H/hc|v_r,v_s>$$

$$= (1/4)K_{rrss}[(v_r+1)(v_r+2)v_s(v_s-1)]^{\frac{1}{2}} \qquad (14)$$

The Darling-Dennison resonances are essential for a correct representation of the bond dissociation anharmonicity in polyatomic molecules. The coefficient K_{rrss} may also be related to the anharmonicity constant x_m. For two interacting bond oscillators, as in the H_2O molecule (or for the CH stretching vibrations of CH_2Cl_2), the relations are

$$x_{11} = (1/4)x_{13} = x_{33} = (1/4)K_{1133} = x_m/2 \qquad (15)$$

Relations of this kind are conveniently called "x,K relations" [9]. They become more complicated for three or more interacting bond vibrations (as in NH_3, CH_4, etc.), where indeed there are a number of different Darling-Dennison resonances; the corresponding relations have been discussed for various molecules by Mills and Robiette [7] and Mills and Mompean [9], and for the case of benzene by Lehmann [10]. They are approximate rather than exact relations, because they are based on the approximation that the only important anharmonicity is that associated with bond stretching, and that the mixing of bending coordinates into the stretching vibrations is unimportant. Nonetheless they are remarkably successful in practice, as shown by the results in Figure 6 (taken from Mills and Robiette) which compares the observed anharmonic constants for H_2O and H_2S determined many years ago by Benedict and coworkers [11] and by Allen and coworkers [12], with values constrained to fit the x,K relations in equations (15).

Finally it should be emphasized that the two ways of setting up the hamiltonian matrix, in either local mode basis functions or normal mode basis functions, are mathematically exactly equivalent; they are two different representations of the same problem. This is emphasized by Figure 7, which shows as an example the two matrices for V = 4 (with the x,K relations already imposed in the normal mode matrix). The only parameters appearing are ω_m, x_m, and λ. The two matrices have identical eigenvalues. Their

Test of the x, K relations for H_2O and H_2S

$$x_{11} = \tfrac{1}{4} x_{13} = x_{33} = \tfrac{1}{4} K_{1133} = \tfrac{1}{2} x_m$$

		H_2O		D_2O	
	cm^{-1}	calc	obs	calc	obs
H_2O	x_{11}	-42.0	-42.6	-22.2	-22.6
$x_m = -84$ cm^{-1}	x_{13}	-168.0	-165.8	-88.9	-87.1
	x_{33}	-42.0	-47.6	-22.2	-26.1
	K_{1133}	-168.0	-155.0	-88.9	-81.2

		H_2S		D_2S	
	cm^{-1}	calc	obs	calc	obs
H_2S	x_{11}	-24.0	-25.1	-12.4	-
$x_m = -48$ cm^{-1}	x_{13}	-96.0	-94.7	-49.5	-
	x_{33}	-24.0	-24.0	-12.4	-
	K_{1133}	-96.0	-91.4	-49.5	-

Fig.6   A test of the x,K relations

474

Normal mode basis H_2O,   V = 4

$(4,0)$	$(3,1)$	$(2,2)$	$(1,3)$	$(0,4)$
$4\omega+14x+4\lambda$	0	$\sqrt{6}\,x$	0	0
0	$4\omega+17x+2\lambda$	0	$3x$	0
$\sqrt{6}\,x$	0	$4\omega+18x$	0	$\sqrt{6}\,x$
0	$3x$	0	$4\omega+17x-2\lambda$	0
0	0	$\sqrt{6}\,x$	0	$4\omega+14x-4\lambda$

Local mode basis H_2O,   V = 4

$[4,0]$	$[3,1]$	$[2,2]$	$[1,3]$	$[0,4]$
$4\omega+20x$	2λ	0	0	0
2λ	$4\omega+14x$	$\sqrt{6}\,\lambda$	0	0
0	$\sqrt{6}\,\lambda$	$4\omega+12x$	$\sqrt{6}\,\lambda$	0
0	0	$\sqrt{6}\,\lambda$	$4\omega+14x$	2λ
0	0	0	2λ	$4\omega+20x$

Fig.7   The hamiltonian matrices for H_2O, V=4, in the two alternative sets of basis functions

eigenvectors differ by the transformation that relates the two sets of basis functions. In the case of two equivalent oscillators this transformation is in fact identical to the transformation between the v_x, v_y and the v, l basis functions for a two dimensional harmonic oscillator, as has been shown by Lehmann [13] and by Kellman [14]. For three or more symmetrically related bond oscillators the transformation is more complex, but the two representations remain exactly equivalent [7,14].

5.   VIBRATIONAL BAND INTENSITIES

Although it is not possible to discuss here all aspects of vibrational overtone spectroscopy, some mention of the problem of understanding band intensities is required. Intensities are related to understanding the dipole moment surface as a function of the vibrational coordinates, and in general this is a complex problem. However two facts stand out from the experimental point of view. The first is that the intensities of V - 0 electric dipole transitions from the ground vibrational state drop rapidly with increasing V, and that the most intense bands are almost invariably associated with the overtones of hydrogen stretching vibrations. The second is that even amongst the hydrogen stretching overtones only a few of the many possible bands appear to carry appreciable intensity: the most intense bands are invariably the lowest energy bands of the group for each value of V, corresponding to a truly localized excitation, represented by the basis functions [V,0,0, ...]

Our understanding of these observations must await a more detailed analysis of the dipole moment function, but an approximate understanding may be obtained simply by noting that if the dipole moment could be regarded as a sum of essentially independent bond contributions, each of which is a (slowly varying) function of the bond length, then in the local mode basis only the truly localized functions, in which all the excitation is in one bond, will have a non-zero dipole transition moment to the ground state. The mixing of the local mode basis functions by the interbond coupling terms redistributes the intensity among all the eigenstates that have strong contributions from these basis functions, and this model appears to be reasonably successful in reproducing the observed relative intensities amongst the set of levels of a given V.

6. EFFECTS OF FERMI RESONANCE, AND RESULTS ON DICHLOROMETHAN

It is a familiar observation in vibrational spectroscopy that a hydrogen bending or deformation vibration has a fundamental frequency approximately half that of the hydrogen stretching modes. It may then happen that cubic anharmonic interaction terms in the potential couple the stretching mode to the overtone of the bending mode. The terms concerned will be of the type

$$V(\text{anharmonic}) \;=\; k_{sbb}q_s q_b^2 \tag{16}$$

where s denotes a stretching and b a bending vibration. The effect, known as Fermi resonance, leads to a mixing of the corresponding wavefunctions and a pushing apart of the corresponding energy levels. The effect occurs not only at the fundamental level, ν_s with $2\nu_b$, but it is repeated with increasing strength between overtones, where the interacting states form Fermi polyads of increasing dimension at the higher levels of excitation (for example the states $2\nu_s$, $\nu_s+2\nu_b$, and $4\nu_b$ form an interacting Fermi triad). For this simplest example the situation is illustrated in Figure 8.

The Fermi resonance coupling matrix elements increase with increasing excitation. At the same time the degree of near degeneracy between the "unperturbed" states that are coupled together may also change. For example it may be that the ν_s level is appreciably higher in energy than the $2\nu_b$ level, but owing to the rapid convergence of stretching overtones compared with bending overtones it may be that $4\nu_s$ is lower than $3\nu_s+2\nu_b$. Such an effect is shown diagrammatically in Figure 8.

Dichloromethane is a molecule which shows exactly these effects, and Figures 9 to 12 show spectra of this molecule that we have taken recently at Reading [15,16]. The two stretching vibrations involved are associated with the two C-H bonds, which give rise to the fundamental bands ν_1(type b, A_1 species in C_{2v}) and ν_6(type c, B_1 species); the deformation mode involved in the Fermi resonance is the CH_2 scissoring mode ν_2(type b, A_1 species). We have modelled the problem by extending the vibrational hamiltonian matrices described in Section 4 to include all the Fermi polyads appropriate to each value of the total vibrational excitation quantum number $V = V(\text{stretch}) + \tfrac{1}{2}V(\text{bend})$. In the normal mode basis the off diagonal matrix elements arise from both Darling-Dennison interactions (K_{1166}) and from Fermi resonance interactions (k_{122}); in the local mode basis they arise from interbond coupling (λ) and Fermi resonance. Once again the

Fermi resonance ν_{str} with $2\nu_{bend}$

... the effect of differing anharmonicity

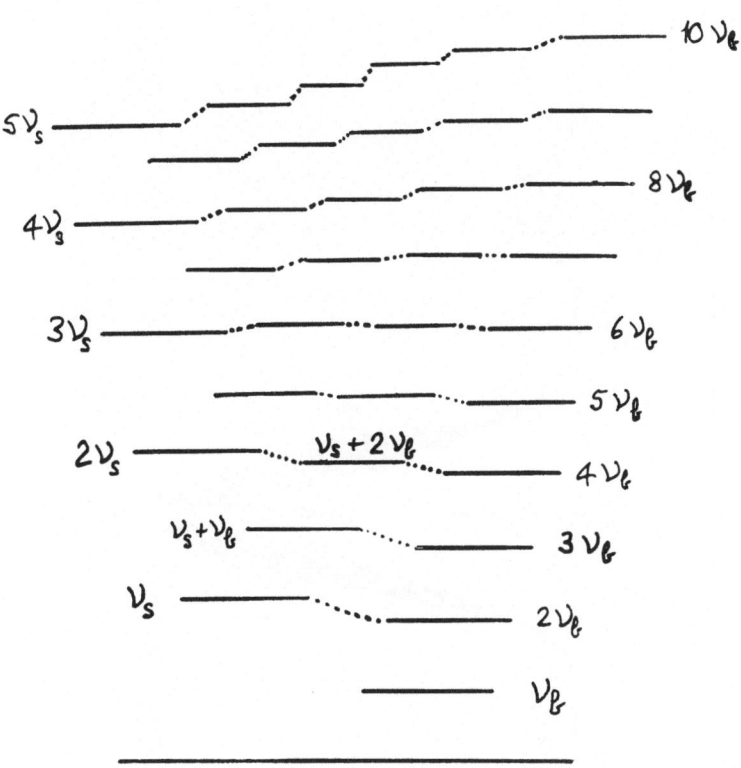

Fig.8   The effects of Fermi resonance

478

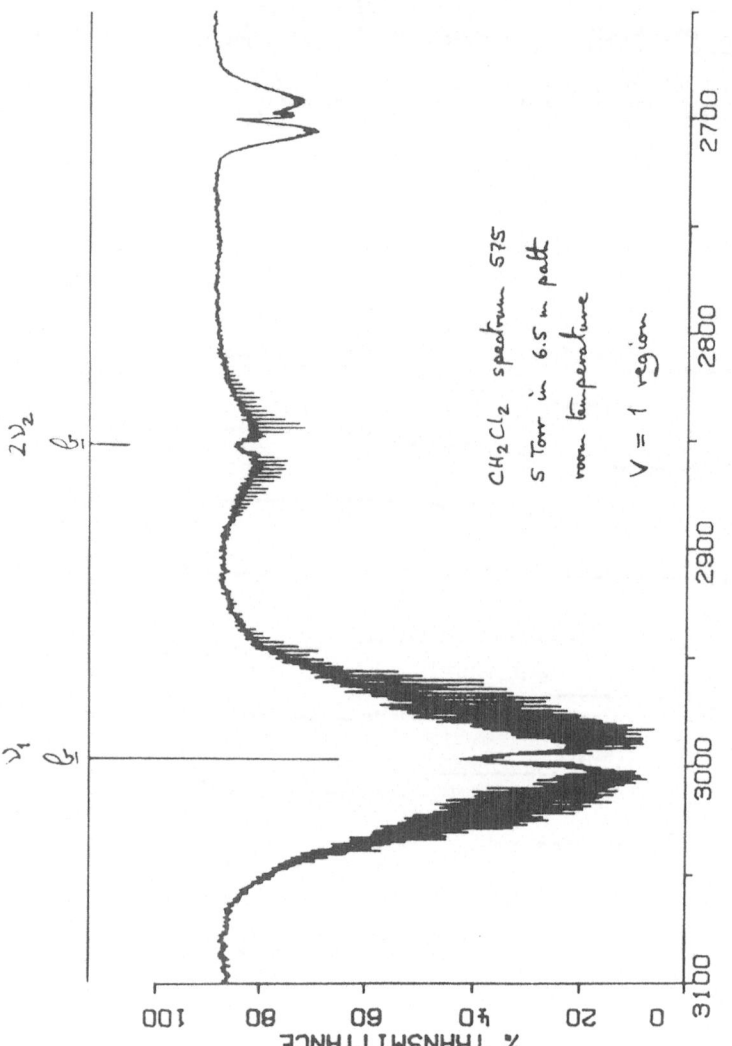

Fig.9 Observed and calculated spectra of CH_2Cl_2 in the V=1 region

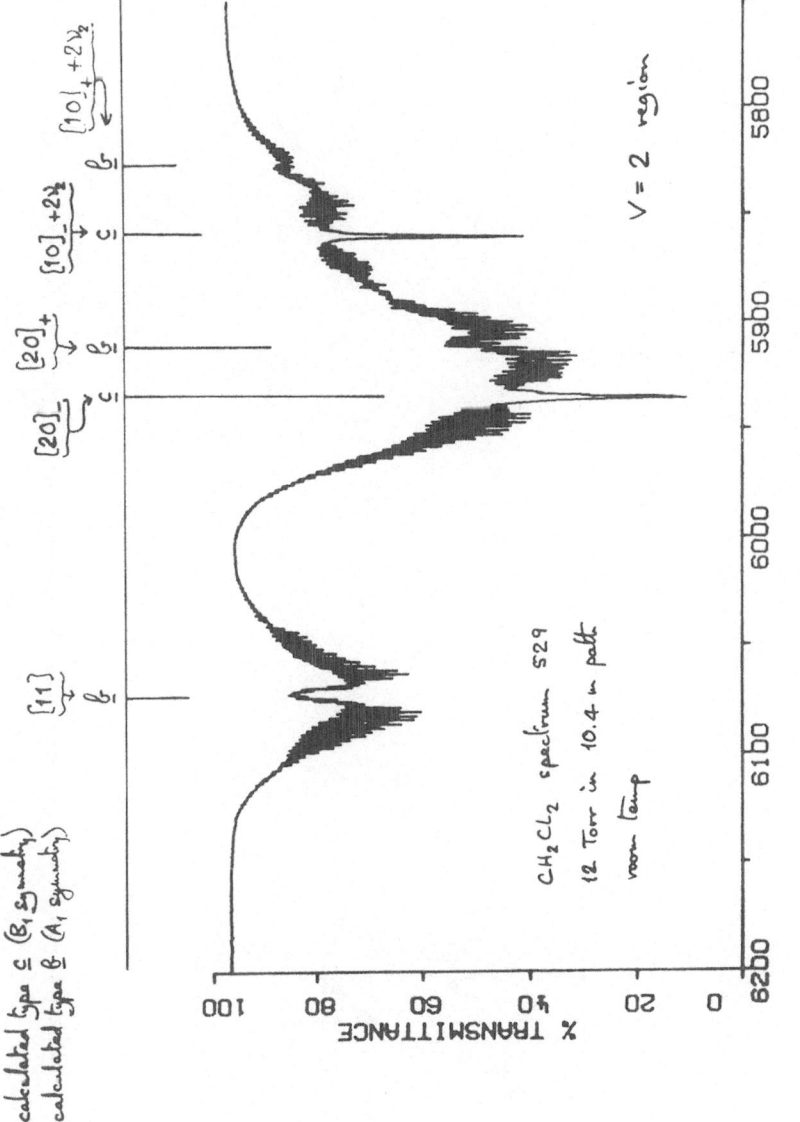

Fig.10 Observed and calculated spectra of CH_2Cl_2 in the V=2 region

480

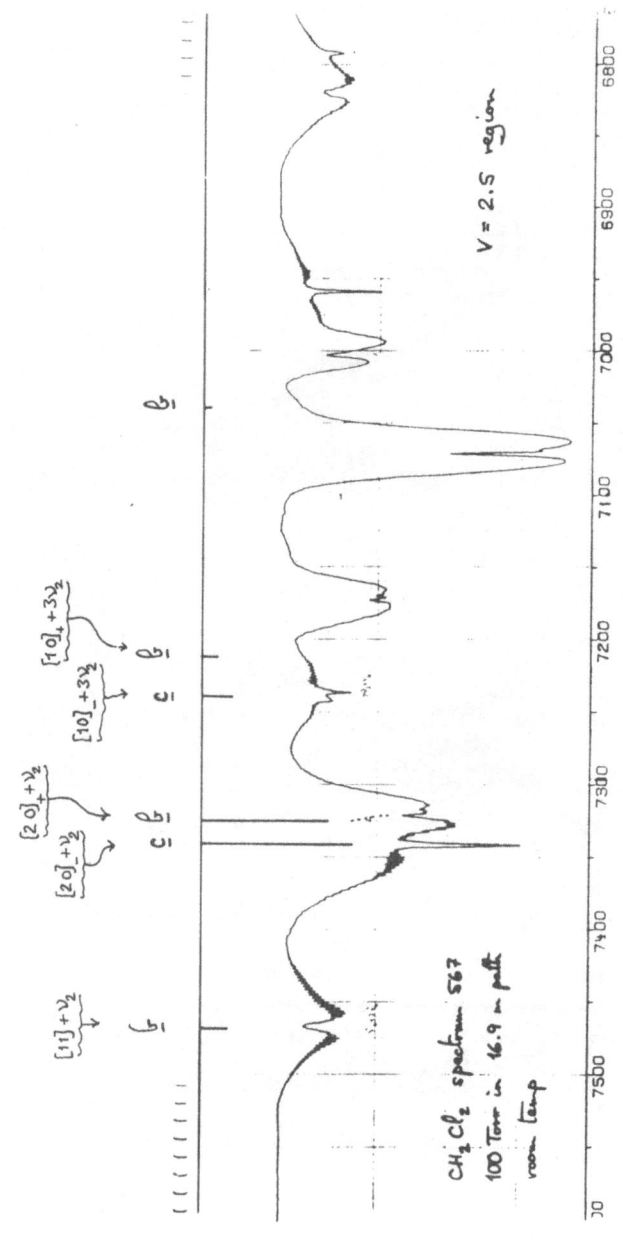

Fig.11 Observed and calculated spectra of CH_2Cl_2 in the V=2.5 region

481

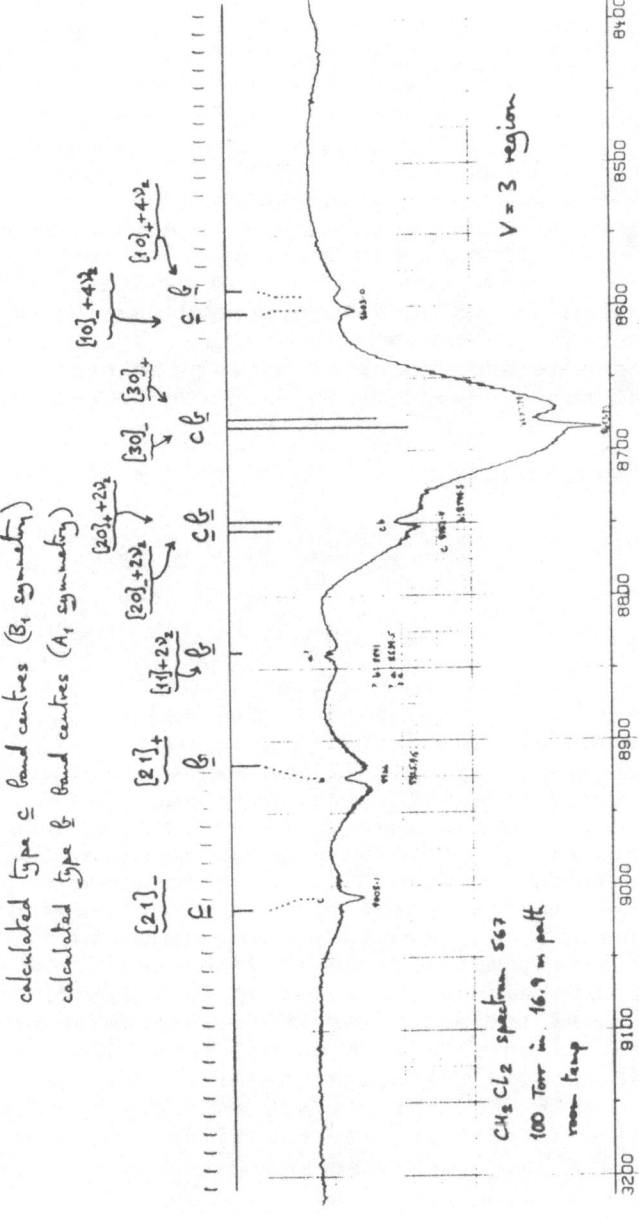

Fig.12 Observed and calculated spectra spectra of CH_2Cl_2 in the V=3 region

two alternative basis sets give identical results.

We have fitted 18 observed bands with a six parameter model, and obtained a fit to the observed band centres good to a few cm^{-1} in every case. The calculated band centres are shown alongside the observed bands that appear in Figures 9 to 12, and the parameters obtained from the least squares fit are shown in Figure 13. Figure 14 shows the vibrational hamiltonian matrices for V=2, in the normal mode basis functions, by way of example. Note that the Fermi interacting levels cross over between V=1 and V=3. At V=1, around 3000 cm^{-1}, ν_1 and ν_6 are more than 100 cm^{-1} above $2\nu_2$; but at V=3, around 8700 cm^{-1}, the $[3,0]_\pm$ levels are about 70 cm^{-1} below the $[2,0]_\pm + 2\nu_2$ levels. The labels are determined by examining the eigenvectors for the dominant basis functions in each eigenfunction.

7. NICKEL TETRACARBONYL

To provide one further illustration of the way in which this model works, and to show that the model is not restricted to a discussion of hydrogen stretching vibrations, we may consider the overtones of the carbonyl stretching vibrations of the nickeltetracarbonyl molecule, $Ni(CO)_4$. The carbonyl stretching normal modes are ν_1 (A_1 species in T_d, singly degenerate) and ν_5 (F_2 species, triply degenerate). The necessary x,K relations have been given by Mills and Robiette [7]; they relate the seven anharmonic and Darling-Dennison constants x_{11}, x_{15}, x_{55}, G_{55}, T_{55}, K_{1155}, and K_{1555} to the single anharmonic constant x_m. There are no detectable Fermi resonances since all the other fundamental vibrations are at least a factor 3 lower in frequency. Thus the entire carbonyl overtone spectrum can be modelled in terms of only 3 parameters, namely ω_m, x_m, and λ. The results of fitting the 12 observed overtones up to V=4 [17] are shown in Figures 15 and 16; as for dichloromethane an excellent fit is obtained. A normal mode labelling has been adopted in Figure 16, although the mixing of the basis functions is such that this is not entirely appropriate. It should also be noted that only the F_2 species vibrational overtone levels have been observed, because only these are dipole active in combination with the ground vibrational state.

8. DISCUSSION

The conclusions of this review are that local mode effects occur in overtones of stretching vibrations in

dichloro
methane

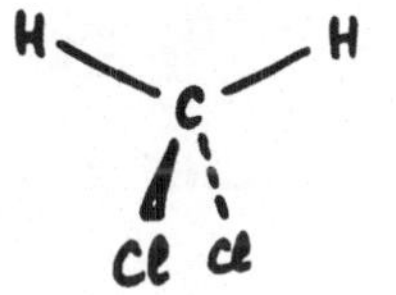

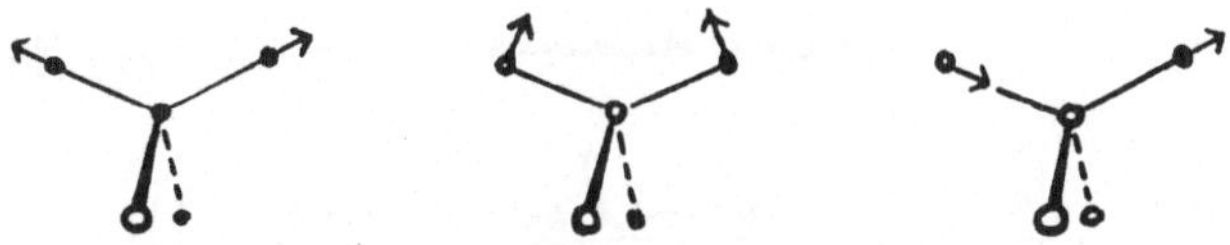

ν_1 ν_2 ν_6

Baggott, Caldow, Law and Mills (to be published)
observe and fit 18 bands, up to 9000 cm^{-1},
with a local mode + Fermi resonance model [rms deviation of 1.9 cm^{-1}]
obtaining the following parameters:

CH stretch: ω_m = 3158.1 cm^{-1}

x_m = -62.0 cm^{-1}

λ = -30.3 cm^{-1}

$$\left\{ \begin{array}{l} \omega_1 = \omega_m + \lambda \ , \quad \omega_6 = \omega_m - \lambda \\ x_{11} = \frac{1}{4} x_{16} = x_{66} = \frac{1}{4} K_{1166} = \frac{1}{2} x_m \end{array} \right\}$$

CH$_2$ scissor: ω_2 = 1463.1 cm^{-1}

x_{22} = -3.9 cm^{-1}

x_{12} = x_{26} = -21.3 cm^{-1}

Fermi resonance: k_{122} = 50.1 cm^{-1}

Fig.13   Parameters in the vibrational hamiltonian for CH$_2$Cl$_2$

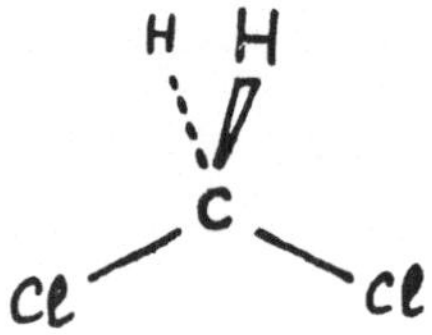

$\nu_1 = CH_2$ sym str

$\nu_2 = CH_2$ α-bend (scissor)

$\nu_6 = CH_2$ asym str

Fermi resonance: $V/hc = k_{122} q_1 q_2^2$

Matrix for $V=2$ in normal mode basis $|\nu_1, \nu_6 ; \nu_2\rangle$

A_1 matrix

	$2\nu_1$	$2\nu_6$	$\nu_1 + 2\nu_2$	$4\nu_2$
	$|2,0;0\rangle$	$|0,2;0\rangle$	$|1,0;2\rangle$	$|0,0;4\rangle$
	$G(2\nu_1)$	K_{1166}	$\frac{1}{\sqrt{2}}k_{122}$	0
		$G(2\nu_6)$	0	0
			$G(\nu_1+2\nu_2)$	$\frac{\sqrt{6}}{2}k_{122}$
				$G(4\nu_2)$

B_1 matrix

	$\nu_1+\nu_6$	$\nu_6+2\nu_2$
	$|1,1;0\rangle$	$|0,1;2\rangle$
	$G(\nu_1+\nu_6)$	$\frac{1}{2}k_{122}$
		$G(\nu_6+2\nu_2)$

Fig.14 Hamiltonian matrices for V=2 in CH_2Cl_2

$$\boxed{Ni \, (CO)_4}$$

parameters :
$$\omega_m = 2090.7 \ (1.0) \ cm^{-1}$$
$$x_m = -12.9 \ (0.4) \ cm^{-1}$$
$$\lambda = +20.2 \ (0.2) \ cm^{-1}$$

$$\omega_1 = \omega_m + 3\lambda \qquad \omega_5 = \omega_m - \lambda$$

$$\left. \begin{aligned} x_{11} &= \tfrac{1}{4} x_{15} = \tfrac{5}{9} x_{55} \\ &= -\tfrac{5}{3} G_{55} = -5 T_{55} \\ &= \tfrac{1}{4} K_{1155} = \tfrac{1}{16} K_{1555} \end{aligned} \right\} = \tfrac{1}{4} x_m$$

Fig.15 Parameters in the vibrational hamiltonian
determined by a least squares fit to the overtone
levels of nickeltetracarbonyl

Ni(CO)$_4$			obs	calc	obs−calc
V=1	ν_1	A_1	2125.0	2125.4	−0.4
	ν_5	F_2	2044.5	2044.6	−0.1
V=2	$2\nu_1$	A_1		4244.9	
	$2\nu_5$	E		4088.6	
	$2\nu_5$	A_1		4069.2	
	$\nu_1+\nu_5$	F_2	4159.4	4159.2	+0.2
	$2\nu_5$	F_2	4073.3	4074.3	−1.0
V=3	$3\nu_1$	A_1		6358.8	
	$\nu_1+2\nu_5$	E		6188.8	
	$\nu_1+2\nu_5$	A_1		6179.3	
	$3\nu_5$	A_1		6084.3	
	$2\nu_1+\nu_5$	F_2	6269	6267.7	+1.3
	$\nu_1+2\nu_5$	F_2	6185	6181.0	+4.0
	$3\nu_5$	F_2(a)	6123	6121.0	+2.0
	$3\nu_5$	F_1		6108.1	
	$3\nu_5$	F_2(b)	6078	6070.1	−1.1
V=4	$3\nu_1+\nu_5$	F_2	8377	8370.3	+6.7
	$2\nu_1+2\nu_5$	F_2	8275	8280.2	−5.2
	$\nu_1+3\nu_5$	F_2(a)		8207.9	
	$\nu_1+3\nu_5$	F_2(b)	8184	8184.6	−0.6
	$4\nu_5$	F_2(a)		8115.5	
	$4\nu_5$	F_2(b)	8063	8064.8	−1.8

Fig.16 Observed and calculated vibrational overtones for Ni(CO)$_4$, in cm^{-1}

polyatomic molecules whenever bond-dissociation anharmonicity dominates interbond coupling effects between symmetrically related bond stretching coordinates. A simple quantum mechanical model for the effects may be constructed with only three key parameters: the harmonic vibration wavenumber of the bond vibrator ω_m (ω_m = 3706 cm^{-1} for H_2O), an interbond coupling constant λ (λ = -50 cm^{-1} for H_2O), and the familiar Morse anharmonicity constant x_m that would be appropriate to a diatomic molecule (x_m = -85 cm^{-1} for H_2O). Typical values of the constant x_m for symmetry related bonds in various small polyatomic molecules, so far as they are known at present, are shown in Figure 17. These constants can be used to estimate the bond dissociation energy using equation (2). This is equivalent to a Birge-Sponer extrapolation, which is known to be of questionable reliability; there is a general tendency for this calculation to overestimate the bond dissociation energy, owing to deviations from the Morse potential function in real molecules. Nonetheless the results obtained give an interesting insight into relative bond strengths which compare well with values determined by thermochemical arguments.

The model described here is based on a series of approximations, but it has nonetheless been found to be quantitatively successful in fitting the observed vibrational levels of a considerable number of small molecules [6,7,9,15,16,17,18]. The most obvious approximations concern the use of perturbation theory and the implicit use of harmonic basis functions in deriving the vibrational hamiltonian matrices, and the neglect of matrix elements coupling different values of the total vibrational excitation quantum number V. However when the same approximations are applied to a diatomic Morse oscillator they yield the exact solution, as observed by Mills and Robiette [7], and some similar cancellation of approximations holds for the polyatomic examples. Another more serious problem concerns the mixing with other vibrations; this has been discussed here only in connection with Fermi resonance, but its importance in other ways is only just being investigated.

Models of vibrational overtone energy levels and wavefunctions are of great importance in discussing intravibrational relaxation and molecular dynamics in polyatomic molecules, particularly in connection with photochemically induced reactions. There seems little doubt that highly vibrationally excited molecules always arrive in the lowest energy states for each value of V, corresponding to the truly localized excitation labelled here as [V,0,0, ...]. The mixing of these states through various

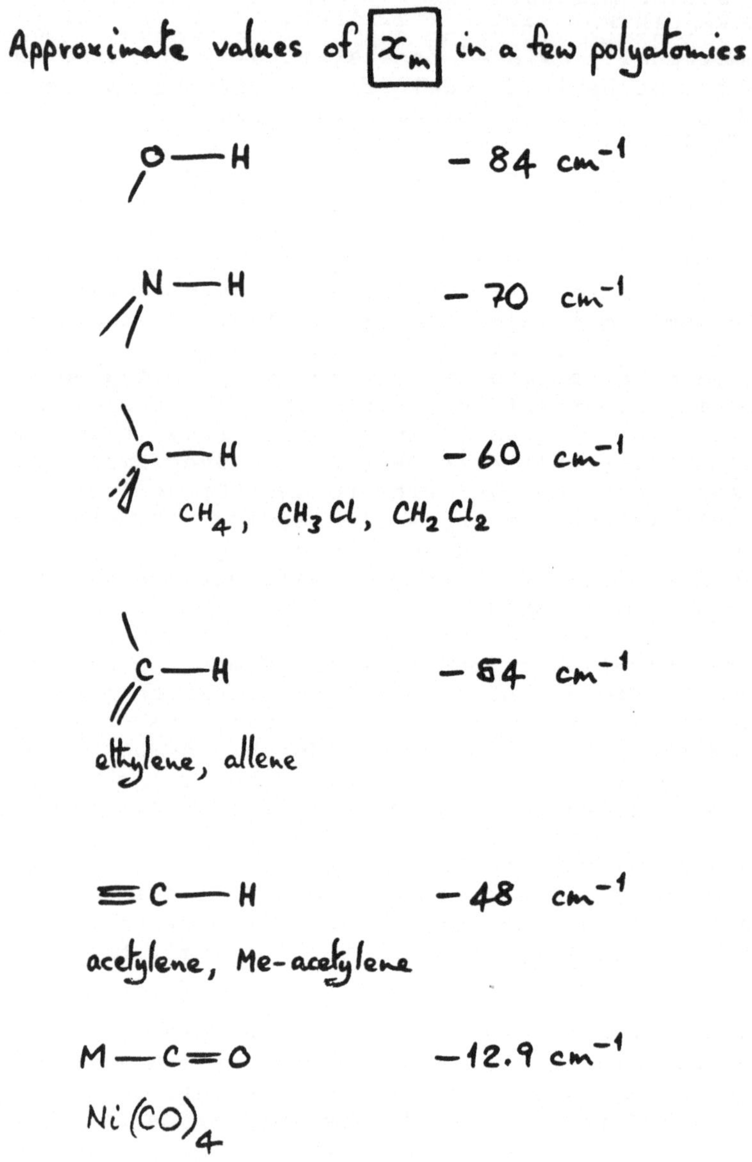

Approximate values of $\boxed{x_m}$ in a few polyatomics

O—H -84 cm^{-1}

N—H -70 cm^{-1}

C—H -60 cm^{-1}

CH_4, CH_3Cl, CH_2Cl_2

C—H -54 cm^{-1}

ethylene, allene

≡C—H -48 cm^{-1}

acetylene, Me-acetylene

M—C≡O -12.9 cm^{-1}

Ni(CO)$_4$

Fig.17 Approximate values of x_m for various molecules

kinds of resonance with the skeletal vibrations of the molecule then play a crucial role in the redistribution of energy that precedes a chemical reaction.

9.   ACKNOWLEDGEMENTS

I am particularly indebted to my colleagues at Reading, Dr J E Baggott and Dr G L Caldow, and to my former colleague Dr A G Robiette, for collaboration and discussion that has led to the development of my own understanding of this subject.   I am also indebted to the Science and Engineering Research Council who have continued to support our work in this field over a number of years.

REFERENCES

[1]   J E Baggott, H J Clase and I M Mills,
        Spectrochim. Acta A42, 319-334 (1986).
[2]   J E Baggott, D W Law, P D Lightfoot and I M Mills,
        J. Chem. Phys. 85, 5414-5429 (1986).
[3]   W Knippers, K van Helvoort, S Stolte and J Reuss,
        Chem. Phys. 98, 1-6 (1985).
[4]   E B Wilson, J C Decius and P C Cross, Molecular
        Vibrations, McGraw Hill, New York 1955.
[5]   I M Mills, Harmonic and Anharmonic Force Field
        Calculations, Spec. Per. Reports of the Chem. Soc.
        No 33, Vol 1 on Theoretcical Chemistry
        (Ed. R N Dixon), The Chemical Society, London, 1974.
[6]   M S Child and R T Lawton, Faraday Discussions of the
        Royal Society of Chemistry, 71, 273-285 (1981).
[7]   I M Mills and A G Robiette,
        Molecular Physics 56, 743-765 (1985).
[8]   B T Darling and D M Dennison,
        Phys. Rev. 57, 128-139 (1940).
[9]   I M Mills and F J Mompean,
        Chem. Phys. Letts. 124, 425-431 (1986).
[10]   K K Lehmann, J. Chem. Phys. 84, 6524-6525 (1986).
[11]   W S Benedict, N Gailar and E K Plyler,
        J. Chem. Phys. 24, 1139-1165 (1956)
[12]   H C Allen and E K Plyler,
        J. Chem. Phys. 25, 1132-1136 (1956).
[13]   K K Lehmann, J. Chem. Phys. 79, 1098-1099 (1983).
[14]   M E Kellman, J. Chem. Phys. 83, 3843-3858 (1985).
[15]   J E Baggott, D W Law and I M Mills,
        Molecular Physics (in press, April 1987)
[16]   J E Baggott, G L Caldow, D W Law and I M Mills, to be
        published.
[17]   I M Mills, Molecular Physics (in press, April 1987).
[18]   A Amrein, H R Dubal and M Quack,
        Molecular Physics 56, 727-735 (1985)

INTERSTELLAR MASERS AND STAR FORMATION IN THE GALAXY

GISBERT WINNEWISSER
I. Physikalisches Institut, Universität zu Köln
5000 Köln
West Germany.

ABSTRACT. Interstellar molecular lines have produced spectacular examples of non-equilibrium excitation, culminating in strong maser action of OH, SiO, H_2O and CH_3OH. Within the past 10 years it was recognized that maser emission is associated with star formation.

After a short historical overview of the history of astrophysical spectroscopy some general properties of star forming regions will be discussed. The bipolar flow region L1228 will be shown as an example and a theoretical model for bipolar flow mechanism is proposed.

1. INTRODUCTION

The detection of the molecular species CH, CH^+, and CN in interstellar clouds in the mid-thirties marked the beginning of the science of interstellar molecules. Although this date lies more than 50 years back, it is really only within the last 15 years that it has become clear that about 50% of the interstellar matter is in molecular form. Even more recent is the recognition that molecular clouds are widespread within the Galaxy, that they are associated with dark clouds of dust which obscure the radiation at optical wavelength, and that they are the sites of active star formation. Within the last years the evidence has grown to the point that one believes star formation can only take place in molecular clouds and that maser action is associated both with the formation of young stars and with evolved, dying stars. Maser action has come to be a "signpost" for the detection and study of the physical conditions and kinematics of the gas around and near very young and very old stars.

Interstellar masers reveal themselves by an extremely intense emission of radiation (brightness temperatures in

A. C. P. Alves et al. (eds.), Frontiers of Laser Spectroscopy of Gases, 491–508.
© 1988 by Kluwer Academic Publishers.

excess of 10^{12}K) which is often variable on a time scale of weeks to years and are often highly polarized. The maser emission is concentrated on a very small frequency interval (typically in the kilo-Hz region) and emanates spatially from very localized spots (typically of stellar dimensions) of fairly high density (in excess of 10^{7-9} cm^{-3}). Most of these features can be recognized in the H_2O maser spectra from the source S106. (Fig. 1)

The interstellar molecular radiation is received predominantly from sources within our Galaxy, although extremely strong maser action as well as thermal molecular radiation has been detected from many other galaxies. Thus most of the interstellar radiation we receive with our telescopes today has begun its travel towards the earth long before this beautiful University of Coimbra was founded almost 700 years ago.

It is appropriate at this special session in honour of Professor Dr. Fernando Pinto-Coelho to take an historical perspective of the main developments of interstellar spectroscopy at large and then describe to which specific areas of knowledge interstellar molecules have contributed. I will then discuss briefly several examples of recent research which are thought to be typical of star forming regions associated with maser emission. I am well aware of the very limited and biased selection I have chosen in view of the fantastic data available.

I would, however, like to mention that there are several excellent reviews, discussing the various physical processes of interstellar masers (Elitzur 1982; Reid and Moran 1981) and the physics of molecular clouds in the galaxy (Spitzer 1976; Winnewisser et al. 1979). Aspects of interstellar chemistry and the determinations of interstellar abundances from molecular line observations have been given by various authors (Duley and Williams 1984; Winnewisser and Herbst 1987).

2. HISTORICAL NOTES

2.1 Early History of Spectroscopy

The discovery that the sun and the stars are made up of the same chemical elements as the Earth is probably one of the earliest and greatest achievements of spectroscopy and its application to astronomical objects. In the last century the detailed and thorough work of Joseph von Fraunhofer whose 200th birthday has just occured (6. March 1787), led in 1814 to the discovery of more than 700 sharp absorption lines in the spectrum of the sun (Fraunhofer 1817) and in 1823 to the discovery of the strongest of these lines also in several stars. The identification of

S 106 $\alpha_{50} = 20^h 25^m 32^s 75$
H$_2$O - Maser $\delta_{50} = 37° 12' 54''$

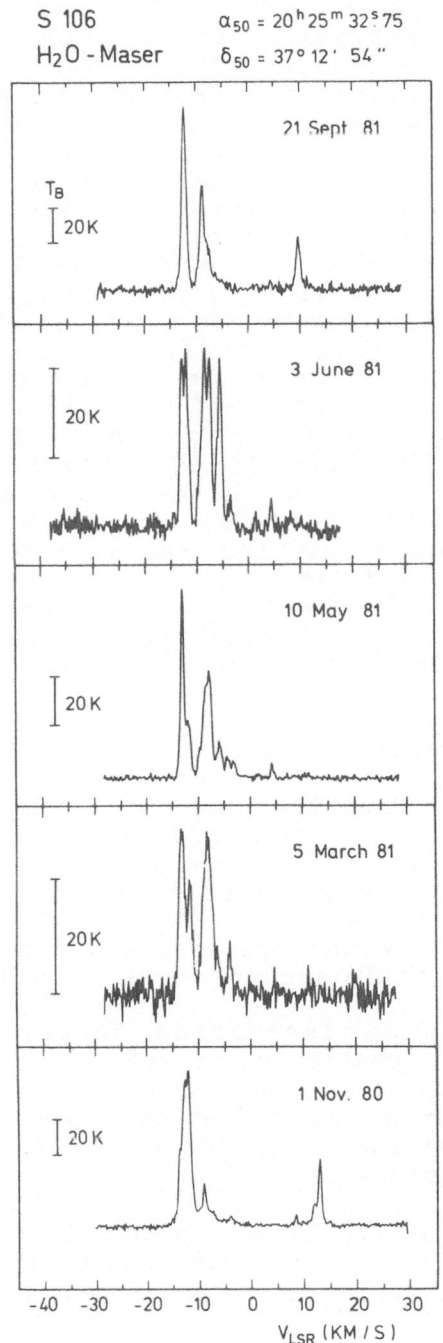

T$_B$
$\rfloor$ 20K

21 Sept 81

$\rfloor$ 20K

3 June 81

$\rfloor$ 20K

10 May 81

$\rfloor$ 20K

5 March 81

$\rfloor$ 20K

1 Nov. 80

-40 -30 -20 -10   0   10   20   30
V$_{LSR}$ (KM / S)

Fig. 1

Time variability observed for the H$_2$O maser in S106. The spectrum of Sept., 21,81 and Nov. 1 1980 show some similarity which each other. A double-peaked profile is recognizable. (from Stutzki, Ungerechts, and Winnewisser: 1982, Astron. Astrophys. 111, 201).

the lines, now named after him, the Fraunhofer-lines was put on a solid footing in the years 1859 - 1861 by the then newly established spectrum analysis introduced by R.W. Bunsen and G.R. Kirchhoff, Professors of Chemistry and Physics respectively at the University of Heidelberg. The Fraunhofer lines corresponded to laboratory spectra of substances which apparently were also present in the solar atmosphere. This detection marks the time and place of the establishment of spectroscopy as an analytical tool with an unparalleled power to study both physical and chemical processes in cosmic dimensions. The power of this new tool "spectroscopy" was already recognized by Kirchhoff and Bunsen and its future application summarized in two points: (i) Analysing the light emitted by heavenly bodies (and they meant stars) will open "... an entirely untrodden field, stretching far beyond the limits of the earth, or even of our solar system". The second application was seen in the "almost inconceivable" sensitivity of spectroscopy which will lead to the detection of hitherto unknown elements. Many new elements, e.g. Cs, Rb, He were subsequently detected by their atomic spectra. A detailed understanding of the spectra and their associated energy level diagrams had to await the advent of quantum mechanics. However, the early spectroscopic techniques by Kirchhoff and Bunsen have been extended and refined over the years.

2.2 Recent History of Interstellar Molecules

The basic spectroscopic technique applied by these early investigators of matching frequencies or wavelengths between spectral lines from stars and those obtained in the laboratory to secure a spectroscopic identification has remained in use up to the present day. Now, however, considerably improved accuracy and the availability of almost the entire electromagnetic spectrum are the basic present-day tools.

The early spectroscopic discoveries pertained essentially to atomic stellar spectra and it was not until 1904 that the first indication of matter dispersed between the stars was deduced from the non-participation of the Na-D absorption lines in the periodic Doppler shift of the delta-Orionis close binary star system.

These observations marked the discovery of interstellar matter although it really was in the thirties when the concept of interstellar gas and dust was begun to be appreciated. The discovery of the electronic spectra of the three diatomic radicals CH, CH$^+$ and CN, seen as sharp absorption features against the continuum spectra of bright background stars, furnished first evidence that the interstellar medium was not devoid of molecules (Swings and

Rosenfeld 1937, McKellar 1940, Douglas and Herzberg 1941).
A summary of the presently known interstellar molecules
detected by optical techniques in diffuse interstellar
clouds has been given by Winnewisser et al. 1979.

Simultaneously with these sharp features broad
interstellar absorption features were discovered which have
now become known as the diffuse interstellar bands (see
e.g. Herbig, 1975). The source of the diffuse interstellar
bands has remained a mystery for over 50 years. Many
detailed suggestions have been made, none of which have
been proven. It is, however, undisputed that these
absorptions are caused by some constituent of the general
interstellar medium, molecular gas or dust or a combination
of both.

3.0 INTERSTELLAR MOLECULAR SPECTROSCOPY

The next leap forward in our knowledge about the
interstellar medium came about 1970 with the discovery of
many polyatomic molecules. Today we know that the
interstellar medium contains a startling number of
molecules (see Table I), including those which had hitherto
not even been detected in the laboratory. These molecules
are normally detected in the gas phase in ·the microwave,
millimeter- and submillimeter wave-, and far infrared
region of the electromagnetic spectrum via characteristic
spectral emission and/or absorption lines. These
wavelengths are sufficiently long to be largely unaffected
by the extinction properties of the interstellar dust at
shorter wavelengths, notably the optical region. Therefore
molecular transitions in the radio region normally probe
deep into the interior of "dense interstellar clouds".

Besides being of interest by themselves, interstellar
molecules have become essential tools for astronomers,
physicists and chemists interested in the study of the
general properties of the interstellar medium. Areas which
have been deeply influenced by the observation of
interstellar molecules and where substantial new insight
into physical and chemical processes have been gained can
be placed into four large groups:

(i) Before the advent of studies on interstellar
molecules the knowledge of the distribution of neutral
matter on a galactic scale, i.e. galactic structure, had
been dependent on stellar statistics and since 1951 on
measurements of the 21cm hyperfine line of neutral
hydrogen. However, since the early 70′s it has become
evident that the galactic distribution of interstellar
molecular clouds can be traced out by the most abundant and
widely distributed molecules. Although H_2 is the most
abundant interstellar molecule and indeed constitutes 90%

Interstellar Molecules (july 87)

2	3	4	5	6	7	8	9	10	11	13
H_2	H_2O	NH_3								
OH	H_2S									
SO	N_2H^+									
SO^+	SO_2									
SiO	HNO									
SiS	NaOH(?)									
NO	H_2D^+									
NS										
HCl										
PN										
CH^+	HCN	H_2CO	HC_3N	CH_3OH	HC_5N	$HCOOCH_3$	HC_7N	$CH_3C_5N(?)$	HC_9N	$HC_{11}N$
CH	HNC	HNCO	C_4H	CH_3CN	CH_3CCH	CH_3C_3N	$(CH_3)_2O$	$(CH_3)_2CO$		
CN	C_2H	H_2CS	H_2CNH	CH_3SH	CH_3NH_2		CH_3CH_2OH			
CO	C_2S	HNCS	H_2C_2O	NH_2CHO	CH_3CHO		CH_3CH_2CN			
CS	SiCC	C_3N	NH_2CN	H_2CCH_2	H_2CCHCN		CH_3C_4H			
CC	HCO	C_3H(lin)	HCOOH	C_5H	C_6H					
	HCO^+	C_3H(ring)	(CH_4)							
	$HOC^+(?)$	C_3O	(SiH_4)							
	OCS	C_3S	C_3H_2							
	HCS^+	$HOCO_2^+$								
	CCS	(HCCH)								
		$HCNH^+$								

( ) : circumstellar identifications

(?) : uncertain identifications

of the total gas density, it is of limited use for the surveys of low temperature clouds. This is due to its widely spaced energy levels (the first excited para-H_2 rotational state occurs at over 350 cm^{-1} above the ground state, which puts the lowest rotational quadrupole transition (J = 2 -> 0) into the far infrared at 28 micron). Consequently high excitation requirements are needed to populate the energy levels. The second most abundant molecule, CO, and its different isotopic species with its low excitation requirements and closely spaced rotational levels, paired with a small permanent electric dipole moment, are perfectly suited to be tracer molecules for the neutral matter in interstellar clouds, as long as one assumes that the CO/H_2 ratio is practically constant at 10^{-4}, which occurs if most of the carbon in the clouds is tied up as CO. The lowest rotational transition of the ubiquitous tracer molecule CO, the J = 1 -> 0 transition at 115 GHz, is therefore being extensively used for general survey purposes, ranging from mapping the details of individual molecular clouds, to surveying the large scale galactic and extragalactic distribution of neutral matter.

The first composite CO survey of the entire Milky Way has recently been completed by Thaddeus and his group (Dame et al. 1987). The main structural features of the Galaxy, the so called molecular ring, i.e. the molecular clouds in the inner spiral arms are clearly seen as an intense ridge of CO radiation.

(ii) Knowledge of the physics and chemistry of molecular clouds has been gained. Molecular transitions have for the first time yielded precise information on a large variety of cloud parameters, such as extent, size, density, total mass of the clouds, temperature distribution within the cloud, and the various excitation mechanisms. It has been recognized, for example, that certain giant molecular clouds such as the well-studied Orion cloud (OMC-1 or Orion Molecular Cloud 1) have masses approaching 100,000 times that of the sun.

Furthermore, the molecular spectra reveal a plethora of dynamical effects, which are otherwise practically impossible to obtain. Molecular spectra have made close investigations possible of the mechanisms which govern star formation processes. One example is the discovery that in molecular clouds gravitational collapse and subsequent heating leads to the birth of young stars. Associated with the star formation are phenomena such as bipolar flow of material in the neighbourhood of young stars, shock fronts emanating from newly born stars and maser emission (H_2O, OH, SiO, CH_3OH) associated with these active regions, as well as the interaction of newly born stars with their maternal molecular clouds. Highly excited rotational molecular lines of many species and vibration-rotation lines of H_2 and CO are the essential tools for probing these hot but confined regions embedded in many molecular clouds.

(iii) Substantial new information has also been obtained concerning the mass loss associated not only with young stars but also with old stars. Stars near the end of their stellar life cycle shed large amounts of stellar processed material (typically 10^{-4} to $10^{-5} M_\odot$ /year) back into space. This is evidenced by the thick circumstellar envelopes covering these old stars -notably the so-called carbon-stars, of which IRC+10216 is the best studied.

(iv) Finally we have gained totally new insight into the composition and chemistry of molecular clouds, the galaxy as a whole, and other galaxies.

4. INTERSTELLAR MASERS: strong deviations from LTE

The four molecules OH, SiO, H_2O, and CH_3OH show in one or more transitions extremely strong maser action. Only very recently has methylalcohol been added to the list of strong maser molecules (Batrla et al. 1987). Its intensity rivals that of the well known OH or H_2O masers. For the observational studies of star formation OH, H_2O and CH_3OH masers are well suited. Probably with the one exception of the source Orion A SiO masers occur in red giants and supergiants. Fig. 3 shows a series of SiO spectra towards Orion A which reveal the complex structure within the double-peaked line profiles and their time variation.

Weak masers have been detected for a number of molecules such as NH_3, H_2CO, SiS, HCN ($v_2=2$). Level inversions have been reported for HC_3N, HC_5N and others.

Masers are found in two types of sources, (i) those which are connected with star forming regions near embedded infrared sources (mainly H_2O masers) or near compact HII(H+)-regions (OH, CH_3OH) and (ii) in stellar sources, i.e. in the atmospheres of red-giant stars (i.e. OH, H_2O, SiO). The emission shows the typical double-peaked line profiles shown for the case of SiO.

Strong maser radiation is a manifestation of extreme deviations of molecular level populations from thermal equilibrium. In localized interstellar regions where the level population is inverted by one and/or several pump mechanisms radiation is amplified through stimulated processes.

The intensity of maser radiation and its dependence on the molecular and level population parameters is best analysed by the concept of radiation brightness temperature T_B and solution of the radiative transfer equation (see for general discussion e.g. Winnewisser et al., 1979; maser specific reference Elitzur, 1982). The intensity of

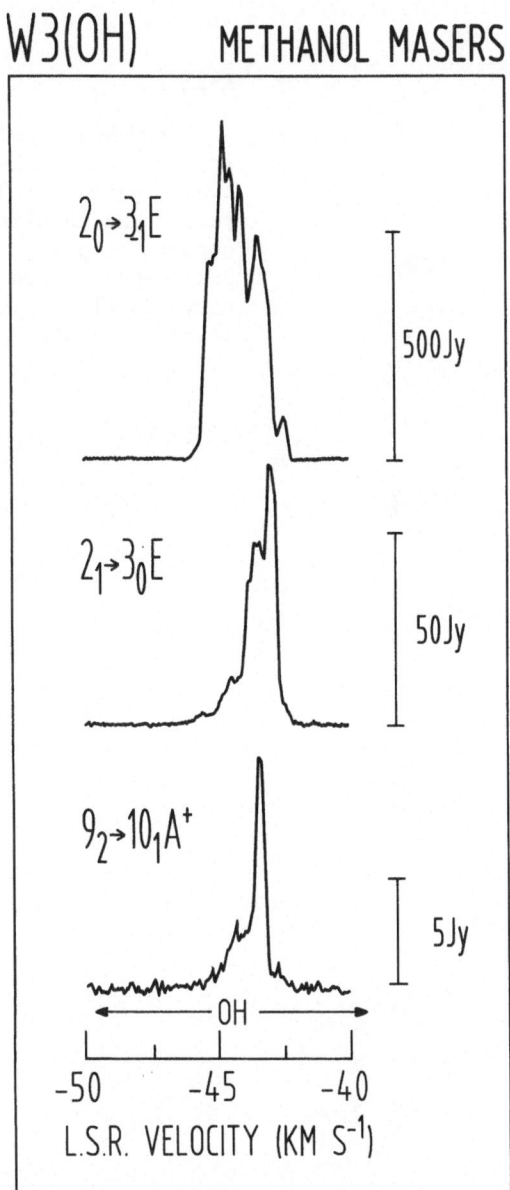

<u>Fig. 2</u>   The newly discovered methanol maser. (2_0-3_{-1}, E) by Batrla et al 1987).

500

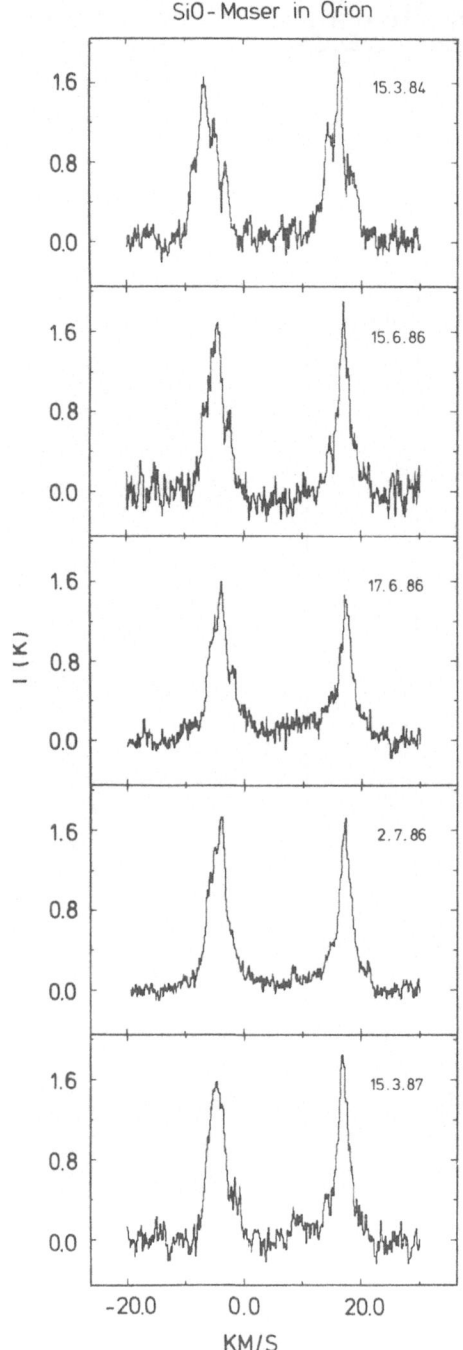

SiO-Maser in Orion

15.3.84

15.6.86

17.6.86

2.7.86

15.3.87

I (K)

-20.0   0.0   20.0

KM/S

Fig. 3

SiO–Maser emission
from Orion A, taken
over a period of 3
years. The complex
structure in the
double-peaked profiles
vary. No clear pattern
is recognizable.

radiation is defined as the energy flux carried by the radiation per unit frequency interval and solid angle. Assuming a blackbody radiation distribution the total intensity or "brightness" (as called by radioastronomers) $B_v(T)$ is a function of the brightness temperature T_B as defined by Planck's radiation law. In the Rayleigh-Jeans limit, the Planck function can be expressed as

$$kT_B = (c^2/2v^2)I_v$$

showing that the brightness temperature T_B is directly proportional to the emitted intensity of the source $I_v = B_v(T_B)$. The brightness temperature T_B is related to the line excitation temperature at the source T_{ex} (which is defined through the level population of the upper n_u and lower n_l level involved in the transition, i.e. $n_u/n_l = (g_u/g_l)\exp(-hv/kT_{ex})$ and a possible continuum background brightness T_c via the solution of the radiative transfer equation:

$$T_B = T_{ex}(1-\exp(-\tau)) + T_c\exp(-\tau)$$

where τ is the optical depth, i.e. the integrated absorption coefficient $\tau = \int k d v$. For maser radiation to occur $T_{ex}<0$, and $|\tau| >1$. This means that the absorption factor $\exp(-\tau)$ turns into an amplification factor. $|\tau|$ is referred to as "gain". Brightness temperatures T_B up to 10^{15}K have been derived, after the small sizes of the maser sources were observed with interferometric techniques.

Our understanding of the inversion mechanism has remained rather limited, although various pump mechanisms have been proposed such as radiative pumps, collisional pumps and chemical pumps (for a complete discussion see Elitzur 1982). Despite this uncertainty observations of interstellar maser lines have lead to an interesting number of astrophysical applications: Maser lines (i) serve as a "signpost" showing where by some mechanism energy is pumped into the gaseous interstellar medium. It is now known that masers occur in those regions of interstellar clouds where star formation is taking place. (ii) allow very accurate position determinations to be made. Due to their high line intensity their position and angular size of the emitting region can be measured by interferometric techniques. (iii) contain a large amount of kinematic information. Maser emission spots for most sources are clustered. Thus proper motion measurements together with the spatial information can be used to estimate the distances to the emitting sources. This technique has been applied for the ORION-KL and W51 source employing H_2O maser emission (Genzel et al. 1981) with the following distance results 480±80 pc and 7 ± 1.5 kpc respectively. (iv) The double peaked 1612 MHz profiles seperated by about 30 km/s of the OH maser in infrared stars allow determination of the expansion velocity of the envelope. In addition for these OH-IR-stars one believes that the OH-masers are radiatively pumped, based on the fact that the 1612 MHz maser emission does vary in phase with the far infrared luminosity at 53 and 35 micron ($^2\pi_{3/2}(J=3/2)$ -> $^2\pi_{1/2}(J=3/2$ or $J=5/2$ respectively). A summary of masers in circumstellar shells is given by Walmsley 1987.

Aside from molecular maser emission of OH, H_2O and CH_3OH there exist several additional indicators of star formation in molecular clouds:
(i) Infrared sources embedded in molecular clouds. A newly born star heats its surroundings by ultraviolet photon emission which is absorbed mainly by the dust and then reradiated in the infrared region. Thus one of the important results of the IRAS-(Infrared Astronomical Satellite) mission was the creation of the Iras-Point-Source-Catalogue (IPSC) with 246000 entries(see e.g. Israel 1985). It contains 144000 stars, 22000 galaxies, 33000 infrared cirrus components and 35000 infrared galactic components including those sources which are newly born stars embedded in dense molecular clouds.

Thus a comparison of the IR-emission of these sources with the molecular emission (mainly CO J = 1 - 0) allows a detailed study of star forming regions and their environment. Such a comparison has been made recently for the dark cloud complex L1228 using data obtained with the Cologne 3-m-radiotelescope at Gornergrat, which has at 115 GHz the same spatial resolution as the IRAS-maps. The blue- and red-shifted line profiles of this source show beautifully the bipolar-outflow associated with the formation of young stars. Fig. 4 shows an integrated CO J = 1->0 contour map of L1228 as well as high- and low-velocity maps around the central area of the exciting star.
(ii) The high-velocity outflows which manifest themselves either in broad wing emission which is often bipolar in nature. Velocities between 50 < v < 200 km/s and mass flows of $10^{-8}M_{\odot}$ < m < $10^{-2}M_{\odot}$/year are found. Fig. 5 shows a broad-wing-emission profile of CO towards the Orion molecular cloud.
(iii) The shock-excited molecular gas, i.e. the vibrationally excited H_2 and CO. Hot NH_3(e.g. J,K = (10,9), (11,9)...) is seen in star forming regions.

Since the first observational evidence of bipolar flows in L1551 (Snell et al. 1980) was found an intensive effort has been made both observationally and some extent also theoretically to understand the nature and cause of the bipolar-flows (see e.g. Lada 1985). The existence of bipolar flows is probably the most direct evidence for

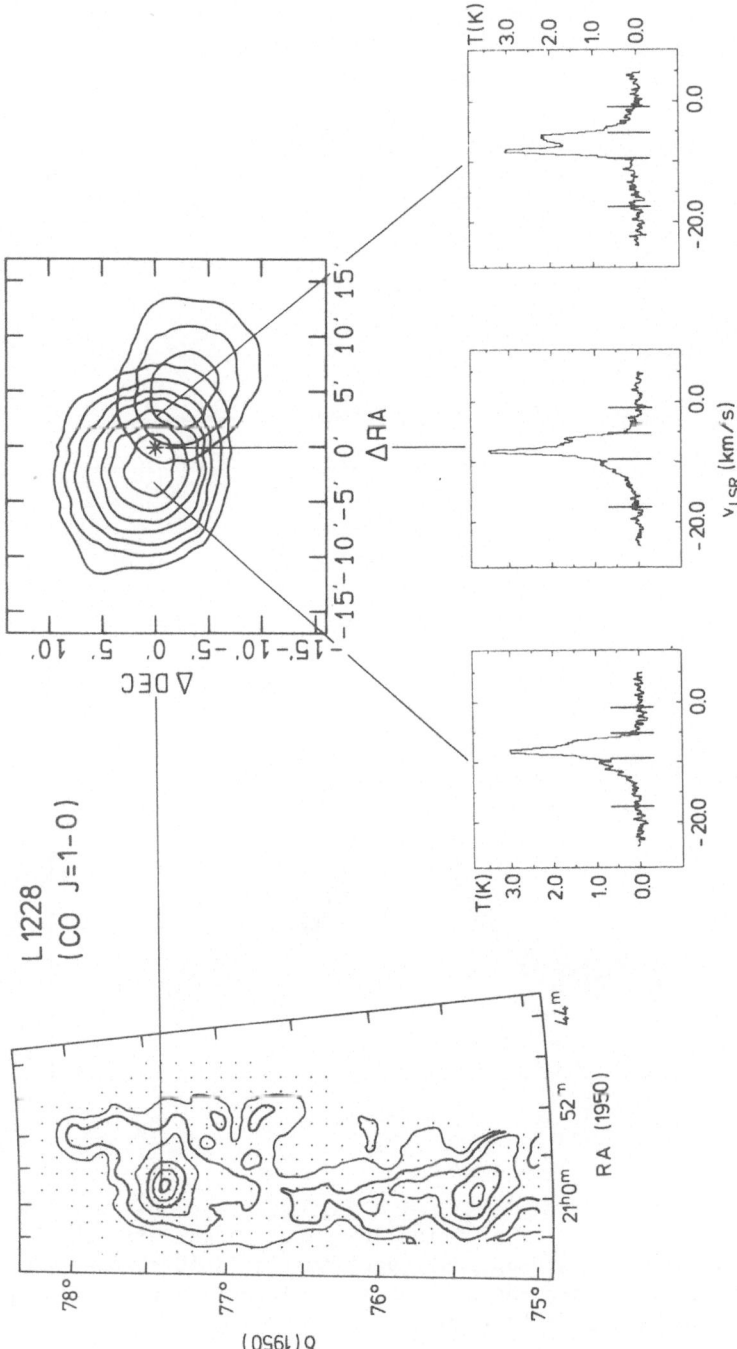

Fig. 4   Bipolar outflow source L1228 (see text). The exciting star is indicated. (Data from L. Haikala).

504

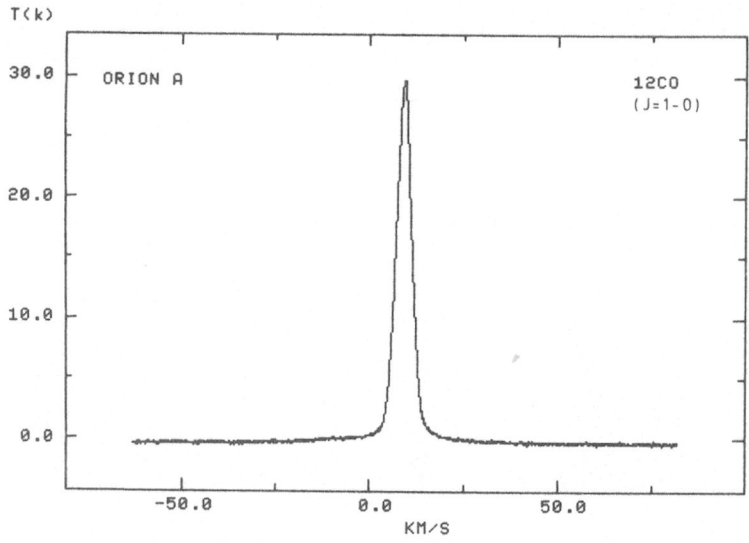

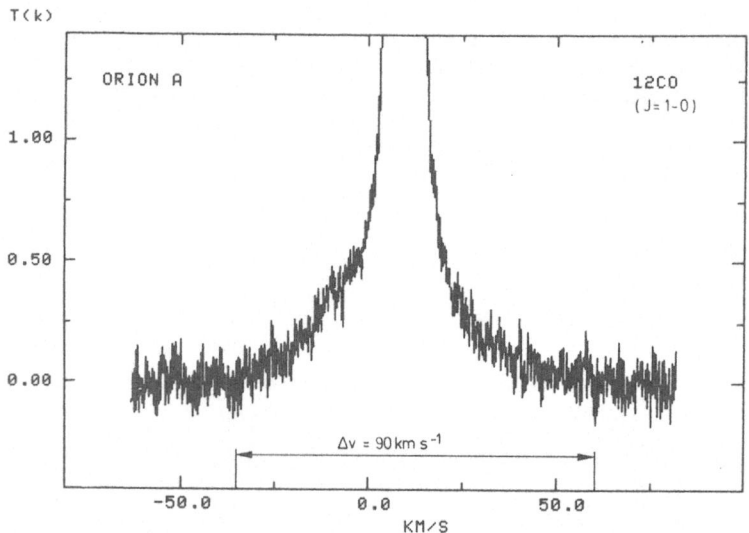

Fig. 5   CO J = 1-0 spectrum of Orion. Data were taken with the high
         resolution acousto-optical spectrometer (2048 channels,
         resolution 30 kHz/channel) of the Cologne 3-m-radiotelscope
         at Gornergrat, Switzerland

star-formation. It is now undisputed that bipolar flows are
associated with high velocity water vapour maser sources,
with vibrationally excited H_2 and sometimes with optically
visible jets.

In an attempt to contribute theoretically towards an
understanding of the physics of bipolar-flows, we have
developed a model aimed at explaining the major
observational features of outflow sources. The initial
scenario required by the model is that of a central stellar
object embedded in a parental cloud of gas and dust forming
a halo whose shape deviates slightly from spherical
symmetry, an assumption which should be met in practically
all realistic cases.

Different gravitational attraction of the
nonsymmetrical halo leads to a disc-like structure and
initiates a mass inflow towards the central object. This
radial disk-like inflow draws the main bulk of its mass
from a basically non-rotating halo surrounding a rotating
disk structure. On its way towards the central stellar
object, the inflowing matter traverses with freefall
velocity a mass gap between the inner rim of the rotating
disk and the outer fringes of the central object. Upon
reaching the neighborhood of the equator of the stellar
object, the inflow divides itself into two parts: the inner
part with mass flux $\dot{M}_1$ penetrates a small ring-like strong
shock region and mixes finally with the surface layer of
the stellar object, while the outer part of the mass flux
$\dot{M}_2$, which is small compared to $\dot{M}_1$, is deflected, by passing
the shock region and moving towards the poles. The model is
shown schematically in Fig.6. The kinetic energy carried by
$\dot{M}_1$ is released in the form of radiation at a zone around
the equator. A small part of the radiation energy is
absorbed above the "radiation zone" by the circumventing
material $\dot{M}_2$, which becomes the outflowing material at the
poles. Details of the calculations are given by Yue et al
1987.

The radial inflow is a transient phenomenon lasting
only about $(2-3) \times 10^4$ years if we assume a typical radial
scale of 10^4 AU for the non-rotating halo which supplies the
radial disk flow. When the non-rotating halo is exhausted,
the radial inflow and the well-collimated bipolar outflow
stop, leaving behind a rotating faint disk around the
stellar object.

Thus, the four major observational features associated
with bipolar outflows,

(i) its bipolar geometry,
(ii) the high velocity of the outflowing material of
order of 100 km/s at very large distance ($10^3 - 10^4$
AU),
(iii) the high degree of collimation and
(iv) the time scale of about 10^4 years

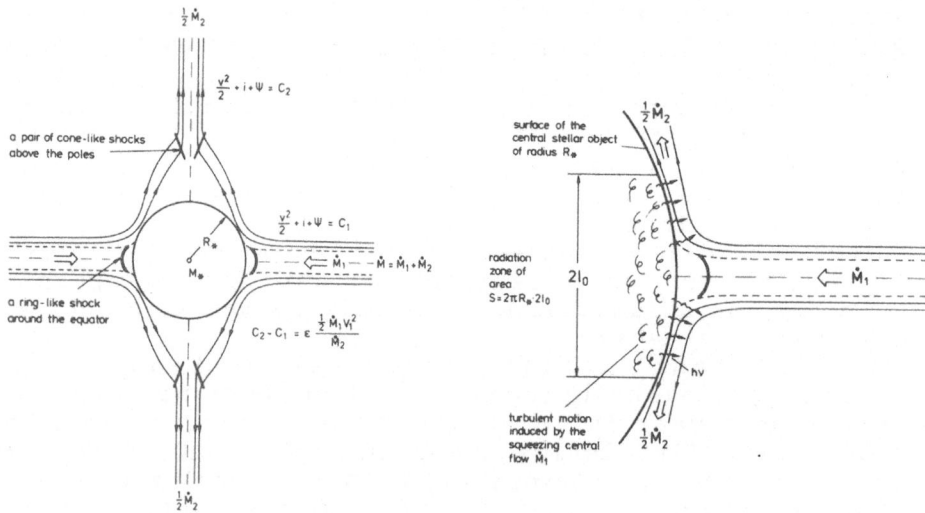

Fig. 6

Bipolar jets are induced by the radial disk like inflow surrounding the equatorial region of the stellar object. (left figure). The radial inflow is divided into two flows: the central major flow passes a shock region and mixes with the surface material of the stellar object. The outer flow bypasses the central object and forms the bipolar outflow after absorbing some radiation energy and passing through a second shock region. The Bernoulli equations determine the energy balance. A detailed picture of the equatorial region is given on the right hand drawing.

can all be explained by the proposed dynamical model. The interaction with the local environment into which the bipolar flow is finally ejected determines its observed geometrical shape. Thus bipolar-flows are a very common but transient phenomenon associated for about 10^4 years with the very early stages of star-formation.

5. FUTURE PERSPECTIVES

It seems appropriate to conclude this short discussion of interstellar masers and star-formation with the question of the problems left open. It is clearly one of the most fascinating areas of astronomical research to study regions of star-formation and hopefully find additional tracers and criteria for the investigation of these regions. The question of the interaction between a newly born star and its parental molecular cloud is one of the most fascinating areas of future research. The study of this detailed problem has to be complemented by the conduct of unbiased surveys for star-forming regions in molecular clouds to enhance the statistical material with the aim to understand more about the possible evolutionary track of a proto-star on its way to the main sequence.

These problems can be tackled by combining data from interferometers operating in the millimeter-wave region and large and small single dish instruments which operate throughout the millimeter- and into the submillimeter-wave region. With more high-spatial resolution data at hand one may then also hope to shed more light on the long standing question of the existence of other planetary systems other than our own, and finally on the details of planetary formation in the wake of stellar birth.

6. REFERENCES

Batrla, W., Matthews, H.E., Menten, K.M., and Walmsley, C.M.:   1987, Nature **326**, 49.
Dame, T.M., Ungerechts, H., Cohen, R.S., de Geus, E., Grenier,   A., May, J., Murphy, D.C., Nyman, L.-A., and Thaddeus, P.:   1987, Astrophys J., (in press).
Douglas, A.E. and Herzberg, G.: 1941, Astrophys. J.,**94**,381.
Duley, W.W. and Williams, D.A.: 1984, **Interstellar Chemistry**,Academic Press, New York.
Elitzur, M.: 1982, Rev. Mod. Phys., **54**, 1225.
Genzel, R., Reid, M.J., Moran, J.M., Downes, D.: 1981, Astrophys.J., **244**, 884; ibid **247**.
Fraunhofer, J.v.: 1817, Ann. d. Physik, **56**, 264.
Hartmann, J.: 1905, Astrophys. J., **21**, 389.
Herbig, G.H.: 1975, Astrophys. J., **196**, 129.
Israel, F.P.: 1985, Light on Dark Matter. Reidel Publ. Comp.
Kirchhoff, G. and Bunsen, R.: 1860, Ann. d. Phys. u. Chemie, **110**, 160; 1861, ibid **113**, 337.
Lada, Ch.J.: 1985, Ann. Rev. Astron. Astrophys. **23**, 267.
McKellar, A.: 1940, Publ. Astron. Soc. Pacific, **52**, 189.
Reid, M.J. and Moran, J.M.: 1981, Ann. Rev. Astron. Astrophys., **19**, 231.
Snell, R.L., Loren, R.B., Plambeck, R.R.: 1980, Astrophys. J. **239**, L17.
Spitzer, L.: 1978, Physical Processes in the Interstellar Medium,   Wiley, New York.
Swings, P. and Rosenfeld, L. 1937.: Astrophys. J. **86**, 483.
Walmsley, C.M. 1987.: IAU Symposium: Astrochemistry **369**, Reidel Publ. Comp.
Winnewisser, G., Churchwell, E. and Walmsley, C.M.: 1979, Modern Aspects of Microwave Spectroscopy, p.313, (edited by G.W. Chantry), Academic Press, New York.
Winnewisser, G. and Herbst, E.: 1987, Topics in Current Chemistry, Springer Verlag, Berlin Heidelberg, **139**, 121.
Yue, Z.Y., Zhang, B. and Winnewisser, G.: 1987, Astron. Astrophys. (submitted).

512